Wildlife Monitoring

Luciano Martins Verdade •
Rafael de Andrade Moral
Editors

Wildlife Monitoring

Integrating Conservation and Innovation in Human-Altered Landscapes

Springer

Editors
Luciano Martins Verdade
Wildlife Monitoring Consultancy
Campinas, São Paulo, Brazil

Rafael de Andrade Moral
Department of Mathematics and Statistics
Maynooth University
Maynooth, Co. Kildare, Ireland

ISBN 978-3-032-05820-1 ISBN 978-3-032-05821-8 (eBook)
https://doi.org/10.1007/978-3-032-05821-8

This Springer imprint is published by the registered company Springer Nature Switzerland AG
The registered company address is: Gewerbestrasse 11, 6330 Cham, Switzerland

To Bob and Cadu.
—LMV

To my wife Sonya and our daughters Eabhí and Éala, with the hopes this book will sow the seeds of a world where preserving wildlife is as natural as breathing, and the planet answers not with applause, but with clean air, full rivers, and life in abundance.
—RAM

Foreword

The editors summarise this book's aim as: to propose a long-term protocol for wildlife monitoring that can operate under real-world constraints, while still yielding meaningful ecological inferences. This can be seen in contrast to the standard focus of ecological research that has been on the natural environment, while here the attention is on the humanly managed settings of agriculture and silviculture. Of course, there is a lot to be learnt from the existing work in terms of data collection, appropriate models, and statistical analyses, and how inferences and predictions can be made on important matters of species conservation and management. However, moving to overtly managed environments poses a whole new range of challenges, and it is the intention of the work presented here to address some of these issues and redress the previously somewhat narrow and controlled focus of ecological research. The collected chapters cover a diverse range of topics, including the opportunities and challenges for data collection, specific statistical techniques tailored to the various settings, and a general discussion of the issues in the modelling and analysis of such data. As such, the book will be a useful resource to ecologists, agricultural scientists, statisticians, and data scientists working on the humanly managed environment. It will complement the many existing books on wildlife ecology and hopefully provide a stimulus for the synthesis and development of new approaches taking advantage of the rapidly expanding tools for data collection and capture.

The authors come from a diversity of different backgrounds and specialities, both methodological and technical, and are grounded in a fascinating range of substantive application areas. A primary basic theme in many of the papers is data collection, whether it be direct, in terms of traps and photo/observer identification, and indirect making use of vestiges (tracks, scats, etc.), along with remote sensing. Data quality is always an issue of concern, while data availability can range from scarce to almost overabundant. Important considerations addressed concern spatial and temporal scales; these can become more challenging when trying to synthesise data from multiple sources, where differing data quality often also needs to be accounted for. As is often the case, there can be a tension between expensive, planned, limited high-quality expert observer data and cheaper more plentiful indirect opportunistic data, where in the extreme one can think about making use of citizen science data. There are many issues addressed in the various papers and both guidance and suggestions for much further work. There are many future challenges for both data storage, processing, and analysis with the likely explosion of data over time, particularly with methods for autonomous data collection.

Most of the chapters also address issues of data analysis, including some tutorial-like material on new and developing techniques, including Bayesian hierarchical modelling, flexible distribution modelling, where the focus is not just on the mean but also on the variance and other parametric aspects of distributional shape, and an introduction to matrix modelling and control theory. However, the approach taken by the authors in discussing modelling and analysis is not a dry, methodological one, but rather an applied perspective firmly rooted in the substantive application and focused on answering significant questions of ecological interest. As ever in this age of machine learning and artificial intelligence there is the contrast between black-box algorithmic approaches and explicit (statistical) models, with trade-offs between performance and interpretability. There is no simple answer to this and it can depend on the goals of the study. Here, with the emphasis on ecological studies in multifunctional agricultural landscapes, there is the key question on how to use studies to balance between the demands of production and conservation. Simple models and predictions for abundance and coverage may no longer suffice, instead one wants to inform conservation policy decisions and land management for agriculture and, maybe also, ecotourism. Some of the studies reported here show that conservation strategies, interventions, and policy changes, even when planned, can have surprising and significant unintended consequences.

There is much to be gained for the reader of this book. It includes something for everyone and provides a taster of what may be possible in many different settings. One cannot but fail to be stimulated by the diversity and scope of the approaches described and, importantly, applied in real and fascinating areas. The material is supported by extensive references in each chapter, as well as pointers to appropriate software tools (generally in R) for model fitting, validation, and presentation. Without a doubt this collection is a great resource and a snapshot of the topic as it stands. However, it is not a cookbook of methods and simple answers, and it is now up to the reader/researcher to apply the ideas to their own work and contribute to the extension of these techniques and the scope of their application in the real world to inform policies for conservation and agriculture.

Enjoy the material and be inspired!

Emeritus Professor of Statistics
University of Galway
Galway, Ireland

John Hinde

Acknowledgments The São Paulo Research Foundation (FAPESP # 2017/01304-4), National Council for Scientific and Technological Development (CNPQ # 312049/2015-3), and the Brazilian Agência Nacional de Energia Elétrica (ANEEL # PD-07427-0225/2025 and 0064-1036/2014) supported the research in this book.

The co-editors are thankful to the University of São Paulo and Maynooth University, for providing them with the academic freedom to pursue the research contained in this book.

Contents

Part I
Conceptual Innovations

Chapter 1
Why Should We Monitor Wildlife and Its Habitat in Agricultural Landscapes?

Luciano Martins Verdade and Rafael de Andrade Moral

Abstract Despite their innumerous and profound environmental impacts agricultural landscapes provide complementary conservation to a relevant part of the biota, including endangered, damaging, and valuable species, as well as populations which do not fit any of those categories. Monitoring is the major demand of wildlife management in human-altered environments. However, like any other field of knowledge, biodiversity monitoring still has conceptual, technological, and societal limitations. Conceptual framework still needs improvement in the number and nature of indicators, as well as the spatiotemporal dimensions of patterns and processes to be monitored. In addition, technological innovations are still needed for both sampling and analytical methodology, including statistical models and algorithms designed to accommodate low detectability, many heterogeneous sources of variation, spatiotemporal autocorrelation, and data scarcity. Although wildlife monitoring in multifunctional agricultural landscapes can provide a direct measure of their conservation value, there are still some societal constraints regarding the tangibility of such value, as well as who should pay for it. This realism is central to this book's broader aim: to propose a long-term protocol for wildlife monitoring that can operate under real-world constraints, while still yielding meaningful ecological inferences.

Keywords Wildlife monitoring · Biodiversity · Biocomplexity · Human-altered environments · Conceptual framework · Innovation · Societal constraints

L. Martins Verdade (✉)
Wildlife Monitoring Consultancy, Campinas, São Paulo, Brazil
e-mail: luciano.verdade@wildlifemgmtco.com

R. A. Moral
Department of Mathematics and Statistics, Maynooth University, Maynooth, Co. Kildare, Ireland

L. Martins Verdade, R. A. Moral (eds.), *Wildlife Monitoring*,
https://doi.org/10.1007/978-3-032-05821-8_1

1 Wildlife in Agricultural Landscapes

No country in the world is able to provide conservation to its whole biota only in conservation units such as National Parks and Ecological Reserves. Consequently, agricultural landscapes provide complementary conservation to a relevant part of the biota despite their innumerous and profound environmental impacts (Verdade et al. 2014a). However, a fundamental difference between National Parks and agricultural landscapes is that the former has the primary mission of biological conservation, whereas the latter has the primary mission of biological production. The first step to improve the conservation value of agricultural landscapes would be to formally recognize its second but also fundamental mission of biological conservation, which gives them the status of multifunctional landscapes (Martinelli et al. 2010). The second step would be the implementation of wildlife management based on the decision-making process regarding the possible necessity of human intervention to remediate—or at least mitigate—possible negative anthropogenic impacts (Verdade et al. 2016).

According to Caughley (1994), there are four possible human interventions regarding wildlife management: increase a depleted population, decrease an excessive population, reach the maximum sustainable yield, or "do nothing but keep an eye on it". These categories correspond to the main fields related to applied evolutionary ecology, respectively, biological conservation, damage control, sustainable use, and monitoring (Verdade et al. 2014b). Their ecological components usually refer to ecological processes like demography and use of space, whereas their evolutionary approach refers to the use of population—not the species—as the unit of management. The combination of both can lead to the application of distinct interventions to distinct populations of the same species. In addition, such a combination also allows the conservation of distinct lineages of the same species; therefore, preserving distinct selective pressures and their respective evolutionary processes (Verdade et al. 2014b).

Agricultural landscapes may have endangered, damaging, valuable, and populations which do not fit any of those categories. Notwithstanding, the demands of such categories are quite distinct. For instance, in a large country like Brazil there are thousands of endangered species, hundreds of valuable ones, tens of damaging ones, but millions which do not fit any of these categories. Consequently, biodiversity monitoring is the major demand we have in terms of wildlife management specially in human-altered environments (Fig. 1.1). That is why we should monitor wildlife and its habitats in agricultural landscapes.

Biodiversity monitoring has been significantly evolving in the last decades (e.g., Sutherland 2006; Lindenmeyer and Likens 2010; Magurran and McGill 2011; Magnusson et al. 2013). However, it is still biased towards long-term ecological research instead of the decision-making process regarding wildlife management. Despite clear overlaps both initiatives have distinct goals and, consequently, distinct *modus operandi*. Long-term ecological programs usually aim at improving our knowledge about ecosystem structure and functioning, whereas biodiversity mon-

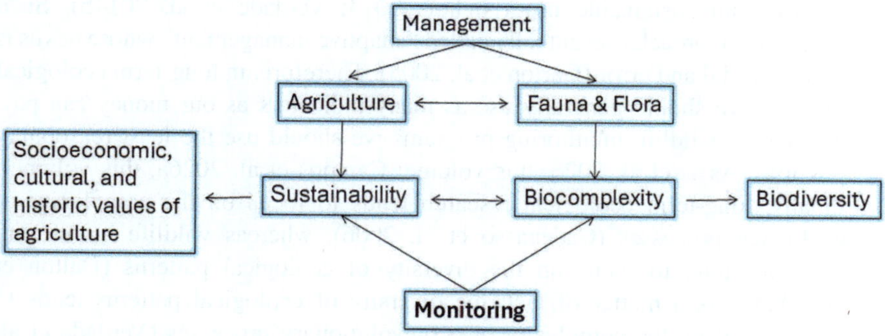

Fig. 1.1 Rationale for a long-term crossing scale biodiversity monitoring program on multifunctional agricultural landscapes. (Adapted from Verdade et al. (2022))

itoring programs aim at improving our capacity to provide biological conservation, sustainable use, and/or wildlife damage control. However, like any other field of knowledge, biodiversity monitoring still has conceptual, technological, and societal challenges (Verdade et al. 2022), as follows.

2 Challenges in Wildlife Monitoring

2.1 *Conceptual Challenges*

The following conceptual questions can be considered as priority to the establishment of biodiversity monitoring programs:

(a) **What should indicators indicate?** The use of indicators is a usual and useful reductionism in wildlife monitoring due to our incapacity to monitor everything, everywhere, all the time. In the academic world indicators tend to be specialist specific, as specialists tend to perceive responses to environmental change faster and clearer on species familiar to them than on others. Actually, all living species respond somehow to environmental changes. To detect their responses, we just need to be able to read their signs . . . Notwithstanding, pragmatically, a good indicator is one which clearly responds to environmental changes ahead of other species, before their impacts are irreversible. Just like the canaries used by former coal miners to indicate toxic levels of carbon monoxide (Pollock 2016).

(b) **How many indicators are too many?** Long-term ecological research tends to have a holistic approach required to understand complex systems such as ecosystems and communities (Pezzini et al. 2012). On the contrary, as stated above, wildlife monitoring programs tend to have a pragmatic approach to support the decision-making process regarding the possible necessity of direct or indirect human interventions concerning biological conservation, damage control, and sustainable use (Nichols 2014; Verdade et al. 2014b). Such pragmatic approach is eventually called "adaptive management" whose nexus is based on trial and error (Garton et al. 2005). Therefore, in long-term ecological research we should generally use as many indicators as our money can pay, whereas in wildlife monitoring programs we should use the least reasonably possible (Assis et al. 2026, this volume; Campos et al. 2026a, this volume). Finally, long-term ecological research tends to focus on the complexity of ecological processes (Cadenasso et al. 2006), whereas wildlife monitoring programs tend to focus on the diversity of ecological patterns (Dalton et al. 2023). As a matter of fact, the diversity of ecological patterns tends to be moulded by the complexity of eco-evolutionary processes (Verdade et al. 2014b). Therefore, the decision-making process driven by wildlife monitoring programs would be significantly improved if based on the complexity of eco-evolutionary processes (Campos et al. 2026b, this volume), even when society pursues specific patterns of biological diversity based on affective memory (e.g., Schama 2004). Ironically, the only principle that wildlife monitoring programs should borrow from long-term ecological research is usually not taken into account...

(c) **How relevant is the 3rd spatial dimension for spatial heterogeneity?** The Theory of Island Biogeography (MacArthur and Wilson 1967) and later the field of Landscape Ecology (Forman 1995) can be considered the current conceptual basis to understand spatial heterogeneity of species distribution and abundance. Both are predominantly based on a two-dimensional approach of space. Such an approach is likely based on the distinct order of magnitude between area (i.e., a combination between width and length) and height of native vegetation. However, the role of height as a third spatial dimension in terms of habitat quality (i.e., resource availability) can be as relevant as area itself (e.g., Bayliss 1987), specially in multifunctional agricultural landscapes (Vásquez-Uribe et al. 2026, this volume).

Multifunctionality can only be achieved by agricultural landscapes on a land-sharing approach with remaining areas of native (re)vegetation kept interspersed over a plantation matrix (Verdade et al. 2016). The land-sparing approach, in contrast, proposes monofunctional landscapes, either for biological production (i.e., large areas of intense agriculture) or biological conservation (i.e., conservation units) (Green et al. 2005). The Brazilian Forest Code (BFC) adopted the land-sharing approach decades before such a debate (Rother et al. 2023). In addition, its conservative yet pragmatic approach tends to cover both controversial conservation strategies from the former "several small vs. single large" (SLOSS) conservation units' debate (Soulé and Simberloff 1986): single large conservation areas represented by its Legal Reserves (LR) as well as several small conservation areas represented by its Areas of Permanent Protection (APP) (Ab'Saber 2011). However, although visionary the BFC has been systematically questioned by a relevant part of the Brazilian agribusiness (Metzger et al. 2010).

In multifunctional agricultural landscapes the pulse of plant biomass dramatically varies between the plantation and the conservation areas (e.g., APP and LR in Brazil). The former presents a low spatial heterogeneity but a high temporal heterogeneity, which is directly related to the agricultural landscape's production value (e.g., productivity and revenue). On the contrary, the latter presents a high spatial heterogeneity and a low temporal heterogeneity, which is directly related to the agricultural landscapes conservation value (e.g., sequestered carbon, biodiversity, and biocomplexity) (Deodoro et al. 2026, this volume; Vásquez-Uribe et al. 2026, this volume).

(d) **How relevant is temporal heterogeneity?** In the short term we usually detect sampling processes, whereas in the medium term we tend to detect ecological processes, and in the long term we tend to detect evolutionary processes (Preston 1960). Thus, wildlife monitoring programs should operate across spatial scales, from local to regional, with medium term frequency of data collection and long-term program operations, preferably permanent for as long as agricultural operations last (Verdade et al. 2014a, 2022). The data generated by such programs need proper statistical treatment to ensure proper inferences are made (Mimnagh et al. 2026b, this volume; Stasinopoulos et al. 2026, this volume; Fokianos 2026, this volume; Shimadzu 2026, this volume; Bassani et al. 2026, this volume).

2.2 *Methodological and Technological Challenges*

In general wildlife counting methods are like looking for a needle in a haystack with the needle usually hiding from you or occasionally attacking you. In addition, it is common to use methods developed for temperate pristine environments in tropical anthropic ones (Penteado et al. 2014), which usually results in low precision and unknown accuracy (Verdade et al. 2013, Magnusson 2014, Robim et al. 2026, this volume).

The lack of specialists and field personnel familiarized enough with certain taxonomic groups is another challenge for the implementation of wildlife monitoring programs in tropical areas with high diversity (Magnusson 2001). Such lack of knowledge appears to be correlated with a decrease in scientific production on natural history and a concomitant increase in scientific production on species distribution modelling based on secondary data (Magnusson and Mourão 1995). Such a technicism tends to be more successful to impulse academic carriers than to improve the decision-making process related to wildlife monitoring (Keville et al. 2017).

It is possible to associate technological innovation with natural history knowledge to improve this decision-making process. Molecular markers are genetic tools that can be used to determine species composition, population density, and diet of carnivores (Jacobus 2026, this volume). Passerine songbirds can also be taped and identified by AI and machine learning to determine species composition and abundance (Gibbons et al. 2026, this volume). Species can also be detected by vestiges (e.g., tracks and marks), but their identification also requires expertise and familiarity with the focused species. In short, technological constraints can be related to sampling and analytical methodology.

2.2.1 Rethinking Tools for Wildlife Monitoring in Human-Altered Landscapes

Traditional monitoring approaches, such as live-trapping, radio-collaring, and intensive field surveys, while informative, are often invasive, expensive, and logistically constrained (Verdade et al. 2013). These methods can be difficult to implement over long time frames or large spatial extents. As a result, they can fall short in delivering generalizable and long-term insights needed for effective wildlife management (especially in human-altered environments). This brings about the need for tools that allow for the collection and analysis of wildlife monitoring data. It is important to stress, however, that the tools should be developed to accommodate the different data structures that can be found.

The analytical methodologies presented in Part III of this book emerge from this principle. Rather than assuming ideal sampling conditions, we tackle the reality of incomplete, noisy, scarce, and heterogeneous data. We present statistical models and algorithms that have been designed to accommodate low detectability, many heterogeneous sources of variation, spatiotemporal autocorrelation, and data scarcity. This realism is central to the book's broader aim: to propose a long-term protocol for wildlife monitoring that can operate under real-world constraints (Verdade et al. 2026, this volume), while still yielding meaningful ecological inferences.

It is important to emphasise that no single model or technique can adequately address all monitoring challenges. There is already a plethora of material in the statistical and ecological literature addressing different methodological approaches, including reviews of commonly used techniques (e.g., Haslett et al. 2022), and tutorials on the analyses of specific types of data and implementation of software (e.g., Fatoretto et al. 2018; Bachl et al. 2019; Moral et al. 2023). Moreover, models can be not only mathematical/deterministic or stochastic/statistical in nature, but also mechanistic, which reflect empirical knowledge (e.g., Pollard et al. 2020), or fully predictive (based on black-box machine learning tools, e.g., Palma et al. 2025). Therefore, the material provided in this book does not aim to be exhaustive. Much on the contrary, we intend to offer a selection of methods, ranging from hierarchical models (e.g., Mimnagh et al. 2026a, b, this volume) to time series (Fokianos 2026, this volume) and mathematical dynamic models (Mason 2026, this volume), that reflect both current advances and promising directions. Each chapter is framed around a set of key problems rather than a fixed solution, which should be regarded as an open forum that invites readers to critically evaluate and adapt techniques to their own monitoring needs.

2.2.2 Data-Driven Approaches

One of the unifying themes of this book is the development of monitoring practices that are non-invasive, inexpensive, and scalable. These three qualities are not only desirable but necessary if we are to monitor millions of species across different spatial and temporal scales. To achieve this, we present methods that emphasise indirect sampling techniques, such as acoustic monitoring (Gibbons et al. 2026, this volume), camera trapping (Campos et al. 2026a, this volume; Bassani et al. 2026, this volume), and vestige collection (Assis et al. 2026, this volume; Campos et al. 2026b, this volume; Azevedo et al. 2026, this volume) coupled with remote sensing (Deodoro et al. 2026, this volume), as primary data sources. These methods, when compared to traditional capture-recapture studies, distance sampling and unmarked counts, minimise ecological disturbance as well as the financial burden, while still allowing for the estimation of biodiversity indices, wildlife abundance, and community dynamics, including trophic structures.

A foundational methodology that recurs throughout Part III is Bayesian hierarchical modelling (for which an introduction is provided by Mimnagh et al. 2026b, this volume). This framework allows us to account for observation error, latent ecological states, and complex data dependencies. Most importantly, it is a very efficient framework to quantify uncertainty about variables of interest. Especially when dealing with scarce data (which demand, in contrast, an abundance of assumptions), Bayesian modelling allows for the incorporation of prior knowledge and the appropriate quantification of uncertainty. This is particularly valuable in human-modified landscapes, where ecological signals can be confounded by anthropogenic noise (Ford 2000).

The integration of artificial intelligence further illustrates the potential of modern data workflows. For instance, machine learning models for species identification, including convolutional neural networks trained on images or soundscapes, can significantly reduce the burden of manual data processing, both in terms of human and financial resources. These tools also allow for real-time monitoring systems that can automatically classify species and detect anomalies and long-term ecological shifts. It is important, however, to stress that a model is only useful when it appropriately describes the data generating mechanisms. In this context, techniques to assess model adequacy, goodness-of-fit, and to select specific models for inferential tasks is crucial to test hypotheses appropriately and obtain reliable conclusions (McInerney et al. 2026, this volume).

2.3 Societal Challenges

Wildlife monitoring in multifunctional agricultural landscapes can provide a direct measure of their conservation value (Verdade et al. 2014a, 2016). We can advocate that any commodity produced in a landscape with a resilient trophic chain is more valuable than a similar one produced in a landscape where most of the biodiversity is extinct. However, most of this value is intangible (Abercrombie and Verdade 1995). An attempt to make it tangible is to call ecosystem services the ecological processes made by some wild species, such as pollination of domesticated species or predation of their natural enemies (Tan et al. 2020). Such an economic approach tends to transform values into market prices (Costanza 1991), although it brings the risk of devaluation of wild species as a result of unpredictable market crises. Not to mention market prices vary due to intangible values like, for example, the lack of confidence on a new president of a rich and powerful country (Taleb 2005).

Knowledge and memory can be used to produce tangible goods by interconnectivity or simply sharing of the imaginary (Harari 2024). However, both are based on intangible human characteristics. Wildlife and its habitat are relevant parts of humans' landscape memory (Schama 2004; Verdade et al. 2016). Local human knowledge can help establish wildlife monitoring programs (Islas et al. 2024) and spread their generated knowledge at regional scales, enhancing human culture (Webb 2014).

Poor wildlife governance tends to limit wildlife management including monitoring programs (Verdade et al. 2022). Practical questions regarding the use of conservation units (Robim et al. 2026, this volume) and the reintroduction of wild species in agricultural landscapes (Verdade and Moral 2026, this volume) usually require a functional wildlife governance. In such a context, wildlife monitoring is fundamental to improving wildlife management on the edge between biological production and conservation. However, a final question remains: Who is going to pay the bill? Verdade et al. (2026, this volume) presents a cost-effective standardized monitoring protocol of wildlife and its habitat especially useful for human-altered environments. The combination of data on biodiversity, biocomplexity, carbon sequestration with commodities' production might significantly reduce its cost for the agro-industrial sectors in charge of land use in agricultural and silvicultural multifunctional landscapes.

3 Future Innovation Opportunities

Our objectives are not to provide only a methodological toolkit. In addition to novel and extended methodologies, as well as tutorials for practitioners, we intend to provide a space for conceptual experimentation. Many of the chapters include thought exercises, theoretical extensions, and techniques that aim to stimulate dialogue and innovation. These elements are intentional. They reflect our belief that progress in wildlife monitoring solutions will come not only from refining existing tools, but also from re-imagining what is possible within data-rich, resource-limited, and ecologically complex scenarios.

One such example is our inclusion of control theory perspectives (Mason 2026, this volume), which is an approach more commonly associated with engineering and robotics. We introduce readers to the concept of observers in population models, with the aims of expanding interdisciplinary approaches to wildlife monitoring.

Another example involves distributional regression (Stasinopoulos et al. 2026, this volume) and mean-variance modelling (Moral et al. 2026, this volume), which according to our own anecdotal evidence are less familiar to many field ecologists and offer a powerful statistical framework to accommodate flexible distributions of response variables. These chapters challenge conventional assumptions and invite readers to consider how insights from other domains can advance wildlife monitoring, and how advanced statistical tools can be used at the service of answering scientific questions and quantifying evidence for or against specific hypotheses.

There is a need to advance both a methodological toolkit and a conceptual agenda, with constant feedback between the two, i.e., conceptual advancements guide new data-driven methodologies, which in turn can be employed to test hypotheses that will update concepts and paradigms (Popovic et al. 2024). In such a context, interdisciplinarity is key, alongside practical implementation. Only under such a framework is it possible to redefine how we observe, understand, and protect wildlife in a rapidly changing world.

Acknowledgements This study was partially funded by Fundação de Amparo à Pesquisa do Estado de São Paulo (FAPESP, Proc. No. 2017/01304-4).

References

Ab'Saber AN (2011) Evolving from a forest code to a biodiversity code. Biota Neotrop 10(4):331–336

Abercrombie CL, Verdade LM (1995) Dinâmica populacional de crocodilianos: elaboração e uso de modelos. In: Larriera A, Verdade LM (eds) Conservación y Manejo de los Crocodylia de America Latina, vol 1. Fundación Banco Bica, Sao Tomé, pp 33–55. (ISBN 950-9632-21-X)

Assis CK, Moral RA, Piña CI, Verdade LM (2026) Estimating the number of groups and population density of social species based on vestiges in human-altered environments: the case study of collared-peccaries (*Pecari tajacu*). In: Martins Verdade L, Moral RA (eds) Wildlife monitoring: integrating conservation and innovation in human-altered landscapes. Springer, Cham

Azevedo FCC, Moral RA, Verdade LM (2026) Wild boar introduction might result in the selection of larger pumas in Central-South Brazil. In: Martins Verdade L, Moral RA (eds) Wildlife monitoring: integrating conservation and innovation in human-altered landscapes. Springer, Cham

Bachl FE, Lindgren F, Borchers DL, Illian JB (2019) Inlabru: an R package for Bayesian spatial modelling from ecological survey data. Methods Ecol Evol 10(6):760–766

Bassani VF, Forace GC, Rodrigues DFL, Lourenço MFR, Timo TPC, Sanches A, Mimnagh N, Moral RA, Ferreira IEP (2026) The influence of fragmented landscapes on population density studies using camera trapping. In: Martins Verdade L, Moral RA (eds) Wildlife monitoring: integrating conservation and innovation in human-altered landscapes. Springer, Cham

Bayliss P (1987) Kangaroo dynamics. In: Caughley G, Shepherd N, Short J (eds) Kangaroos: their ecology and management in the sheep rangelands of Australia. Cambridge University Press, Cambridge, UK, pp 119–134

Cadenasso ML, Pickett STA, Grove JM (2006) Dimensions of ecosystem complexity: heterogeneity, connectivity, and history. Ecol Complex 3(1):1–12

Campos BM, Charters JD, Piña CI, Moral RA, Verdade LM (2026a) Estimating population density of Neotropical terrestrial mammals in human-altered environments based on camera-trapping. In: Martins Verdade L, Moral RA (eds) Wildlife monitoring: integrating conservation and innovation in human-altered landscapes. Springer, Cham

Campos CB, Gheler-Costa C, Esteves CF, Rech-Fernandes TC, Rosalino LM, Moral RA, Piña CI, Verdade LM (2026b) Sampling scats in trails: trophic approach in wildlife surveys. In: Martins Verdade L, Moral RA (eds) Wildlife monitoring: integrating conservation and innovation in human-altered landscapes. Springer, Cham

Caughley G (1994) Directions in conservation biology. J Anim Ecol 63:215–244

Costanza R (1991) Ecological economics: the science and management of sustainability. Columbia University Press, New York

Dalton DT, Berger V, Adams V, Botha J, Halloy S, Kirchmeir H, Sovinc A, Steinbauer K, Švara V, Jungmeier M (2023) A conceptual framework for biodiversity monitoring programs in conservation areas. Sustainability 15(8):6779

Deodoro SC, Fealy R, McCarthy T (2026) The use of remote sensing data for above-ground biomass estimation. In: Martins Verdade L, Moral RA (eds) Wildlife monitoring: integrating conservation and innovation in human-altered landscapes. Springer, Cham

Fatoretto MB, Moral RA, Demétrio CGB, de Pádua CS, Menarin V, Rojas VMA, D'Alessandro CP, Delalibera I Jr (2018) Overdispersed fungus germination data: statistical analysis using R. Biocontrol Sci Tech 28(11):1034–1053

Fokianos K (2026) A tutorial on observation-driven count time series models applied to ecology. In: Martins Verdade L, Moral RA (eds) Wildlife monitoring: integrating conservation and innovation in human-altered landscapes. Springer, Cham

Ford ED (2000) Scientific method for ecological research. Cambridge University Press, Cambridge, UK

Forman RT (1995) Land mosaics: the ecology of landscapes and regions. Cambridge University Press, Cambridge, UK

Garton EO, Ratti JT, Giudice JH (2005) Research and experimental design. In: Braun CE (ed) Techniques for wildlife investigations and management. The Wildlife Society, Bethesda, pp 43–71

Gibbons A, Donohue I, Parnell A (2026) Monitoring with machines: a review of computational bioacoustics. In: Martins Verdade L, Moral RA (eds) Wildlife monitoring: integrating conservation and innovation in human-altered landscapes. Springer, Cham

Green RE, Cornell SJ, Scharlemann JP, Balmford A (2005) Farming and the fate of wild nature. Science 307(5709):550–555

Harari YN (2024) Nexus: a brief history of information networks from the stone age to AI. Random House, New York

Haslett J, Parnell AC, Hinde J, Moral RA (2022) Modelling excess zeros in count data: a new perspective on modelling approaches. Int Stat Rev 90(2):216–236

Islas CA, Verdade LM, Seixas CS (2024) Indications of changes in hunting culture in the central-south region of Brazil in the last 25 years: a systematic literature review. Biota Neotrop 24(1):e20231531. https://doi.org/10.1590/1676-0611-BN-2023-1531

Jacobus AP (2026) Molecular ecology techniques for wildlife monitoring integrating conservation and innovation in human-altered landscapes. In: Martins Verdade L, Moral RA (eds) Wildlife monitoring: integrating conservation and innovation in human-altered landscapes. Springer, Cham

Keville MP, Nelson CR, Hauer FR (2017) Academic productivity in the field of ecology. Ecosphere 8(1):e01620. https://doi.org/10.1002/ecs2.1620

Lindenmeyer DB, Likens GE (2010) Effective ecological monitoring. CSIRO Publishing, Collingwood

MacArthur RH, Wilson EO (1967) The theory of Island biogeography. Princeton University Press, Princeton

Magnusson WE (2001) Standard errors of survey estimates: what do they mean? Neotrop Primates 9(2):53–54

Magnusson WE (2014) Uncertainty and the design of in-situ biodiversity-monitoring programs. Nat Conserv 8:77–94. https://doi.org/10.3897/natureconservation.8.5929

Magnusson WE, Mourão GM (1995) Safe options for the management of crocodilians. CSG Newsl 14(4):3–5

Magnusson WE, Braga-Neto R, Pezzini F, Baccaro F, Bergallo H, Penha J, Rodrigues D, Verdade LM, Lima A, Albernaz AL, Hero J-M, Lawson B, Castilho C, Drucker D, Franklin E, Mendonça F, Costa F, Galdino G, Castley G, Zuanon J, do Vale J, Campos L, Luizão R, Cintra R, Barbosa RI, Lisboa A, Koblitz RV, Mendes-Pontes AR, da Cunha CN (2013) Biodiversidade e monitoramento ambiental integrado. Áttema Editorial, Santo André. 352p. (ISBN 978-85-65551-04-5)

Magurran AE, McGill BJ (2011) Biological diversity: frontiers in measurement and assessment. Oxford University Press, Oxford, UK

Martinelli LA, Joly CA, Nobre CA, Sparovek G (2010) A falsa dicotomia entre a preservação da vegetação natural e a produção pecuária. Biota Neotrop 10:323–330

Mason O (2026) Control theoretic perspectives on wildlife monitoring: a tutorial introduction to observers for population models. In: Martins Verdade L, Moral RA (eds) Wildlife monitoring: integrating conservation and innovation in human-altered landscapes. Springer, Cham

McInerney R, Valiquette E, Reis SF, Moral RA (2026) Statistical model adequacy, goodness-of-fit, and model selection. In: Martins Verdade L, Moral RA (eds) Wildlife monitoring: integrating conservation and innovation in human-altered landscapes. Springer, Cham

Metzger JP, Lewinsohn TM, Joly CA, Verdade LM, Martinelli LA, Rodrigues RR (2010) Brazilian law: full speed in reverse? Science 329:276–277

Mimnagh N, Ferreira IEP, Verdade LM, Moral RA (2026a) Counting animals we can't see: the triple poisson model for vestiges. In: Martins Verdade L, Moral RA (eds) Wildlife monitoring: integrating conservation and innovation in human-altered landscapes. Springer, Cham

Mimnagh N, Moral RA, Prado EB (2026b) An introduction to Bayesian hierarchical modelling applied to wildlife monitoring. In: Martins Verdade L, Moral RA (eds) Wildlife monitoring: integrating conservation and innovation in human-altered landscapes. Springer, Cham

Moral RA, Vishwakarma R, Connolly J, Byrne L, Hurley C, Finn JA, Brophy C (2023) Going beyond richness: modelling the BEF relationship using species identity, evenness, richness and species interactions via the DImodels R package. Methods Ecol Evol 14(9):2250–2258

Moral RA, Santin M, Verdade LM (2026) Mean-variance modelling of Neotropical passerine bird songs. In: Martins Verdade L, Moral RA (eds) Wildlife monitoring: integrating conservation and innovation in human-altered landscapes. Springer, Cham

Nichols J (2014) The role of abundance estimates in conservation decision-making. In: Verdade LM, Lyra-Jorge MC, Piña CI (eds) Applied ecology and human dimensions in biological conservation. Springer, Heidelberg, pp 117–131. https://doi.org/10.1007/978-3-642-54751-5_8. (ISBN 978-3-642-54750-8)

Palma GR, Mello RF, Godoy WA, Engel E, Lau D, Markham C, Moral RA (2025) Forecasting insect abundance using time series embedding and machine learning. Eco Inform 85:102934

Penteado M, Silva WR, Verdade LM (2014) Wildlife surveys in agricultural landscapes: birds. In: Verdade LM, Lyra-Jorge MC, Piña CI (eds) Applied ecology and human dimensions on biological conservation. Springer, Heidelberg, pp 149–158. https://doi.org/10.1007/978-3-642-54751-5_10. (ISBN 978-3-642-54750-8)

Pezzini FFPH, Melo A, Oliveira DMS, Amorim RX, Figueiredo FOG, Drucker DP, Rodrigues FRO, Zuquim G, Sousa TEL, Costa FRC, Magnusson WE, Sampaio AF, Lima AP, Garcia ARM, Manzatto AG, Nogueira A, Costa CP, de Araújo Barbosa CE, Castilho CBCV, Cunha CN, Freitas CG, Cavalcante CO, Brandão DO, Rodrigues DJ, Santos ECPR, Baccaro FB, Ishida FY, Carvalho FA, Moulatlet GM, Guillaumet J-LB, Pinto JLPV, Schietti J, Vale JD, Belger L, Verdade LM, Pansonato MP, Nascimento MT, Santos MCV, Cunha MS, Arruda R, Barbosa RI, Romero RL, Pansini S, Pimentel TP (2012) Biodiversity & ecology. J Div Biodiv Evol Ecol Plants 4:265–274. https://doi.org/10.7809/b-e.00083

Pollard CP, Griffin CT, Moral RA, Duffy C, Chuche J, Gaffney MT, Fealy RM, Fealy R (2020) phenModel: a temperature-dependent phenology/voltinism model for a herbivorous insect incorporating facultative diapause and budburst. Ecol Model 416:108910

Pollock C (2016) The canary in the coal mine. J Avian Med Surg 30(4):386–391

Popovic G, Mason TJ, Drobniak SM, Marques TA, Potts J, Joo R, Altwegg R, Burns CCI, McCarthy MA, Johnston A, Nakagawa S (2024) Four principles for improved statistical ecology. Methods Ecol Evol 15(2):266–281

Preston FW (1960) Time and space and the variation of species. Ecology 41(4):611–627

Robim MJ, Villani JP, Menezes DS, Barros FC, Souza HA, Toledo MCB, Piña CI, Moral RA, Verdade LM (2026) The influence of rafting on the use of riparian Atlantic Forest by middle to large terrestrial mammals. In: Martins Verdade L, Moral RA (eds) Wildlife monitoring: integrating conservation and innovation in human-altered landscapes. Springer, Cham

Rother DC, Romanelli JP, Rodrigues RR (2023) Historical trajectory of restoration practice and science across the Brazilian Atlantic Forest. Restor Ecol 31(8):e14041

Schama S (2004) Landscape and memory. Harper Perennial, London

Shimadzu H (2026) Temporal turnover applications in wildlife monitoring. In: Martins Verdade L, Moral RA (eds) Wildlife monitoring: integrating conservation and innovation in human-altered landscapes. Springer, Cham

Soulé ME, Simberloff D (1986) What do genetics and ecology tell us about the design of nature reserves? Biol Conserv 35(1):19–40

Stasinopoulos M, Bastiani F, Burke K, Merder J (2026) Distributional regression techniques for wildlife management. In: Martins Verdade L, Moral RA (eds) Wildlife monitoring: integrating conservation and innovation in human-altered landscapes. Springer, Cham

Sutherland WJ (2006) Ecological census techniques: a handbook. Cambridge University Press, Cambridge

Taleb N (2005) The black swan: the impact of the highly improbable. Random House, New York

Tan PY, Zhang J, Masoudi M, Alemu JB, Edwards PJ, Grêt-Regamey A, Richards DR, Saunders J, Song XP, Wong LW (2020) A conceptual framework to untangle the concept of urban ecosystem services. Landsc Urban Plan 200:103837

Vásquez-Uribe LC, Toledo RM, Moral RA, Pivello VR, Verdade LM (2026) Relationship between above-ground biomass of native vegetation and the abundance of small mammals in agro-silviculutral landscapes. In: Martins Verdade L, Moral RA (eds) Wildlife monitoring: integrating conservation and innovation in human-altered landscapes. Springer, Cham

Verdade LM, Moral RA (2026) Rewilding agricultural landscapes: conceptual, technological, and societal constraints. In: Martins Verdade L, Moral RA (eds) Wildlife monitoring: integrating conservation and innovation in human-altered landscapes. Springer, Cham

Verdade LM, Moreira JR, Ferraz KMPMB (2013) Counting capybaras. In: Moreira JR, Ferraz KMPMB, Herrera EA, Macdonald DW (eds) Capybara: biology, use and conservation of an exceptional neotropical species. Springer, New York, pp 357–370. https://doi.org/10.1007/978-1-4614-4000-0_22. (ISBN 978-1-4614-3999-8)

Verdade LM, Lyra-Jorge MC, Piña CI (2014a) Redirections in conservation biology. In: Verdade LM, Lyra-Jorge MC, Piña CI (eds) Applied ecology and human dimensions in biological conservation. Springer, Heidelberg, pp 3–17. https://doi.org/10.1007/978-3-642-54751-5_1. (ISBN 978-3-642-54750-8)

Verdade LM, Penteado M, Gheler-Costa C, Dotta G, Rosalino LM, Pivello VR, Piña CI, Lyra-Jorge MC (2014b) The conservation value of agricultural landscapes. In: Verdade LM, Lyra-Jorge MC, Piña CI (eds) Applied ecology and human dimensions in biological conservation. Springer, Heidelberg, pp 91–102. https://doi.org/10.1007/978-3-642-54751-5_6. (ISBN 978-3-642-54750-8)

Verdade LM, Gheler-Costa C, Lyra-Jorge MC (2016) The multiple facets of agricultural landscapes. In: Gheler-Costa C, Lyra-Jorge MC, Verdade LM (eds) Biodiversity in agricultural landscapes of Southeastern Brazil. De Gruyter, Berlin, pp 2–13. https://doi.org/10.1515/9783110480849-003

Verdade LM, Bianchi RC, Galetti PM Jr, Pivello VR, Silva WR, Uezu A (2022) Biodiversity management and research in multifunctional landscapes. Biota Neotrop 22(spe):e20221407. https://doi.org/10.1590/1676-0611-BN-2022-1407

Verdade LM, Toledo RM, Pivello V, Jacobus AP, Moral RA (2026) Long-term spatial-crossing scale wildlife monitoring: a protocol for agricultural and silvicultural landscapes of Central-south Brazil. In: Martins Verdade L, Moral RA (eds) Wildlife monitoring: integrating conservation and innovation in human-altered landscapes. Springer, Cham

Webb G (2014) Wildlife conservation: in the belly of the beast. Charles Darwin University Press, Darwin

Chapter 2
Estimating the Number of Groups and Population Density of Social Species Based on Vestiges in Human-Altered Environments: The Case Study of Collared-Peccaries (*Pecari tajacu*)

Cecília Kruszynski, Rafael de Andrade Moral, Thais R. Diniz-Reis, Carlos I. Piña, and Luciano Martins Verdade

Abstract We estimated the population density of collared peccaries (*Pecari tajacu*) in a landscape composed of *Eucalyptus* plantations as a matrix and remnants of native vegetation in southeast Brazil. To do so, we estimated the number of groups based on a combination of the detection of tracks in line transects and home ranges (minimum and maximum) from a literature review. We then estimated group size based on camera trapping. As a result, we found 6–10 local groups with seven to nine individuals each with an estimated population density between 0.96 and 2.06 individuals/km^2. This is approximately one-half of the density described for the species in Brazilian Pantanal, where it has an intermediate home range size in comparison to the minimum and maximum values used in this study. The present results suggest that: (a) the present density estimate is reliable, despite its relatively smaller sampling effort than traditional methods; (b) despite of negative impacts related to agriculture intensification, silvicultural landscapes may be relevant for the conservation of collared peccaries and the maintenance of functional trophic processes in human-altered environments of southeast Brazil, and (c) the possible

C. Kruszynski
Wiley VCH, Berlin, Germany

R. A. Moral
Department of Mathematics and Statistics, Maynooth University, Maynooth, Co. Kildare, Ireland

T. R. Diniz-Reis
Universidade de São Paulo, CENA, Piracicaba, São Paulo, Brazil

C. I. Piña
CICyTTP-CONICET Entre Ríos/UADER, España, Diamante, Entre Ríos, Argentina

L. Martins Verdade (✉)
Wildlife Monitoring Consultancy, Campinas, São Paulo, Brazil
e-mail: luciano.verdade@wildlifemgmtco.com

L. Martins Verdade, R. A. Moral (eds.), *Wildlife Monitoring*,
https://doi.org/10.1007/978-3-032-05821-8_2

impact of the introduction of wild boars on collared peccaries should be prioritized by future studies in silvicultural landscapes of southeast Brazil.

Keywords Wildlife management · Forestry · Neotropics · Monitoring

1 Introduction

The collared peccary (*Pecari tajacu*) can be more tolerant to habitat degradation and human hunting pressure than the white-lipped peccary (*Tayassu pecari*) (Altrichter 2005; Dotta and Verdade 2011). However, habitat fragmentation has resulted in its decline in the Atlantic Forest (Mendes et al. 2023). Their conservation should be prioritized because they are fundamental agents in seed dispersion, mostly for palm trees (Keuroghlian et al. 2004; Reis et al. 2006) and one of the largest prey of top carnivores, such as pumas (*Puma concolor*) and jaguars (*Panthera onca*) (Martins et al. 2008; Mazzolli et al. 2008; Polisar et al. 2003; Pontes and Chivers 2007). Monitoring their population size should, therefore, be pursued to guide the decision-making process concerning poaching enforcement, and species conservation.

Collared peccaries are distributed from southwest United States to northeast Argentina (Oliver 1993). However, their actual distribution, especially in South America, is not well delimited because of the lack of data about the species occurrence in remote areas (Sowls 1997; Desbiez et al. 2012). The species lives in distinct environments such as tropical forests, dry and humid savannas, Chaco, and deserts (Oliver 1993). Its home range is usually constant throughout the year (Keuroghlian and Eaton 2008a). Group size varies from 2 to 50 individuals and seems to be influenced by rain frequency, hunting pressure, and forest fragmentation (Sowls 1997). In the Amazon, it is rare to find more than ten individuals per group (Peres 1996). However, groups can split into subgroups and/or smaller associations along the day and reunite to rest during the night (Sowls 1997; Robinson and Eisenberg 1985; Keuroghlian and Eaton 2008a). Reproduction seems to occur throughout the year, without a pronounced breeding season, with one or two cubs per female after approximately 145 days of gestation period (Sowls 1997).

The species can be found in silvicultural landscapes of the state of São Paulo in southeastern Brazil (Timo et al. 2015) and other human-modified environments (Rocha et al. 2023). However, the collared peccaries face three major challenges: the environmental impacts caused by agricultural intensification, poaching, and the local introduction of wild boars (Desbiez et al. 2012; Pereira et al. 2021).

Although relevant, hunting pressure is not well studied in Brazil due to the lack of professional secrecy and confidentiality to biologists, which prevents us from knowing its actual dimension (Verdade and Seixas 2013). Moreover, the species is consumed by rural communities in southeast Brazil (Morais et al. 2022) as well as in other countries of Latin America (Redford and Robinson 1987; Figueiredo and Sousa 2023).

Wild boar is considered one of the worst invasive species in the world (Barrios-Garcia and Ballari 2012). Its presence results in possible competition for resources, predation, and the introduction of infectious diseases in other wild pigs (Herrera et al. 2005; Nogueira-Filho et al. 2009; Miller et al. 2017; McClure et al. 2018). However, the possible impact (either positive or negative) of the control of wild boars on peccaries is still largely unknown.

The planted area of *Eucalyptus* is of 7.3 million hectares in Brazil (ABRAF 2022). The paper and cellulose industry follows the Brazilian Environmental Law relatively well due to the certification process of the Forest Stewardship Council (FSC) (Verdade et al. 2014a, 2016a). In addition, the *Eucalyptus* plantations are used as habitat by a relevant number of medium to large mammals, such as the maned wolf (*Chrysocion brachiurus*), ocelot (*Leopardus guttulus*), coati (*Nasua nasua*), deer (*Mazama* sp.), armadillo (*Dasypus novemcinctus* and *Cabassous unicinctus*), giant anteater (*Myrmercophaga tridactyla*), and the collared peccary (Mello 2005; Dotta and Verdade 2007; Timo et al. 2015; Campos et al. 2018). Certified silvicultural landscapes can, therefore, be considered relevant for wildlife conservation in Southeastern Brazil as an effective example of multifunctional landscapes (Bodmer et al. 1997; Martinelli et al. 2010; Verdade et al. 2014a, 2016b).

A trophic approach on wildlife monitoring might be useful to determine the complexity of its trophic ecology and its conservation value beyond simple patterns of biological diversity (Campos et al. 2026, this volume). Its presence possibly enhances the local habitat carrying capacity of top predators such as pumas (*P. concolor*) (Fragoso 1994; Polisar et al. 2003; Altrichter and Boaglio 2004; Keuroghlian et al. 2004; Pontes and Chivers 2007; Martins et al. 2008; Mazzolli et al. 2008; Nava 2008) or even (controversially) the jaguar (*Panthera onca*) (Da Rosa et al. 2016; Verdade et al. 2016b, 2018; Hegel and Marini 2018), which would enhance the conservation value of silvicultural landscapes (Verdade et al. 2014b, 2016a).

Distance methods have been successfully used in wildlife surveys and monitoring worldwide in the last decades (Nichols 2014). However, its assumption that the presence of observers does not affect the permanence of animals on the transect lines can barely work for hunted species like the collared peccary (e.g., Verdade 1996; Potratz et al. 2024). This is due to the avoidance behavior of those species to the observer's presence due to likely previous experiences with poachers (Fragoso et al. 2016; Silva et al. 2022). In addition, wildlife use of agricultural landscapes tends to be biased as their matrix is usually considered as non-habitat (Verdade et al. 2014a). In the present study, we adapted the method proposed by Fragoso et al. (2016) to estimate group size and population density of collared peccaries based on sights and vestiges (Krebs 2006: 364–365) in a silvicultural landscape of southeast Brazil.

2 Study Area

This study was carried out at two silvicultural farms: Fazenda Três Lagoas (3.242 ha, 23°22′0″–23°20′41″S/48°28′00″– 48°27′57″W) and Fazenda Arca (1.123 ha, 23°20′0″–23°18′51″S/ 48°27′30″–48°28′20″W). They are both located in the municipality of Angatuba, south-central region of the state of São Paulo in Southeast Brazil (Fig. 2.1). Fazenda Três Lagoas consists of a matrix of *Eucalyptus* plantations (2224 ha) established between August 2006 and December 2007. The farm also has 587 ha of Legal Reserve (LR) and 270 ha of Areas of Permanent Protection (APP) with remnants of native vegetation and abandoned pasture. In Fazenda Arca, there was also the replacement of pasture by 722 ha *Eucalyptus* plantations in 2008 with 294 ha of APP and 67 ha of LR (Martin et al. 2012). The total study area was then 4365 ha (i.e., 43.65 km^2).

In both farms, APP and LR have remnants of native vegetation composed of a transition between Cerrado and semideciduous Atlantic Forest in varying stages of regeneration (Athayde et al. 2015, 2016) following the Brazilian Forest Code (BFC) (Metzger et al. 2010; Lopes et al. 2016). The present study was carried out during the first 7-year commercial cycle of the *Eucalyptus* plantations in both farms.

3 Methodology

3.1 Sampling Methodology

We searched for direct sights and vestiges (e.g., tracks, scats, and marks in trees and on the ground) on two transect lines at least 1 km away from each other on the two local drains—southwest (SW) and northeast (NE)—in eight campaigns every other week from July to October 2012. The SW transect was 8 km long (4.95 km in *Eucalyptus* and 3.05 km in APP/LR) and the NE transect was 12 km long (7.84 km in *Eucalyptus* and 4.16 km in APP/LR; Fig. 2.1), totaling 160 km (i.e., 8 campaigns × 20 km total transect lines). Walking speed was approximately 1 km/h with brief stops every 50 m for better viewing and listening of the environment (see Cullen Jr. et al. 2009 for methodological assumptions).

Nineteen camera traps were deployed in both farms, on dirt roads, and places where we previously found vestiges of animals during November 2012. Corn was placed at camera stations to serve as bait for the peccaries. Photos recorded were used to estimate the peccaries' group sizes (Gonçalves et al. 2024).

3.2 Analytical Methodology

The sampling effort efficiency was checked by an adapted collector's curve between the number of direct and indirect animals' detections and the distance covered by the observer. We fitted a Bayesian nonlinear asymptotic model to better account for the uncertainty via credible intervals. The model is

$$Y_t \sim \mathrm{N}\left(\mu_i, \sigma^2\right)$$

$$\mu_t = \alpha\left(1 - \exp\left\{-t \times \exp\left(\beta\right)\right\}\right)$$

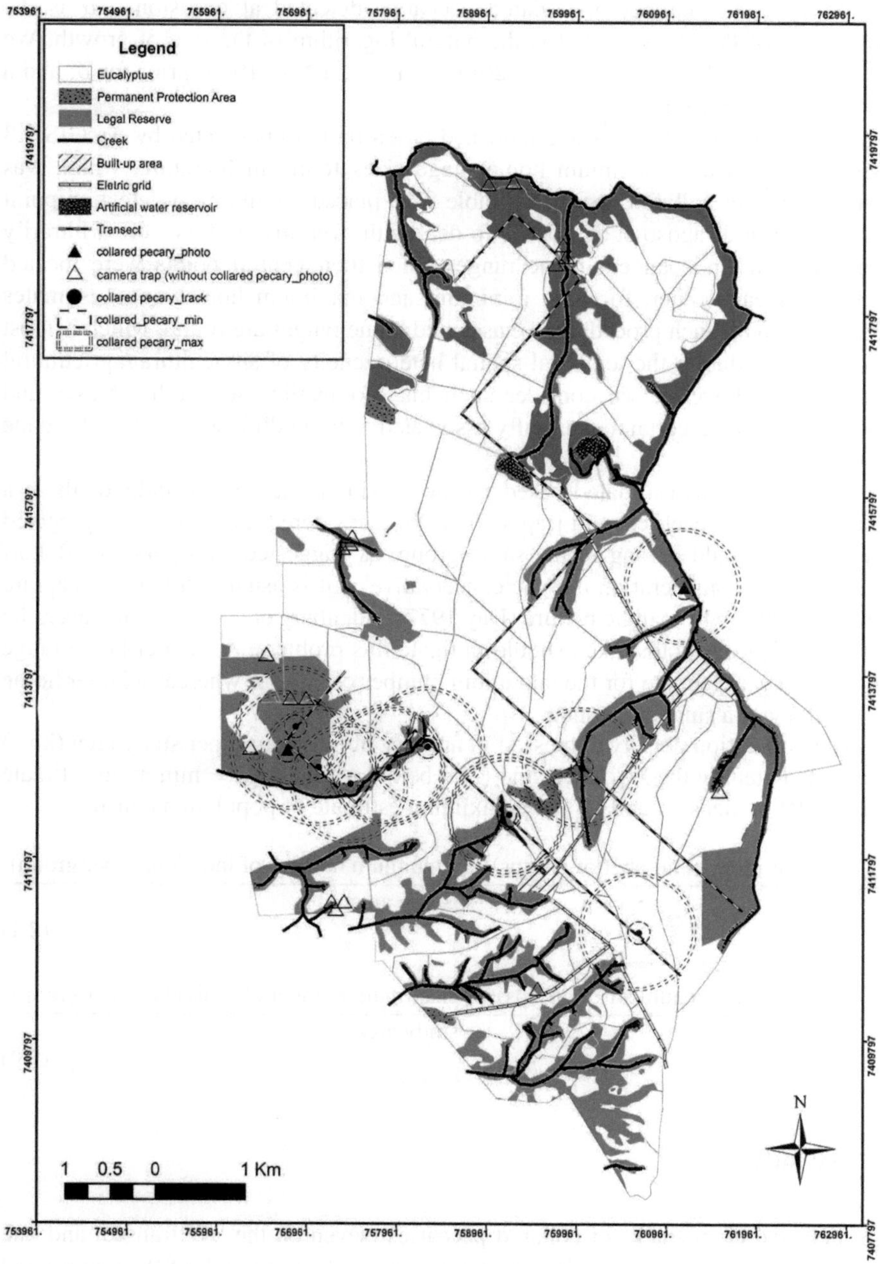

Fig. 2.1 Farms Três Lagoas and Arca (respectively at south and north), in Angatuba municipality, state of São Paulo, southeast Brazil (Datum Corrego Alegre). Location of vestiges and sights of collared peccaries, and their minimum and maximum home range buffers

where y_i is the number of collared peccaries detected at occasion t, α is the asymptote of the curve, and β is the natural logarithm of the rate of growth. We assumed a weakly informative U(0,20) prior for α, a N(0, 1000) prior for β, and a half-Cauchy prior for σ.

The number of groups was established based on buffers created by ArcGIS 9.3 with minimum and maximum home range sizes found in literature, which was 0.4–12 km^2 for collared peccaries (Table 2.1), placed around each detection point (adapted from Fragoso et al. 2016). To deal with overlaps, buffers were arbitrarily considered as independent home ranges when their central points were located outside of each other, for both minimum and maximum home range estimates (Fig. 2.2). With such procedure, we assumed home ranges are round, which is most likely biased due to the temporal-spatial heterogeneity of silvicultural/agricultural landscapes. However, we consider such bias compatible with other biases and methodological uncertainties usually associated with wildlife surveys (see Verdade et al. 2013).

Group size was estimated based on the maximum number of individuals in a single photo for each camera trap station. We considered that animals aggregated by the bait would belong to the same group, as suggested by Byers and Bekoff (1981). Such consideration may be conservative as it is usually difficult to capture the whole group in a single picture (Day 1977; Oldenburg et al. 1985). However, the large number of photos taken should mitigate this problem. A smaller home range yields a larger estimate for the maximum number of groups, whereas a larger home range induces a smaller estimate.

The population density expressed in number of individuals per study area (km^2) was calculated by the Eqs. (2.1) and (2.2) below where D_{min} = minimum estimate of population density, and D_{max} = maximum estimate of population density:

$$D_{\min} = \frac{(\text{minimum number of groups}) \times (\text{minimum number of individuals per group})}{\text{study area}} \quad (2.1)$$

$$D_{\max} = \frac{(\text{maximum number of groups}) \times (\text{maximum number of individuals per group})}{\text{study area}} \quad (2.2)$$

4 Results

We detected eight tracks of collared peccaries (seven on the SW transect and one on the NE transect) and two individuals in camera-traps (one the SW transect and one on the NE transect) (Table 2.2). The adapted collector's curve showed that the present sampling effort reached 64.5% of the estimated number of detections (asymptote posterior mean = 15.50; actual = 10) (Fig. 2.3). Based on the buffers placed around each detection, we estimated between 6 and 10 groups of collared peccaries (Fig. 2.1).

Table 2.1 Group size and home range of collared peccaries

Group size	Home range (km^2)	Population density (ind/km^2)	Habitat	Hunting	Study area	Country	Reference
3–12	–	0.062–0.91	Chaco	Yes	Chaco province	Argentina	Altrichter (2005)
5–15	0.4–8	–	Desert	–	Arizona and Texas	United States	Jennings and Harris (1953), Knipe (1957), Neal (1959), Minnamon (1962), Schweinsburg (1971), Bigler (1974), Sowls (1974), Bissonette (1976)
12–36	10–12	2–7.2	Forest	Yes	Amazon	Brazil	Fragoso (1994)
3–20	1		Forest and savanna	Yes	Maracá Island	Brazil	Fragoso (1999)
5–50	1.57–2.43	6.17–95.54	Forest	–	Saint-Eugène	French Guiana	Judas and Henry (1999)
7–22	1.23–3.05	2.8–8.9	Cerrado	Yes	São Paulo	Brazil	Keuroghlian et al. (2004)
14–25	2	–	Forest	–	Barro Colorado Hato Massaguaral	Panama Venezuela	Robinson and Eisenberg (1985)
4–16	2	4–6	Wetland	Yes	Pantanal	Brazil	Oliveira-Santos et al. (2011)
3–10	2–3	3–15	Forest	No	La Selva	Costa Rica	Suarez (1993)
4–11	1.96	2–5.61	Chaco		Paraguayan Chaco	Paraguay	Taber et al. (1994)
25	3.88	6.44	Desert	–	Roosevelt Lake	United States	Zervanos and Hadley (1973)
	1.26–2.22 1.88–2.55	12.5 25		Yes No	Scott Site Texas Shaffer Site Texas	United States	Ellisor and Harwell (1969)
3.43	–	2.09	Forest	No	Uauaçu Lake Amazonia	Brazil	Haugaasen and Peres (2005)
2–8	1–3.6	2.6–15.5	Semiarid	–	Texas	United States	Gabor and Hellgren (2000)
	–	2.25–8.2	Forest	No	Manu NP	Peru	Endo et al. (2010)

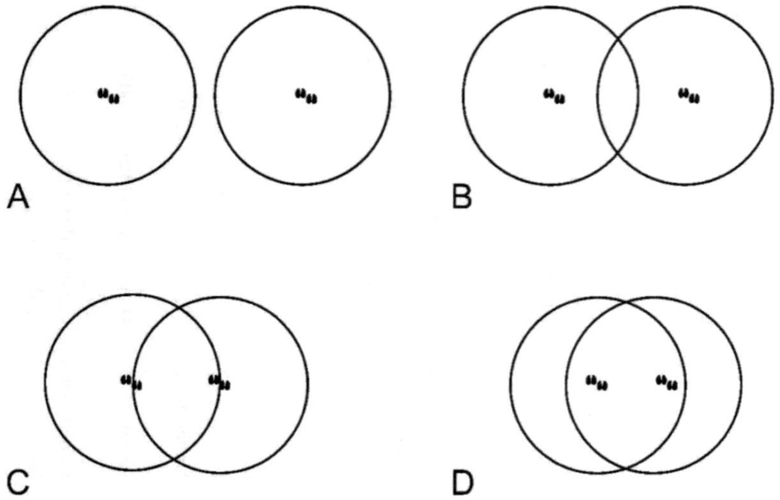

Fig. 2.2 Criteria for the differentiation of home ranges (i.e., buffers) among groups based on vestiges (e.g., tracks) found in the trail. (**a, b**) are considered as two distinct groups with occasionally some spatial overlap, whereas (**c, d**) are considered the same group sampled twice. Criteria are described in the text

Table 2.2 Number of detections of collared peccary and wild boar in the study area

	Transect line		Total
Detection type	SW	NE	
Camera trap	1	1	2
Track	7	1	8
Total	8	2	10

In addition, we took 1471 photos of collared peccaries in two of the 19 camera-trap stations (one in each transect). The photographic records suggest that the collared peccary groups contain between seven and nine individuals (Fig. 2.4). Thus, the number of individuals in the study area was estimated to be between 42 (i.e., 6 groups × 7 individuals/group) and 90 (i.e., 10 groups × 9 individuals/group). Considering the total area of 43.65 km^2 for both farms together, we estimated a density of collared peccaries between 0.96 and 2.06 individuals/km^2. We detected the local presence of exotic wild boars by tracks (N = 20), sights (N = 2), and photographs (N = 2). However, we did not estimate their abundance as the species had just recently arrived in the study area with anecdotal reports of high population growth from local residents.

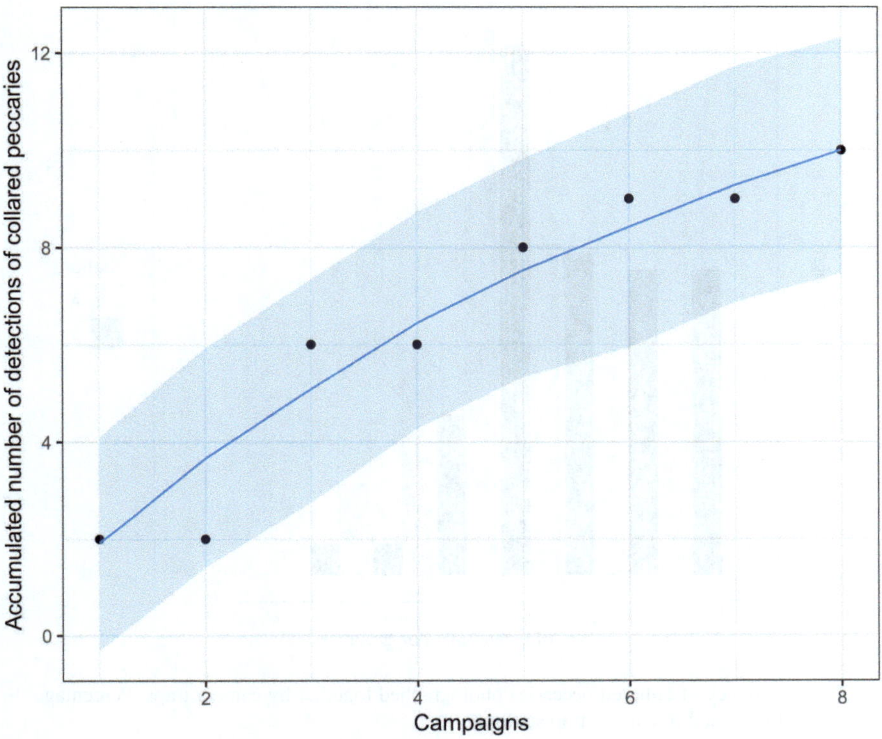

Fig. 2.3 Observed accumulated number of detections of collared peccaries (black dots) and fitted Bayesian nonlinear model (blue curve), alongside 95% credible intervals for the true mean number of detections (shaded ribbon)

5 Discussion and Conclusion

The sampling effort was enough to detect approximately 65% of the total number of vestiges, which is slightly below the range reached by comparable previous local surveys of medium to large terrestrial mammals (between 71.5% and 95.2% by Timo et al. (2015), and between 83.1% and 91.1% by Campos et al. (2018)). These values mean that the collared peccary abundance might have been underestimated by two-thirds of their actual population density, which can be considered comparable to other studies (Verdade et al. 2013).

The collared peccary appears to use *Eucalyptus* plantations like other species of mammals (Martin et al. 2012; Rosalino et al. 2013; Timo et al. 2015), birds (Millan et al. 2015), and amphibians (Lopes et al. 2016). However, they appear to dwell at the conservation areas (APP and RL) nearby, as described in previous studies (Altrichter and Boaglio 2004; De Araújo et al. 2008; Kasper et al. 2007; Licona et al. 2011). Such a pattern stresses the importance of conservation areas as part of silvicultural landscapes.

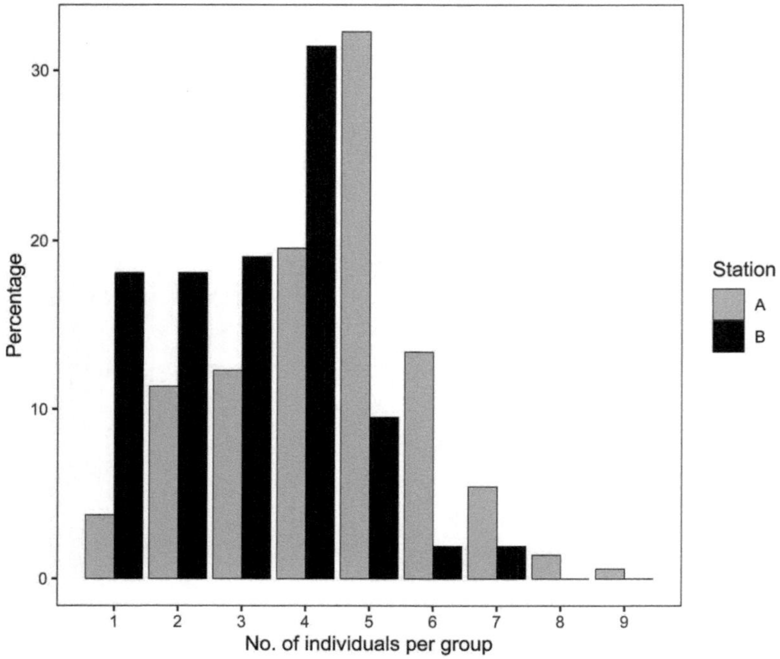

Fig. 2.4 Frequency of collared peccaries photographed together by camera traps. Percentage: % of photos/films. *A* and *B*: camera trap stations

The group sizes are compatible with those described in studies carried out in similar landscapes (Table 2.1). However, local population density is approximately half of those found for the species in Brazilian Pantanal, which may be due to habitat quality as the species appears to be resistant to relatively high hunting pressure (Cullen Jr. et al. 2000; Hill et al. 2003; Parry et al. 2009; Endo et al. 2010; Hellgren et al. 2012). However, hunting pressure can affect the behavior of collared peccaries (Keuroghlian and Eaton 2008b) and other mammals (Verdade 1996), making them avoid human presence. This may result in an underestimation of their population density by distance-based methods (Fragoso et al. 2016). The use of vestiges can compensate for such effect and generate more realistic estimates of species distribution (Hill et al. 1997; Stander et al. 1997; Stander 1998; Longoria and Weckerly 2007; Luzar et al. 2011; Munari et al. 2011) and population size (Fragoso et al. 2016), even with a relatively small sampling effort, such as in the present study. Although conservative, the present estimated density may potentially be more accurate than those obtained from distance methods based on direct animals' sight because it is not biased by the avoidance behavior of the species towards the observers.

The expansion of the exotic wild boar current distribution may result in the following threats to the collared peccaries: direct predation (Baber and Coblentz 1986; Castro 2001), competition for food resources (Gabor and Hellgren 2000), and the introduction of exotic parasites and pathogens (Galetti et al. 2016; Santos et al.

2016; Rostami et al. 2017; McClure et al. 2018; Pereira et al. 2021). An effective control of the wild boar should be prioritized to mitigate such impacts (Barrios-Garcia and Ballari 2012). Nonetheless, such a strategy would require the existence of an effective Wildlife Service to provide law enforcement in the hunting process to prevent illegal hunting of peccaries and other native species (Verdade 2004). Such Wildlife Service should also act on the possible conflict between wild boars and farmers related to agricultural damage (Marchini 2014).

The collared peccary presence in silvicultural landscapes may not only enhance biodiversity but also increase the complexity of the local trophic structure and dynamics (as suggested by Verdade et al. 2014a), despite the possible effects of agriculture intensification (Verdade et al. 2015). The possible interactions between wild boars and collared peccaries in such a circumstance are still unclear and should, therefore, be prioritized in future studies. The use of stable isotopes (C^{13} and N^{15}; Marques et al. 2014) and DNA (Palomares and Begoña 2014; Nascimento et al. 2024) might be useful to improve our knowledge of their trophic ecology.

Silvicultural landscapes should be included in management plans for the conservation of collared peccaries. In addition, the study of the wild boar ecology should be prioritized to assess its real impact on native wildlife and their habitats and to establish effective measures for their control when necessary. This requires an effective link between ecological research and environmental public policy (e.g., Martinelli et al. 2010).

The present study provides a relatively low-cost method to estimate the population density of collared peccaries that can be adapted to other species of medium to large terrestrial mammals. Although the methods used to estimate population density are very simple, they allow us to obtain crucial information with regards to the decision-making process related to wildlife management in silvicultural landscapes. More complex methods can be used, such as the one developed by Mimnagh et al. (2026, this volume), who revisit the dataset presented here alongside other case studies using a Bayesian hierarchical modelling framework to estimate abundance based on vestige counts.

Acknowledgments This study was financed by the Biota Program of *Fundação de Amparo à Pesquisa do Estado de São Paulo São Paulo* (FAPESP) (Grant Nos. 2006/06954-4, 2017/01304-4). LMV held a Productivity Fellowship from Conselho Nacional de Desenvolvimento Científico e Tecnológico (CNPq Proc. No. 312049/2015-3), CK had a Research Initiation Fellowship from FAPESP (Grant No. 2011/06697-8). Jose Fragoso gave us precious suggestions on sampling and analytical methodology.

Conflict of Interest The Wiley VCH publisher declares no involvement or association with the content of this publication.

References

ABRAF (2022) Anuário Estatístico da ABRAF 2022: ano base 2021. Associação Brasileira de Produtores de Florestas Plantadas – Brasília

Altrichter M (2005) The sustainability of subsistence hunting of peccaries in the Argentine Chaco. Biol Conserv 126:351–362

Altrichter M, Boaglio GI (2004) Distribution and relative abundance of peccaries in the Argentine Chaco: associations with human factors. Biol Conserv 116:217–225

Athayde EA, Cancian LF, Verdade LM, Morellato LPC (2015) Functional and phylogenetic diversity of scattered trees in agricultural landscape: implications for conservation. Agric Ecosyst Environ 199:272–281

Athayde EA, Cancian LF, Morellato LPC (2016) The value of agricultural landscape for tropical trees. In: Gheler-Costa C, Lyra-Jorge MC, Verdade LM (eds) Biodiversity in agricultural landscapes of southeastern Brazil. De Gruyter, Berlin, pp 87–111

Baber DW, Coblentz BE (1986) Density, home range, habitat use and reproduction in feral pigs on Santa Catalina Island. J Mammal 67:512–525

Barrios-Garcia MN, Ballari SA (2012) Impact of wild boar (*Sus scrofa*) in its introduced and native range: a review. Biol Invasions 14:2283–2300

Bigler WJ (1974) Seasonal movements and activity patterns of the collared peccary. J Mammal 55:851–855

Bissonette JA (1976) The relationship of resource quality and availability to social behavior and organization in the collared peccary. Unpublished Ph.D. dissertation, University of Michigan, Ann Arbor

Bodmer RE, Eisenberg JF, Redford KH (1997) Hunting and the likelihood of extinction of amazonian mammals. Conserv Biol 11:460–466

Byers JA, Bekoff M (1981) Social, spacing, and cooperative behavior of the collared peccary, *Tayassu tajacu*. J Mammal 62:767–785

Campos BM, Charters J, Verdade LM (2018) Diversity and distribution patterns of medium to large mammals in a silvicultural landscape in southeastern Brazil. iForest 11:802–808. https://doi.org/10.3832/ifor2721-011

Campos CM, Gheler-Costa C, Esteves CF, Rech-Fernandes TC, Rosalino LM, Moral RA, Piña CI, Verdade LM (2026) Sampling scats on trails: trophic approach in wildlife surveys in silvicultural landscapes. In: Martins Verdade L, Moral RA (eds) Wildlife monitoring: integrating conservation and innovation in human-altered landscapes. Springer, Cham

Castro J (2001) Estudo da população de javali (Sus Scrofa Linnaeus, 1758) no Parque Natural da Serra da Estrela. Available in: http://portal.icn.pt/NR/rdonlyres/80C3A8C8-6A74-46A5-A98673296A2E591E/0/PNSEJavali_Populacao_2001.pdf. Accessed 7 May 2013

Cullen L Jr, Bodmer RE, Valladares-Padua C (2000) Effects of hunting in habitat fragments of the Atlantic forests, Brazil. Biol Conserv 95:49–56

Cullen L Jr, Rudran R, Valladares-Pádua C (2009) Métodos de estudos em biologia da conservação e manejo da vida silvestre. UFPR, Parana

Da Rosa CA, Puertas F, Galetti M (2016) The feral pig as prey for jaguars: a reply to the 'Letter from the Conservation Front Line' by Verdade et al. Anim Conserv 20:111–112

Day GI (1977) Climate and its relationship to survival of young javelina. Arizona Game and Fish Dep. P-R Final Rep. W-78-R, Phoenix

De Araújo RM, De Souza MB, Ruiz-Miranda CR (2008) Densidade e tamanho populacional de mamíferos cinegéticos em duas Unidades de Conservação do Estado do Rio de Janeiro, Brasil. Iheringia Ser Zool 98:391–396

Desbiez ALJ, Keuroghlian A, Beisiegel BM, Medici EP, Gatti A, Pontes ARM, Campos CB, Tófoli CF, Moraes EA Jr, Azevedo FC, Pinho GM, Cordeiro JLP, Santos TS Jr, Morais AA, Mangini PR, Flesher K, Rodrigues LF, Almeida LB (2012) Avaliação do risco de extinção do cateto *Pecari tajacu* Linnaeus, 1758, no Brasil. Biodiv Bras 3:74–83

Dotta G, Verdade LM (2007) Trophic categories in a mammal assemblage: diversity in an agricultural landscape. Biota Neotrop 7:287–292

Dotta G, Verdade LM (2011) Medium to large-sized mammals in agricultural landscapes of South-Eastern Brazil. Mammalia 75:345–352. https://doi.org/10.1515/MAMM.2011.049

Ellisor JE, Harwell WF (1969) Mobility and home range of collared peccary in Southern Texas. J Wildlife Manag 33:425–427

Endo W, Peres CA, Salas E, Mori S, Sanchez-Vega J-L, Shepard GH, Pacheco V, Yu DW (2010) Game vertebrate densities in hunted and nonhunted forest sites in Manu National Park, Peru. Biotropica 42:251–261

Figueiredo MT, Sousa LC (2023) Evaluating the effectiveness of community-based conservation programs for collared peccaries. Conserv Biol 37(1):200–211

Fragoso JMV (1994) Large mammals and the community dynamics of an Amazonian Rain Forest. Ph.D. thesis, University of Florida, 210 pp

Fragoso JMV (1999) Perception of scale and resource partitioning by peccaries: Behavioral causes and ecological implications. J Mammal 80(3):993–1003

Fragoso JMV, Levi T, Oliveira LFB, Luzar JB, Overman H, Read JM, Silvius KM (2016) Line transect surveys underdetect terrestrial mammals: implications for the sustainability of subsistence hunting. PLoS One 11(4):e0152659

Gabor TM, Hellgren EC (2000) Variation in peccary populations: landscape composition or competition by an invader? Ecology 81:2509–2524

Galetti M, Pedrosa F, Keuroghlian A, Sazima I (2016) Liquid lunch – vampire bats feed on invasive feral pigs and other ungulates. Front Ecol Environ 14(9):505–506

Gonçalves FM, Barros LM, Silva JA (2024) Advances in camera trapping: implications for monitoring collared peccary populations. Wildlife Res Technol 12(1):77–89

Haugaasen T, Peres CA (2005) Mammal assemblage structure in amazonian flooded and unflooded forests. J Trop Ecol 21:133–145

Hegel CGZ, Marini MA (2018) Large felids as predators of wild boars in the Atlantic forest: reconciling Verdade et al. and Rosa et al. Anim Conserv 21:363–364

Hellgren EC, Synatzske DR, Oldenburg PW, Guthery FS (2012) Demography of a collared peccary population in South Texas. J Wildlife Manag 59:153–163

Herrera HM, Norek A, Freitas TPT, Rademaker V, Fernandes O, Jansen AM (2005) Domestic and wild mammals' infection by *Trypanosoma evansi* in a pristine area of the Brazilian Pantanal region. Parasitol Res 96:121–126

Hill K, Padwe J, Bejyvagi C, Bepurangi A, Jakugi F, Tykuarangi R, Tykuarangi T (1997) Impact of hunting on large vertebrates in the Mbaracayu Reserve, Paraguay. Conserv Biol 11:1339–1353

Hill KIM, Mcmillan G, Fariña R (2003) Hunting-related changes in game encounter rates from 1994 to 2001 in the Mbaracayu Reserve, Paraguay. Conserv Biol 17:1312–1323

Jennings WS, Harris JT (1953) The collared peccary in Texas. F.A. Report Series No. 12. Texas Game and Fish Commission, Austin

Judas J, Henry O (1999) Seasonal variation of home range of collared peccary in tropical rain forests of French Guiana. J Wildlife Manag 63:546–552

Kasper CB, Mazim FD, Soares JBG, de Oliveira TG, Fabián ME (2007) Composição e abundância relativa dos mamíferos de médio e grande porte no Parque Estadual do Turvo, Rio Grande do Sul, Brasil. Ver Bra Zool 24:1087–1100

Keuroghlian A, Eaton DP (2008a) Fruit availability and peccary frugivory in an isolated Atlantic forest fragment: effects on peccary ranging behavior and habitat use. Biotropica 40:62–70

Keuroghlian A, Eaton DP (2008b) Importance of rare habitats and riparian zones in a tropical forest fragment: preferential use by *Tayassu pecari*, a wide-ranging frugivore. J Zool 275:283–293

Keuroghlian A, Eaton DP, Longland WS (2004) Area use by white-lipped and collared peccaries (*Tayassu pecari* and *Tayassu tajacu*) in a tropical forest fragment. Biol Conserv 120:411–425

Knipe T (1957) Javelina in Arizona, Wildlife bulletin no. 2. Arizona Game and Fish Department, Phoenix

Krebs CJ (2006) Mammals. In: Sutherland WJ (ed) Ecological census techniques: a handbook, 2nd edn. Cambridge University Press, Cambridge, pp 351–369

Licona M, McCleery R, Collier B, Brightsmith DJ, Lopez R (2011) Using ungulate occurrence to evaluate community-based conservation within a biosphere reserve model. Anim Conserv 14:206–214

Longoria MP, Weckerly FW (2007) Estimating detection probabilities from sign of collared peccary. J Wildlife Manag 71:652–655

Lopes PC, Marques TS, Martin PS, Gheler-Costa C, Verdade LM (2016) Patterns of leaf-litter amphibian diversity in a silvicultural landscape of southeastern Brazil. In: Gheler-Costa C, Lyra-Jorge MC, Verdade LM (eds) Biodiversity in agricultural landscapes of southeastern Brazil. De Gruyter, Berlin, pp 226–242

Luzar JB, Silvius KM, Overman H, Giery ST, Read JM, Fragoso JMV (2011) Large-scale environmental monitoring by indigenous peoples. Bioscience 61:771–781

Marchini S (2014) Who's in conflict with whom? Human dimensions of the conflicts involving wildlife. In: Verdade LM, Lyra-Jorge MC, Piña CI (eds) Applied ecology and human dimensions in biological conservation. Springer, Heidelberg, pp 189–209. (ISBN 978-3-642-54750-8)

Marques TS, Bassetti LAB, Lara NRF, Araújo MS, Piña CI, Camargo PB, Verdade LM (2014) Isotopic discrimination factors (Δ13C and Δ15N) between tissues and diet of the Broad-snouted Caiman (*Caiman latirostris*). J Herpetol 48:332–337

Martin PS, Gheler-Costa C, Lopes PC, Rosalino LM, Verdade LM (2012) Terrestrial non-volant small mammals in agro-silvicultural landscapes of southeastern Brazil. Forest Ecol Manag 282:185–195

Martinelli LA, Joly CA, Nobre CA, Sparovek G (2010) A falsa dicotomia entre a preservação da vegetação natural e a produção agropecuária. Biota Neotrop 10:1–9

Martins R, Quadros J, Mazzolli M (2008) Hábito alimentar e interferência antrópica na atividade de marcação territorial do *Puma concolor* e *Leopardus pardalis* (Carnivora: Felidae) e outros carnívoros na Estação Ecológica de Juréia-Itatins, São Paulo, Brasil. Ver Bras Zool 25:427–435

Mazzolli M, Ludwig M, Hammer A (2008) Qualidade de ambiente para a onça-pintada, puma e jaguatirica na Baía de Guaratuba, Estado do Paraná, utilizando os aplicativos Capture e Presence. Atlantica 21:105–117

McClure ML, Burdett CL, Farnsworth ML, Sweeney SJ, Miller RS (2018) A globally-distributed alien invasive species poses risks to United States imperiled species. Sci Rep 8:5331

Mello A (2005) Distribuição da mastofauna de médio e grande porte em um mosaico florestal. M.S. dissertation, Universidade do Vale do Rio do Sinos, 52 pp

Mendes AR, Oliveira VP, Santos LF (2023) Habitat fragmentation and its effects on the home range of *Pecari tajacu* in the Atlantic Forest. Biotropica 55(1):150–164

Metzger JP, Lewinsohn TM, Joly CA, Verdade LM, Martinelli LA, Rodrigues RR (2010) Brazilian law: full speed in reverse? Science 329:276–277

Millan CH, Develey PF, Verdade LM (2015) Stand-level management practices increase occupancy by birds in exotic *Eucalyptus* plantations. Forest Ecol Manag 336:174–182

Miller RS, Sweeney SJ, Slootmaker C, Grear DA, Di Salvo PA, Kiser D, Shwiff SA (2017) Cross-species transmission potential between wild pigs, livestock, poultry, wildlife, and humans: implications for disease risk management in North America. Sci Rep 7:7821

Mimnagh N, Ferreira IEP, Verdade LM, Moral RA (2026) Counting animals we can't see: A triple Poisson model for scarce vestige data. In: Martins Verdade L, Moral RA (eds) Wildlife monitoring: integrating conservation and innovation in human-altered landscapes. Springer, Cham

Minnamon PS (1962) The home range of the collared peccary, Pecari tajacu (Mearns), in the Tucson Mountains. M.S. thesis, University of Arizona, Tucson

Morais BHS, Cardoso DL, Costa JS, Mayor P, Albuquerque NI, Chisté RC, Guimarães DAA (2022) Use of wildlife as an alternative protein source: Collared peccary meat. Meat Sci 192:108895

Munari DP, Keller C, Venticinque EM (2011) An evaluation of field techniques for monitoring terrestrial mammal populations in Amazonia. Mamm Biol 76:401–408

Nascimento EF, Pinheiro MS, Lopez GG (2024) Use of environmental DNA (eDNA) to assess collared peccary populations in fragmented habitats. Environ DNA 6(1):23–35

Nava AFD (2008) Espécies sentinelas para a Mata Atlântica: as conseqüências epidemiológicas da fragmentação florestal no Pontal do Paranapanema, São Paulo. Doctoral thesis, Faculdade de Medicina Veterinária e Zootecnia, Universidade de São Paulo, 147 pp

Neal BJ (1959) A contribution on the life history of the collared peccary in Arizona. Am Midl Nat 61:177–190

Nichols JD (2014) The role of abundance estimates in conservation decision-making. In: Verdade LM, Lyra-Jorge MC, Piña CI (eds) Applied ecology and human dimensions in biological conservation. Springer, Heidelberg, pp 117–131

Nogueira-Filho SLG, Nogueira SSC, Fragoso JMV (2009) Ecological impacts of feral pigs in the Hawaiian Islands. Biodivers Conserv 18:3685–3686

Oldenburg PW, Ettestad PJ, Grant WE, Davis E (1985) Structure of collared peccary herds in South Texas: spatial and temporal dispersion of herd members. J Mammal 66:764–770

Oliveira-Santos LGR, Dorazio RM, Tomas WM, Mourão G, Fernandez FAS (2011) No evidence of interference competition among the invasive feral pig and two native peccary species in a neotropical wetland. J Trop Ecol 27:557–561

Oliver WLR (1993) Pigs, peccaries and hippos: status survey and conservation action plan, Monographic series: IUCN/SSC action plans for the conservation of biological diversity. IUCN, Gland, p 202

Palomares F, Begoña A (2014) The use of molecular tools in ecological studies of mammalian carnivores. In: Verdade LM, Lyra-Jorge MC, Piña CI (eds) Applied ecology and human dimensions in biological conservation. Springer, Heidelberg, pp 105–116. (ISBN 978-3-642-54750-8)

Parry L, Barlow J, Peres CA (2009) Hunting for sustainability in tropical secondary forests. Conserv Biol 23:1270–1280

Pereira GL, Martinez EQ, Ribeiro JD (2021) Impact of invasive Sus scrofa on native peccary populations in South American ecosystems. Biol Invasions 23(3):679–693

Peres CA (1996) Population status of white-lipped *Tayassu pecari* and collared peccaries *T. tajacu* in hunted and unhunted Amazonian forests. Biol Conserv 77:115–123

Polisar J, Maxit I, Scognamillo D, Farrell L, Sunquist ME, Eisenberg JF (2003) Jaguars, pumas, their prey base, and cattle ranching: ecological interpretations of a management problem. Biol Conserv 109:297–310

Pontes ARM, Chivers DJ (2007) Peccary movements as determinants of the movements of large cats in Brazilian Amazonia. J Zool 273:257–265

Potratz EJ, Holt RD, Brown JS (2024) Ecology of fear: acclimation and adaptations to hunting by humans. Sustainability 16(3):1216

Redford KH, Robinson JG (1987) The game of choice: patterns of indians and colonist hunting in the neotropics. Am Anthropol 89:650–667

Reis NR, Peracchi AL, Pedro WA, Lima IP (2006) Mamíferos do Brasil. Sociedade Brasileira de Zoologia, Londrina

Robinson JG, Eisenberg JF (1985) Group size and foraging habits of the collared peccary *Tayassu tajacu*. Society 66:153–155

Rocha V, Santos AC, Moreira LR (2023) A meta-analysis of collared peccary diet and foraging behavior in modified landscapes. J Mammal 104(2):485–498

Rosalino LM, Martin PS, Gheler-Costa C, Lopes PC, Verdade LM (2013) Allometric relations of neotropical small rodents (Sigmodontinae) in anthropogenic environments. Zool Sci 30:585–590

Rostami A, Riahi SM, Fakhri Y, Saber V, Hanifehpour H, Valizadeh S, Gholizadeh M, Pouya RH, Gamble HR (2017) The global seroprevalence of *Toxoplasma gondii* among wild boars: a systematic review and meta-analysis. Vet Parasitol 244:12–20

Santos LMJF, Farias NAR, Oliveira PA, Cademartori BG, Ramos TS, Oliveira FC, Ruas JL (2016) Presence of *Toxoplasma gondii* infection in wild board in Southern Brazil. Sch J Agric Vet Sci 3:238–241

Schweinsburg RE (1971) The home range, movements and herd integrity of the collared peccary. J Wildlife Manag 35:455–460

Silva EJ, Costa HP, Alvarez MR (2022) Current trends in the population dynamics of collared peccaries in Brazilian agroforestry systems. J Wildl Manag 86(2):345–359

Sowls LK (1974) Social behavior of the collared peccary, *Dicotyles tajacu*, L. In: Geist V, Walther F (eds) The behavior of ungulates and its relation to management, vol 24. IUCN Pub. Series, Morges, pp 144–165

Sowls LK (1997) Javelinas and other peccaries: their biology, management, and use, 2nd edn. Texas A&M University Press

Stander PE (1998) Spoor counts as indices of large carnivore populations: the relationship between spoor frequency, sampling effort and true density. J Appl Ecol 35:378–385

Stander PE, Ghau I, Tsisaba D, Vi OI (1997) Tracking and the interpretation of spoor: a scientifically sound method in ecology. J Zool 242:329–341

Suarez IMT (1993) Ecología de los grupos de sainos (Tayassu tajacu) y daños que ocasionan en los cultivos vecinos a la Estación Biológica La Selva. M.S. thesis, Universidad Nacional de Costa Rica, San Jose

Taber AB, Doncaster CP, Neris NN, Colman F (1994) Ranging behaviour and activity patterns of two sympatric peccaries, *Catagonus wagneri* and *Tayassu tajacu*, in the Paraguayan Chaco. Mammalia 58:61–71

Timo TPC, Lyra-Jorge MC, Gheler-Costa C, Verdade LM (2015) The effect of plantation age on the use of *Eucalyptus* stands by medium to large-sized wild mammals in Southeastern Brazil. iForest 8:108–113

Verdade LM (1996) The influence of hunting pressure on the social behavior of vertebrates. Rev Bras Biol 56:1–13

Verdade LM (2004) A exploração da fauna silvestre no Brasil: jacarés, sistemas e recursos humanos. Biota Neotrop 4(2):1–12

Verdade LM, Seixas CS (2013) Confidencialidade e sigilo profissional em estudos sobre caça. Biota Neotrop 13:21–23

Verdade LM, Moreira JR, Ferraz KMPMB (2013) Counting Capybaras. In: Moreira JR, Ferraz KMPMB, Herrera EA, Macdonald DW (eds) Capybara: biology, use and conservation of an exceptional neotropical species. Springer, New York, pp 357–370. (ISBN 978-1-4614-3999-8)

Verdade LM, Penteado M, Gheler-Costa C, Dotta G, Rosalino LM, Pivello VR, Piña CI, Lyra-Jorge MC (2014a) The conservation value of agricultural landscapes. In: Verdade LM, Lyra-Jorge MC, Piña CI (eds) Applied ecology and human dimensions in biological conservation. Springer, Heidelberg, pp 91–102. (ISBN 978-3-642-54750-8)

Verdade LM, Lyra-Jorge MC, Piña CI (2014b) Redirections in conservation biology. In: Verdade LM, Lyra-Jorge MC, Piña CI (eds) Applied ecology and human dimensions in biological conservation. Springer, Heidelberg, pp 3–17. (ISBN 978-3-642-54750-8)

Verdade LM, Piña CI, Rosalino LM (2015) Biofuels and biodiversity: challenges and opportunities. Environ Dev 15:64–78

Verdade LM, Gheler-Costa C, Lyra-Jorge MC (2016a) The multiple facets of agricultural landscapes. In: Gheler-Costa C, Lyra-Jorge MC, Verdade LM (eds) Biodiversity in agricultural landscapes of southeastern Brazil. De Gruyter, Berlin, pp 2–13

Verdade LM, Palomares F, Couto HTZ, Polizel JL (2016b) Recent land use changes and the expansion of an exotic potential prey: a possible redemption for Atlantic forest jaguars? Anim Conserv 19:209–211

Verdade LM, Palomares F, Couto HTZ, Polizel JL (2018) Land-use change and an exotic potential prey for the jaguar: a reply to da Rosa et al. Anim Conserv 21:89–90

Zervanos SM, Hadley NF (1973) Adaptational biology and energy relationships of the collared peccary (*Tayassu tajacu*). Ecology 54:759–774

Chapter 3
Estimating Population Density of Neotropical Terrestrial Mammals in Human-Altered Environments Based on Camera-Trapping and Secondary Data on Home Range

Bruna M. Campos, Jacob D. Charters, Carlos I. Piña, Rafael de Andrade Moral, and Luciano Martins Verdade

Abstract Wildlife abundance estimates can be necessary to understand ecological processes that determine the patterns of diversity, crucial to monitoring programs. Human-modified environments, such as agricultural and silvicultural landscapes can act as a complement to conservation units (e.g., National Parks), despite their numerous environmental impacts. In this study we estimated population densities (ind.km^{-2}) of medium to large terrestrial mammals in a silvicultural landscape in southeast Brazil, based on a combination of camera-trapping and home range data from literature review with the following results: *Mazama gouazoubira* (0.06), *Pecari tajacu* (0.94–1.31), *Sus scrofa* (0.06–0.09), *Cerdocyon thous* (0.06), *Chrysocyon brachyurus* (0.03–0.06), *Leopardus guttulus* (0.06), *Puma concolor* (0.03–0.06), *Eira barbara* (0.03), *Nasua nasua* (0.94–1.88), *Cabassous unicinctus* (0.09), *Dasypus novemcinctus* (0.13), *Lepus europaeus* (0.09), *Sylvilagus brasiliensis* (0.01–0.03), *Mymercophaga tridactyla* (0.09–0.13), *Tamandua tetradactyla* (0.09), and *Cuniculus paca* (0.03). These estimates are similar to previous studies on Neotropical canids and felids, even those carried out in pristine environments, but

B. M. Campos
Universidade de São Paulo, CENA, Piracicaba, São Paulo, Brazil

J. D. Charters
Kimberley Land Council. Wunggurr, Gibb River Station, Broome, WA, Australia

C. I. Piña
CICyTTP-CONICET Entre Ríos/UADER, España, Diamante, Entre Ríos, Argentina

R. A. Moral
Department of Mathematics and Statistics, Maynooth University, Maynooth, Co. Kildare, Ireland

L. Martins Verdade (✉)
Wildlife Monitoring Consultancy, Campinas, São Paulo, Brazil
e-mail: luciano.verdade@wildlifemgmtco.com

L. Martins Verdade, R. A. Moral (eds.), *Wildlife Monitoring*,
https://doi.org/10.1007/978-3-032-05821-8_3

species with smaller home-ranges appear to be underestimated. The present results indicate that the methodology is applicable to wildlife monitoring in human-altered environments, especially agricultural/silvicultural landscapes, in which environmental certification requires periodical wildlife surveys. Future studies should prioritize local estimates of species home-range and habitat use, as well as the sampling design and effort to improve species detectability.

Keywords Wildlife management · Monitoring · Silvicultural landscapes · Abundance

1 Introduction

Management actions are fundamentally linked to the knowledge of population processes (Caughley 1994; Nichols 2014), because they indicate the status of wildlife populations and their changes through time and space. Understanding these changes is key to assess ecological processes such as seed dispersion, pollination, predator-prey relationships, and human-wildlife conflicts (Gaston and Fuller 2007; Gaston 2011; Nichols 2014). Much of the world's tropical areas are dominated by human-modified environments, such as agricultural and silvicultural landscapes (McNeely and Scherr 2003). It has been shown that designated conservation areas are insufficient to conserve biodiversity as they are generally not well connected and too small to maintain sufficient populations of wide-ranging species, such as top predators and grazing ungulates (Chape et al. 2003; Rodrigues et al. 2004).

Previous studies have shown that some agricultural and silvicultural landscapes are able to sustain a relevant biodiversity, including medium to large mammals. Many species use these areas not only as protective corridors to move between native habitats, but also as primary habitat in which they obtain food and shelter (Rosalino et al. 2009; Dotta and Verdade 2011; Martin et al. 2012; Penteado et al. 2016; Verdade et al. 2020). Agricultural and silvicultural landscapes can help conserve biodiversity and should be managed as multifunctional landscapes (Vandermeer and Perfecto 1997; Martinelli et al. 2010) with a primary mission of biological production, and a secondary—but still fundamental—mission of biological conservation (Verdade et al. 2014a). Consequently, any changes in the agricultural landscapes may influence the carrying capacity of local wildlife populations and thus species richness and abundance within them (Fahrig et al. 2011).

Once land use changes take place, local species populations may be affected. Thus, the primary management strategy should be monitoring local wildlife populations to quickly detect variation in ecological processes and, consequently, their change in patterns of biological diversity (Caughley 1994; Verdade et al. 2014b). Such a demand is, at least partially, attended by environmental certification such as the Forest Stewardship Council (FSC) in silvicultural landscapes, which demands periodical wildlife surveys (Schulte-Herbrüggen and Davies 2006).

Wildlife survey methods based on distance from the observer (Buckland et al. 2001, 2004) tend to underestimate population density of hunted species such as deer and peccaries (Fragoso et al. 2016). In addition, even consolidated methods tend to have low precision and unknown accuracy (Verdade et al. 2013). On the other hand, occupancy modelling that provides presence-absence data can be used to monitor species and reasonably estimate population density by detection points (Royle and Nichols 2003; MacKenzie et al. 2006) whenever home range data from similar environments are available (Fragoso et al. 2016). This is the aim of the present study: to estimate population density of medium to large-sized mammals based on a combination between camera-trapping and home range data from literature review.

2 Study Area

This study was carried out at Fazenda Três Lagoas (32.42 km^2, 23°22′00″—23°20′41″S/48°28′00″—48°27′57″W), which is located within the Alto Paranapanema river, in Angatuba municipality, São Paulo, southeast Brazil (Fig. 3.1). Local climate is classified as subtropical (Cwa, according to Köppen classification).

From the late nineteenth Century to the late 1920's, Angatuba municipality was predominatly composed of cotton and then coffee plantations, which were later replaced by cattle raising (Lisboa and Prado 2019). At Fazenda Três Lagoas, 22.2 km^2 of *Eucalyptus* were planted between 2006 and 2007. Conservation areas were allocated within the farm according to the Brazilian Forest Code, delimiting 8.8 km^2 as Legal Reserve and 3.36 km^2 for Areas of Permanent Protection, according to the Brazilian Forest Code (Sparovek et al. 2012). Currently, this farm is composed of abandoned pastures, remnants of native vegetation in various stages of regeneration (i.e., Cerrado *stricto senso*, "cerradão", and semidecidual forest), and *Eucalyptus* plantations, forming a silvicultural landscape (Athayde et al. 2016).

3 Methodology

3.1 Sampling Methodology

The sampling method was based on the Rapid Assessment of Long-term Ecological Research (RAPELD) protocol used by the Brazilian Research Program on Biodiversity (PPBio), which is a spacial grid composed of sampling units 1 km apart from each other (Magnusson et al. 2005, 2014). Five sampling units were randomly chosen from each habitat type within the farm (i.e., *Eucalyptus*, native vegetation, and abandoned pasture), in a total of 15 sampling units, constituting a stratified sampling scheme within the grid described above (Fig. 3.1).

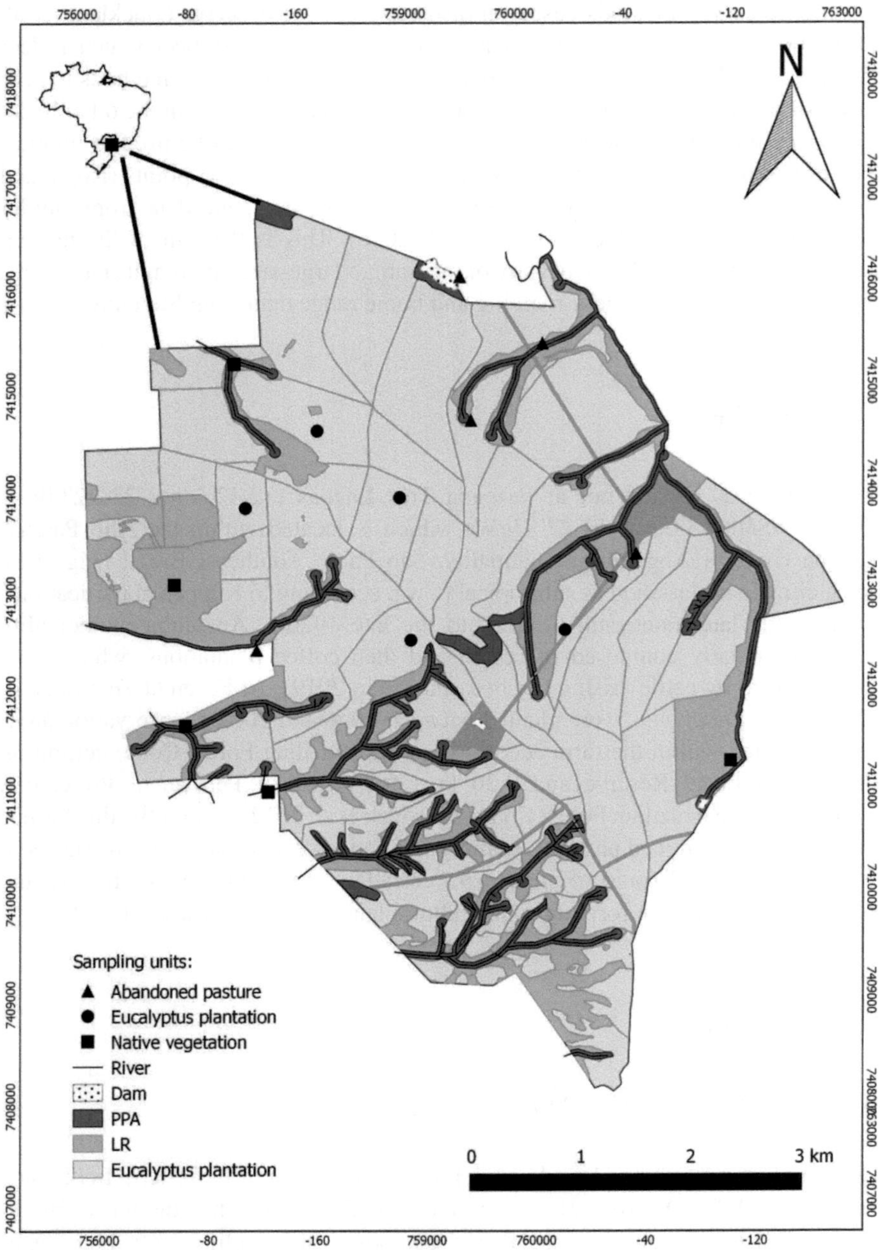

Fig. 3.1 Fazenda Três Lagoas, in Angatuba municipality, São Paulo, Brazil. PPA area of permanent protection, LR legal reserve. Datum Córrego Alegre

Each sampling unit had a passive infrared digital camera Bushnell Trophy Cam® installed approximately 40 cm above ground facing south. The sampling period encompassed 12 monthly campaigns of 1-week duration each (7 days.month^{-1}), from July 2015 to July 2016 (except in May 2016) during the second 7-year production cycle of the *Eucalyptus* plantation. Cameras were arranged to record videos 24 h.day^{-1} for 15 s, with 1-min intervals between them to facilitate species identification. Batteries were checked and replaced when necessary, before each campaign. Because we aimed to capture the medium to large mammals' species assemblage instead of one target species, no baits or lures were placed in front of the camera traps, since they can bias the results by attracting or repelling different species (Foster and Harmsen 2012).

3.2 *Analytical Methodology*

From the camera trap records, we were able to register the following information from middle to large sized mammals: location, species (or the lowest identifiable taxon), number of individuals, and capture date and time. Consecutive records of the same species at the same camera were deemed independent when there was at least 1 h interval between them, unless multiple individuals could be distinguished by appearing simultaneously in the same video (O'Brien et al. 2003).

Population density was estimated by a method adapted from Rowcliffe et al. (2008) and Fragoso et al. (2016), based on species' detection points (e.g., tracks, scats, camera trapping, sightings) and the placement of round buffers relative to the species home range, based on literature review. Fragoso et al. (2016) used average home range values, whereas we used minimum and maximum values previously described in environments as similar as possible to the ones from the present study. As a result, minimum and maximum population densities (D_{min} and D_{max}, respectively, in ind./km^2) were estimated according to the formulas below (adapted from Kruszynski et al. 2026, this volume), considering the area of Fazenda Três Lagoas (32.42 km^2) as the focused area of management unit:

$$D_{\text{min}} = \frac{(\text{minimum number of groups}) \times (\text{minimum number of individuals per group})}{32.42}$$

$$D_{\text{max}} = \frac{(\text{maximum number of groups}) \times (\text{maximum number of individuals per group})}{32.42}$$

A literature review was carried out to determine the minimum and maximum comparable home ranges for each medium to large mammal species detected in the area (Table 3.1). The following criteria were used to select the data to create home range buffers:

Table 3.1 Medium to large mammals' home range sizes review

Species	Home range (km^2)		Methodology	Habitat type	Country	Reference
	Minimum	Maximum				
Mozama gouazoubira	0.33[a]	0.98[a]	MCP	*Cerradão* and semideciduous forest	Brazil	Neto (2016)
	0.03 (male)	0.14 (male)	Observations	Mature secondary forests, shrubs and brush, and grassland	Argentina	Black-Décima (2000)
	0.03 (female)	0.1 (female)	Observations	Mature secondary forests, shrubs and brush, and grassland	Argentina	Black-Décima (2000)
Pecari tajacu	1.02[a]	2.87[a]	MCP (collars)	Atlantic forest	Brazil	Keuroghlian et al. (2004)
	2.3			Pantanal	Brazil	Desbiez et al. (2010)
	1.23	3.05		Cerrado	Brazil	Keuroghlian et al. (2004)
Sus scrofa	2.5 (female)[a]	33.7 (male)[a]	MCP	Variable	Australia	Caley (1997)
	1.8 (female)	2.3 (male)	MCP	Forest and rice field	EUA	Wood and Brenneman (1980)
Cerdocyum thous	0.3 (female)[a]	11.1 (male)[a]	MCP	Disturbed semi-deciduous Atlantic Forest, Cerrado and *Eucalyptus* plantation	Brazil	Michalski et al. (2006)
	2.8	8.14	MCP	Atlantic forest	Brazil	Oliveira (2002)
	1.55	2.63	MCP	Chato-Chiquitano transitional dry forest	Bolivia	Maffei and Taber (2003)
Chrysocyon brachyurus	15.09[a]	67.69[a]	MCP	Cerrado	Brazil	Jácomo et al. (2009)
	39	58	MCP	Pantanal	Brazil	Trolle et al. (2006b)
	75 ± 34.6		MCP	Cerrado	Brazil	Carvalho and Vasconcellos (1995)

Leopardus guttulus	5.28 (male)[a]	10.30 (female)[a]	MCP	Rice and soy crops, cattle ranching, native forest	Brazil	Kasper et al. (2012)
	2 (female)	8 (male)	MCP	Subtropical forest fragments and agriculture	Brazil	Oliveira (2008)
Puma concolor	18.9[a]	75.5[a]	MCP	Atlantic Forest and *Eucalyptus* plantation	Brazil	Mazzolli (2010)
	41	428	MCP	Cerrado	Brazil	Silveira (1999)
	32	155	MCP	Pantanal	Brazil	Crawshaw and Quigley (1984)
Eira barbara	2.25 (female)[a]	9 (female)[a]	MCP	Venezuelan llanos	Venezuela	Sunquist et al. (1989)
	5.3 (female)		MCP	Disturbed semi-deciduous Atlantic forest, Cerrado and *Eucalyptus* plantation	Brazil	Michalski et al. (2006)
Nasua nasua	2.2[a]	7.55[a]	MCP	Cerrado	Brazil	Trovati et al. (2010)
	2.78	5.88	MCP	Atlantic forest	Brazil	Beisiegel and Mantovani (2006)
	1.2	1.5	MCP	Pantanal	Brazil	Rocha (2006)
Cabassous unicinctus	1.01			Cerrado	Brazil	Encarnação (1987)
	0.12	1.75	MCP	Pantanal	Brazil	Desbiez et al. (2018)
Dasypus novemcinctus	0.11[a]	1.38[a]	MCP		EUA	Layne and Glover (1977)
	0.02	0.2	Review			Loughry and McDonough (1998)
	2.03		MCP		EUA	Fitch et al. (1952)
Lepus europaeus	0.7	2.68	MCP	Agricultural landscape	France	Reitz and Léonard (1994)
	0.2[a]	0.63[a]	MCP	Agricultural landscape	Germany	Rühe and Hohmann (2004)

(continued)

Table 3.1 (continued)

Species	Home range (km^2)		Methodology	Habitat type	Country	Reference
	Minimum	Maximum				
Sylvilagus brasiliensis[b]	0.01[a]	0.03[a]	Harmonic-mean and core-area		EUA	Forys and Humphrey (1996)
Myrmecophaga tridactyla	6.93 (female)[a]	10.80 (male)[a]	MCP	Cerrado	Brazil	Miranda (2004)
	2.74 (male)	3.67 (female)	MCP	Cerrado	Brazil	Shaw et al. (1987)
	4 (male)	7.5 (male)	MCP	Pantanal, forest patches, savanna, scrub savanna, and seasonally flooded grasslands	Brazil	Medri and Mourão (2005)
Tamandua tetradactyla	1.06 (female)[a]		MCP		Brazil	Trovati and Brito (2009)
	3.8		MCP	Venezuelan llanos	Venezuela	Montgomery and Lubin (1977)
	1		MCP	Serra da Mesa, Cerrado	Goiás, Brazil	Rodrigues et al. (2008)
Cuniculus paca	0.0149 (male)[a]	0.0344 (female)[a]	Fixed kernel	Tropical lowland wet forest	Costa Rica	Beck-King et al. (1999)
	0.875–2.049 (male)	0.507–86.7 (female)	MCP	Mosaic (agroecosystems, wetlands, shrubland and forest)	Belize	Harmsen et al. (2018)
	3.5	11.8				
	0.01	0.03	MCP	Rural agricultural community in a montane tropical forest	Beni, Bolivia	Benavides et al. (2017)

[a]Selected home range for density estimate
[b]*Sylvilagus palustris hefneri*

1. Home range of the same species within similar landscapes (i.e. agricultural/silvicultural landscapes, Atlantic Forest or Cerrado biome, southeast Brazil);
2. Home ranges assessed with radio collars; and,
3. The most recent studies available.

To determine minimum and maximum number of groups of each species (respectively, $N_{group\ min}$ and $N_{group\ max}$), each detection point was used as a central point for minimum and maximum home range buffers generated on the QGis software. When minimum and maximum home range data were unavailable, a single value was used. The minimum and maximum home range buffers around detection points estimated maximum and minimum number of groups, respectively. Overlaps were considered as such whenever detection points were exclusive (i.e., the detection point is outside the neighboring buffer) (Fig. 3.2). Minimum and maximum number of individuals per group were established according to the minimum and maximum number of individuals of each species captured in a single video. The present method may be limited by the uncertainties concerning the use of hypothetical home-range estimates and the limitations concerning the use of camera-traps (Table 3.2). However, such uncertainties tend to be comparable to those from traditional wildlife survey methods (Verdade et al. 2013).

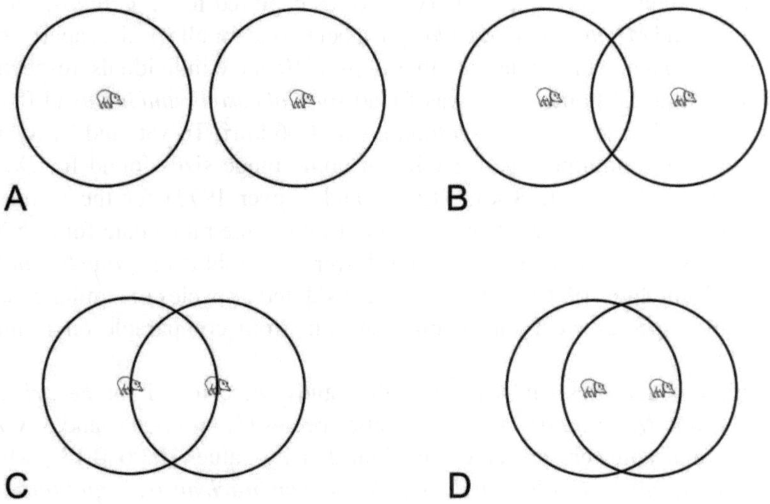

Fig. 3.2 Criteria for the differentiation of home ranges (i.e., buffers) among groups based on camera-trapping. (**a**) and (**b**) are considered as two distinct groups with occasionally some spatial overlap, whereas (**c**) and (**d**) are considered the same group sampled twice. Criteria are described in the text

Table 3.2 Uncertainties associated to the methodology used in this study and possible future developments

Uncertainty	Reference	Future development
Lack of accurate home range data from studies conducted within similar environments		Local home-range estimates
Use of buffers as round home-range	Saïd et al. (2009)	Local studies on habitat use
The assumption that overlaps occurs only when a buffer does not include another buffer's detection point	Fieberg and Kochanny (2005)	Local studies on dispersion
Uncertainties associated with camera trapping	Royle and Nichols (2003); O'Brien et al. (2003); Burton et al. (2015); Meek et al. (2015)	Studies on sampling design and effort

4 Results

Sixteen taxa could be identified to the species level from the camera trap records. *Leopardus* sp. and *Coendou* sp. were unable to be identified to the species level and were therefore excluded from analyses. The number of individuals per group did not include parent-offspring temporary bond (as captured for *L. guttulus*, *Mazama gouazoubira* and *Myrmecophaga tridactyla*), being one for all species, apart from *N. nasua* and *P. tajacu*, which consisted in groups of 10 and 6 individuals, respectively.

A single value of home range was found for *Cabassous unicinctus* (1.01 km^2, Encarnação 1987) and *Tamandua tetradactyla* (1.06 km^2, Trovati and Brito 2009). We then used the minimum and maximum home range sizes found for *Dasypus novemcinctus* (0.11 and 1.38 km^2, Layne and Glover 1977) for the former and kept the single value for the latter. We did not find home range data for *Sylvilagus brasiliensis*. We then used data from the lower keys rabbit (*S. palustris hefneri*) (Forys and Humphrey 1996), which can be considered a species of similar behavior. For the other species we found applicable data from comparable environments (Table 3.1).

The highest densities (ind.km^{-2}) of this study were found for *Pecari tajacu* (0.94–1.31) and *N. nasua* (0.94–1.88). Exotic species (*L. europaeus* and *S. scrofa*), armadillos, and anteaters presented medium density values (0.06–0.13), whereas *Mazama gouazoubira*, *Cerdocyon thous*, *Chrysocyon brachyurus*, *L. guttulus*, *Puma concolor*, *Eira barbara*, *S. brasiliensis*, and *Cuniculus paca* presented the lowest densities (up to 0.06) (Table 3.3, Fig. 3.3).

Considering the maximum estimated density values, the omnivores, herbivores and formivores were the most abundant groups, whereas frugivores, carnivores and insectivores were the least. Considering the minimum estimated density values, the variation among functional groups considerably decreased. In this study the species with the largest home ranges (i.e., *C. brachyurus* and *P. concolor*) presented density

Table 3.3 Minimum and maximum population density of medium to large-sized mammals

Order	Family	Species (Scientific name)	Species (Common name)	Functional group	Population Density (ind.km^{-2})		# of possible groups	
					Min	Max	Min	Max
Artiodactyla	Cervidae	*Mazama gouazoubira*	Grey brocket	Herbivore	0.06	0.06	2	2
	Suidae	*Sus scrofa*[a]	Wild boar	Omnivore	0.06	0.09	2	3
	Taysuidae	*Pecari tajacu*	Collared peccary	Frugivore	0.94	1.31	5	7
Carnivora	Canidae	*Cerdocyon thous*	Crab-eating fox	Omnivore	0.06	0.06	2	2
		Chrysocyon brachyurus	Maned wolf	Omnivore	0.03	0.06	1	2
	Felidae	*Leopardus guttulus*	Southern tiger cat	Carnivore	0.06	0.06	2	2
		Puma concolor	Puma	Carnivore	0.03	0.06	1	2
	Mustelidae	*Eira barbara*	Tayra	Omnivore	0.03	0.03	1	1
	Procyonidae	*Nasua nasua*	South American coati	Omnivore	0.94	1.88	3	6
Cingulata	Dasypodidae	*Cabassous unicinctus*	Southern naked-tailed armadillo	Formivore	0.09	0.09	3	3
		Dasypus novemcinctus	Nine-banded armadillo	Insectivore	0.13	0.13	4	4
Lagomorpha	Leporidae	*Lepus europaeus*[a]	European hare	Herbivore	0.09	0.09	3	3
		Sylvilagus brasiliensis	Tapeti	Herbivore	0.03	0.03	1	1
Pilosa	Myrmecophagidae	*Myrmecophaga tridactyla*	Giant anteater	Formivore	0.09	0.13	3	4
		Tamandua tetradactyla	Southern tamandua	Formivore	0.09	0.09	3	3
Rodentia	Cuniculidae	*Cuniculus paca*	Aguti	Frugivore	0.03	0.03	1	1

Groups of all species included one individual, except *N. nasua* and *P. tajacu* whose group size was estimated to be up to 10 and 6 individuals, respectively, based on images obtained from cameras

[a]Exotic species

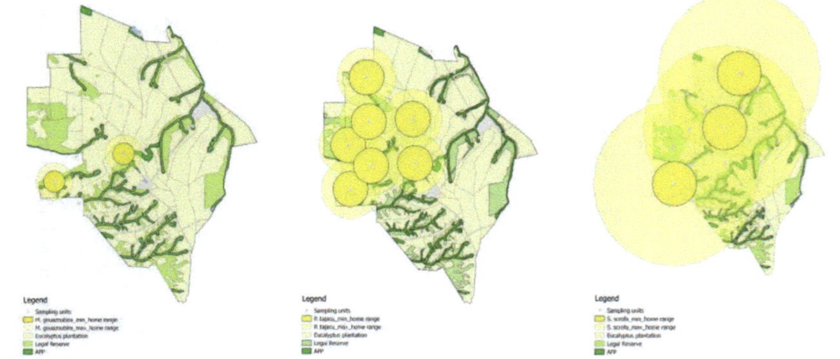

a) Artiodactyla: *M. gouazoubira* ($N_{group\ min.}$=2; $N_{group\ max.}$=2); *P. tajacu* ($N_{group\ min.}$=5; $N_{group\ max.}$=7) and *S. scrofa* ($N_{group\ min.}$=2; $N_{group\ max.}$=3)

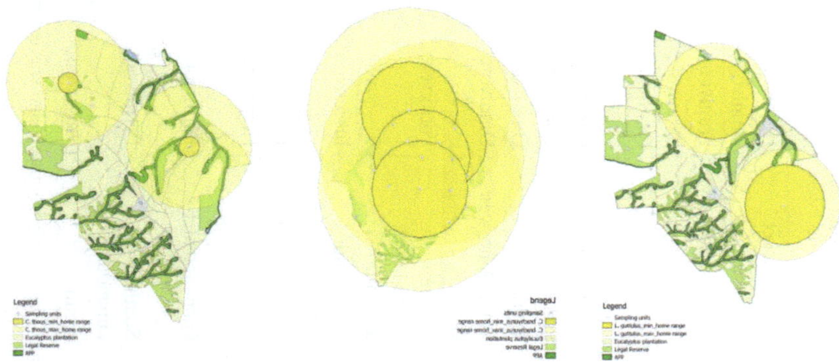

Fig. 3.3 Minimum ($N_{group\ min}$) and maximum ($N_{group\ max}$) home range buffers of medium to large Neotropical mammals in a silvicultural landscape of Southeast Brazil. (**a**) Artiodactyla: *M. gouazoubira* ($N_{group\ min} = 2$; $N_{group\ max} = 2$); *P. tajacu* ($N_{group\ min} = 5$; $N_{group\ max} = 7$) and *S. scrofa* ($N_{group\ min} = 2$; $N_{group\ max} = 3$). (**b**) Carnivora: *C. thous* ($N_{group\ min} = 2$; $N_{group\ max} = 2$); *C. brachyurus* ($N_{group\ min} = 1$; $N_{group\ max} = 2$); *L. guttulus* ($N_{group\ min} = 2$; $N_{group\ max} = 2$); *P. concolor* ($N_{group\ min} = 1$; $N_{group\ max} = 2$); *E. barbara* ($N_{group\ min} = 1$; $N_{group\ max} = 1$); *N. nasua* ($N_{group\ min} = 3$; $N_{group\ max} = 6$). (**c**) Cingulata: *C. unicinctus* ($N_{group\ min} = 3$; $N_{group\ max} = 3$) and *D. novencinctus* ($N_{group\ min} = 4$; $N_{group\ max} = 4$). (**d**) Lagomorpha: *L. europaeus* ($N_{group\ min} = 3$; $N_{group\ max} = 3$); *S. brasiliensis* ($N_{group\ min} = 1$; $N_{group\ max} = 1$). (**e**) Pilosa: *M. tridactyla* ($N_{group\ min} = 3$; $N_{group\ max} = 4$) and *T. tetradactyla* ($N_{group\ min} = 3$; $N_{group\ max} = 3$). (**f**) Rodentia: *C. paca* ($N_{group\ min} = 1$; $N_{group\ max} = 1$)

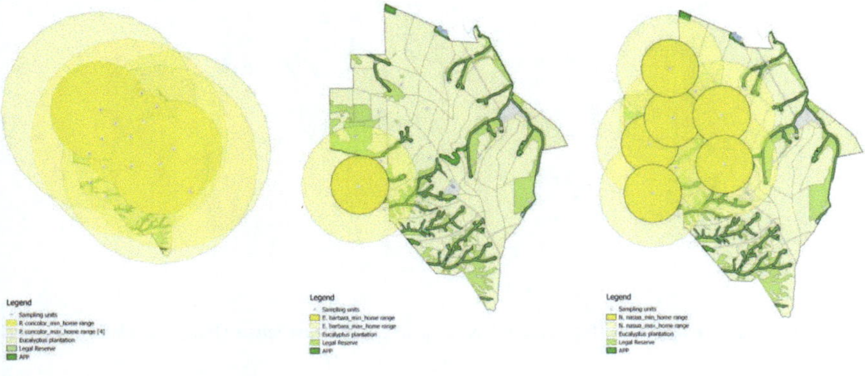

b) Carnivora: *C. thous* ($N_{group\ min.}$=2; $N_{group\ max.}$=2); *C. brachyurus* ($N_{group\ min.}$=1; $N_{group\ max.}$=2); *L. guttulus* ($N_{group\ min.}$=2; $N_{group\ max.}$=2); *P. concolor* ($N_{group\ min.}$=1; $N_{group\ max.}$=2); *E. barbara* ($N_{group\ min.}$=1; $N_{group\ max.}$=1); *N. nasua* ($N_{group\ min.}$=3; $N_{group\ max.}$=6)

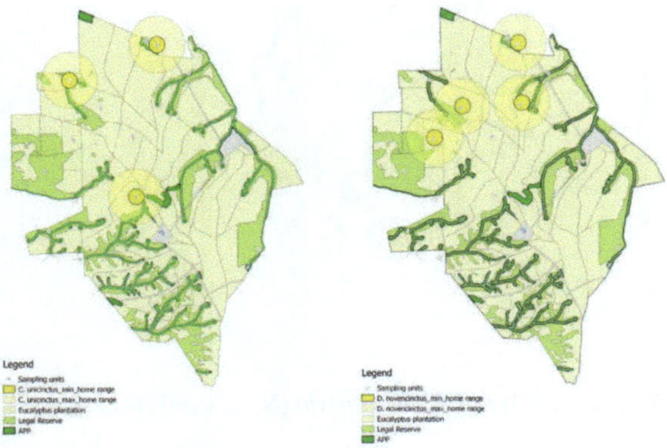

c) Cingulata: *C. unicinctus* ($N_{group\ min.}$=3; $N_{group\ max.}$=3) and *D. novencinctus* ($N_{group\ min.}$=4; $N_{group\ max.}$=4)

Fig. 3.3 (continued)

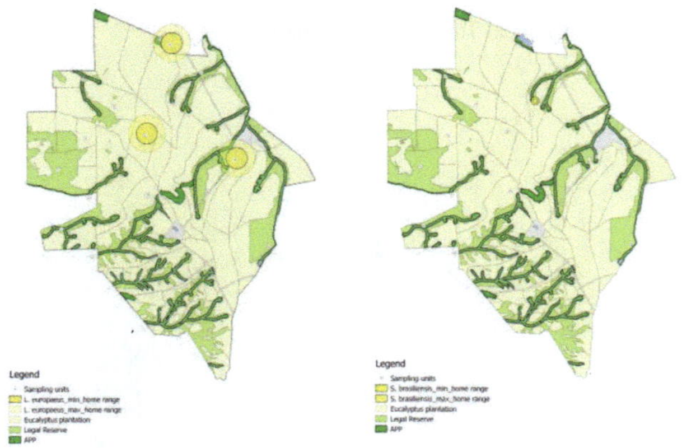

d) Lagomorpha: *L. europaeus* ($N_{group\ min.}$=3; $N_{group\ max.}$=3); *S. brasiliensis* ($N_{group\ min.}$=1; $N_{group\ max.}$=1)

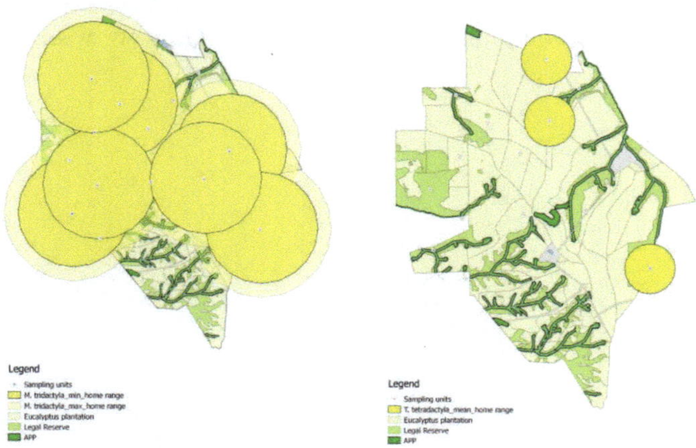

e) Pilosa: *M. tridactyla* ($N_{group\ min.}$=3; $N_{group\ max.}$=4) and *T. tetradactyla* ($N_{group\ min.}$=3; $N_{group\ max.}$=3)

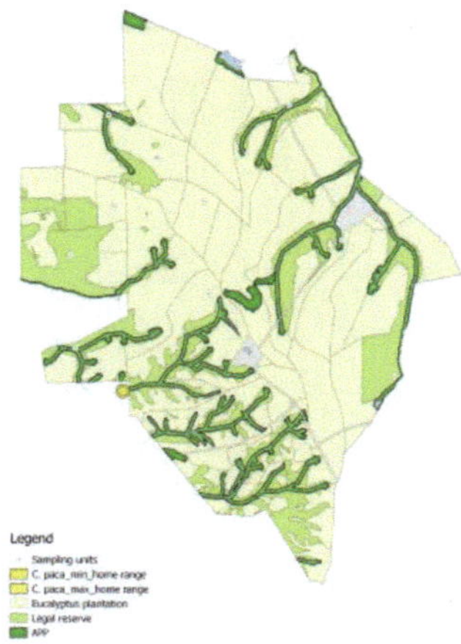

Fig. 3.3 (continued)

estimates quite like previous studies carried out in comparable environments. However, species with smaller home ranges presented smaller density estimates when compared to similar studies in the literature (Table 3.4).

5 Discussion and Conclusion

In this study, we detected at least 16 species of medium to large terrestrial mammals. Similar studies in silvicultural landscapes of São Paulo state, Brazil, detected from 15 to 23 species. However, for a similar sampling period (i.e., 12 monthly campaigns), Dotta and Verdade (2011) detected 23 species, more than 44% than the present study. This is likely due to the relatively low efficiency of camera-trapping when the goal is individual identification (Caravaggi et al. 2017). Individualization, whenever possible, can be extremely useful to both demographic and behavioural-ecological studies. On the former it can be useful to determine sex ratio, age-distribution and population density (Trolle and Kéry 2003).

Agricultural landscapes may be relevant to biological conservation (Verdade et al. 2014a). However, land use changes and intensification might result in relevant impacts to wildlife (Verdade et al. 2015). Ideally, such impacts should be detected early to help decision making processes to be effective (Verdade et al. 2014b). In such context, camera-trapping has the following strengths:

(a) Species identification: Videos and photos usually allow for more accurate species identification than vestiges, such as tracks and scats;
(b) Individual identification: Natural marks and spotted pelage might effectively help individual identification in videos and photos (Trolle and Kéry 2003); and,
(c) Operationality: Local personnel can be easily trained to operate a grid of stationary camera-traps in agricultural landscapes.

On the other hand, the use of camera-trapping in wildlife monitoring under systematic design may have the following weaknesses:

(d) Relatively high cost: The cost of equipment may be relatively high depending on the number of camera traps required and local available budget;
(e) Equipment loss by robbery: In certain areas, camera traps can be stolen; and,
(f) Detectability of medium-large terrestrial mammals can be relatively low by camera traps in a systematic design: There is a trade-off between the distance between cameras on a grid and the area effectively sampled by the grid, which tends to reduce the probability of animals to be captured by the cameras (Burton et al. 2015).

For the reasons above, the use of vestiges might be complementary to camera-traps or even replace them in some circumstances (e.g., "e" and "f" above). In addition, the sampling of vestiges can be done on already existing trails in agricultural landscapes (e.g., Palomares et al. 2018), which can facilitate its operationality. However, the search and identification of wildlife vestiges (e.g., tracks, scats and marks in trees) require some level of expertise, whose training may be longer and more expensive than the training required for operating camera traps.

Table 3.4 Comparative density estimates of medium to large Neotropical mammals

Functional group	Species	Density estimate (ind.km^{-2})		References
		This study	Traditional methods	
Carnivores	*Leopardus gutulus*	0.06	0.07–0.13	Oliveira-Santos et al. (2012)
	Puma concolor	0.03–0.06	0.0068–0.03; 0.062–0.065	Eisenberg et al. (1979); Mazzolli (2010)
Formivores	*Myrmecophaga tridactyla*	0.09–0.13	0.14–0.34	Desbiez and Medri (2010); Desbiez et al. (2010)
	Cabassous unicinctus	0.09	0.14–0.27	Bonato et al. (2008)
Frugivores	*Cuniculus paca*	0.03	3.5–6.9	Huanca-Huarachi et al. (2011); Leuchtenberger et al. (2018)
	Pecari tajacu	0.94–1.31	5.48 ± 1.62	Desbiez et al. (2010)
Herbivores	*Mazama gouazoubira*	0.06	2.63–23	Desbiez et al. (2010); Ferreguetti et al. (2015)
Omnivores	*Chrysocyon brachyurus*	0.03	0.029–0.044	Trolle et al. (2006a, b)
	Cerdocyon thous	0.06	0.55	Desbiez et al. (2010)
	Eira barbara	0.03	0.48	Desbiez et al. (2010)
	Nasua nasua	0.94–1.88	10.5 ± 1.88	Desbiez et al. (2010)
	Sus scrofa	0.06–0.09	0.01–52	Saunders and McLeod (1999); Melis et al. (2006)

The use of home range data from the literature helps to obtain simple estimates of population density. The main advantages of such a procedure are its low cost and the relative speed of obtaining the estimates. This approach would not be recommended for, however, endangered species on National Parks when individualization should be pursued to increase precision and accuracy in population estimates, but it can be rather useful in agricultural landscapes, considering that they are predominantly occupied by generalist species, which do not require high accuracy on abundance estimates to be managed (Nichols 2014; Verdade et al. 2014a). However, as species adapt to human-altered environments (e.g., Rosalino et al. 2014; Potratz et al. 2024), their space use might significantly vary in response to land use change. In addition, home-ranges are not actually round as the species tend to use habitats distinctively. Thus, to improve the use of secondary data on home range, future studies should prioritize the use of space (i.e., habitat use and home range) in agricultural and silvicultural landscapes.

The present population density estimates are similar to those generated by traditional methods to *P. concolor*, *C. brachyurus* and *C. thous*, but considerably smaller to other species (Table 3.4). These three species have the largest home ranges among all from the present study (Table 3.3). This suggests that the present sampling design is adequate to species with relatively large home ranges, but insufficient to those with smaller ones. Two conclusions can then be taken: (a) the present method is viable to estimate population density of medium-large terrestrial mammals in human-modified environments on the Neotropics; and, (b) the sampling design should be adapted to improve estimates of species with smaller home-ranges, which could be experimentally tested with sub-grids (e.g., cameras 100 $\times$ 100 m apart from each other) within larger grids (e.g., cameras 1000 $\times$ 1000 m apart from each other) in distinct spatial combinations.

The use of the present method in larger scale might be successful if its intrinsic uncertainties are treated (Table 3.2). The lack of accurate home range data might be mitigated by the development of future studies on local home-range estimates. In addition, the use of round buffers as home-range might be improved with future studies on habitat use and modelling, for instance, based on occupancy/detectability (e.g., Royle and Nichols 2003) or probability of occurrence (e.g., Ferraz et al. 2009).

The assumption that overlaps occur only when a buffer does not include another buffer's detection point should be tested as hypothesis on future studies on dispersion (sensu Caughley 1994). In this case, spatial modelling might also be useful. Finally, the uncertainties associated with camera trapping may be accommodated by carrying out studies on sampling design and effort to improve detectability of local species (Ferreira et al. 2026, this volume); and by using vestiges in complement (or replacement) of camera traps, and utilizing distance-based methods, or hierarchical Bayesian approaches (Mimnagh et al. 2026, this volume).

Acknowledgements Fazenda Três Lagoas provided logistics for this study. Tiago S. Marques cooperated at field work. This study was financed by the Biota Program of *Fundação de Amparo à Pesquisa do Estado de São Paulo São Paulo* (FAPESP) (Grant Nos. 2006/06954-4, 2017/01304-4). This study was based on the use of camera-traps without the capture or handling of animals, which does not require field study permissions in Brazil.

References

Athayde EA, Cancian LF, Morellato LPC (2016) The value of agricultural landscape for tropical trees. In: Gheler-Costa C, Lyra-Jorge MC, Verdade LM (eds) Biodiversity in agricultural landscapes of southeastern Brazil. De Gruyter Open, Berlin, pp 87–103

Beck-King H, Helversen OV, Beck-King R (1999) Home range, population density, and food resources of *Agouti paca* (Rodentia: Agoutidae) in Costa Rica: a study using alternative methods. Biotropica 31:675–685. https://doi.org/10.1111/j.1744-7429.1999.tb00417.x

Beisiegel BM, Mantovani W (2006) Habitat use, home range and foraging preferences of *Nasua nasua* in a pluvial tropical Atlantic forest area. J Zool 269:77–87. https://doi.org/10.1111/j.1469-7998.2006.00083.x

Benavides C, Arce A, Pacheco LF (2017) Home range and habitat use by pacas in a montane tropical forest in Bolivia. Acta Amazon 47:227–236. https://doi.org/10.1590/1809-4392201603163

Black-Décima P (2000) Home range, social structure, and scent marking behavior in brown brocket deer (*Mazama gouazoubira*) in a large enclosure. Mastozoología Neotropical 7(1):5–14

Bonato V, Martins EG, Machado G, Da-Silva CQ, dos Reis SF (2008) Ecology of the armadillos *Cabassous unicinctus* and *Euphractus sexcinctus* (Cingulata: Dasypodidae) in a Brazilian Cerrado. J Mammal 89:168–174. https://doi.org/10.1644/06-MAMM-A-187.1

Buckland ST, Anderson DR, Burnham KP, Laake JL, Borchers DL (2001) Introduction to distance sampling. Oxford University Press, Oxford, UK

Buckland ST, Anderson DR, Burnham KP, Laake JL, Borchers DL (2004) Advanced distance sampling. Oxford University Press, Oxford, UK

Burton AC, Neilson E, Moreira D, Ladle A, Steenweg R, Fisher JT, Bayne E, Boutin S (2015) Wildlife camera trapping: a review and recommendations for linking surveys to ecological processes. J Appl Ecol 52:675–685. https://doi.org/10.1111/1365-2664.12432

Caley P (1997) Movements, activity patterns and habitat use of feral pigs (*Sus scrofa*) in a tropical habitat. Wildl Res 24:77–87. https://doi.org/10.1071/WR94075

Caravaggi A, Banks PB, Burton AC, Finlay CM, Haswell PM, Hayward MW, Rowcliffe MJ, Wood MD (2017) A review of camera trapping for conservation behaviour research. Remote Sens Ecol Conservation 3(3):109–122. https://doi.org/10.1002/rse2.48

Carvalho CT, Vasconcellos LEM (1995) Disease, food and reproduction of the maned wolf—Chrysocyon brachyurus (Illiger) (Carnivora, Canidae) in Southeast Brazil. Rev Bras Zool 12:627–640. https://doi.org/10.1590/S0101-81751995000300018

Caughley G (1994) Directions in conservation biology. J Anim Ecol 63:215–244. https://doi.org/10.2307/5542

Chape S, Blyth S, Fish L, Fox P, Spalding M (2003) United Nations list of protected areas. IUCN – World Conservation Union, Gland

Crawshaw PG, Quigley HB (1984) A ecologia do jaguar ou onça-pintada no Pantanal. Instituto Brasileiro de Desenvolvimento Florestal (IBDF), Brasília

Desbiez ALJ, Medri ´iM (2010) Density and habitat use by giant anteaters (*Myrmecophaga tridactyla*) and southern tamanduas (*Tamandua tetradactyla*) in the Pantanal wetland, Brazil. Edenata 11:4–10. https://doi.org/10.1896/020.011.0102

Desbiez ALJ, Bodmer RE, Tomas WM (2010) Mammalian densities in a neotropical wetland subject to extreme climatic events. Bi tropica 42:372–378. https://doi.org/10.1111/j.1744-7429.2009.00601.x

Desbiez ALJ, Massocato GF, Kluyber D, Fernandes-Santos RC (2018) Unraveling the cryptic life of the southern naked-tailed armadillo, *Cabassous unicinctus squamicaudis* (Lund, 1845), in a neotropical wetland: home range, activity pattern, burrow use and reproductive behaviour. Mamm Biol 91:95–103. https://doi.org/10.1016/j.mambio.2018.02.006

Dotta G, Verdade LM (2011) Medium to large-sized mammals in agricultural landscapes of South-Eastern Brazil. Mammalia 75:345–352. https://doi.org/10.1515/MAMM.2011.049

Eisenberg JF, O'Connell MA, August PV (1979) Density, productivity, and distribution of mammals in two Venezuela habitats. In: Eisenberg JF (ed) Vertebrate ecology in the northern Neotropics. Smithsonian Institution Press, Washington, DC, pp 187–207

Encarnação CD (1987) Contribuição à ecologia dos tatus (Xenarthra, Dasypodidae) da Serra da Canastra. Minas Gerais. Dissertation, Universidade Federal do Rio de Janeiro

Fahrig L, Baudry J, Brotons L, Burel FG, Crist TO, Fuller RJ, Sirami C, Siriwardena GM, Martin JL (2011) Functional landscape heterogeneity and animal biodiversity in agricultural landscapes. Ecol Lett 14:101–112. https://doi.org/10.1111/j.1461-0248.2010.01559.x

Ferraz KMPMB, Peterson AT, Scachetti-Pereira R, Vettorazzi CA, Verdade LM (2009) Distribution of capybaras in an agroecosystem, Southeastern Brazil, based on ecological niche modeling. J Mammal 90:189–194. https://doi.org/10.1644/07-MAMM-A-338.1

Ferreguetti AC, Tomás WM, Bergallo HG (2015) Density, occupancy, and activity pattern of two sympatric deer (*Mazama*) in the Atlantic Forest, Brazil. J Mammal 96:1245–1254. https://doi.org/10.1093/jmammal/gyv132

Ferreira IEP et al (2026) The influence of fragmented landscapes on population density studies using camera trapping. In: Martins Verdade L, Moral RA (eds) Wildlife monitoring: integrating conservation and innovation in human-altered landscapes. Springer, Cham

Fieberg J, Kochanny CO (2005) Quantifying home-range overlap: the importance of the utilization distribution. J Wildl Manag 69:1346–1359. https://doi.org/10.2193/0022-541X(2005)69[1346:QHOTIO]2.0.CO;2

Fitch HS, Goodrum P, Newman C (1952) The armadillo in the southwestern United States. J Mammal 33:21–37. https://doi.org/10.2307/1375638

Forys EA, Humphrey SR (1996) Home range and movements of the lower keys marsh rabbit in a highly fragmented habitat. J Mammal 77:1042–1048. https://doi.org/10.2307/1382784

Foster RJ, Harmsen BJ (2012) A critique of density estimation from camera-trap data. J Wildl Manag 76:224–236. https://doi.org/10.1002/jwmg.275

Fragoso JM, Levi T, Oliveira LF, Luzar JB, Overman H, Read JM, Silvius KM (2016) Line transect surveys underdetect terrestrial mammals: implications for the sustainability of subsistence hunting. PLoS One 11:e0152659. https://doi.org/10.1371/journal.pone.0152659

Gaston KJ (2011) Common ecology. Bioscience 61:354–362. https://doi.org/10.1525/bio.2011.61.5.4

Gaston KJ, Fuller RA (2007) Commonness, population depletion and conservation biology. Trends Ecol Evol 23:14–19. https://doi.org/10.1016/j.tree.2007.11.001

Harmsen BJ, Wooldridge RL, Gutierrez SM, Doncaster CP, Foster RJ (2018) Spatial and temporal interactions of free-ranging pacas (*Cuniculus paca*). Mammal Res 63:161–172. https://doi.org/10.1007/s13364-017-0350-0

Huanca-Huarachi G, Herrera JC, Noss AJ (2011) Population density and habitat use of the paca (*Cuniculus paca*) in the north of the Amboró-Carrasco conservation complex. Ecologia en Bolivia 46:4–13

Jácomo ATA, Kashivakura CK, Ferro C, Furtado MM, Astete SP, Tôrres NM, Sollmann R, Silveira L (2009) Home range and spatial organization of maned wolves in the Brazilian grasslands. J Mammal 90:150–157. https://doi.org/10.1644/07-MAMM-A-380.1

Kasper CB, Bastazini VAG, Soares JBG, Freitas TRO (2012) Abundance of Conepatus chinga (Carnivora, Mephitidae) and other medium-sized mammals in grasslands of southern Brazil. Iheringia, Série Zoologia 102:303–310. https://doi.org/10.1590/S0073-47212012000300009

Keuroghlian A, Eaton DP, Longland WS (2004) Area use by white-lipped and collared peccaries (*Tayassu pecari* and *Tayassu tajacu*) in a tropical forest fragment. Biol Conserv 120:411–425. https://doi.org/10.1016/j.biocon.2004.03.016

Kruszynski C, Moral RA, Diniz-Reis T, Piña C, Verdade LM (2026) Estimating the number of groups and population density of social species based on vestiges in human-altered environments: The case study of collared-peccaries (*Pecari tajacu*). In: Martins Verdade L, Moral RA (eds) Wildlife monitoring: integrating conservation and innovation in human-altered landscapes. Springer, Cham

Layne JN, Glover D (1977) Home range of the armadillo in Florida. J Mammal 58:411–413. https://doi.org/10.2307/1379340

Leuchtenberger C, de Oliveira ES, Cariolatto LP, Kasper CB (2018) Activity pattern of medium and large sized mammals and density estimates of Cuniculus paca (Rodentia: Cuniculidae) in the Brazilian Pampa. Braz J Biol 78:697–705. https://doi.org/10.1590/1519-6984.174403

Lisboa MAM, Prado BHS (2019) Fazenda das Areias e Fazenda da Conquista: História. Memoria e Cultura, RV Content, SP

Loughry WJ, McDonough CM (1998) Spatial patterns in a population of nine-banded armadillos (*Dasypus novemcinctus*). Am Midl Nat 140:161–169. https://doi.org/10.1674/00030031(1998)140[0161:SPIAPO]2.0.CO;2

MacKenzie DI, Nichols JD, Royle JA, Pollock KH, Bailey LL, Hines J (2006) Occupancy estimation and modeling. Academic, Oxford

Maffei L, Taber AB (2003) Area de acción, actividad y uso de hábitat del zorro patas negras, *Cerdocyon thous* Linnaeus, 1776 (Carnivora: Canidae) en un bosque seco. Mastozoología Neotropical 10:154–160

Magnusson WE, Lima AP, Luizão R, Luizão F, Costa FRC, Castilho CV, Kinupp VF (2005) RAPELD: a modification of the gentry method for biodiversity surveys in long-term ecological research sites. Biota Neotrop 5(2):19–24. https://doi.org/10.1590/S1676-06032005000300002

Magnusson WE, Lawson B, Baccaro F, de Castilho CV, Castley JG, Costa F, Drucker DP, Franklin E, Lima AP, Luizão R, Mendonça F, Pezzini F, Schietti J, Toledo JJ, Tourinho A, Verdade LM, Hero J-M (2014) Multi-taxa surveys: integrating ecosystem processes and user demands. In: Verdade LM, Lyra-Jorge MC, Piña CI (eds) Applied ecology and human dimensions in biological conservation. Springer, Heidelberg, pp 177–187

Martin PS, Gheler-Costa C, Lopes PC, Rosalino LM, Verdade LM (2012) Terrestrial non-volant small mammals in agro-silvicultural landscapes of Southeastern Brazil. Forest Ecol Manag 282:185–195. https://doi.org/10.1016/j.foreco.2012.07.002

Martinelli LA, Joly CA, Nobre CA, Sparovek G (2010) A falsa dicotomia entre a preservação davegetação natural e a produção agropecuária. Biota Neotrop 10:323–330. https://doi.org/10.1590/S1676-06032010000400036

Mazzolli M (2010) Mosaics of exotic forest plantations and native forests as habitat of pumas. Environ Manag 46:237–253. https://doi.org/10.1007/s00267-010-9528-9

McNeely JA, Scherr SJ (2003) Ecoagriculture: strategies for feed the world and save wild biodiversity. Island Press, Washington, DC

Medri MI, Mourão G (2005) Home range of giant anteaters (*Myrmecophaga tridactyla*) in the Pantanal wetland, Brazil. J Zool 266:365–375. https://doi.org/10.1017/S0952836905007004

Meek PD, Ballard GA, Fleming PJ (2015) The pitfalls of wildlife camera trapping as a survey tool in Australia. Aust Mammal 37:13–22. https://doi.org/10.1071/AM14023

Melis C, Szafrańska PA, Jędrzejewska B, Bartoń K (2006) Biogeographical variation in the population density of wild boar (*Sus scrofa*) in western Eurasia. J Biogreogr 33:803–811. https://doi.org/10.1111/j.1365-2699.2006.01434.x

Michalski F, Crawshaw PG Jr, Oliveira TG, Fabia ME (2006) Notes on home range and habitat use of three small carnivore species in a disturbed vegetation mosaic of Southeastern Brazil. Mammalia 70:52–57. https://doi.org/10.1515/MAMM.2006.004

Mimnagh N, Ferreira IEP, Verdade LM, Moral RA (2026) Counting animals we can't see: a triple count model for scarce vestige data. In: Martins Verdade L, Moral RA (eds) Wildlife monitoring: integrating conservation and innovation in human-altered landscapes. Springer, Cham

Miranda GHB (2004) Ecologia e conservação do tamandua-bandeira (*Myrmecophaga tridactyla*, Linnaeus, 1758) no Parque Nacional das Emas. Dissertation, Universidade de Brasilia

Montgomery GG, Lubin YD (1977) Prey influences on movements of neotropical anteaters. In: Phillips RL, Jonkel C (eds) Proceedings of the 1975 predator symposium. University of Montana Press, Missoula, Montana, pp 103–131

Neto FG (2016) Ecologia do veado-catingueiro (*Mazama gouazoubira*, Fischer 1814) no Pantanal. Dissertation, Universidade Federal do Paraná

Nichols JD (2014) The role of abundance estimates in conservation decision-making. In: Verdade LM, Lyra-Jorge MC, Piña CI (eds) Applied ecology and human dimensions in biological conservation. Springer, Heidelberg, pp 117–131

O'Brien TG, Kinnaird MF, Wibisono HT (2003) Crouching tigers, hidden prey: Sumatran tiger and prey populations in a tropical forest landscape. Anim Conserv 6:131–139. https://doi.org/10.1017/S1367943003003172

Oliveira ENC (2002) Ecologia alimentar e área de vida de carnívoros da Floresta Nacional de Ipanema, Iperó, SP (Carnivora: Mammalia). Dissertation, Universidade Estadual de Campinas

Oliveira TG (2008) Leopardus tigrinus. In: Machado ABM, Drummond GM, Paglia AP (eds) Livro Vermelho da Fauna Brasileira Ameaçada de Extinção. Fundação Biodiversitas, Curitiba, Paraná, pp 787–788

Oliveira-Santos LGR, Graipel ME, Tortato MA, Zucco CA, Cáceres NC, Goulart FV (2012) Abundance changes and activity flexibility of the oncilla, Leopardus tigrinus (Carnivora: Felidae), appear to reflect avoidance of conflict. Zoologia-Curitiba 29:115–120. https://doi.org/10.1590/S1984-46702012000200003

Palomares F, González-Borrajo N, Chávez C, Rubio Y, Verdade LM, Monsa R, Harmsen B, Adrados B, Zanin M (2018) Scraping marking behaviour of the largest neotropical felids. PeerJ 6:e4983. https://doi.org/10.7717/peerj.4983

Penteado M, Yamashita C, Marques TS, Verdade LM (2016) Bird diversity in relation to land use in agricultural landscapes of southeastern Brazil. In: Gheler-Costa MC, Lyra-Jorge MC, Verdade LM (eds) Biodiversity in agricultural landscapes of southeastern Brazil. De Gruyter Open, Berlin, pp 243–268

Potratz EJ, Holt RD, Brown JS (2024) Ecology of fear: acclimation and adaptations to hunting by humans. Sustainability 16(3):1216. https://doi.org/10.3390/su16031216

Reitz F, Léonard Y (1994) Studies on the European hare. Characteristics of European hare *Lepus europaeus* use of space in a French agricultural region of intensive farming. Acta Theriol 39:143–157

Rocha FL (2006) Áreas de uso e seleção de habitats de três espécies de carnívoros de médio porte na fazenda Nhumirim e arredores, Pantanal da Nhecolândia, MS. Dissertation, Universidade Federal de Mato Grosso do Sul

Rodrigues ASL, Andelman SJ, Bakarr MI, Boitani L, Brooks TM, Cowling RM, Fishpool LDC, da Fonseca GAB, Gaston KJ, Hoffmann M, Long JS, Marquet PA, Pilgrim JD, Pressey RL, Schipper J, Sechrest W, Stuart SN, Underhill LG, Waller RW, Watts MEJ, Yan X (2004) Effectiveness of the global protected area network in representing species diversity. Nature 428:640–643. https://doi.org/10.1038/nature02422

Rodrigues FHG, Medri IM, Miranda GHB, Alves CC, Mourao G (2008) Anteater behavior and ecology. In: Vizcaino SF, Loughry WJ (eds) The biology of the Xenarthra. University Press of Florida, Gainesville, pp 257–268

Rosalino LM, do Rosario J, Santos-Reis M (2009) The role of habitat patches on mammalian diversity in cork oak agroforestry systems. Acta Oecol 35:507–512. https://doi.org/10.1016/j.actao.2009.03.006

Rosalino LM, Verdade LM, Lyra-Jorge MC (2014) Adaptation and evolution in changing environments. In: Verdade LM, Lyra-Jorge MC, Piña CI (eds) Applied ecology and human dimensions in biological conservation. Springer, Heidelberg, pp 53–71

Rowcliffe JM, Field J, Turvey ST, Carbone C (2008) Estimating animal density using camera traps without the need for individual recognition. J Appl Ecol 45(4):1228–1236. https://doi.org/10.1111/j.1365-2664.2008.01473.x

Royle JA, Nichols JD (2003) Estimating abundance from repeated presence–absence data or point counts. Ecology 84(3):777–790. https://doi.org/10.1890/0012-9658(2003)084[0777:EAFRPA]2.0.CO;2

Rühe F, Hohmann U (2004) Seasonal locomotion and home-range characteristics of European hares (*Lepus europaeus*) in an arable region in Central Germany. Eur J Wildl Res 50:101–111. https://doi.org/10.1007/s10344-004-0049-9

Saïd S, Gaillard J-M, Widmer O, Débias F, Bourgoin G, Delorme D, Roux C (2009) What shapes intra-specific variation in home range size? A case study of female roe deer. Oikos 118:1299–1306. https://doi.org/10.1111/j.1600-0706.2009.17346.x

Saunders G, McLeod S (1999) Predicting home range size from the body mass or population densities of feral pigs, *Sus scrofa* (Artiodactyla: Suidae). Aust J Ecol 24:538–543. https://doi.org/10.1046/j.1442-9993.1999.01001.x

Schulte-Herbrüggen B, Davies G (2006) Wildlife conservation and tropical timber certification. ZSL Conservation Report No. 6. Zoological Society of London, London

Shaw JH, Machado-Neto J, Carter TS (1987) Behavior of free-living giant anteaters (*Myrmecophaga tridactyla*). Biotropica 19:255–259. https://doi.org/10.2307/2388344

Silveira L (1999) Ecologia e conservação dos mamíferos carnívoros do Parque Nacional das Emas, Goiás. Dissertation, Universidade Federal de Goiás

Sparovek G, Berndes G, Barretto AGDOP, Klug ILF (2012) The revision of the Brazilian Forest act: increased deforestation or a historic step towards balancing agricultural development and nature conservation? Environ Sci Pol 16:65–72. https://doi.org/10.1016/j.envsci.2011.10.008

Sunquist ME, Sunquist F, Daneke DE (1989) Ecological separation in a Venezuelan llanos carnivore community. In: Redford K, Eisenberg J (eds) . Advances in neotropical mammalogy, Sandhill Crane Press, Gainsville, FL, pp 197–232

Trolle M, Kéry M (2003) Estimation of ocelot density in the Pantanal using capture-recapture analysis of camera-trapping data. J Mammal 84:607–614. https://doi.org/10.1644/1545-1542(2003)084<0607:EOODIT>2.0.CO;2

Trolle M, Bissaro MC, Prado HM (2006a) Mammal survey at a ranch of the Brazilian Cerrado. In: Hawksworth DL, Bull AT (eds) Vertebrate conservation and biodiversity. Springer, Dordrecht, pp 1205–1211

Trolle M, Noss AJ, De Lima ES, Dalponte JC (2006b) Camera-trap studies of maned wolf density in the Cerrado and the Pantanal of Brazil. In: Hawksworth DL, Bull AT (eds) Vertebrate conservation and biodiversity. Springer, Dordrecht, pp 1197–1204

Trovati RG, Brito BA (2009) Nota sobre deslocamento e área de uso de Tamandúa-mirim (*Tamandia tetradactyla*) translocado no Cerrado brasileiro. Neotrop Biol Conservation 4(3):144–149. https://doi.org/10.4013/nbc.2009.43.04

Trovati RG, Brito BA, Duarte JMB (2010) Habitat use and home range of the brown-nosed coati, *Nasua nasua* (Carnivora: Procyonidae) in the Brazilian Cerrado. Rev Biol Trop 58:1069–1077

Vandermeer J, Perfecto I (1997) The agroecosystem: a need for the conservation biologist's lens. Conserv Biol 11:591–592. https://doi.org/10.1046/j.1523-1739.1997.07043.x

Verdade LM, Moreira JR, Ferraz KMPMB (2013) Counting capybaras. In: Moreira JR, Ferraz KMPMB, Herrera EA, Macdonald DW (eds) Capybara: biology, use and conservation of an exceptional neotropical species. Springer, New York, pp 357–370

Verdade LM, Lyra-Jorge MC, Piña CI (2014a) Redirections in conservation biology. In: Verdade LM, Lyra-Jorge MC, Piña CI (eds) Applied ecology and human dimensions in biological conservation. Springer, Heidelberg, pp 3–17

Verdade LM, Penteado M, Gheler-Costa C, Dotta G, Rosalino LM, Pivello VR, Piña CI. Lyra-Jorge MC. 2014b. The conservation value of agricultural landscapes. In: Verdade LM, Lyra-Jorge MC, Piña CI (eds) Applied ecology and human dimensions in biological conservation. Springer, Heidelberg, pp. 91–102

Verdade LM, Piña CI, Rosalino LM (2015) Biofuels and biodiversity: challenges and opportunities. Environ Dev 15:64–78

Verdade LM, Moral RA, Calaboni A, Amaral MVSG, Martin PS, Amorim LS, Gheler-Costa C, Piña CI (2020) Temporal dynamics of small mammals' in *Eucalyptus* plantations in Southeast Brazil. Glob Ecol Conserv 24:e01217. https://doi.org/10.1016/j.gecco.2020.e01217

Wood GW, Brenneman RE (1980) Feral hog movements and habitat use in South Carolina. J Wildl Manag 44:420–427. https://doi.org/10.2307/3807973

Part II
Technological Innovations I: Indicators and Sampling Methodology

Chapter 4
Sampling Scats on Trails: Trophic Approach in Wildlife Surveys in Silvicultural Landscapes

Cláudia B. Campos, Carla Gheler-Costa, Carolina F. Esteves, Tatiane C. Rech-Fernandes, Luís Miguel Rosalino, Rafael de Andrade Moral, Carlos I. Piña, and Luciano Martins Verdade

Abstract In this study we used a trophic approach to survey mesopredators (i.e., canids and felids) and small mammals (i.e., rodents and marsupials) in silvicultural landscapes. The small mammals were captured by pitfall traps and detected on the feaces of canids and felids by the presence of their remaining undigested hair and bones. Canids and felids were surveyed using line transects. The study was carried out at two farms devoted to *Eucalyptus* production in São Paulo, Brazil, which have a combination of mixed Ombrophyle forest and Cerrado under varying stages of conservation as Areas of Permanent Protection and Legal Reserve, according to the Brazilian Forest Code. We detected 12 terrestrial species of small mammals and 20 of medium to large mammals, most of them generalists. The species richness and relative abundance of these mammals were similar to those found in previous studies carried out in silvicultural landscapes in Southeast Brazil. Despite the intense anthropic history of local land use, mammalian trophic ecology is relatively

C. B. Campos · C. F. Esteves
Instituto Pró-Carnívoros, Atibaia, São Paulo, Brazil

C. Gheler-Costa
Ecologia Aplicada: pesquisa, ensino e serviços ambientais, Bauru, São Paulo, Brazil

T. C. Rech-Fernandes
TCR Assessoria Ambiental, Cotia, São Paulo, Brazil

L. M. Rosalino
cE3c – Center for Ecology, Evolution and Environmental Change & CHANGE – Global Change and Sustainability Institute, Faculdade de Ciências, Universidade de Lisboa, Lisboa, Portugal

R. A. Moral
Department of Mathematics and Statistics, Maynooth University, Maynooth, Co. Kildare, Ireland

C. I. Piña
CICyTTP-CONICET Entre Ríos/UADER, España, Diamante, Entre Ríos, Argentina

L. Martins Verdade (✉)
Wildlife Monitoring Consultancy, Campinas, São Paulo, Brazil
e-mail: luciano.verdade@wildlifemgmtco.com

L. Martins Verdade, R. A. Moral (eds.), *Wildlife Monitoring*,
https://doi.org/10.1007/978-3-032-05821-8_4

complex, with tens of predator-prey interactions in which canids appeared to have a less specialized diet compared to felids, which is a usual pattern. The trophic approach might be useful for monitoring wildlife in agricultural and silvicultural landscapes instead of the usual patterns of biodiversity since trophic interactions reflect the diversity of biological patterns as well as the complexity of the trophic process which tend to shape them.

Keywords Wildlife management · Monitoring · Small mammals · Mesopredators · Biocomplexity

1 Introduction

Species richness has been used as a major criterion for conservation purposes in pristine environments at global and regional scales (Mittermeier et al. 2004). However, at local scales, transition areas can support more species than former ecosystems, but with lower carrying capacity and consequently higher chance of local extinction (Macdonald et al. 2007). Canids and felids are possibly the most effective local terrestrial predators in agricultural and silvicultural landscapes of Southeast Brazil (Dotta and Verdade 2007) as they usually present a relatively high diversity (Dotta and Verdade 2011; Campos et al. 2018). In general, felids are considered more specialist than canids in terms of diet (Macdonald et al. 2010). However, even errant domestic dogs (*Canis lupus familiaris*) and domestic cats (*Felis catus*) can act as effective predators in human-modified environments (Campos et al. 2007).

An extensive niche-overlap occurs among *Leopardus guttulus*, *L. pardalis*, and *L. wiedii*, whereas *Herpailurus yagouaroundi* has the largest dietary niche breadth among small to medium felids of Southern Brazil (Wang 2002; Rocha-Mendes et al. 2010; Seibert et al. 2015). In addition, diet overlap between *Cerdocyon thous* and *Chrysosyum brachyurus* may vary significantly among regions (Juarez and Marinho-Filho 2002; Bueno and Motta-Junior 2004; Jácomo et al. 2004). Nevertheless, small mammals are a staple food for both canids and felids in Southeast Brazil (Dotta and Verdade 2009; Bianchi et al. 2011; Verdade et al. 2011). However, in general only 40–50% of their original assemblage remains in silvicultural landscapes in that region (Gheler-Costa et al. 2012; Martin et al. 2012). Such reduction in prey diversity may lead to an increase in dietary niche overlap among mesopredators.

By hypothesis, species interactions can be more effective than species richness to characterize ecosystem structure and function as they are related to the evolutionary-ecological processes that determine the patterns of species abundance and distribution (Jones and Safi 2011; Verdade et al. 2014b). Particularly in special in human-modified environments, the loss of complexity of ecological processes may lead to the loss of diversity of ecological patterns (May et al. 2007; Loeuille et al. 2013; Heleno et al. 2014; Sanders et al. 2018).

Brazil is one of the largest *Eucalyptus* producers in the world with more than 7.3 million hectares planted (ABRAF 2022). Because *Eucalyptus* plantations are homogeneous forests, landscapes whose matrices are *Eucalyptus* commercial plantations can be called silvicultural landscapes. The paper industry (major user of *Eucalyptus* production) usually follows the Brazilian Forest Code (BFC, Federal Law No. 12.651/2012) including the maintenance of Areas of Permanent Protection (APP) and Legal Reserve (LR) due to the market-driven social-environmental certification process, especially by the Forest Stewardship Council (FSC) (Burger et al. 2005). According to the BFC, approximately 20–30% of silvicultural landscapes must be maintained as APP and LR in the region originally covered by Atlantic Forest (Metzger et al. 2010), including native vegetation in various levels of integrity, from abandoned pastures to relatively well conserved forest, with a considerable wildlife diversity (Millan et al. 2015; Lopes et al. 2016), which improves conservation value of silvicultural landscapes (Verdade et al. 2014a). However, human-altered environments may experience a process of homogenization in local biodiversity, favoring the persistence of generalist species (Magnusson 2006).

In the present study we assessed the diversity of non-volant terrestrial mammals and their trophic interactions in silvicultural landscapes of São Paulo state, Southeast Brazil. In addition, we explore the potential application of a trophic approach in wildlife surveys as a measure of both biodiversity (i.e., species composition and relative abundance of predators and their prey) and biocomplexity (i.e., the complexity of their trophic interactions).

2 Study Area

This study was carried out at Fazenda Ibiti (24°10′ S, 49°15′ W) (11,594 ha) and Fazenda Santa Isabel (24°10′ S, 49°05′ W) (5840 ha). The former is located at Itararé municipality, with 65% of its area covered by *Eucalyptus* plantations, 25% by mixed Ombrophile forest, 5.6% by Cerrado, 3.9% by secondary growth forest and 0.5% by human constructions; whereas the latter is located at Itapeva municipality with 63% of its area covered by *Eucalyptus* plantations and 37% by Cerrado. The altitude in both areas averages between 700 and 750 m (Fig. 4.1). The local climate is sub-tropical, with an average annual mean temperature of 25 °C.

3 Methodology

3.1 Sampling Methodology

We carried out 10 bimonthly three-day sampling campaigns for small mammals and nine for medium to large mammals between August 2004 and February 2006.

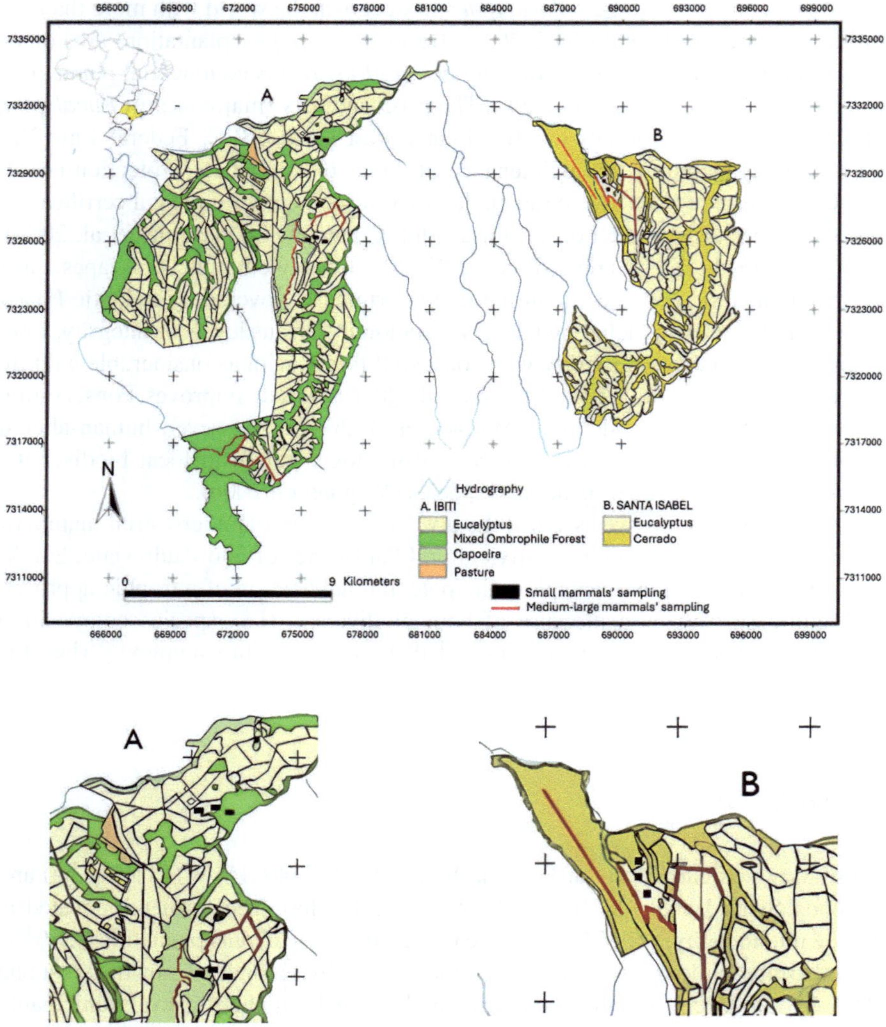

Fig. 4.1 Land use composition with location of pit-fall traps and linear transects at Fazenda Ibiti (zoomed at **a**) and Fazenda Santa Isabel (zoomed at **b**) farms in São Paulo State, Brazil (*Datum*: Córrego Alegre)

By that time, the *Eucalyptus* plantations were on their late 7-year cycle (i.e., between 4 and 6 years of age), with trees reaching 18 m high on average. We used nine sampling units distributed over *Eucalyptus* plantations and three distinct types of native vegetation: mixed ombrophile forest, secondary forest and Cerrado) (Table 4.1, Fig. 4.1). The selection of sampling units were established according to the following *a priori* criteria: (a) Largest area of native vegetation cover patches of each farm; (b) Lowest area-perimeter ratio; and (c) Minimum distance of 5 km between sampling sites, to reduce spatial autocorrelation.

Table 4.1 Sampling units' spatial distribution

Sampling unit	Environment
1	Mixed ombrophile forest 500 m from the border of *Eucalyptus* plantation
2	Border between mixed ombrophile forest and *Eucalyptus* plantation
3	*Eucalyptus* plantation 500 m from the border of mixed ombrophile forest
4	Secondary forest 500 m from the border of *Eucalyptus* plantation
5	Border between secondary forest and *Eucalyptus* plantation
6	*Eucalyptus* plantation 500 m from the border of secondary forest
7	Cerrado 500 m from the border of *Eucalyptus* plantation
8	Border between the Cerrado and *Eucalyptus* plantation
9	*Eucalyptus* plantation 500 m from the border of Cerrado

Small rodents and marsupials were collected by pitfall-trapping (Cechin and Martins 2000; Lyra-Jorge and Pivello 2001). We used nine sampling units, each of them composed by a Y-shaped structure with sixteen 100 litre buckets, spaced 10 m apart: five along the 60 m Y-branches and one at the center, totaling 144 buckets (Fig. 4.1, Table 4.1). The total sampling effort was 4320 bucket-nights (i.e., 10 monthly campaigns × 3 nights per campaign × 9 sampling units × 16 buckets per sampling units). Bottom holes were made in the buckets to facilitate drainage. In addition, a Styrofoam plate and a water container were left inside each bucket to respectively prevent drowning and dehydration of captured animals. Each trap was checked twice a day during field campaigns (Sutherland 2006). All captured animals were individually marked with a subcutaneous microchip (Trovan™), sexed, and identified according to their morphometrics and pelage color characteristics (Gheler-Costa et al. 2012; Martin et al. 2012). Capture and handling were performed aseptically to prevent pathogen transmission according to international standards (Gannon and Sikes 2011). The animals were then released at the capture site.

We considered medium to large sized terrestrial mammals' species with adult weight above 1 kg. We established nine 3-km transect lines, at least 5 km apart from each other, also distributed over *Eucalyptus* plantations and three distinct types of native vegetation: mixed ombrophile forest, secondary forest and Cerrado) (Table 4.1, Fig. 4.1). The total distance covered was 243 km (i.e., 9 monthly campaigns × 9 transect lines × 3 km per transect line). All direct observations (live or dead animals) and tracks of medium to large non-volant terrestrial mammal species were considered a record of detection (Wemmer et al. 1996) and identified, according to their morphometrics (Becker and Dalponte 2013). The taxonomic nomenclature of the species followed the guidelines of the Brazilian Society of Mammalogy (Abreu et al. 2023).

To reduce uncertainty, we grouped species with records identified by similar signs in the lowest possible taxonomic level (Carvalho et al. 2015). Thus, deer (*Subulo gouazoubira* and *Mazama americana*) were grouped as Cervidae as their detection has only been done by tracks (Angeli et al. 2014). In addition, the Southern tiger cat (*Leopardus guttulus*) and margay (*L. wiedii*) were grouped as *Leopardus* spp. for diversity analyses but kept separately for diet analyses. This is because the footprints of both species are indistinguishable, however these species can be distinguished in terms of their fur found in their own scats.

The diet of canids and small to medium felids was assessed by triage of items found in the feces collected along the transect lines (Dotta and Verdade 2007). Undigested items (e.g., seeds, bones, teeth, and hair) were manually screened using a stereoscope and tweezers. The species of mesopredators and their mammal preys were identified using cuticle imprints and medullar patterns of hair samples found in the scats. We used a light microscope to analyze hair (cuticle/medulla) comparing them with pre-existing classification keys (Quadros and Monteiro-Filho 2006; Martin et al. 2009). Seeds and other plant material were identified by specialists. Dubious identifications of both scats and vestiges were excluded from the analyses.

3.2 *Analytical Methodology*

We considered species richness as the asymptote of the rarefaction curve generated by the mathematical relation between sampling effort and the number of species detected, using a Bootstrap procedure to estimate uncertainty. This analysis was carried out using the EstimateS package (Colwell 2006). The number of small mammal specimens captured in pitfall traps per sampling unit was used as a measure of species relative abundance. Trapping success was considered as the number of captured individuals divided by the number of bucket-nights, in percentage points (i.e., multiplied by 100). For medium to large sized mammals, each detection was considered as an occurrence. We assumed that the most abundant species would be widely dispersed and therefore be found more frequently than the less abundant ones (Sutherland 2006). Since transect sampling was performed on unpaved trails, we assumed that the variation in detectability of tracks and scats was negligible (Harmsen et al. 2010). A species ordination was established by Multidimensional Scaling analysis, carried out in R Software (R Core Team 2020).

We quantified diet items by their frequency of occurrence, which is expressed as the percentage of scats containing each item (Breuer 2005). Then we fitted ordination curves of the diet items frequency of occurrence, considering only small rodents and marsupials as food items for canids and small to medium-sized felids. Since diet items' frequencies are positive, continuous, and right-skewed, we fitted a gamma generalized linear model with the log link function, as explained below.

Let Y_{ij} be the observed predation frequency for the i-th predator and j-th diet item. We assumed that $Y_{ij} \sim \text{Gamma}(\mu_{ij}, \phi)$, where the mean is modelled with a different linear regression per predator on the scale of the linear predictor, i.e.,

$$\log\left(\mu_{ij}\right) = \beta_{0i} + \beta_{1i} \times s_j$$

where s_j is the diet item rank (starting from 0 to a maximum of 35) and the dispersion ϕ is modeled with a single intercept, common to all predator species. Model goodness-of-fit was assessed using half-normal plots with a simulated envelope (Moral et al. 2017). This analysis was carried out using R software (R Core Team 2020).

We fitted ordination curves for each species and compared their intercepts and slopes using likelihood-ratio tests. We used the Bonferroni correction for the significance level, since we are performing 15 pairwise comparisons simultaneously; hence our adjusted significance level for each pairwise comparison is $p = 0.05/15 = 0.0033$. The slopes comprise the relationship between the number of diet items and their frequency of occurrence, which can be considered a complexity index (adapted from Jost (2006) and Ulanowicz et al. (2014)), analogous to β-diversity (Zar 1996: 353–356).

4 Results

We detected 32 taxa of non-volant terrestrial mammals in the study area. We captured 215 individuals of 12 species of small mammals (93% of the estimated species richness detected by pitfall traps) as follows: eight rodents (Rodentia), four marsupials (Didelphimorphia), and two armadillos (Cingulata) (Table 4.2). Trapping success was 3.3% in Capoeira, 4.1% in Cerrado, and 7.6% in mixed Ombrophile forest, with 5% as the average for the whole study area. *Oligoryzomys flavescens* was the most abundant species, being present in all environments, although *Necromys lasiurus*, *Rhipidomys mastacalis* and *Calomys tener* can also be considered relatively abundant (Table 4.2). None of the detected species is considered threatened.

We also detected 20 taxa of medium to large sized mammals in transects (approximately 87% of the estimated richness) from 197 occurrences (i.e., tracks and sightings) as follows: Carnivora (10 spp.), Cingulata (3 spp.), Cetartiodactyla (2 spp.), Lagomorpha (2 spp.), Perissodactyla (1 sp.), Pilosa (1 sp.), and Primates (1 sp.) (Table 4.2). *Dasypus novemcinctus* and *D. septemcinctus* have also been captured by pitfall traps. However, they were only considered in the analyses of medium to large mammals, unless stated otherwise.

The more frequently detected taxa of medium to large mammals were Cervidae (N = 55), *Chrysocyon brachyurus* and *Myrmecophaga tridactyla* (N = 31 each). Three species have some degree of threat, according to the International Union for Conservation of Nature and Natural Resources (IUCN) Red List of Threatened Species: *C. brachyurus* is considered near threatened (Paula and DeMatteo 2015), whereas *M. tridactyla* and *Tapirus terrestris* are considered vulnerable (Miranda et al. 2014; Varela et al. 2019).

Approximately 16% (5/32) of the species of terrestrial mammals were only detected in native vegetation, 9% (3/32) only on the *Eucalyptus* plantations and 6% (2/32) only on the border between these two environments (Table 4.2). There appears to be a certain distinction between native vegetation (mixed Ombrophyle Forest, Cerrado and Capoeira) and *Eucalyptus* plantations (including their borders)

Table 4.2 Non-volant terrestrial mammals at Fazenda Ibiti and Fazenda Santa Isabel, Southeast Brazil

Order	Species	Common names	Cerrado			Mixed Ombrophyle forest			Capoeira			
			Ma	Bo	Na	Ma	Bo	Na	Ma	Bo	Na	Total
Carnivora[1]	*Canis lupus familiaris*	Domestic dog		1				3				4
	Cerdocyon thous	Crab-eating fox		3		1					1	5
	Chrysocyon brachyurus	Maned wolf	4	6	1	6	4		4	4	2	31
	Eira barbara	Tayra	1	2	2		2			1		8
	Herpailurus yagouaroundi	Jaguarundi		2		1	1			1		5
	Leopardus pardalis	Ocelot			1			4	1		1	7
	Leopardus sp.	Wild cats			1						2	3
	Nasua nasua	Coati	1									1
	Puma concolor	Puma	1	3	1		1		1	3	2	12
	Procyon cancrivorus	South American raccoon	1	2		1						4
Cetartiodactyla[1]	Dicotyles *tajacu*	Collared peccary	1		2					5	3	11
	Mazama americana[a]	Red-brocket deer	9	7	5	9	5	2	7	7	4	55
	Subulo gouazoubira[a]	Brown brocket deer										
Cingulata[2]	*Dasypus novemcinctus*[1]	Nine-banded armadillo		1(1)		1		(1)			1(1)	6
	Dasypus septemcinctus[1]	Seven-banded armadillo	(1)	1	(1)						(1)	4
	Euphractus sexcintus	Six-banded armadillo			2							2
Didelphimorfia[3]	*Didelphis albiventris*	White-eared opossum							1			1
	Cryptonanus agricolai	Agricola's gracile opossum	1	3	1	1			1			7

	Gracilinanus microtarsus	Brazilian gracile opossum	4	1						1		6
	Monodelphis kunsi	Pygmy short-tailed opossum			1							1
Lagomorpha[1]	*Lepus europaeus*	European hare									1	1
	Sylvilagus brasiliensis	Brazilian cottontail				5				3		8
Perissodactyla[1]	*Tapirus terrestris*	Tapir					2		1			3
Pilosa[1]	*Myrmecophaga tridactyla*	Giant anteater	6	5	2	1	4		4	6	3	31
Primates[1]	*Alauatta guariba clamitans*	Southern brown howler						2				2
Rodentia[3]	*Akodon montensis*	Montane grass mouse		1	1			1	1		8	12
	Bibimys labiosus	Large-lipped crimson-nosed rat		1								1
	Calomys tener	Delicate Vesper Mouse	3	1	1	8	7	1	2			23
	Kannabateomys amblyonyx	Atlantic bamboo rat		1								1
	Necromys lasiurus	Hairy-tailed bolo mouse	6		3	1		12			11	33
	Oligoryzomys flavescens	Yellow pygmy rice rat	4	9	7	29	33	1	6	1	7	97
	Oligoryzomys nigripes	Black-footed pygmy rice rat	1								1	2
	Rhipidomys mastacalis	Long-tailed Climbing Mouse	1	2	3	5	7	2	4		1	25
Total			45	53	35	69	66	29	33	32	50	412

Matrix core of the *Eucalyptus* stands (Ma), border between *Eucalyptus* stands and native vegetation (Bo), and core of the native vegetation (Na). The numbers represent the frequency of occurrence, i.e., the number of sights or vestiges detected by an observer in transects (Carnivora, Cetartiodactyla, Lagomorpha, Perissodactyla, Pilosa, and Primates), and an index of abundance, i.e., the number of individuals captured in pitfall traps (Didelphimorfia and Rodentia), or both with the abundance index in parentheses (Cingulata)

Sampling methods: [1]Transects; [2]Both methods (pitfall traps in parentheses); [3]Pitfall traps

[a]*Mazama americana* and *Subulo gouazoubira* are grouped as Cervidae, as their tracks' identification can be dubious

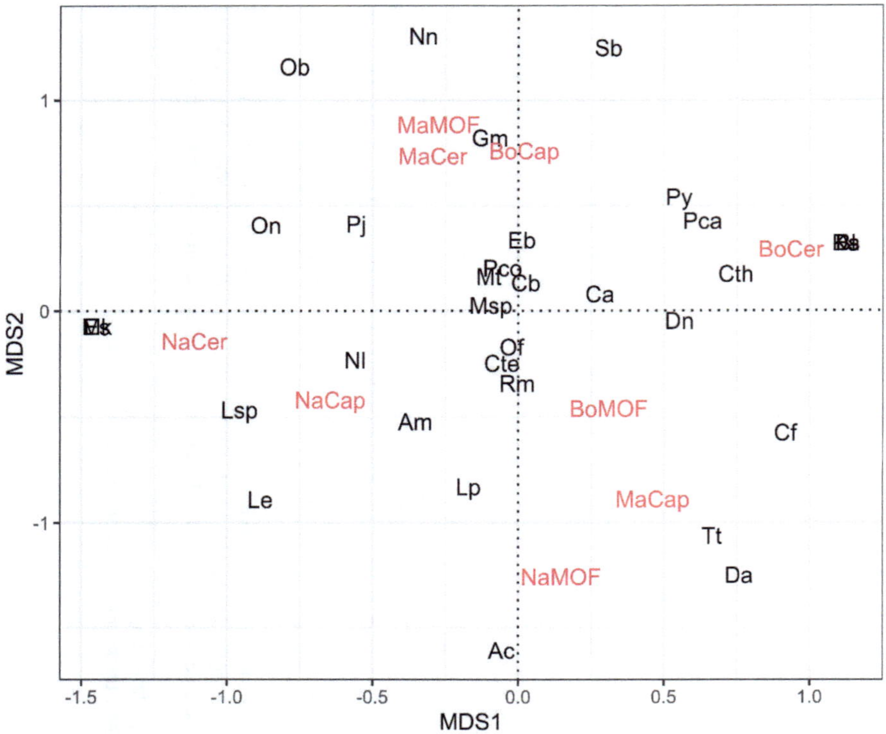

Fig. 4.2 Multidimensional Scaling analysis between habitats and non-volant terrestrial mammals on a silvicultural landscape of southeastern Brazil. Habitats: Cerrado (**MaCer**: Matrix; **BoCer**: Border; **NaCer**: Native vegetation); Mixed Ombrophyle Forest (**MaMOF**: Matrix; **BoMOF**: Border; **NaMOF**: Native vegetation); and Capoeira (**MaCap**: Matrix; **BoCap**: Border; **NaCap**: Native vegetation). *Akodon montensis* (Am), *Bibimys labiosus* (Bl), *Calomys tener* (Cte), *Cryptonanus agricolai* (Ca), *Didelphis albiventris* (Da), *Gracilinanus microtarsus* (Gm), *Kannabateomys amblyonyx* (Ka), *Monodelphis kunsi* (Mk), *Necromys lasiurus* (Nl), *Oligoryzomys flavescens* (Of), *Oligoryzomys nigripes* (*On*), *Rhipidomys mastacalis* (Rm); Medium to Large-sized species: *Alouatta guariba clamitans* (Ac), *Canis lupus familiaris* (Cf), *Cerdocyon thous* (Cth), Cervidae (Msp), *Chrysocyon brachyurus* (Cb), *Dasypus novemcinctus* (Dn), *Dasypus septemcinctus* (Ds), *Eira barbara* (Eb), *Euphractus sexcintus* (Es), *Leopardus* sp. (Lsp), *Leopardus pardalis* (Lp), *Lepus europaeus* (Le), *Myrmecophaga tridactyla* (Mt), *Nasua nasua* (Nn), *Dicotyles tajacu* (Pt), *Procyon cancrivorus* (Pca), *Puma concolor* (Pco), *Herpailurus yagouaroundi* (Py), *Sylvilagus brasiliensis* (Sb), *Tapirus terrestris* (Tt). Records from *Dasypus novemcinctus* and *D. septemcinctus* from both methods (pitfall traps and traces) have been pooled together

in terms of species composition. Such distinction is captured by Component 2 of the Multidimensional Scaling analysis (Fig. 4.2). Howler monkeys (*Alouatta* guariba *clamitans*), ocelots (*Leopardus pardalis*) and, surprisingly, errant domestic dogs (*C. lupus familiaris*) have been clearly associated with the mixed Ombrophyle Forest. However, many species tend to use all landscape elements including the *Eucalyptus* plantations.

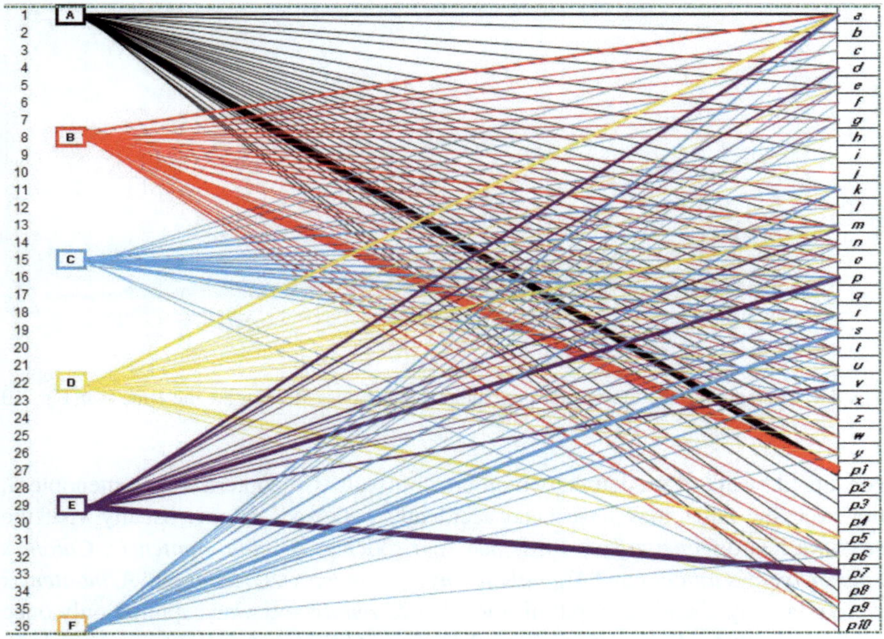

Fig. 4.3 Trophic relations among non-volant terrestrial mammals at silvicultural landscapes in Southeast Brazil. Line thickness proportional to number of occurrences. Predators: A (*C. brachyurus*), B (*C. thous*), C (*H. yagouaroundi*, D: *L. pardalis*, E (*L. wiedii*), F (*L. gutulus*). Diet items: a (Coleoptera), b (Dermaptera), c (Diptera), d (Hemiptera), e (Hymenoptera), f (Isoptera), g (Lepidoptera), h (Orthoptera), i (Acari), j (Stylommatophora), *k* (*Ameiva ameiva*), *l* (*Cercosaura*), m (Aves), *n* (*Akodon montensis*), *o* (*Bibimys labiosus*), *p* (*Calomys tener*), *q* (*Cavia aperea*), *r* (*Clyomys laticeps*), *s* (*Holochilus brasiliensis*), *t* (*Kannabateomys amblyonyx*), *u* (*Necromys lasiurus*), *v* (*Oligoryzomys sp.*), *w* (*Rhipidomys mastacalis*), *x* (*Didelphis albiventris*), *y* (*Sylvilagus brasiliensis*), *z* (*Dasypus novemcinctus*), p1 (*Butia microspadix*), p2 (*Campomanesia pubescens*), p3 (*Citrus* sp.), p4 (*Curcubita* sp.), p5 (*Eucalyptus* sp.), p6 (*Poaceae* sp.), p7 (*Psidium cattleyanum*), p8 (*Psychotria* sp.), p9 (*Solanum lycocarpum*), p10 (*Solanum* sp.)

We collected 143 fecal samples from two species of canids (*C. brachyurus*, N = 45, and *Cerdocyon thous*, N = 37) and four species of felids (*Leopardus gutulus*., N = 13; *L. pardalis*, N = 13; *L. wiedii*, N = 17; and *Herpailurus yagouaroundi*, N = 18). In total, we detected 68 predator-prey interactions between mesopredators (i.e., canids and felids) and small mammals (i.e., lagomorphs, rodents, marsupials, and xenarthrans) (Fig. 4.3). Only eight interactions (11.8%) had a frequency of occurrence above 10% (even though, below 30%). Therefore, almost 90% of them had a frequency of occurrence below 5% (Fig. 4.4).

In total, we identified 35 items in the mesopredators' diets based on their fecal analysis: all 35 items were found in canids, whereas 28 were found in felids, including 13 species of small mammals. The two species of canids had 30 and 31 diet items each, whereas the four species of felids had between 14 (*L. guttulus*) and 20 (*L. pardalis* and *H. yagouaroundi*) (Table 4.3). Eight out of 35 items were

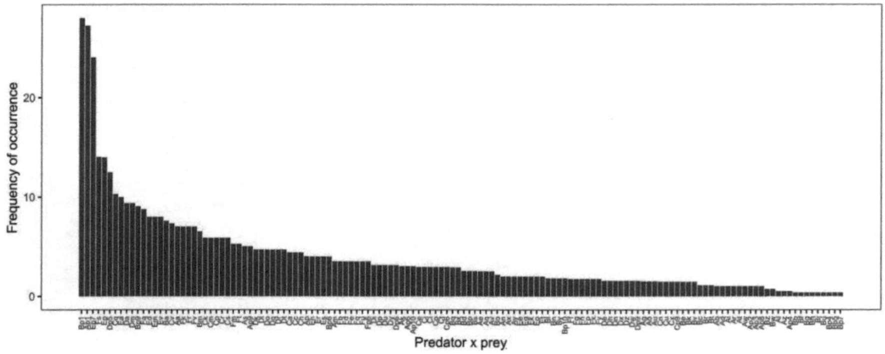

Fig. 4.4 Ordination curve of predator-prey interactions. Codes at x axis are the same as in Fig. 4.3

consumed by all six predators: two orders of insects (Coleoptera and Hymenoptera; however, the latter may have been accidently collected, as ants usually visit the scats), a lizard (*Ameiva* sp.), birds and four rodents (*Akodon montensis*, *Calomys tener*, *Clyomys laticeps* and *Kannabateomys amblyomyx*). *C. tener* and *A. montensis* were relatively abundant in pitfall traps, but *K. amblyonyx* was captured only once during this study. Six species of small rodents (*Bibimys labiosus*, *Cavia aperea*, *Holochilus brasiliensis*, *Necromys lasiurus*, *Olygorysomys* spp., and *Rhipidomys mastacalis*) and a native grass (Poaceae) were consumed by at least five out of the six mesopredators of the present study. However, *C. aperea*, *Clyomys laticeps,* and *H. brasiliensis* were not captured by pitfall traps. On the other hand, three marsupials (*Cryptonanus agricolai*, *Gracilinanus microtarsus*, and *Monodelphis kunsi*) and an armadillo (*Dasypus septemcinctus*) were captured by pitfall traps but were not detected in the predators' scats (Table 4.3, Fig. 4.3). Fruits (*Solanum lycocarpum*, *Butia microspadix*, *Psidium cattleyanum*, *Cucurbita* sp. and *Psicotria* sp.) predominated in canids' diets, but they were also present in felids' diets. In addition, insects (Dermaptera, Diptera, Hemiptera, Lepidoptera and Orthoptera, as well as Coleoptera and Hymenoptera) were common in canids' diet, but rare in the scats of felids (Table 4.3).

Canids and felids consumed a relatively high number of items in their diet including plants (i.e., fruits and grass) and animals (i.e., vertebrates and invertebrates) (Fig. 4.4). When considering only the small rodents, the intercepts and slopes for *H. yagourundi* and *L. guttulus* differed from the canid species' ones, while the slopes for *L. wiedii* also differed when compared to the canid species' ones (Figs. 4.5 and 4.6). Although the felid species' intercepts did not differ within their group, the slope for *L. pardalis* differed from the other felids, while being statistically equal to the canids' intercepts and slopes (Figs. 4.5 and 4.6).

Table 4.3 Frequency of occurrence of prey items consumed by the six predator species sampled in this study

			Predators						N
Taxon	Diet items	Common names	*C. brachyurus*	*C. thous*	*L. guttulus*	*L. pardalis*	*L. wiedii*	*H. jagouaroundi*	
Rodentia	*Akodon montensis*	Montane grass mouse	11.1	10.8	15.4	7.7	5.9	5.6	12
	Bibimys labiosus	Large-lipped crimson-nosed rat	13.3	8.1	7.7	23.1		11.1	1
	Calomys tener	Delicate vesper mouse	15.6	21.6	53.8	15.4	5.9	22.2	23
	Cavia aperea	Brazilian guinea pig	17.8	13.5		23.1	11.8	38.9	
	Clyomys laticeps	Broad-headed spiny rat	6.7	5.4	15.4	23.1	23.5	22.2	
	Holochilus brasiliensis	Web-footed marsh rat	15.6	27.0	15.4		47.1	22.2	
	Kannabateomys amblyonyx	Atlantic bamboo rat	6.7	5.4	7.7	23.1	17.6	11.1	1
	Necromys lasiurus	Hairy-tailed bolo mouse	2.2	8.1		15.4	11.8	5.6	33
	Oligoryzomys spp.	Pygmy rice rat	8.9	13.5	30.8		23.5	27.8	99
	Rhipidomys mastacalis	Long-tailed climbing mouse	17.8	10.8		7.7	5.9	5.6	25
Cingulata	*Dasypus novemcinctus*	Nine-banded armadillo	11.1	10.8		15.4	5.9		3
	Dasypus septemcinctus[1]	Seven-banded armadillo							3
	Didelphis albiventris	White-eared opossum	4.4	2.7		7.7			1
	Cryptonanus agricolai	Agricola's gracile opossum							7
Didelphimorfia	*Gracilinanus microtarsus*	Brazilian gracile opossum							6
	Monodelphis kunsi	Pygmy short-tailed opossum							1
Lagomorpha	*Sylvilagus brasiliensis*	Tapeti	4.4	5.4		7.7			
Aves	Aves	Birds	40.0	18.9	30.8	46.2	17.6	22.2	
Reptiles	*Ameiva ameiva*	Common ameiva lizard	8.9	16.2	7.7	23.1	23.5	22.2	
	Cercosaura sp.	Long-tailed little lizard	11.1	2.7		23.1		11.1	
Invertebrates	Acari	Mite		5.4		15.4		11.1	
	Coleoptera	Beetle	46.7	37.8	38.5	46.2	29.4	16.7	

(continued)

Table 4.3 (continued)

			Predators						N
Taxon	Diet items	Common names	*C. brachyurus*	*C. thous*	*L. guttulus*	*L. pardalis*	*L. wiedii*	*H. jagouaroundi*	
	Dermaptera	Earwigs	2.2	5.4			11.8		
	Diptera	Flies	6.7					16.7	
	Hemiptera	Cicadas	15.6	5.4	30.8		11.8		
	Hymenoptera	Ants	8.9	13.5	7.7	7.7	11.8	11.1	
	Isoptera	Termites	2.2				5.9		
	Lepidoptera	Butterflies	2.2	8.1	7.7		5.9		
	Orthoptera	Grasshoppers	2.2			7.7		16.7	
	Stylommatophora sp.	Common Garden Snail	2.2						
Plants	*Butia microspadix*	Butia palm	33.3	32.4					
	Campomanesia pubescens	Guabiroba		5.4					
	Citrus sp.	Citrus		5.4					
	Curcubita sp.	Cucurbi	2.2	2.7					
	Poaceae sp.	Grass	24.4	16.2		15.4	11.8	11.1	
	Psidium cattleyanum	Cherry guava	2.2		7.7				
	Psychotria sp.	Wild coffee		5.4					
	Solanum lycocarpum	Wolf apple	55.6	27.0		7.7		5.6	
	Solanum sp.	Solanaceous plant	11.1	16.2					

N number of individuals collected during sampling procedures (see sampling methodology)

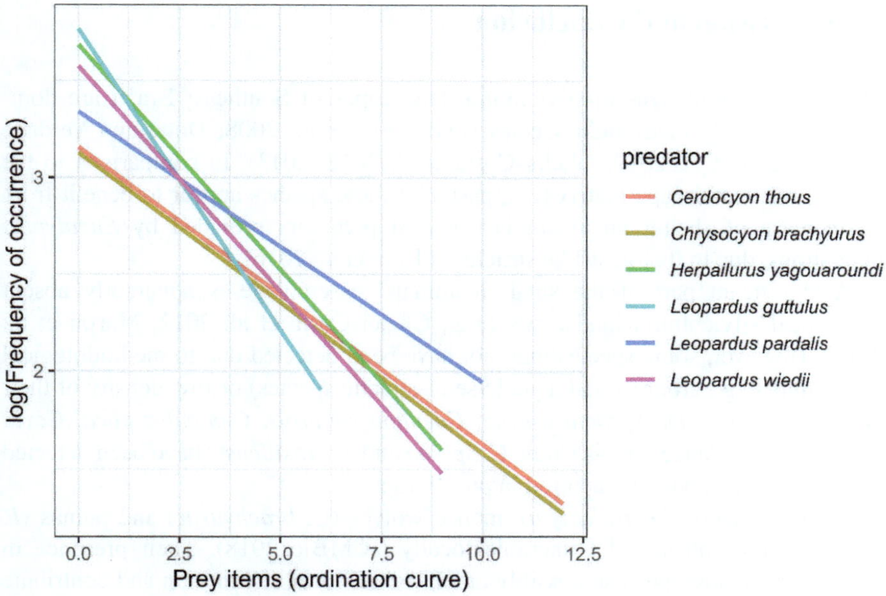

Fig. 4.5 Estimated linear predictors from the gamma generalized linear models fitted to the ordination curves of frequency of occurrence of prey items for six distinct predators

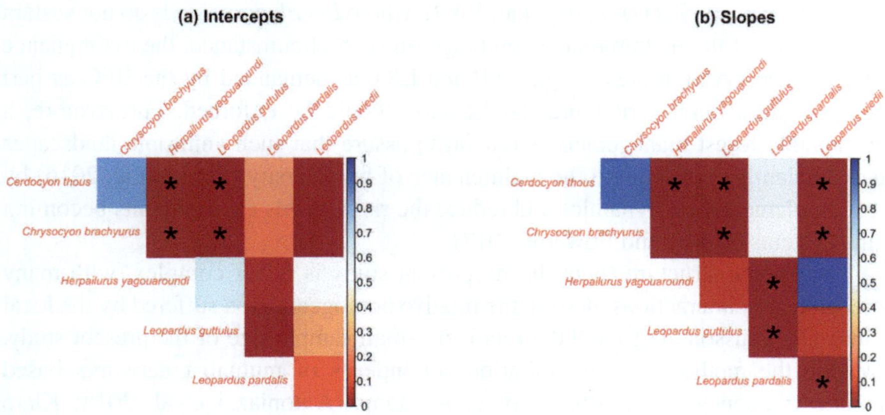

Fig. 4.6 Multiple comparisons between intercepts (**a**) and slopes (**b**) estimated by the gamma generalized linear models fitted to the ordination curves of frequency of occurrence of prey items for six distinct predators. The color shade represents the p-value, while asterisks (*) represent significance at a global level of 0.05, after performing Bonferroni correction (i.e., p-values less than 0.05/15 = 0.0033)

5 Discussion and Conclusion

Mammals' assemblages in silvicultural landscapes of Southeast Brazil are dominated by widely distributed species (Lyra-Jorge et al. 2008; Dotta and Verdade 2011; Martin et al. 2012; Gheler-Costa et al. 2012, 2013). In comparison to the usual former landscape matrix (i.e., pastures) these species appear to benefit from an increase of shelter and resources against predators provided by *Eucalyptus* plantations, due to its forest-like structure (Timo et al. 2016).

A significant part of the small mammals' assemblage is apparently absent from local silvicultural landscapes (e.g., Gheler-Costa et al. 2012; Martin et al. 2012). However, some species may not have been detected due to methodological constraints (e.g., arboreal and aquatic/semi-aquatic species) or low-density of their populations (e.g., *Dasyprocta azarae*, *Coendou spinosus*, *Cuniculus paca*, *Cavia aperea*). For instance, in this study, *C. aperea* and *S. brasiliensis* have been detected in predator scats, but not captured in pitfall traps.

Giant anteaters (*M. tridactyla*), maned-wolves (*C. brachyurus*) and pumas (*P. concolor*) are considered vulnerable locally (ICMBio 2018). Their presence in silvicultural landscapes can possibly amplify their local distribution and contribute to the discussion of the possible complementary role of these landscapes in wildlife conservation (Brockerhoff et al. 2013). Their detection in silvicultural landscapes can possibly amplify their local distribution. The present results suggest that *Eucalyptus* plantations should be considered as habitat for local generalist terrestrial mammals (e.g., Timo et al. 2015). However, a process of landscape complementation may be occurring (Dunning et al. 1992), where *Eucalyptus* stands do not sustain *per se* most of the mammalian assemblage. In such circumstance, the maintenance of local conservation areas (e.g., APP and LR), as demanded by the BFC as part of agricultural and silvicultural landscapes, should be enforced. Furthermore, a sustainable forest management is crucial to assure that such anthropic landscapes can efficiently contribute to the maintenance of biodiversity (Pirard et al. 2016) by its complementation dynamics and reduce the risks of this environments becoming sink systems (Gilroy and Edwards 2017).

The trophic structure found in the present study is rather complex, with many predator-prey interactions, despite the relative homogenization suffered by the local fauna (Magnusson 2006) and the relatively small sample size of the present study. Despite the methodological limitations of indexes of mutualist networks based on the frequency of occurrence of diets' items (Antoniazzi et al. 2018; Klare et al. 2011), the slope of the gamma generalized linear model with the log link function can be considered as a biocomplexity index related to diet specialization among predators. Canids appear to be less specialized than felids in terms of small mammals' predation (either by the intercept or the slope). In general, felids are considered more specialist predators than canids (Macdonald et al. 2010). However, an opposite pattern has been found on a former study for errant domestic dogs and cats on a human-modified environment (Campos et al. 2007). Actually, a momentary concentration of a few items on a considered specialist diet may be only seasonal.

In such a case the trophic process can have low complexity in short-term but high in medium- to long-term as dominant prey items may vary along the time. The relatively short-term of the present study may not allow us to detect such a change. This possibility alone would be a good reason to consider a trophic approach at monitoring wildlife programs, in special in silvicultural landscapes where the main plantation and the surrounding areas of native vegetation have distinct pulses of biomass production (Holt 2008; Verdade et al. 2014a).

The presence of remaining native trees inside the matrix, including palms, which is enforced by environmental certification process of FSC (Athayde et al. 2015), results on an increase in bird diversity on *Eucalyptus* plantations (Millan et al. 2015). Such procedure should be stimulated in other agroindustrial sectors, such as the ethanol/sugar industry (Verdade et al. 2015). Birds are the most diverse group of vertebrates in agricultural/silvicultural landscapes in Southeast Brazil, including more than 200 species (Penteado et al. 2016). The present results suggest that at least some of them are relevant prey for canids and felids, whereas others may compete with mammalian carnivores. As a matter of fact, the species diversity of birds of prey is comparable to the Carnivora in these environments (Penteado et al. 2016). Future studies should prioritize species identification of birds species both as prey and as predators in agricultural/silvicultural landscapes of the Neotropics.

Insects are the most diverse terrestrial animal group in the Neotropics (Winkler et al. 2018). Their relevance as prey of vertebrate species in pristine ecosystems has been extensively described (e.g., Vieira and De Moraes 2003; Bader et al. 2015). Nevertheless, their role on the food web of agricultural/silvicultural landscapes also remains mostly neglected. Future qualitative and quantitative trophic studies should prioritize the evaluation of the role of insects on the food web of agricultural/silvicultural landscapes of the Neotropics.

Future studies should also prioritize the development of wildlife friendly silvicultural practices (e.g., Medeiros et al. 2009) to improve habitat quality of the main plantation and thus increase β-diversity of silvicultural landscapes (Verdade et al. 2014a). In addition, the present results might guide future programs on the reintroduction of locally extinct species. Jaguars, tapirs, and white-lipped peccaries should be prioritized as they would significantly enrich the local food web as the biggest predator and its staple prey in terms of biomass (Hayward et al. 2016; Verdade et al. 2016a). Finally, the small mammals' diversity might be possibly enhanced with the enforcement of the BFC and the consequent recovery of native vegetation areas inside agricultural/silvicultural landscapes (Gheler-Costa et al. 2012; Verdade et al. 2014a, 2016b) with a consequent improvement of the local food web as well. A possible rewilding program of agricultural and silvicultural landscapes (Verdade and Moral 2026, this volume) might be effectively monitored by the trophic approach presented and discussed here.

Acknowledgements This study was funded by Fundação de Amparo à Pesquisa do Estado de São Paulo (FAPESP, Grants # 2006/60954-4 and 2017/01304-4). LMR had fellowships from Fundação para a Ciência e a Tecnologia (FCT), Fundo Social Europeu (III Quadro Comunitário de Apoio) (SFRH/BPD/14435/2003) and from FAPESP (Grant # 2011/00408-4) and was financially supported by the cE3c (DOI: 10.54499/UIDB/00329/2020), through national funds, and the co-funding by the FEDER, within the PT2020 Partnership Agreement and Compete 2020. LMV held Research Productivity Fellowships from CNPq (Grant # 300480/2008-3, 309468/2011-6, and 312049/2015-3). The animal study protocol was approved by the Ethics Committee of Escola Superior de Agricultura "Luiz de Queiroz" of the University of São Paulo (Proc. # 02027.003862/2004-53). The animals were captured under licenses issued by IBAMA (Lic. # 050/04 and 144/05). RIPASA Papel e Celulose gave access to the study areas at Parque Ibiti. and André Henrique S. Fernandes and Jacob D. Charters helped with graphical support.

References

ABRAF (2022) Anuário Estatístico da ABRAF 2022: ano base 2021. Associação Brasileira de Produtores de Florestas Plantadas – Brasília

Abreu EF, Casali D, Costa-Araújo R, Garbino GST, Libardi GS, Loretto D, Loss AC, Marmontel M, Moras LM, Nascimento MC, Oliveira ML, Pavan SE, Tirelli FP (2023) Lista de Mamíferos do Brasil (2023-1) [Data set]. Zenodo. https://doi.org/10.5281/zenodo.10428436

Angeli T, Oliveira ML, Duarte JMB (2014) Differentiation of deer species of the genus *Mazama* by track morphometry. Stud Neotrop Fauna Environ 49:199–203

Antoniazzi R, Dáttilo W, Rico-Gray V (2018) A useful guide of main indices and software used for ecological networks studies. In: Dáttlio W, Rico-Gray V (eds) Ecological networks in the tropics: an integrative overview of species interactions from some of the most species-rich habitats on earth. Springer International Publishing, Cham, pp 185–196

Athayde EA, Cancian LF, Verdade LM, Morellato LPC (2015) Functional and phylogenetic diversity of scattered trees in agricultural landscape: implications for conservation. Agric Ecosyst Environ 199:272–281

Bader E, Jung K, Kalko EK, Page RA, Rodriguez R, Sattler T (2015) Mobility explains the response of aerial insectivorous bats to anthropogenic habitat change in the neotropics. Biol Conserv 186:97–106

Becker M, Dalponte JC (2013) Rastros de mamíferos silvestres brasileiros: um guia de campo, 3rd edn. Technical Books, Rio de Janeiro

Bianchi RDC, Rosa AF, Gatti A, Mendes SL (2011) Diet of margay, *Leopardus wiedii*, and jaguarundi, *Puma yagouaroundi*, (Carnivora: Felidae) in Atlantic rainforest, Brazil. Zoologia (Curitiba) 28(1):127–132

Breuer T (2005) Diet choice of large carnivores in Northern Camerron. Afr J Ecol 43:97–106

Brockerhoff EG, Jactel H, Parrotta JA, Ferraz SFB (2013) Role of eucalypt and other planted forests in biodiversity conservation and the provision of biodiversity-related ecosystem services. For Ecol Manag 301:43–50. https://doi.org/10.1016/j.foreco.2012.09.018

Bueno ADA, Motta-Junior JC (2004) Food habits of two syntopic canids, the maned wolf (*Chrysocyon brachyurus*) and the crab-eating fox (*Cerdocyon thous*), in southeastern Brazil. Rev Chil Hist Nat 77(1):5–14

Burger D, Hess J, Lang B (2005) Forest certification: an innovative instrument in the service of sustainable development? GTZ, Eschborn

Campos CB, Esteves CF, Ferraz KMPMB, Crawshaw PG Jr, Verdade LM (2007) Diet of free-ranging cats (*Felis catus*) and dogs (*Canis familiares*) in a suburban and rural environment of Southeastern. Zoology 273:14–20

Campos BM, Charters J, Verdade LM (2018) Diversity and distribution patterns of medium to large mammals in a silvicultural landscape in southeastern Brazil. iForest 11:802–808. https://doi.org/10.3832/ifor2721-011

Carvalho WD, Rosalino LM, Dalponte JC, Santos B, Adania CH, Esbérard CEL (2015) Can footprints of small and medium sized felids be distinguished in the field? Evidence from Brazil's Atlantic Forest. Trop Conserv Sci 8:760–777

Cechin SZ, Martins M (2000) Eficiência de armadilhas de queda (pitfall traps) em amostragens de anfíbios e répteis no Brasil. Rev Bras Zool 17:729–740
Colwell RK (2006) EstimateS: statistical estimation of species richness and shared species from samples. Version 8. http://viceroy.eeb.uconn.edu/estimates. Accessed 20 Mar 2016
Dotta G, Verdade LM (2007) Trophic categories in a mammal assemblage: diversity in an agricultural landscape. Biota Neotrop 7:287–292
Dotta G, Verdade LM (2009) Felids in an agricultural landscape in São Paulo, Brazil. CatNews 51:22–25
Dotta G, Verdade LM (2011) Medium to large-sized mammals in agricultural landscapes of southeastern Brazil. Mammalia 75:345–352
Dunning JB, Danielson BJ, Pulliam HR (1992) Ecological processes that affect populations in complex landscapes. Oikos 65:169–175
Gannon WL, Sikes RS (2011) Guidelines of the American Society of Mammalogists for the use of wild mammals in research. J Mammal 92:235–253
Gheler-Costa C, Vettorazzi CA, Pardini R, Verdade LM (2012) The distribution and abundance of small mammals in agroecosystems of southeastern Brazil. Mammalia 76:185–191
Gheler-Costa C, Sabino-Santos G Jr, Amorim LS, Rosalino LM, Figueiredo LTM, Verdade LM (2013) The effect of pre-harvest fire on the small mammal assemblage in sugarcane fields. Agric Ecosyst Environ 171:85–89
Gilroy JJ, Edwards DP (2017) Source-sink dynamics: a neglected problem for landscape-scale biodiversity conservation in the tropics. Curr Landsc Ecol Rep 2:51–60. https://doi.org/10.1007/s40823-017-0023-3
Harmsen BJ, Foster RJ, Silver S, Ostro L, Doncaster CP (2010) Differential use of trails by Forest mammals and the implications for camera-trap studies: a case study from Belize. Biotropica 42(1):126–133. https://doi.org/10.1111/j.1744-7429.2009.00544.x
Hayward MW, Kamler JF, Montgomery RA, Newlove A, Rostro-García S, Sales LP, Van Valkenburgh B (2016) Prey preferences of the jaguar *Panthera onca* reflect the post-Pleistocene demise of large prey. Front Ecol Evol 3:148
Heleno R, Garcia C, Jordano P, Traveset A, Gómez JM, Blüthgen N, Memmott J, Moora M, Cerdeira J, Rodríguez-Echeverría S, Freitas H, Olesen JM (2014) Ecological networks: delving into the architecture of biodiversity. Biol Lett 10:20131000
Holt R (2008) Theoretical perspectives on resource pulses. Ecology 89(3):671–681
ICMBio. Instituto Chico Mendes de Conservação da Biodiversidade (2018) Livro vermelho da fauna brasileira ameaçada de extinção: Volume II – Mamíferos, 1st edn. ICMBio/MMA, Brasília
Jácomo AT, Silveira L, Diniz-Filho JAF (2004) Niche separation between the maned wolf (*Chrysocyon brachyurus*), the crab-eating fox (*Dusicyon thous*) and the hoary fox (*Dusicyon vetulus*) in central Brazil. J Zool 262(1):99–106
Jones KE, Safi K (2011) Ecology and evolution of mammalian biodiversity. Philos Trans R Soc Lond B Biol Sci 366:2451–2461
Jost L (2006) Entropy and diversity. Oikos 113:363–375
Juarez KM, Marinho-Filho J (2002) Diet, habitat use, and home ranges of sympatric canids in central Brazil. J Mammal 83(4):925–933
Klare U, Kamler JF, Macdonald DW (2011) A comparison and critique of different scat-analysis methods for determining carnivore diet. Mammal Rev 41(4):294–312
Loeuille N, Barot S, Georgelin E, Kylafis G, Lavigne C (2013) Eco-evolutionary dynamics of agricultural networks: implications for sustainable management. In: Woodward G, Bohan DA (eds) Ecological networks in an agricultural world. Academic, London, pp 339–435
Lopes PC, Marques TS, Martin PS, Gheler-Costa C, Verdade LM (2016) Patterns of leaf-litter amphibian diversity in a silvicultural landscape of southeastern Brazil. In: Gheler-Costa C, Lyra-Jorge MC, Verdade LM (eds) Biodiversity in agricultural landscapes of southeastern Brazil. De Gruyter, Berlin, pp 226–242
Lyra-Jorge MC, Pivello VR (2001) Combining live trap and pitfall to survey terrestrial small mammals in savanna and forest habitats, in Brazil. Mammalia 65:524–530

Lyra-Jorge MC, Ciocheti G, Pivello VR (2008) Carnivore mammals in a fragmented landscape in northeast of São Paulo State, Brazil. Biodivers Conserv 17:1573–1580

Macdonald DW, Tattersall FH, Service KM, Firbank LG, Feber RE (2007) Mammals, agri-environment schemes and set-aside – what are the putative benefits? Mammal Rev 37:259–277

Macdonald DW, Loveridge AJ, Nowell K (2010) Dramatis personae: an introduction to the wild felids. In: Macdonald DW, Loveridge AJ (eds) Biology and conservation of wild felids. Oxford University Press, Oxford, UK, pp 3–58

Magnusson WE (2006) Homogeneização biótica. In: Rocha CFD, Bergallo HG, van Sluys M, Alves MAS (eds) Biologia da conservação: essências. RiMa Editora, São Carlos, pp 211–229

Martin PS, Gheler-Costa C, Verdade LM (2009) Microestruturas de pelos de pequenos mamíferos não-voadores: chave para identificação de espécies de agroecossistemas do estado de São Paulo, Brasil. Biota Neotrop 9:233–241

Martin PS, Gheler-Costa C, Lopes PC, Rosalino LM, Verdade LM (2012) Terrestrial non-volant small mammals in agro-silvicultural landscapes of Southeastern Brazil. For Ecol Manag 282:185–195

May RM, Crawley MJ, Sugihara G (2007) Communities: patterns. In: May RM, McLean A (eds) Theoretical ecology: principles and applications. Oxford University Press, Oxford, UK, pp 111–131

Medeiros GD, Silva E, Martins SV, Neves-Feio R (2009) Diagnóstico da fauna silvestre em empresas florestais brasileiras. Rev Árvore 33:93–100

Metzger JP, Lewinsohn TM, Joly CA, Verdade LM, Martinelli LA, Rodrigues RR (2010) Brazilian law: full speed in reverse? Science 329:276–277

Millan CH, Develey PF, Verdade LM (2015) Stand-level management practices increase occupancy by birds in exotic *Eucalyptus* plantations. For Ecol Manag 336:174–182

Miranda F, Bertassoni A, Abba AM (2014) Myrmecophaga tridactyla. In: The IUCN red list of threatened species: e.T14224A47441961. IUCN, Cambridge, UK

Mittermeier RA, Gil PR, Hoffman M, Pilgrim J, Brooks T, Mittermeier CG, Lamoureux J, Fonseca GAB (2004) Hotspots revisited. Earth's biologically richest and most endangered terrestrial ecoregions. Conservation International, Washington, DC

Moral RA, Hinde J, Demétrio CG (2017) Half-normal plots and overdispersed models in R: the hnp package. J Stat Softw 81:1–23

Paula RC, DeMatteo K (2015) Chrysocyon brachyurus. In: The IUCN red list of threatened species 2015: e. T4819A88135664. IUCN, Cambridge, UK

Penteado M, Yamashita C, Marques TS, Verdade LM (2016) Bird diversity in relation to land use in agricultural landscapes of southeastern Brazil. In: Gheler-Costa C, Lyra-Jorge MC, Verdade LM (eds) Biodiversity in agricultural landscapes of southeastern Brazil. De Gruyter, Berlin, pp 243–268. https://doi.org/10.1515/9783110480849-017

Pirard R, Dal Secco L, Warman R (2016) Do timber plantations contribute to forest conservation? Environ Sci Pol 57:122–130. https://doi.org/10.1016/j.envsci.2015.12.010

Quadros J, Monteiro-Filho ELA (2006) Revisão conceitual, padrões microestruturais e proposta nomenclatória para os pelos-guarda de mamíferos brasileiros. Zoologia 23:279–296

R Core Team (2020) R: a language and environment for statistical computing. R Foundation for Statistical Computing, Vienna. URL https://www.R-project.org

Rocha-Mendes F, Mikich SB, Quadros J, Pedro WA (2010) Feeding ecology of carnivores (Mammalia, Carnivora) in Atlantic forest remnants, southern Brazil. Biota Neotrop 10(4):21–30

Sanders D, Thébault E, Kehoe R, Frank van Veen FJ (2018) Trophic redundancy reduces vulnerability to extinction cascades. Proc Natl Acad Sci 115(10):2419–2424. https://doi.org/10.1073/pnas.1716825115. Epub 2018 Feb 21. PMID: 29467292; PMCID: PMC5878001

Seibert JB, de Oliveira Moreira D, Mendes SL, Gatti A (2015) Diet of two sympatric felids (*Leopardus guttulus* and *Leopardus wiedii*) in a remnant of Atlantic forest, in the montane region of Espírito Santo, southeastern Brazil. Bol do Mus Biol Mello Leitão 37(2):193–200

Sutherland WJ (2006) Ecological census techniques – a handbook. Cambridge University Press, Cambridge, UK

Timo TPC, Lyra-Jorge MC, Gheler-Costa C, Verdade LM (2015) The effect of plantation age on the use of *Eucalyptus* stands by medium to large-sized wild mammals in Southeastern Brazil. iForest 8:108–113

Timo TPC, Costa JC, Gheler-Costa C, Lyra-Jorge MC, Verdade LM (2016) Wildlife and landscape geometry in silvicultural landscapes. In: Gheler-Costa C, Lyra-Jorge MC, Verdade LM (eds) Biodiversity in agricultural landscapes of southeastern Brazil. De Gruyter, Berlin, pp 293–327

Ulanowicz RE, Holt RD, Barfield M (2014) Limits on ecosystem trophic complexity: insights from ecological network analysis. Ecol Lett 17:127–113. https://doi.org/10.1111/ele.12216

Varela D, Flesher K, Cartes JL, de Bustos S, Chalukian S, Ayala G, Richard-Hansen C (2019) Tapirus terrestris. In: The IUCN red list of threatened species 2019: e.T21474A45174127. https://doi.org/10.2305/IUCN.UK.2019-1.RLTS.T21474A45174127.en. Downloaded on 12 March 2020

Verdade LM, Moral RA (2026) Rewilding agricultural landscapes: conceptual, technological, and societal constraints. In: Martins Verdade L, Moral RA (eds) Wildlife monitoring: integrating conservation and innovation in human-altered landscapes. Springer, Cham

Verdade LM, Rosalino LM, Gheler-Costa C, Pedroso NM, Lyra-Jorge MC (2011) Adaptation of mesocarnivores (Mammalia: Carnivora) to agricultural landscapes of Mediterranean Europe and Southeastern Brazil: a trophic perspective. In: Rosalino LM, Gheler-Costa C (eds) Middle-sized carnivores in agricultural landscapes. Nova Science Publishers, New York, pp 1–38

Verdade LM, Penteado M, Gheler-Costa C, Dotta G, Rosalino LM, Pivello VR, Lyra-Jorge MC (2014a) The conservation value of agricultural landscapes. In: Verdade LM, Lyra-Jorge MC, Piña CI (eds) Applied ecology and human dimensions in biological conservation. Springer, Heidelberg, pp 91–102

Verdade LM, Lyra-Jorge MC, Piña CI (2014b) Redirections in conservation biology. In: Verdade LM, Lyra-Jorge MC, Piña CI (eds) Applied ecology and human dimensions in biological conservation. Springer, Heidelberg, pp 3–17

Verdade LM, Piña CI, Rosalino LM (2015) Biofuels and biodiversity: challenges and opportunities. Environ Dev 15:64–78. https://doi.org/10.1016/j.envdev.2015.05.003

Verdade LM, Gheler-Costa C, Lyra-Jorge MC (2016a) The multiple facets of agricultural landscapes. In: Gheler-Costa C, Lyra-Jorge MC, Verdade LM (eds) Biodiversity in agricultural landscapes of southeastern Brazil. De Gruyter, Berlin, pp 2–13

Verdade LM, Palomares F, Couto HTZ, Polizel JL (2016b) Recent land use changes and the expansion of an exotic potential prey: a possible redemption for Atlantic forest jaguars? Anim Conserv 19:209–211

Vieira EM, De Moraes DA (2003) Carnivory and insectivory in neotropical marsupials. Predators with pouches: the biology of carnivorous marsupials. In: Jones M, Dickman C, Archer M (eds) Predators with pouches: the biology of carnivorous marsupials. CSIRO Publishing, Collingwood, pp 271–284

Wang E (2002) Diets of ocelots (*Leopardus pardalis*), margays (*L. wiedii*), and oncillas (*L. tigrinus*) in the Atlantic rainforest in southeast Brazil. Stud Neotrop Fauna Environ 37(3):207–212

Wemmer C, Kunz TH, Lundie-Jenkins G, McShea WJ (1996) Mammalian sign. In: Wilson DE (ed) Measuring and monitoring biological diversity: standard methods for mammals. Smithsonian Institution Press, Washington, DC, pp 157–176

Winkler I, Scheffer SJ, Lewis ML, Ottens KJ, Rasmussen AP, Gomes-Costa GA, Santillan LMH, Condon MA, Forbes AA (2018) Anatomy of a neotropical insect radiation. BMC Evol Biol 18(1):30. https://doi.org/10.1186/s12862-018-1146-9

Zar JH (1996) Biostatistical analysis. Prentice-Hall International, London

Chapter 5
Molecular Ecology Techniques for Wildlife Monitoring

Ana Paula Jacobus

Abstract Molecular ecology techniques have transformed wildlife monitoring and conservation practices by providing innovative tools to assess biodiversity and ecosystem health, particularly in human-altered landscapes. Here, we explore key molecular methodologies, including environmental DNA (eDNA), DNA barcoding, metabarcoding, and genomic approaches, showcasing their application in conservation biology. We demonstrate how these tools address challenges such as habitat fragmentation, climate change, and anthropogenic disturbances, by emphasizing integration with traditional ecological methods. Key applications include rapid biodiversity assessments, species identification, monitoring cryptic and hybrid populations, and evaluating genetic diversity and adaptive potential. We present case studies ranging from urban biodiversity monitoring in São Paulo, Brazil, using eDNA to assess amphibian populations, to genomic studies on pollinator health in agricultural landscapes in the Midwest USA. We also discuss ethical considerations, data complexity, and accessibility challenges while presenting emerging innovations such as portable DNA sequencers, machine learning for data analysis, synthetic biology for ecosystem restoration, and global data-sharing platforms to promote open science. We advocate a synergistic approach to conservation, integrating molecular tools with traditional methods such as telemetry, camera trapping, and citizen science. These integrated strategies enable real-time biodiversity monitoring, support habitat restoration, and inform sustainable management of ecosystems. In the future, molecular ecology techniques will play an increasingly critical role in conservation planning, offering scalable, precise, and actionable insights for safeguarding biodiversity in a rapidly changing world.

Keywords Molecular ecology · Environmental DNA (eDNA) · Metabarcoding and biodiversity assessment · Conservation genomics · Biodiversity monitoring

A. P. Jacobus (✉)
Institute of Bioenergy Research-São Paulo State University, São Paulo, Brazil

L. Martins Verdade, R. A. Moral (eds.), *Wildlife Monitoring*,
https://doi.org/10.1007/978-3-032-05821-8_5

1 Introduction

As natural habitats face increasing pressure from human activities, effective wildlife monitoring has become critical for conservation strategies. Traditional field methods—such as direct observation, camera trapping, and telemetry—while valuable, can be labor-intensive and limited in scope. Molecular ecology techniques offer a complementary approach by providing high-resolution insights into species presence, genetic diversity, and ecosystem health, even in difficult-to-access environments.

As human activities such as deforestation, urbanization, agriculture, and industrialization intensify, wildlife habitats are increasingly fragmented and degraded. This intensifies the urgency for effective monitoring to guide conservation efforts. Monitoring wildlife is fundamental for understanding population trends, identifying threats, and shaping proactive management strategies. However, traditional field methods—such as visual observations, camera traps, and telemetry—come with limitations. While these approaches offer valuable insights into animal behavior and distribution, they are often labor-intensive, time-consuming, and geographically restricted. Additionally, they may fail to detect elusive, nocturnal, or cryptic species, limiting the scope of biodiversity assessments, especially in challenging environments like dense forests or aquatic ecosystems (Beja-Pereira et al. 2009; Thomsen and Willerslev 2015).

Molecular ecology techniques have emerged as powerful tools to complement and overcome the limitations of traditional methods. Tools like environmental DNA (eDNA) and DNA metabarcoding enable the detection of species even in low abundance by analyzing traces of genetic material shed into the environment. This provides a non-invasive and cost-effective way to monitor biodiversity across a range of ecosystems. Unlike direct observation, molecular methods can reveal hidden biodiversity and offer insights into community composition without disturbing the habitat or the organisms themselves (Bohmann et al. 2014).

Moreover, genomic techniques provide detailed information about genetic diversity, population structure, and the adaptive potential of species. Such data are crucial for conservation planning, especially in fragmented landscapes, where maintaining genetic flow is fundamental to prevent inbreeding and support species resilience against environmental changes (Frankham 2005; Hanski and Gaggiotti 2004; Broquet and Petit 2009). For example, genomic data have been used to define management units and inform the design of wildlife corridors that reconnect isolated populations, enhancing their long-term viability (Funk et al. 2012).

The integration of these molecular tools with traditional monitoring methods creates a synergistic approach to conservation. For instance, eDNA sampling combined with camera traps can validate the presence of elusive species detected by molecular analysis, while telemetry data offer insights into the movement patterns of genetically identified individuals. This comprehensive approach enhances our ability to track biodiversity in real-time, detect invasive species early, and monitor ecosystem health, ultimately guiding conservation strategies more effectively in an era of rapid environmental change (Thomas et al. 2004; Goldberg et al. 2015).

In this chapter we highlight cutting-edge molecular tools and their application to monitor wildlife populations and biodiversity. We explore how these technologies align with conservation goals, particularly in human-altered landscapes such as urban areas, agricultural zones, and degraded ecosystems.

2 Key Molecular Techniques for Wildlife Monitoring

2.1 eDNA

eDNA involves the detection of species by analyzing genetic material shed into the environment (e.g., water, soil, or air). eDNA allows for non-invasive monitoring of elusive or endangered species and provides information about biodiversity without the need for direct capture or visual confirmation. Over the last 20 years, eDNA has emerged as a powerful tool, capturing traces of free-floating genetic material released from organisms by skin, mucous, saliva, sperm, eggs, secretions, feces, urine, blood, roots, leaves, pollen, fruits, or decaying remains. Any environmental sample now holds a wealth of eDNA, and thanks to advancements in sample processing and sequencing technologies, we can unlock the biological information contained within (Taberlet et al. 2012).

This once-fictional capability, such as one could see in science fiction shows, such as Spock's "Tricoder" in Star Trek, is now reshaping biological research. Increasingly, scientists are using eDNA for species detection and ecological monitoring, significantly reducing or eliminating the need for direct observation or physical collection of organisms (Bohmann et al. 2014). Moreover, these methods are accelerating discoveries, as identifying species no longer depends on prior knowledge of what might be present in a given habitat. Researchers focused on invasive species, community interactions, ecosystem dynamics, and wildlife conservation are particularly poised to benefit from the widespread adoption of eDNA techniques.

2.1.1 Applications in Conservation

One particular application of eDNA is in Biodiversity Assessment, whereby we may use genetic data to build a rapid inventory of ecosystems. eDNA facilitates comprehensive biodiversity assessments by detecting multiple species from a single environmental sample, such as water, soil, or sediment. This method allows researchers to generate detailed species inventories efficiently, even in complex or inaccessible habitats, such as deep parts of the oceans or dense rainforests. For example, eDNA metabarcoding was used to assess fish diversity in freshwater ecosystems, identifying species that were previously undetected by traditional methods (Deiner et al. 2017).

Another strategy relies on non-invasively monitoring endangered and elusive species. eDNA is particularly valuable for tracking rare, endangered, or elusive species that are difficult to observe directly. By analyzing environmental samples, researchers can detect the genetic material of these species without disturbing their habitats. For instance, eDNA has been used to detect the critically endangered great crested newt (*Triturus cristatus*) in freshwater ponds, offering a non-invasive alternative to traditional survey methods (Thomsen et al. 2012).

The impact on conservation planning, using eDNA, allows conservationists to confirm species presence or absence in habitats under consideration for protection, aiding habitat prioritization and restoration efforts. In this situation it is possible by proactive management to detect invasive species in early phases of invasion. This is because eDNA enables the early detection of invasive species at low population densities, allowing for timely interventions before the species establishes and spreads. This capability is crucial for protecting native biodiversity and reducing the economic impact of invasive species. eDNA was employed to monitor the presence of invasive Asian carp in freshwater systems in the United States, detecting their DNA long before physical sightings occurred (Jerde et al. 2011).

By detecting species known to be indicators of environmental health, eDNA can help assess habitat quality and ecosystem conditions. For example, the presence or absence of specific aquatic invertebrates or fish species can indicate water quality or pollution levels. For example, annual time-series analysis of eDNA revealed seasonal dynamics in lake ecosystems, helping to understand the ecological impacts of environmental changes (Bista et al. 2017).

For long-term monitoring, eDNA is ideal for establishing long-term monitoring programs to track changes in species composition and habitat quality over time, offering insights into the effects of climate change, habitat degradation, and human activities. One advantage is that it can be used to predict and evaluate conservation interventions, as well as to track reintroduction success, by detecting the genetic material of reintroduced individuals and assessing their establishment in the habitat. One example of that is when eDNA was used to monitor the reintroduction of the European beaver (*Castor fiber*) in the United Kingdom, confirming the species' presence and habitat use (Harper et al. 2019a).

eDNA also provides measurable outcomes for habitat recovery projects, which allows for the monitoring and evaluation of habitat restoration by analyzing species diversity before and after restoration efforts (Harper et al. 2019b).

Another example of eDNA use is related to tracking species shifts associated with climate change. As climate change alters habitats, eDNA can detect shifts in species distributions, including the arrival of species in new areas or the disappearance of others from their historical ranges. For instance, eDNA revealed changes in the distributions of cold-water fish species in response to warming water temperatures in North American lakes (Leduc et al. 2021).

While eDNA use offers significant advantages for wildlife monitoring and conservation, there are also several limitations which must be addressed to ensure reliable and accurate outcomes. These challenges include issues related to DNA degradation, contamination, and interpretation of results.

2.1.2 DNA Degradation

eDNA is highly susceptible to degradation due to various environmental factors. Exposure to sunlight (UV radiation), high temperatures, and microbial activity can break down DNA molecules, reducing their detectability and reliability over time (Barnes and Turner 2016). This limits the temporal resolution of eDNA, meaning it provides a snapshot of recent biological presence, typically reflecting activity from the past few hours to several days. In aquatic environments, eDNA can degrade rapidly—often within 24–72 h—due to microbial activity, UV radiation, temperature, and water chemistry.

In contrast, eDNA in sediments or soils may persist for weeks under stable conditions, though it still mainly represents recent organisms. In fast-flowing systems like rivers, detection windows may be as short as a few hours, while cold, low-light, or low-microbial environments can extend eDNA persistence to several days or more. As a result, researchers must carefully consider the timing of sample collection and environmental conditions when designing studies and interpreting results (Strickler et al. 2015).

2.1.3 Contamination

Cross-contamination during field sampling, laboratory processing, or even through airborne DNA particles can lead to false positives, where the presence of a species is erroneously detected. For example, DNA from one habitat could inadvertently contaminate samples from another, skewing conservation assessments (Deiner et al. 2017). To decrease this source of errors, strict field protocols (e.g., wearing gloves, using sterilized equipment) and rigorous laboratory practices (e.g., negative controls, decontamination procedures) are necessary (Goldberg et al. 2016).

2.1.4 Interpretation of Results

Inferring species abundance or biomass from eDNA concentrations remains a major challenge. The amount of DNA shed into the environment varies depending on species-specific factors (e.g., size, behavior, physiology) and environmental dynamics (e.g., water flow, sedimentation) (Tillotson et al. 2018). This variability complicates the relationship between eDNA concentrations and actual population metrics. Without standardized methods to account for these factors (Pilliod et al. 2014), eDNA studies may struggle to provide quantitative insights into population size or density (Yates et al. 2019).

2.1.5 Future Directions in eDNA Technology

eDNA is a rapidly evolving field, with advancements poised to address existing challenges and expand its applications. Future developments are expected to enhance the reliability, efficiency, and scope of eDNA methods, making them more integrative and comprehensive. Key areas of innovation include standardization, integration with remote sensing technologies, and the exploration of environmental RNA (eRNA).

One of the critical barriers to the broader adoption and comparability of eDNA studies is the lack of standardized protocols for sampling, extraction, and analysis. Current practices vary widely across studies, leading to inconsistencies in data quality and interpretability. Developing universal guidelines for each stage of eDNA analysis—including field sampling, laboratory workflows, and bioinformatics pipelines—will improve reproducibility and enable more robust cross-study comparisons. Consistent methods will facilitate meta-analyses and large-scale biodiversity assessments, allowing for more accurate tracking of global conservation trends (Goldberg et al. 2016).

The integration of eDNA data with remote sensing technologies, such as satellite imagery and aerial drones, could potentially provide a holistic view of ecosystems (Li et al. 2024) Remote sensing offers spatial and temporal insights into habitat characteristics, while eDNA delivers species-level biodiversity data. Together, these methods can provide a comprehensive understanding of ecosystem dynamics, including changes in species distributions, habitat fragmentation, and the impacts of climate change.

Emerging as a complementary tool to eDNA, eRNA focuses on detecting RNA molecules from environmental samples. Unlike DNA, RNA is less stable but offers a snapshot of active biological processes, such as gene expression (Cristescu 2019). This can provide deeper insights into the functional aspects of ecosystems, including species' metabolic activities and responses to environmental stressors. The potential benefits of eRNA are the identification of species with active populations, distinguishing them from dormant or recently extinct organisms. It may also be used for real-time monitoring of ecosystem health by assessing functional biodiversity, since eRNA enables the assessment of functional biodiversity by revealing actively expressed genes, allowing the identification of ecological guilds such as decomposers, primary producers, and nitrogen fixers based on their metabolic activity in real time (Adamo et al. 2020). Some challenges are related to this approach such as the instability of RNA in the environment necessitates rapid sample preservation and highly sensitive analytical techniques.

Advancements in eDNA technology are shaping the future of conservation biology, offering more integrative, sensitive, and precise tools for biodiversity monitoring. Standardizing methodologies will enhance reproducibility and comparability, while the combination of eDNA with remote sensing technologies will unlock new dimensions in ecological research. Meanwhile, the advent of eRNA use promises to expand our understanding of ecosystems by providing functional insights that complement the species presence data obtained from eDNA. These innovations characterize molecular approaches as indispensable tools for addressing the global biodiversity crisis.

3 Sample Collection Techniques for eDNA Analysis

Sample collection is a critical step in eDNA analysis, as it determines the quality and reliability of the downstream results. Effective sample collection methods ensure the capture of sufficient DNA while minimizing contamination and degradation. The choice of technique depends on the target organism, the type of environment (e.g., aquatic, terrestrial), and the study's objectives. Below, we discuss the primary methods used in eDNA sample collection, their applications, and key considerations.

3.1 Water Sampling

Water sampling is the most common method for eDNA analysis, especially in aquatic ecosystems. It involves collecting water samples from habitats such as rivers, lakes, oceans, or wetlands, where eDNA is suspended in the water column. Water is collected using sterile bottles, syringes, or pumps, typically from the surface or at specific depths. The sample is filtered immediately on-site or transported on ice to the laboratory, where filtration is performed to capture eDNA on a filter membrane. The filtration volume depends on water turbidity; highly turbid water requires smaller volumes to prevent filter clogging. Finally, the samples must be stored properly (e.g., frozen or preserved with ethanol) to prevent DNA degradation (Deiner et al. 2017).

3.2 Soil and Sediment Sampling

Soil and sediment sampling are used to detect eDNA from terrestrial or benthic organisms. These substrates often contain more stable eDNA due to reduced exposure to degrading factors, such as UV light. Soil or sediment samples are collected using sterile tools, such as scoops or corers, and store in sterile containers. Samples may be sieved to remove debris and homogenized before DNA extraction. It can be used to detect the presence of burrowing species, plants, or microorganisms or to monitor the biodiversity of terrestrial organisms, or that of benthic communities in aquatic systems. Sample depth and location are critical, as eDNA distribution may vary vertically and horizontally. However, DNA extraction from soil or sediment can be challenging due to the presence of inhibitors, such as humic acids. To overcome these inhibitors, strategies include using specialized kits with inhibitor removal, adding adsorbents (e.g., PVPP), prewashing samples, diluting extracts, and employing resistant polymerases or PCR additives, such as BSA (Yoccoz et al. 2012).

3.3 Air Sampling

Air sampling is an emerging technique for eDNA analysis, particularly for detecting airborne DNA from terrestrial species, such as insects, birds, or mammals. Air samplers equipped with filters or adhesive surfaces collect airborne particles. The captured material is then analyzed for DNA content. This sampling can be applied to monitor species' presence in caves, forests, or urban areas, as well as to study pollinators or other airborne organisms. Air sampling is highly sensitive to environmental conditions, such as wind speed and humidity. The technique is still under development and is less widely applied when compared to water or soil sampling (Clare et al. 2021).

3.4 Swabbing and Surface Sampling

Swabbing and surface sampling are used in environments where eDNA accumulates on surfaces, such as rocks, vegetation, or infrastructure. Sterile swabs or wipes are used to collect DNA from surfaces. The swabs are stored in buffer solutions or directly processed for DNA extraction. Detecting DNA from organisms interacting with specific surfaces, such as nests, burrows, or feeding sites. Monitoring the spread of pathogens or diseases. The surface contamination and the heterogeneity of DNA distribution can affect results. Swabbing is highly targeted and may miss broader biodiversity signals (Piaggio et al. 2014).

3.5 Stool and Feces Sampling for eDNA Analysis

Stool and feces sampling is a highly informative method for eDNA analysis, providing insights into biodiversity, diet composition, health, and ecological interactions. Fecal material contains DNA from the organism that produced it (host DNA) and from consumed food items (prey DNA), as well as environmental microbes and parasites. This approach is widely used in wildlife monitoring, conservation genetics, and ecological studies.

This approach can be used for monitoring species presence and biodiversity. Feces are often the only trace left by elusive or cryptic species in their habitats. Sampling and analyzing DNA from fecal material can confirm the presence of these organisms without requiring direct observation or capture. One example is the detection of large carnivores, such as tigers (*Panthera tigris*), in forested regions using scat samples (Wilting et al. 2016).

DNA extracted from the predator's feces often includes prey DNA, providing insights into the dietary preferences and ecological roles of species. This is particularly useful for understanding predator-prey dynamics and food web interactions. For instance, Deagle et al. (2009) analyzed fecal samples from wolves (*Canis lupus*) to determine their diet composition and assess impacts on prey populations. They found that DNA-based methods, such as metabarcoding, provided a more accurate and detailed understanding of wolf diets compared to traditional morphological analyses. Specifically, DNA analysis detected a greater diversity of prey species, including those that are difficult to identify through hair or bone fragments alone. This approach revealed that wolves consumed a variety of prey, with significant regional variations in diet composition, reflecting differences in prey availability. The study highlighted the utility of molecular techniques in ecological research, offering insights into predator-prey dynamics and informing conservation strategies.

Fecal DNA allows researchers to assess genetic diversity, population structure, and relatedness within and between populations of a species, aiding conservation and wildlife management. Monitoring genetic diversity in wild gorilla populations using DNA extracted from fecal samples enabled Arandjelovic et al. (2010) to identify individuals, estimate population sizes, and evaluate gene flow between groups, providing critical data for the conservation of these endangered primates.

Another application using DNA from feces can be used to detect Pathogen and contribute to microbiome studies. Feces contain DNA from gut microbiota and pathogens, making it a valuable source for studying host health, disease dynamics, and microbial community structure. Investigating the gut microbiome diversity of orangutans (*Pongo pygmaeus*) from fecal samples helped Amato et al. (2013) understand the effects of habitat disturbance on their health, revealing that environmental changes significantly influence microbial community composition, which in turn may impact the nutritional status and immune function of these critically endangered primates.

For endangered species, fecal sampling provides non-invasive means to monitor populations, assess threats, and help establish conservation strategies. Fecal DNA analysis of snow leopards (*Panthera uncia*) was used by Janečka et al. (2008) to estimate population density in mountainous regions, enabling the identification of individual animals through genetic markers and providing non-invasive, reliable data on population size, spatial distribution, and connectivity essential for conservation planning in remote and rugged habitats.

Advances in DNA sequencing technologies and bioinformatics are expected to improve the resolution and scalability of fecal eDNA analysis. Integration with other molecular tools, such as eRNA, could provide functional insights into the health and ecology of wildlife populations (Pansu et al. 2019).

4 DNA Barcoding and Metabarcoding

DNA barcoding and metabarcoding are molecular techniques that have revolutionized biodiversity assessment and conservation strategies. DNA barcoding uses short, standardized gene regions (such as the COI gene in animals) to identify species. Metabarcoding extends this approach by analyzing mixed environmental samples, providing community-level insights into biodiversity. By leveraging genetic information, these methods enable researchers to identify species, monitor ecosystems, and evaluate biodiversity more accurately and efficiently. When applied to eDNA analysis, DNA barcoding and metabarcoding allow the detection of species from complex environmental samples, such as water, soil, or air, providing non-invasive, scalable, and precise tools for conservation biology.

4.1 Techniques Used for DNA Barcoding and Metabarcoding

DNA barcoding identifies species by sequencing a standardized region of DNA that is highly conserved across taxa but contains enough variation to distinguish between species. While both rely on sequencing DNA, barcoding and metabarcoding differ in scope and application. DNA barcoding identifies individual species, whereas metabarcoding simultaneously analyzes multiple species from a mixed sample.

4.1.1 DNA Barcoding

DNA barcoding identifies species by sequencing a standardized region of DNA that is highly conserved across taxa but contains enough variation to distinguish between species. The mitochondrial cytochrome oxidase I (COI) gene is the most commonly used marker for animals, whereas plants are typically identified using chloroplast regions such as rbcL and matK. DNA Barcoding steps are presented as follows.

DNA is extracted from a tissue sample (e.g., leaf, fin, or hair). A specific DNA region is amplified using polymerase chain reaction (PCR) with universal primers. The amplified region is sequenced using Sanger or next-generation sequencing (NGS) technologies. The resulting sequence is compared against reference databases such as GenBank or the Barcode of Life Data Systems (BOLD) to identify the species. This technique can be used for species identification (e.g. distinguishing morphologically similar or cryptic species), forensic applications (e.g., identifying illegally traded wildlife products), and monitoring species in an ecosystem.

The barcode approach has some advantages, such as high accuracy for identifying single species and have well-established protocols for many taxa. Limitations include the requirement for high-quality DNA from individual samples and challenges when analyzing mixed or environmental samples (Hebert et al. 2003).

Table 5.1 Comparison of DNA barcoding and metabarcoding

Aspect	DNA barcoding	DNA metabarcoding
Purpose	Identifies single species	Analyzes entire communities
Sample type	Individual tissue samples	Mixed or environmental samples
Sequencing	Sanger or NGS	High-throughput sequencing (HTS)
Applications	Species identification and forensics	Biodiversity monitoring and diet studies
Key limitation	Requires isolated DNA	Primer bias and bioinformatics challenges

4.1.2 DNA Metabarcoding

Metabarcoding extends DNA barcoding to analyze entire communities of organisms from a single environmental or bulk sample. It combines DNA barcoding with high-throughput sequencing (HTS) to identify multiple species simultaneously, making it ideal for complex samples such as soil, water, or gut contents (Table 5.1). DNA Metabarcoding steps are presented as follows.

Environmental or bulk samples (e.g., soil, water, or feces) are collected. DNA is extracted using methods tailored to environmental samples to remove inhibitors. Universal primers target a specific DNA region (e.g., COI for animals, 16S rRNA for bacteria, ITS for fungi) to amplify DNA from all species in the sample. Amplified DNA is sequenced using HTS platforms such as Illumina or Nanopore, generating large datasets of sequences. Sequence data are processed, filtered, and compared against reference databases to identify species. The applications of metabarcoding approach involve biodiversity monitoring, assessing species diversity in ecosystems, dietary analysis through fecal samples, and habitat assessment.

The advantages of metabarcoding are simultaneous analysis of multiple species. In addition, it is applicable to environmental and degraded DNA, scalable, and cost-effective for large-scale studies. However, it has the following limitations: primer bias can influence amplification success for some taxa, the incomplete reference databases may limit species identification resolution and require advanced bioinformatics expertise for data analysis (Deiner et al. 2017).

DNA barcoding and metabarcoding are powerful, complementary tools in molecular ecology and conservation biology. While DNA barcoding excels in species-level identification, metabarcoding enables comprehensive biodiversity assessments from complex samples. Together, these techniques provide critical insights into ecosystems, facilitating species monitoring, ecological research, and conservation efforts.

4.2 Applications of DNA Barcoding and Metabarcoding in Conservation

DNA barcoding and metabarcoding are transformative tools in conservation biology, providing precise, scalable, and efficient means to assess biodiversity and monitor ecosystem dynamics. These molecular techniques have opened new avenues for understanding species composition, detecting cryptic biodiversity, and tracking changes in populations and ecosystems over time. Below, we explore three key applications of these approaches in conservation.

4.2.1 Rapid Biodiversity Assessments of Multiple Taxa from a Single Sample

Traditional biodiversity surveys often require labor-intensive methods and species-specific expertise, particularly in diverse ecosystems. DNA metabarcoding offers a revolutionary alternative by enabling rapid and comprehensive biodiversity assessments from a single environmental sample. This technique uses universal primers to amplify DNA from multiple taxa within a sample, followed by high-throughput sequencing to identify species across taxonomic groups. One example is assessing biodiversity in freshwater, marine, and terrestrial ecosystems, also monitoring community structure in habitats under threat from human activities. Metabarcoding of water samples from freshwater ecosystems has been used to simultaneously detect fish, amphibians, and macroinvertebrates, providing a holistic view of aquatic biodiversity (Deiner et al. 2017). By providing a non-invasive and scalable method for biodiversity assessment, metabarcoding accelerates conservation decision-making and helps prioritize habitats for protection.

4.2.2 Identifying Cryptic Species or Hybrid Populations

Cryptic species—those that are morphologically similar but genetically distinct—pose significant challenges to traditional taxonomic identification. DNA barcoding overcomes these limitations by using genetic sequences to distinguish between closely related species, thereby uncovering hidden biodiversity. Additionally, DNA barcoding can detect hybrid populations that result from interbreeding between species, which may have implications for species survival and ecosystem dynamics. Using barcoding/metabarcoding can help to identify cryptic species in biodiversity hotspots, such as tropical rainforests and coral reefs. Also monitoring hybridization events that may threaten genetic integrity or facilitate adaptation. DNA barcoding revealed the existence of multiple cryptic frog species in the Amazon, which were previously misclassified as a single species based on morphology (Vences et al. 2005). Identifying cryptic and hybrid species ensures that conservation plans account for the full range of biodiversity, preventing the inadvertent loss of genetically distinct lineages.

5 Applications in Conservation Using Genomics and Population Genetics

The application of genomics and population genetics has profoundly impacted conservation biology, offering tools to address challenges in preserving species and ecosystems. This section is divided into three critical applications: defining conservation units and identifying populations at risk of inbreeding, informing reintroduction programs by genetic matching, and investigating adaptive potential in populations under anthropogenic pressures.

The integration of genomics and population genetics into conservation biology has revolutionized our ability to monitor, manage, and conserve biodiversity. These tools provide insights into population structure, genetic diversity, and evolutionary processes, enabling evidence-based strategies for species preservation in the face of anthropogenic pressures and climate change. This chapter explores the applications of these methodologies, emphasizing their transformative potential in conservation efforts.

5.1 Defining Conservation Units and Identifying Populations at Risk of Inbreeding

Accurate delineation of conservation units is fundamental for prioritizing and managing biodiversity effectively. Conservation units, such as evolutionary significant units (ESUs) and management units (MUs), are defined based on genetic distinctiveness and evolutionary trajectories (Moritz 1994). Genomic tools enable fine-scale resolution of population structure, facilitating the identification of ESUs and MUs that may otherwise go unnoticed using traditional methods. For example, genomic analysis of the Iberian lynx (*Lynx pardinus*), one of the most endangered felines, revealed critical population subdivisions that informed localized conservation actions (Casas-Marce et al. 2013).

Inbreeding, a consequence of reduced population size and genetic isolation, poses a significant threat to population viability. Whole-genome sequencing (WGS) and single nucleotide polymorphism (SNP) genotyping are now routinely used to calculate inbreeding coefficients, revealing genomic regions with deleterious mutations (Kardos et al. 2015). In the case of Scandinavian wolves (*Canis lupus*), genomic data uncovered high levels of inbreeding, emphasizing the need for genetic management to enhance gene flow (Åkesson et al. 2016).

5.2 Informing Reintroduction Programs by Genetic Matching

Reintroduction programs aim to establish self-sustaining populations in their native habitats, often relying on captive breeding or translocation. Ensuring genetic compatibility between source and recipient populations is crucial to avoid outbreeding depression or genetic swamping. Population genomics provides tools for assessing genetic similarity and adaptive potential, enabling informed decisions on donor populations. One success story is the reintroduction of the Arabian oryx (*Oryx leucoryx*), which involved genetically informed selection of individuals to maximize genetic diversity and minimize inbreeding (Iyengar et al. 2007). Similarly, the genetic rescue of Florida panthers (*Puma concolor coryi*) involved introducing Texas pumas to counteract inbreeding depression, with subsequent genomic analysis confirming the program's effectiveness in increasing heterozygosity and fitness (Johnson et al. 2010).

5.3 Investigating Adaptive Potential in Populations Under Anthropogenic Pressures

Human activities, including habitat fragmentation, climate change, and pollution, exert selective pressures on natural populations. Investigating adaptive potential in these contexts is critical for designing conservation strategies that promote resilience. Genomic studies can identify loci associated with adaptive traits, such as thermal tolerance, resistance to pathogens, or metabolic efficiency. Coral reef ecosystems exemplify the power of genomic approaches in conservation. For instance, studies on *Acropora* corals have identified genetic markers linked to thermal tolerance, providing insights into potential assisted evolution strategies to combat bleaching events (van Oppen et al. 2015). In terrestrial systems, genomic studies on the Soay sheep (*Ovis aries*) have demonstrated how climate change influences genetic selection on body size and survival traits, offering predictive insights into population dynamics under changing environments (Hayward et al. 2011).

Genomics and population genetics have transformed the field of conservation biology, enabling the delineation of conservation units, reducing the risks associated with reintroduction programs, and enhancing our understanding of adaptive potential under anthropogenic pressures. These tools provide a robust framework for addressing contemporary conservation challenges, emphasizing the importance of integrating genomic data into management and policymaking to safeguard biodiversity.

6 Challenges and Limitations of Molecular Techniques in Conservation Biology

The integration of molecular tools, such as genomics and eDNA analyses, has revolutionized conservation biology, providing innovative methods to monitor biodiversity, assess genetic diversity, and guide species management. Despite their transformative potential, these technologies face significant challenges that need to be addressed to ensure their effective application. These challenges include the complexity of data, sampling limitations, ethical considerations, and cost and accessibility barriers.

6.1 Data Complexity

One of the primary challenges of molecular techniques is managing and interpreting the vast amounts of data generated by genomic studies. Technologies like WGS and SNP genotyping, and transcriptomics produce large, complex datasets that require specialized bioinformatic tools and expertise (Allendorf et al. 2010). The ability to store, process, and analyze these data is often concentrated in well-resourced institutions, leaving conservation practitioners in underfunded regions at a disadvantage (Stephens et al. 2015). Furthermore, the rapid evolution of bioinformatic tools brings the need for continuous training and updates, which can strain resources and capacity. Investments in user-friendly analytical platforms and capacity-building initiatives are critical to democratizing access to genomic insights.

6.2 Ethical Considerations

The collection and use of genetic material for conservation purposes raise important ethical issues, particularly in areas of high biodiversity and cultural significance. Biopiracy—the unauthorized use of genetic resources—and issues of data ownership are especially concerning when working in collaboration with indigenous communities or low-income countries (Morgera et al. 2012). Disparities in access to sequencing technology and data repositories further exacerbate these challenges. Ethical guidelines, such as those outlined in the Nagoya Protocol on Access and Benefit-Sharing, must be adhered to in conservation projects to ensure equitable collaboration and respect for local knowledge and resources (Laird et al. 2020).

6.3 Cost and Accessibility

While the cost of molecular techniques has decreased significantly over the past decade, they remain prohibitively expensive for many conservation programs, particularly in low-income regions (Schwartz et al. 2007). Infrastructure requirements, such as high-throughput sequencers and bioinformatic servers, increase financial burden. This disparity limits the application of these tools in regions where conservation needs are most urgent. Portable sequencing technologies, such as the Oxford Nanopore MinION, and international funding collaborations offer promising solutions to reduce costs and increase accessibility (Jansen et al. 2018).

The transformative potential of molecular tools in conservation biology is undeniable, yet their widespread application is hindered by challenges related to data complexity, sampling limitations, ethical considerations, and cost and accessibility. Addressing these issues requires interdisciplinary collaboration, capacity-building efforts, and adherence to ethical frameworks. By overcoming these limitations, molecular techniques can achieve their full potential in protecting biodiversity and guiding conservation strategies in a rapidly changing world.

7 Integrating Molecular and Traditional Ecological Methods: Enhancing Conservation in Human-Altered Environments

The integration of molecular tools with traditional ecological methods offers a powerful framework for addressing complex conservation challenges, particularly in human-altered landscapes. While molecular techniques such as eDNA analysis and genetic sequencing have transformed biodiversity monitoring, they should not replace traditional approaches but rather complement them. Combining these methodologies allows for robust validation of molecular data and the incorporation of ecological, behavioral, and social dimensions into conservation efforts.

7.1 The Complementary Role of Molecular and Traditional Methods: Validating Molecular Data with Traditional Approaches

Molecular tools like eDNA analysis can detect species presence in an area by genetic traces left in the environment. However, these findings often require validation by direct observations or traditional ecological methods to confirm their accuracy and ecological relevance. For example, camera traps can be deployed in tandem with eDNA surveys to provide visual evidence of the species detected. This approach not only verifies the presence of elusive or nocturnal species but also provides additional ecological insights on animal behavior and population dynamics (Taberlet et al. 2012).

7.2 Mapping Movement Patterns Using Telemetry

Molecular data can identify individuals or populations within a species, but traditional telemetry studies remain necessary for understanding movement patterns and habitat use. By integrating molecular tagging (e.g., individual genotyping) with telemetry, conservationists can track the movements of genetically identified individuals across fragmented landscapes. For example, telemetry studies combined with genetic data have been used to assess habitat connectivity for large carnivores like tigers (*Panthera tigris*), informing the design of wildlife corridors in human-dominated regions (Joshi et al. 2013).

7.3 Engaging Communities Through Citizen Science

Citizen science programs offer an opportunity to combine molecular and traditional methods while involving local communities in conservation. Molecular techniques, such as eDNA sampling, can be simplified for citizen participation, enabling widespread data collection across large geographic areas. These programs not only generate valuable data but also foster public awareness and engagement. For instance, the "Great Australian Platypus Search" combined eDNA sampling with community-driven data collection to monitor platypus populations across Australia, enhancing conservation efforts while promoting community stewardship (Griffiths et al. 2021).

7.4 Case Studies in Human-Altered Environments: Monitoring Biodiversity in Agricultural Landscapes

In intensively managed agricultural landscapes, biodiversity monitoring often requires a combination of molecular and traditional methods. A study in Europe combined eDNA analysis of freshwater systems with traditional aquatic macroinvertebrate surveys to assess the impacts of agricultural runoff on biodiversity. The molecular data provided species-level resolution, while traditional surveys offered broader ecological context, revealing the cumulative impacts of habitat degradation on aquatic communities (Hering et al. 2018).

In Brazil, studies by Verdade et al. have highlighted the importance of integrating traditional ecological surveys with genetic tools to assess biodiversity in agricultural landscapes (e.g., Villela et al. 2008; Zucoloto et al. 2021). Their work demonstrated that landscape mosaics, which combine crop fields, pastures, and remnants of native vegetation, can support a surprising level of biodiversity. For instance, in regions dominated by sugarcane cultivation, broad-snouted caiman (*Caiman latirostris*) populations were shown to persist in connected aquatic environments including wetlands and artificial reservoirs, but with relatively levels of gene flow (Verdade et al. 2002; Villela et al. 2008). These findings underscored the critical role of landscape connectivity in maintaining gene flow and biodiversity resilience in agricultural matrices (Verdade et al. 2014).

The integration of molecular tools, such as eDNA and genetic analyses, with field-based surveys provides a comprehensive understanding of how agricultural practices affect biodiversity. These combined approaches offer actionable insights to guide habitat restoration, implement wildlife corridors, and promote sustainable agricultural practices that balance productivity with conservation goals.

7.5 Conserving Urban Wildlife

Urban areas pose unique challenges for biodiversity conservation due to habitat fragmentation and human-wildlife conflicts. Molecular methods, such as genetic analyses of scats or feathers, can identify urban wildlife populations, while camera traps and telemetry studies provide critical information on their movement and behavior. A study on coyotes (*Canis latrans*) in Chicago combined these approaches to understand how urbanization affects genetic connectivity and movement patterns, leading to recommendations for green infrastructure that supports urban wildlife corridors (Gehrt et al. 2011).

The integration of molecular and traditional ecological methods enhances the scope and reliability of conservation efforts, particularly in human-altered landscapes. By combining the strengths of molecular tools with direct observations, telemetry, and community engagement, conservationists can achieve a holistic understanding of ecosystems and develop more effective management strategies. This collaborative approach not only bridges technological advances with ecological traditions but also underscores the importance of inclusive, interdisciplinary conservation practices.

7.6 Urban Biodiversity Monitoring Using eDNA: Insights from São Paulo, Brazil

eDNA sampling detects genetic material shed by organisms into their environment, such as skin cells, mucus, or waste products. This non-invasive method is particularly well-suited for urban settings, where direct observation of species may be hindered by limited access, pollution, or cryptic behavior (Thomsen and Willerslev 2015). By providing species-level identification from environmental samples, eDNA allows for rapid and cost-effective biodiversity assessments, even in degraded or heavily modified habitats (Ficetola et al. 2008).

Urban ecosystems, often perceived as biologically impoverished, can harbor remarkable biodiversity, including species that persist in fragmented and altered habitats. Monitoring this biodiversity is fundamental for informing urban planning and conservation strategies. eDNA has emerged as a powerful tool for detecting species in urban environments, where traditional survey methods may be impractical or ineffective. In São Paulo, Brazil, eDNA analysis of urban streams has provided groundbreaking insights into the local amphibian diversity, challenging prior assumptions and shaping urban conservation policies.

São Paulo, one of the largest megacities in the world, faces significant environmental challenges, including pollution, habitat fragmentation, and urban sprawl.

Urban streams, often overlooked as biodiversity hotspots, were the focus of an eDNA study aimed at assessing amphibian presence. Remarkably, eDNA analysis revealed the presence of amphibian species previously believed to be absent from the region, including some species of conservation concern. This discovery highlighted the resilience of amphibians in urban environments and underscored the importance of these habitats for biodiversity conservation (Siqueira et al. 2020).

The findings also revealed spatial patterns of biodiversity linked to urban green spaces and water quality, providing critical data for urban planners. Streams adjacent to larger green spaces showed higher species richness, emphasizing the role of connected habitats in supporting urban wildlife. These insights have informed initiatives to preserve and expand green corridors and implement pollution reduction strategies in São Paulo.

The use of eDNA in São Paulo has demonstrated its potential to inform urban conservation policies. The data has been instrumental in advocating for the integration of biodiversity considerations into urban development plans. Specifically, the findings have supported the following possibilities:

1. **Green infrastructure planning**: Identification of biodiversity hotspots has informed the design of green corridors and the prioritization of areas for conservation.
2. **Community engagement**: The visibility of these findings has increased public awareness of urban biodiversity and fostered community support for conservation initiatives.
3. **Monitoring and evaluation**: eDNA provides a scalable and repeatable method for tracking biodiversity over time, enabling the assessment of conservation interventions and urban ecosystem health.

The application of eDNA in urban biodiversity monitoring, such as what was carried out in São Paulo, highlights its transformative potential for urban conservation. By revealing hidden patterns of biodiversity and providing actionable data for policymakers, eDNA complements traditional methods and advances the understanding of urban ecosystems. As cities continue to expand, incorporating eDNA into urban planning and conservation strategies will be critical for promoting coexistence between humans and wildlife in the Anthropocene.

8 Genomic Tools in Agricultural Landscapes: Understanding and Managing Biodiversity

Agricultural landscapes play a crucial role in supporting global food production but often come at the cost of habitat fragmentation and biodiversity loss. Pollinators and other wildlife, vital for ecosystem functioning and agricultural productivity, are particularly vulnerable in these environments. Genomic tools are critical to understanding the impacts of agriculture on biodiversity, support management strategies, and promote coexistence between agriculture and conservation.

8.1 Genomic Insights into Pollinator Populations

Pollinators, such as bees, are critical for the reproduction of many crops and wild plants. However, habitat fragmentation, pesticide exposure, and agricultural intensification have significantly threatened their populations. Genomic analyses provide insights into the genetic health, population structure, and connectivity of pollinator populations in agricultural regions.

A notable study in the Midwest of the USA employed genomic tools to assess the genetic diversity of wild bee populations across intensively farmed landscapes (Koh et al. 2016). The analysis revealed reduced genetic diversity in populations near high-intensity agricultural zones, which was strongly associated with pesticide exposure and the loss of flowering plants. These findings underscored the need for interventions to restore pollinator-friendly environments.

In response to these genomic insights, conservation practitioners and agricultural stakeholders implemented measures to enhance habitat connectivity. The establishment of wildlife corridors, such as hedgerows, buffer strips, and flowering cover crops, has supported gene flow among fragmented populations, promoting their resilience and adaptive capacity in the face of agricultural pressures (Dicks et al. 2021).

In Southeast Brazil, studies have demonstrated how landscape mosaics, combining crop fields, pastures, and native vegetation patches, can support both biodiversity and agricultural productivity (Verdade et al. 2016). By using ecological and genomic tools, these researchers have assessed the impacts of agricultural intensification on several species, highlighting the role of agricultural landscapes as critical spaces for biodiversity conservation.

One example of such work is the integration of genomic analyses to study gene flow in mammal populations inhabiting fragmented forest patches within agricultural matrices. These studies have informed management practices such as reforestation and the creation of ecological corridors, ensuring the long-term persistence of species in these human-dominated landscapes.

8.2 Implications for Conservation and Sustainable Agriculture

The integration of genomic tools and ecological research, exemplified by those studies, highlights several key strategies for biodiversity conservation in agricultural landscapes:

1. **Restoration of habitat connectivity**: Genomic data can identify genetic bottlenecks and guide the design of wildlife corridors that support gene flow across fragmented habitats.
2. **Pollinator-friendly practices**: Genomic studies on pollinators inform the implementation of strategies such as reducing pesticide use and planting flowering vegetation to promote pollinator health.

3. **Sustainable landscape management**: Ecological and genomic research together emphasize the importance of landscape mosaics in supporting biodiversity while maintaining agricultural productivity.

Genomic tools have provided critical insights into the impacts of agricultural practices on biodiversity, especially pollinators and other key species. Combined with traditional ecological methods and the innovative approaches, these tools offer a pathway towards more sustainable agricultural landscapes. By integrating science and practice, it is possible to balance the needs of food production with the imperative to conserve biodiversity for future generations.

9 Future Directions: Emerging Innovations in Molecular Ecology

Molecular ecology is poised to revolutionize conservation biology through continuous innovation and cross-disciplinary approaches. As environmental challenges intensify, the development and integration of advanced tools will play a pivotal role in conserving biodiversity, particularly in human-altered environments. The future of molecular ecology lies in its ability to adapt to the dynamic needs of conservation science while fostering collaboration across global networks.

One of the most exciting advancements is the development of portable DNA sequencing technologies, such as the Oxford Nanopore MinION, which enable real-time biodiversity assessments in the field. These tools allow researchers to detect and identify species directly at collection sites, significantly reducing time and logistical barriers (Pomerantz et al. 2018). This portability enhances the ability to monitor remote or inaccessible ecosystems, improving conservation responses to emerging threats.

Machine learning and artificial intelligence (AI) offer transformative potential by automating the identification of species and populations from genetic data. These technologies can analyze vast genomic datasets with high precision, identifying cryptic species or tracking changes in population genetics over time (Schmidt et al. 2021). By streamlining the analysis process, AI applications can accelerate conservation decision-making and improve the accuracy of biodiversity assessments.

Synthetic biology, another frontier in molecular ecology, presents opportunities to restore degraded ecosystems through engineered microbial communities. For instance, synthetic microbes designed to enhance soil fertility or combat invasive species could provide innovative solutions to long-standing ecological challenges (de Lorenzo et al. 2018). While these approaches require rigorous ethical considerations and risk assessments, their potential to rehabilitate ecosystems cannot be overlooked.

Global data-sharing platforms, such as the Earth BioGenome Project and other open science initiatives, are breaking down silos in conservation research. These platforms facilitate collaboration, promote transparency, and democratize access to molecular data, empowering conservation efforts worldwide (Lewin et al. 2018). By uniting researchers, policymakers, and communities, such initiatives foster a global commitment to biodiversity conservation.

An emerging yet underexplored frontier in molecular ecology is the study of rapid evolution—adaptive genetic changes occurring over just a few generations in response to intense anthropogenic pressures (Reznick et al. 2004). Although ecological consequences of environmental disturbances are widely recognized, the evolutionary dimension remains largely overlooked. Agricultural landscapes, for example, impose strong and continuous selective pressures on local fauna and flora through habitat fragmentation, pesticide use, and monoculture cycles. These pressures can drive genetic shifts in surprisingly short timeframes. In Brazil, the commercial rotation of *Eucalyptus* plantations is approximately 6 years. Over just three cycles, more than 50 generations of small mammals such as native rodents may occur—ample time for adaptive genetic changes to arise and spread within populations (Rosalino et al. 2014). Molecular tools, including genome-wide association studies (GWAS), temporal population genomics, and epigenetic profiling, can uncover such evolutionary responses, revealing how species are genetically tracking environmental change. Integrating these approaches into conservation planning is crucial for capturing not just the ecological, but also the evolutionary resilience or vulnerability of species living in human-modified environments (Tellier et al. 2024).

While molecular techniques provide unprecedented insights, their integration with traditional ecological methods remains crucial. Traditional approaches, such as field surveys, telemetry, and citizen science, offer critical context and validation for molecular findings. Together, these methods create a more comprehensive understanding of ecosystems, enabling conservationists to design informed strategies that address complex challenges in human-altered landscapes.

As landscapes continue to change under anthropogenic pressures, molecular tools will become increasingly central to conservation strategies. They provide the means to monitor wildlife, assess genetic health, and guide habitat restoration in ways that were previously unimaginable. From protecting pollinators in agricultural regions to uncovering hidden biodiversity in urban ecosystems, molecular ecology offers a powerful framework for conserving life on Earth.

The future of molecular ecology lies in the seamless integration of innovation, collaboration, and ethical stewardship. By leveraging the potential of portable sequencing, AI-driven analytics, synthetic biology, and global networks, conservationists can address emerging challenges with precision and agility. As we move toward a more sustainable future, molecular tools will not only enhance our ability to protect biodiversity but also inspire new ways of understanding and interacting with the natural world.

References

Adamo M, Voyron S, Chialva M, Marmeisse R, Girlanda M (2020) Metabarcoding on both environmental DNA and RNA highlights differences between fungal communities sampled in different habitats. PLoS One 15(12):e0244682

Åkesson M, Bensch S, Hasselquist D, Hansson B (2016) Inbreeding and genetic rescue in the Scandinavian wolf population. Biol Conserv 191:51–59

Allendorf FW, Luikart G, Aitken SN (2010) Conservation and the genetics of populations. Wiley-Blackwell, Holboken

Amato KR et al (2013) Gut microbial diversity and its relationship with diet in the wild orangutan. PLoS One 8(11):e77177

Arandjelovic M et al (2010) Effective non-invasive genetic monitoring of multiple wild western gorilla groups. Biol Conserv 143(7):1780–1791

Barnes MA, Turner CR (2016) The ecology of environmental DNA and implications for conservation genetics. Conserv Genet 17(1):1–17

Beja-Pereira A, Oliveira R, Alves PC, Schwartz MK, Luikart G (2009) Advancing ecological understandings through technological transformations in noninvasive genetics. Mol Ecol Resour 9(5):1279–1301

Bista I et al (2017) Annual time-series analysis of aqueous eDNA reveals ecologically relevant dynamics of lake ecosystem biodiversity. Nat Commun 8:14087

Bohmann K, Evans A, Gilbert MTP, Carvalho GR, Creer S, Knapp M, Yu DW, de Bruyn M (2014) Environmental DNA for wildlife biology and biodiversity monitoring. Trends Ecol Evol 29(6):358–367

Broquet T, Petit EJ (2009) Molecular estimation of dispersal for ecology and population genetics. Annu Rev Ecol Evol Syst 40:193–216

Casas-Marce M, Soriano L, López-Bao JV, Godoy JA (2013) Genetics at the verge of extinction: insights from the Iberian lynx. Mol Ecol 22(23):5503–5515

Clare EL et al (2021) eDNAir: proof of concept that animal DNA can be collected from air sampling. PeerJ 9:e11030

Cristescu ME (2019) Can environmental RNA revolutionize biodiversity science? Trends Ecol Evol 34(8):694–697

Deagle BE et al (2009) Improving DNA-based diet studies through the use of PCR blocking primers designed to augment species-specific amplification of prey DNA. Mol Ecol Resour 9(5):771–778

Deiner K, Bik HM, Mächler E, Seymour M, Lacoursière-Roussel A, Altermatt F, Creer S, Bista I, Lodge DM, De Vere N, Pfrender ME (2017) Environmental DNA metabarcoding: transforming how we survey animal and plant communities. Mol Ecol 26(21):5872–5895

Dicks LV, Breeze TD, Ngo HT et al (2021) A global-scale expert assessment of drivers and risks associated with pollinator decline. Nat Ecol Evol 5(2):145–152

Ficetola GF, Miaud C, Pompanon F, Taberlet P (2008) Species detection using environmental DNA from water samples. Biol Lett 4(4):423–425

Frankham R (2005) Genetics and extinction. Biol Conserv 126(2):131–140

Funk WC, McKay JK, Hohenlohe PA, Allendorf FW (2012) Harnessing genomics for delineating conservation units. Trends Ecol Evol 27(9):489–496

Gehrt SD, Anchor C, White LA (2011) Home range and landscape use of coyotes in a metropolitan landscape: conflict or coexistence? J Mammal 90(5):1045–1057

Goldberg CS, Strickler KM, Pilliod DS (2015) Moving environmental DNA methods from concept to practice for monitoring aquatic macroorganisms. Biol Conserv 183:1–3

Goldberg CS, Turner CR, Deiner K, Klymus KE, Thomsen PF, Murphy MA et al (2016) Critical considerations for the application of environmental DNA methods to detect aquatic species. Methods Ecol Evol 7(11):1299–1307

Griffiths J, Watson R, Reid S (2021) Citizen science and eDNA: engaging communities in monitoring platypus populations. Conserv Sci Pract 3(5):e461

Hanski I, Gaggiotti OE (2004) Ecology, genetics, and evolution of metapopulations. Elsevier Academic Press, Saint Louis

Harper LR et al (2019a) Robust biological monitoring of ecological restoration using environmental DNA: a case study in reintroduction ecology. Restor Ecol 27(5):1173–1181

Harper LR, Lawson Handley L, Hahn C, Boonham N, Rees HC, Gough KC, Lewis E, Adams IP, Brotherton P, Phillips S, Hänfling B (2019b) Biological monitoring using environmental DNA: a review of tools and applications. Biol Rev 94(5):1223–1243

Hayward AD, Wilson AJ, Pilkington JG et al (2011) Natural selection on a measure of parasite resistance varies across ages and environmental conditions in a wild mammal. J Evol Biol 24(8):1664–1676

Hebert PDN et al (2003) Biological identifications through DNA barcodes. Proc R Soc B Biol Sci 270(1512):313–321

Hering D, Borja A, Carstensen J et al (2018) The European water framework directive at the age of 20: achievements and future directions. Biol Conserv 228:30–40

Iyengar A, Gilbert T, Woodfine T et al (2007) Remnants of ancient genetic diversity preserved within captive groups of scimitar-horned oryx (Oryx dammah). Mol Ecol 16(12):2436–2449

Janečka JE et al (2008) Population monitoring of snow leopards using noninvasive collection of scat samples: a pilot study. Anim Conserv 11(2):92–101

Jansen HJ, Liem ASN, van Schaik CP (2018) Portable sequencing in conservation genetics: promises and pitfalls. Conserv Genet Resour 10(3):455–462

Jerde CL et al (2011) "Silent invasion": a covert threat to aquatic biodiversity by the invasive Asian carp detected with environmental DNA. Biol Conserv 144(8):2175–2179

Johnson WE, Onorato DP, Roelke ME et al (2010) Genetic restoration of the Florida panther. Science 329(5999):1641–1645

Joshi AR, Dinerstein E, Wikramanayake E et al (2013) Connectivity of tiger population in the human-dominated landscape of central India. Ecol Model 268:1–10

Kardos M, Luikart G, Allendorf FW (2015) Measuring individual inbreeding in the age of genomics: marker-based measures are better than pedigrees. Heredity 115(1):63–72

Koh I, Lonsdorf EV, Williams NM et al (2016) Modeling the status, trends, and impacts of wild bee abundance in the United States. Proc Natl Acad Sci 113(1):140–145

Laird SA, Wynberg RP, Rourke M (2020) Access and benefit-sharing as a mechanism for protecting traditional knowledge and biodiversity. Nat Sustain 3(5):399–405

Leduc AOHC et al (2021) Environmental DNA detects range shifts in cold-water fish species in response to climate change. Glob Chang Biol 27(8):1623–1634

Lewin HA, Robinson GE, Kress WJ et al (2018) Earth BioGenome project: sequencing life for the future of life. Proc Natl Acad Sci 115(17):4325–4333

Li Y, Devenish C, Tosa MI, Luo M, Bell DM, Lesmeister DB, Greenfield P, Pichler M, Levi T, Yu DW (2024) Combining environmental DNA and remote sensing for efficient, fine-scale mapping of arthropod biodiversity. Philos Trans R Soc Lond B Biol Sci 379(1904):20230123

de Lorenzo V, Marlière P, Sole R (2018) Bioremediation at a global scale: from the test tube to the field. Curr Opin Biotechnol 53:56–62

Morgera E, Tsioumani E, Buck M (2012) Unraveling the Nagoya protocol: a commentary on the Nagoya protocol on access and benefit-sharing to the convention on biological diversity. Brill, Leiden

Moritz C (1994) Defining "evolutionarily significant units" for conservation. Trends Ecol Evol 9(10):373–375

van Oppen MJH, Oliver JK, Putnam HM, Gates RD (2015) Building coral reef resilience through assisted evolution. Proc Natl Acad Sci 112(8):2307–2313

Pansu J et al (2019) Applying environmental DNA metagenomics to monitor host-parasite interactions in wildlife. Mol Ecol Resour 19(5):1286–1303

Piaggio AJ et al (2014) Detecting invasive species in ecological communities using eDNA. Conserv Biol 28(2):430–439

Pilliod DS, Goldberg CS, Arkle RS, Waits LP (2014) Factors influencing detection of eDNA from a stream-dwelling amphibian. Mol Ecol Resour 14(1):109–116

Pomerantz A, Peñafiel N, Arteaga A et al (2018) Real-time DNA barcoding in a rainforest using nanopore sequencing: opportunities for rapid biodiversity assessments and local capacity building. Gigascience 7(4):giy033

Reznick D, Rodd H, Nunney L (2004) Empirical evidence for rapid evolution. In: Ferrière R, Dieckmann U, Couvet D (eds) Evolutionary conservatio biology. Cambridge University Press, Cambridge, UK, pp 101–118

Rosalino LM, Lyra-Jorge MC, Verdade LM (2014) Adaptation and evolution in changing environments. In: Verdade LM, Lyra-Jorge MC, Piña CI (eds) Applied ecology and human dimensions on biological conservation. Springer, Heidelberg, pp 53–71. https://doi.org/10.1007/978-3-642-54751-5_4

Schmidt C, Hornick T, Mönchgesang S (2021) Machine learning in molecular biodiversity research: opportunities and challenges. Mol Ecol Resour 21(3):676–692

Schwartz MK, Luikart G, Waples RS (2007) Genetic monitoring as a promising tool for conservation and management. Trends Ecol Evol 22(1):25–33

Siqueira T, Alexandrino ER, Prado VHM (2020) Urban stream biodiversity revealed through environmental DNA: implications for amphibian conservation in São Paulo, Brazil. Urban Ecosyst 23(5):941–953

Stephens ZD, Lee SY, Faghri F et al (2015) Big data: astronomical or genomical? PLoS Biol 13(7):e1002195

Strickler KM, Fremier AK, Goldberg CS (2015) Factors influencing eDNA degradation in freshwater ecosystems. Environ Sci Technol 49(24):14032–14039

Taberlet P, Coissac E, Hajibabaei M, Rieseberg LH (2012) Environmental DNA. Mol Ecol 21(8):1789–1793

Tellier A, Hodgins K, Stephan W, Stukenbrock E (2024) Rapid evolutionary adaptation: potential and constraints. Mol Ecol 33:e17350

Thomas CD, Cameron A, Green RE, Bakkenes M, Beaumont LJ, Collingham YC, Erasmus BFN, Ferreira de Siqueira M, Grainger A, Hannah L, Hughes L, Huntley B, van Jaarsveld AS, Midgley GF, Miles L, Ortega-Huerta MA, Peterson AT, Phillips OL, Williams SE (2004) Extinction risk from climate change. Nature 427(6970):145–148

Thomsen PF, Willerslev E (2015) Environmental DNA – an emerging tool in conservation for monitoring past and present biodiversity. Biol Conserv 183:4–18

Thomsen PF et al (2012) Monitoring endangered freshwater biodiversity using environmental DNA. Mol Ecol 21(11):2565–2573

Tillotson HK, Kelly RP, Duda JJ, Hoy M, Kralj J, Quinn TP (2018) Concentrations of environmental DNA (eDNA) reflect spawning salmon abundance at fine spatial and temporal scales. Biol Conserv 220:1–11

Vences M et al (2005) Deciphering amphibian diversity through DNA barcoding: chances and challenges. Philos Trans R Soc Lond B Biol Sci 360(1462):1859–1868

Verdade LM, Zucoloto RB, Coutinho LL (2002) Microgeographic variation in *Caiman latirostris*. J Exp Zool 294(4):387–396

Verdade LM et al (2014) The conservation value of agricultural landscapes. In: Verdade L, Lyra-Jorge M, Piña C (eds) Applied ecology and human dimensions in biological conservation. Springer, Berlin, Heidelberg

Verdade LM, Gheler-Costa C, Lyra-Jorge C (2016) The multiple facets of agriculture landscape Brazil. In: Gheler-Costa C, Lyra-Jorge MC, Verdade LM (eds) biodiversity in agricultural landscapes of southeastern Brazil. De Gruyter Open Poland, Warsaw, pp 2–13

Villela PMS, Coutinho LL, Piña CI, Verdade LM (2008) Macrogeographic genetic variation in broad-snouted caiman (*Caiman latirostris*). J Exp Zool 309A:1–9

Wilting A et al (2016) A scats-based method for identifying the presence of rare carnivores using DNA barcoding. Conserv Genet Resour 8(3):255–265

Yates MC, Glaser D, Post JR, Cristescu ME, Fraser DJ (2019) The relationship between eDNA concentration and biomass in marine fishes. Environ DNA 1(1):141–151

Yoccoz NG et al (2012) DNA from soil mirrors plant taxonomic and growth form diversity. Mol Ecol 21(15):3647–3655

Zucoloto RB, Bomfim GC, Fernandes FMC, Schnadelbach AS, Piña CI, Verdade LM (2021) Effective population size of broad-snouted caiman (*Caiman latirostris*) in Brazil: a historical and spatial perspective. Global Ecol Conserv 28:e01673

Chapter 6
The Use of Remote Sensing Data for Above-Ground Biomass Estimation

Sandra Cristina Deodoro, Rowan Fealy, and Tim McCarthy

Abstract Since the launch of satellites into space in the 1970s, the usage of their data has expanded across various applications, from communication (telecommunication and broadcasting) and navigation (e.g., the Global Positioning System-GPS) to precision agriculture and environmental monitoring. Advances in sensors, instruments, and platforms—including drones—have equipped researchers with new measurements of the physical properties of elements on Earth (e.g., vegetation, soil, and water) leading to improved environmental assessment and monitoring across scales. Earth Observation includes remote sensing technologies, which play a vital role in monitoring vegetation—providing information on vegetation extent, deforestation, phenological stages, crop growth, above-ground biomass, and changes in land use/land cover over time (temporal series) and across different spatial scales (larger or smaller areas). Satellite remote sensing technologies can be further classified into optical/non-optical as well as active and passive systems. Optical, passive remote sensing of vegetation relies on the interaction between plant pigments/cell structure and incident radiation in the electromagnetic spectrum, particularly in the red and near-infrared spectrum. In green vegetation, remote sensing techniques leverage the process whereby chlorophyll absorbs incident energy while cell structure strongly scatters near-infrared radiation to prevent cell damage. Non-optical, active, microwave remote sensing systems transmit radiation from the satellite onto the ground target, in this case biomass such as crops, which will then reflect the radiation as backscatter to the satellite. This reflected backscatter signal is affected by the physical properties, water content and electric potential energy stored in plants (including grass) and tress. Considering that biomass can serve as a proxy for biodiversity, this chapter presents examples of estimating

S. C. Deodoro (✉) · R. Fealy
Irish Climate Analysis and Research Units (ICARUS), Department of Geography, Maynooth University, Maynooth, Ireland
e-mail: sandra.deodoro.2022@mumail.ie

T. McCarthy
National Centre for Geocomputation, Maynooth University, Maynooth, Ireland

L. Martins Verdade, R. A. Moral (eds.), *Wildlife Monitoring*,
https://doi.org/10.1007/978-3-032-05821-8_6

above-ground biomass using remote sensing data and techniques in two contrasting landscapes: forested and managed landscapes.

Keywords Earth observation · GIS · Optical · Microwave · Electromagnetic · Landscapes

1 Introduction

Remote sensing is a technique for obtaining images of objects on the Earth's surface, without any physical contact between the sensor and the object, facilitating their identification, classification, and measurements (area and composition) (Elachi and Van Zyl 2021). Despite being cutting-edge technology, its roots can be traced to the invention of the camera dating back to the end of the nineteenth century. While the development and application of remote sensing using aerial photography largely has its origins in military surveillance during World War I, subsequent developments during the Cold War led to the development of satellite-based remote sensing applications (e.g. measurements of the upper atmosphere) (Cracknell 2018). With the advent of civilian satellite earth observation during the 1970s, it has since been employed in a wide variety of applications (e.g., communication, navigation, urban planning, environmental monitoring, disaster response, and agricultural optimisation), given its prompt and direct way to acquire information on the Earth's surface at different spatial resolutions. This is due to the intrinsic characteristics of the satellite instruments (i.e., platforms and sensors) and their methods of data acquisition (Liu 2015). For example, low-resolution Earth Observation (EO) platforms provide broad-scale perspectives over a large swath width (horizontal distance on the ground covered by the sensor), while higher-resolution sensors capture more resolved and finer details on the ground but with a reduced swath width.

Remote sensing imagery has, and continues to, play a critical role in environmental science and geoscience as it provides continuous and repeat "scans" of the Earth's surface—capturing information from both the surface and subsurface (depth of approximately 50 cm in bare soil conditions with radar satellite)—that can be used to monitor and track changes over time (e.g. land use and land cover (LULC) change information). Furthermore, remote sensing provides data for quantifying biophysical and geophysical variables on both temporal (i.e. geostationary; repeat orbit) and spatial scales, leading to substantially improved understanding of Earth system dynamics (atmosphere, hydrosphere, biosphere and geosphere) and ecosystem resilience (ESA 2023). Importantly, it provides a non-destructive approach, particularly important for monitoring potentially sensitive environments. The data is spatial (as represented by the swath width) and is typically captured at a high temporal resolution (e.g. ~6 days with the Sentinel satellite), with acquired data being made available in near real-time. In addition, data collected by public institutions and agencies are provided for free (e.g. European Space Agency

(ESA), National Aeronautics and Space Administration (NASA)). However, the data requires a level of expertise to process and may be limited by cloud cover (e.g. optical-based sensors) or potential for visual interpretation (e.g. radar or microwave-based sensors). In spite of these limitations, the use of remote sensing offers significant potential for contributing to environmental monitoring and measurement, across scales.

Here we focus on the use and application of remote sensing using both satellites and unmanned aerial vehicles (UAV)/drones for estimating above-ground biomass in different land cover/use contexts, namely in forest, grassland, and agricultural applications to illustrate its potential for contributing to improving our understanding of these human-altered landscapes. As managed landscapes, information concerning grass management, pasture forage, and forest timber production (Zhang et al. 2023), along with other information, can be readily obtained from satellites from these land cover types. Furthermore, considering grassland and forest as contrasting landscapes, they are highlighted here to represent the capability of remote sensing data and techniques for estimating above-ground biomass in light of two specific biophysical properties, and to show how different sensors obtain their data, providing land cover and vegetation information.

The rationale of this chapter is rooted in the tenet that biomass can act as a proxy for biodiversity. It is worth noting that despite highlighting the benefit of a remote sensing approach for biomass estimation, the need for in-situ measurements (e.g. biomass sampling and plant height measurements) remains. Quantitative data collection in the field is essential to fit, train, and evaluate biomass estimation models derived from any means, including remote sensing imagery. However, the combination of both measurement and monitoring approaches outlined here provides an opportunity to develop spatio-temporal models, facilitating estimation across wider spatial scales and over time.

2 The Electromagnetic Spectrum: Obtaining Information from Remote Sensors

There are four components to obtaining measurements with remote sensing: the energy source or electromagnetic energy, the transmission path, the target, and the satellite sensor, which need to be considered for different applications. Among these, the energy source plays an essential role in transmitting information from the target to the sensor (i.e. the received signal, the reflectance, the echo or backscattering).

The electromagnetic spectrum underpins the fundamental principles of remote sensing from designing and engineering the satellites to acquiring data. The electromagnetic spectrum ranges from short wave to long wave lengths, wherein the former includes gamma and x-rays; the latter, microwaves and broadcast radio waves (Figs. 6.1 and 6.2). The portion of the spectrum that the sensor measures

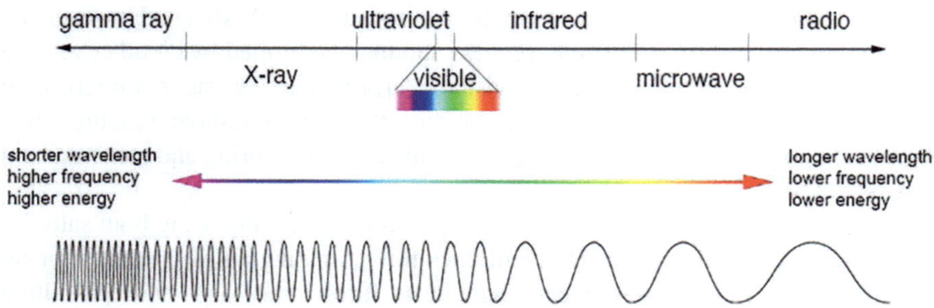

Fig. 6.1 The electromagnetic spectrum and its graphical representation of wavelength (on bottom), frequency and energy/region (on top). (Credit: NASA's Imagine the Universe. https://imagine.gsfc.nasa.gov/science/toolbox/emspectrum1.html)

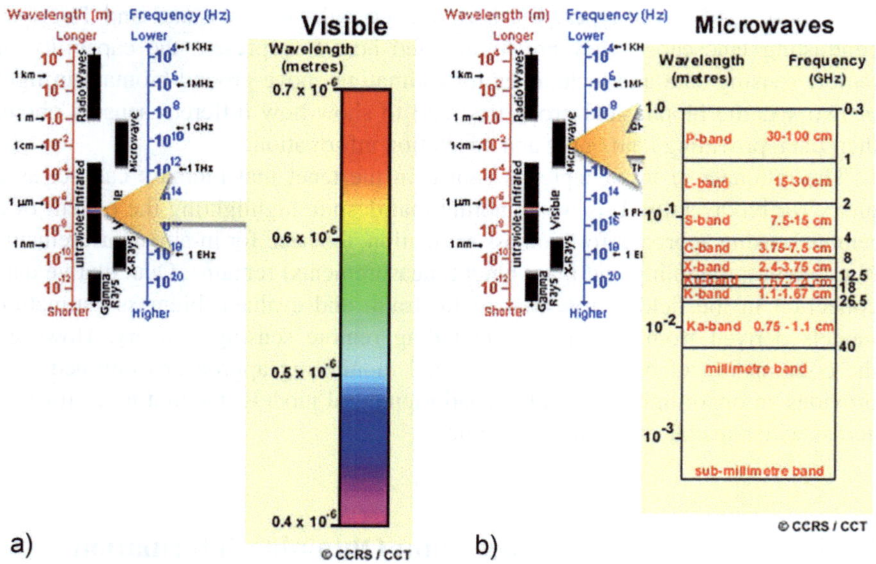

Fig. 6.2 The electromagnetic spectrum, with a focus on the visible region (**a**) and microwave region (**b**). (Source: Canada Centre for Remote Sensing. https://www.nrcan.gc.ca/sites/www.nrcan.gc.ca/files/earthsciences/pdf/resource/tutor/fundam/pdf/fundamentals_e.pdf)

in or sensor bandwidths (i.e., band size in the satellite sensor context) determine the spectral resolution of the remote sensing device or instrument, which, in turn, provides information on the extent to which the electromagnetic energy will interact with elements on the Earth's surface (e.g. vegetation, soil, water, urban structures) and to what depth on that surface.

Electromagnetic radiation transmitted by different objects typically varies by wavelength (e.g the Earth emits in the long wavelength portion of the spectrum, while the Sun predominantly emits in the short wavelength portion). Electromagnetic waves passing through a medium, such as the the atmosphere, undergo scattering and absorption effects. Radiation in shorter wavelengths is more readily scattered than longer wavelengths—associated with the interaction between the radiation wavelength and particles of different sizes in the atmosphere. The areas of the spectrum where atmospheric effects are limited or non-existent are known as atmospheric windows (Walsh 2003; Jensen 2009), that is, where radiation or emitted energy in particular wavelengths can pass through a medium, unimpeded. These "windows" are used in remote sensing to obtain information concerning the Earth's surface and objects on it. In other words, the transmissivity of the terrestrial atmosphere is reliant on wavelength. Optical remote sensing (i.e. optical radiation) refers to the part of the electromagnetic spectrum characterised by the wavelengths from the visible to infrared regions. The visible region, whose wavelengths range from approximately 0.4 to 0.7 μm (400–700 nm[1]), lies in the spectral range that the human eye can perceive (i.e. visible light). The infrared (IR) region ranges from approximately 0.7 to 100 μm (700–100,000 nm) and is divided into reflected IR (0.7–3.0 μm wavelengths) and thermal IR (3.0–100 μm) (Walsh 2003) or thermal energy. Reflected IR (near wave infrared (NIR) lies just beyond the visible range: 0.7–1.4 μm and short wave infrared (SWIR): 1.4–3 μm) is useful for discriminating between healthy and unhealthy vegetation or fallow vegetation, and for distinguishing between vegetation, soil and rocks, whereas thermal IR is employed for monitoring temperature variations in land, water, and ice (Walsh 2003). Data from the visible and infrared portion, specifically the reflected near-infrared (NIR), is important for vegetation analysis as this range is sensitive to photosynthetic vegetation.

Longer wavelengths in the microwave and radio range, referred to as radar, are also of interest to the environmental sciences. Microwave radiation is the longest wavelength used for remote sensing (Walsh 2003) and while microwave or radar data is not visually interpretable and requires a level of expertise to work with, it has a higher information content than data acquired in the optical range.

In addition to optical and microwave/radar, remote sensors can also be classified in terms of energy emitted and/or received (the principle of transmitting and receiving signals, and receiving only). Active sensors emit a pulse or beam towards an object and measure the response back—the received quantity of emitted energy; while passive sensors do not emit radiation but measure the backscatter or solar energy reflected by a target on the Earth's surface. For clarity, we employ the optical remote sensing (visible and infrared) and microwave remote sensing naming conventions in the remainder of this chapter. Table 6.1 summarises the main characteristics of, and differences between, optical and microwave remote sensing.

In remote sensing, the incident radiation, or radiation that interacts with a target, is decomposed into different wavelengths—for example, from visible to short-wave infrared in the optical case—providing quantitative data known as spectral curves or signatures[2] through which information on the Earth's surface can be analysed and interpreted. The quantitative part is the spectral reflectance and refers to the

---

[1] 1 nanometre (nm) is equal to 10^{-9} m; 1 micrometre (μm) is equal to 10^{-6} m.

[2] The reader can refer to Figures 5 and 7 in Deodoro et al. (2024). https://doi.org/10.1016/j.jag.2024.103742.

Table 6.1 Advantages and disadvantages of optical and microwave remote sensing

Type of remote sensing instrument	Characteristic	Device examples[a]	Advantages	Disadvantages
Optical (passive sensor)	Relies on and measures reflected emissions produced by elements on the earth's surface.	Satellite optical sensor[b], spectrometer, radiometer, spectroradiometer, hyperspectral radiometer, imaging radiometer, sounder, accelerometer.	Easy interpretation of the outputs.	Cloudiness is an issue; requires sunlight to generate reflectance from the targets.
Microwave (active sensor)	Emits its own radiation to illuminate a target and measure the reflected or scattered signal.	Radar[b], lidar[b], laser altimeter, sounder (weather), scatterometer	Operates regardless of weather conditions (e.g. cloudiness) and luminosity level; operates day and night; is relatively independent of atmospheric scatterings.	Interpretation and analysis of the outputs are not easy or straightforward.

[a]EOS Data Analytics. https://eos.com/blog/types-of-remote-sensing/
[b]Commonly employed in biomass estimation

amount of reflectance in a specified wavelength range. The vegetation components that are of relevance in remote sensing are plant pigment (e.g., chlorophyll) and cell structure, as they affect the incident electromagnetic radiation. For a detailed illustration of this process, readers can refer to ESA (n.d.: 18).

3 The Importance of Pixel Size and Geographical Scale

The spatial resolution of an image (Fig. 6.3) is one of the pivotal characteristics as it relates to the level of detail that can be resolved in a scene (ESA 2020). In a contextual example from biology conservation, selecting imagery that is capable of mapping a habitat of interest depends largely on the size of the habitat. That is where spatial resolution (pixel size) and image resolution (number of pixels in image) need to be considered, in tandem with spectral resolution (number of bands or wavelength intervals).

Remote sensing instruments represent the Earth's surface as a grid of sampling units, known as pixels (or cells) (Fig. 6.4), which is an abbreviation for picture

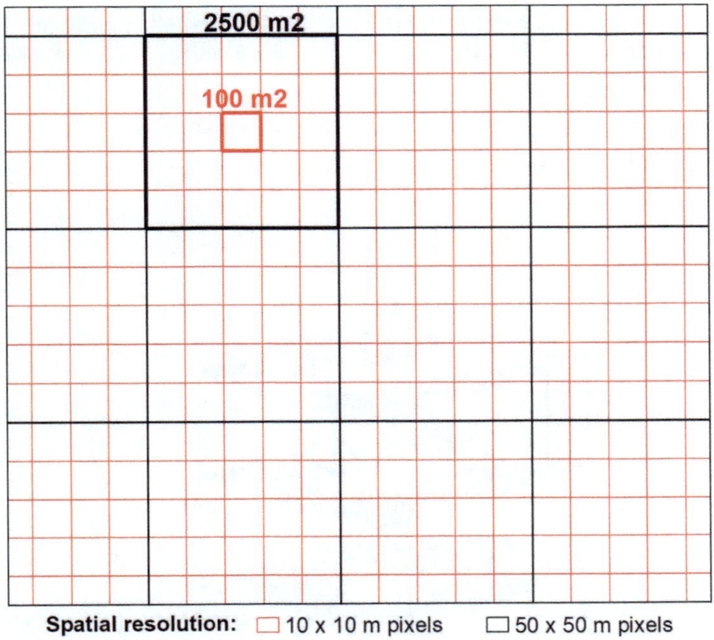

Fig. 6.3 An example of the effect of different spatial resolutions for mapping. The different spatial gridding with pixels at 10 × 10 m or 100 m^2 each on the terrain and at 50 × 50 m or 2500 m^2 each

element. In remote sensing imagery, a pixel is represented as a cell in an array conveying data values—of which the acquired data value represents an integrated value for each pixel (e.g. a single value is representative of the pixel area). Pixel and cell are interchangeable concepts; the former is often used when describing digital imagery (i.e. a two-dimensional square-shaped picture element), whereas cell is often used when describing raster data (e.g. in a Tagged Image File format (.tif) or georeferenced Tagged Image File format (.geotiff). The pixel size roughly equates to the spatial resolution of the sensor. For example, each pixel in a Sentinel-2 image measuring 10 m by 10 m (10 m square on the ground) (Fig. 6.5) means that the smallest linear dimension on the Earth's surface which could be resolved is approximately 10 m. In this example, the Sentinel 2 imagery would be incapable of detecting habitats whose dimensions are less than 10 m by 10 m which would require higher resolution imagery. Thus, to resolve or identify a target or object on the Earth (e.g., a tree), the spatial resolution of the imagery should be smaller than the size of the intended target, typically half the size of the feature, to discern that object (Walsh 2003), while the image resolution should be large enough to determine the spatial extent (e.g., forest).

Image resolution is commonly measured in terms of pixel size by Ground Sample Distance (GSD), which is the ground distance between adjacent pixel centres (ESA 2020) (Fig. 6.6). Since it refers to the size of a single pixel in an image, the lower

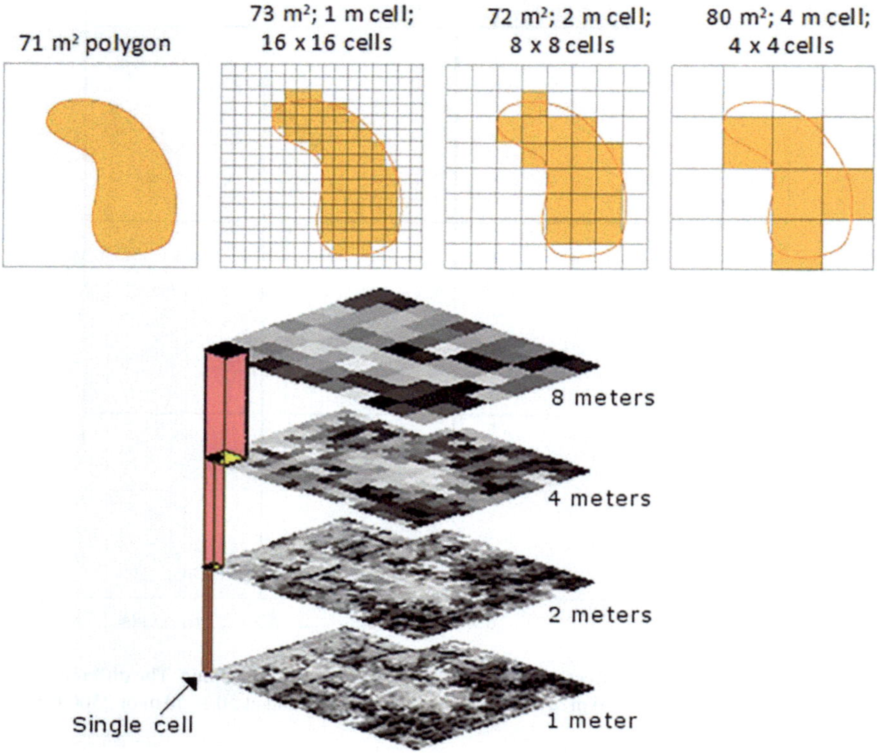

Fig. 6.4 A comparison of small and large cell sizes (on top) and a comparison of spatial resolution to information content (bottom). The red polygons that contour the orange pixels depicted on the top of the figure could be a representation of a patch or habitat. (Source: ArcGIS documentation (ArcGIS 10.8). Cell size of raster data. https://desktop.arcgis.com/en/arcmap/latest/manage-data/raster-and-images/cell-size-of-raster-data.htm)

this number, the finer the detail in the image. Commonly, low-resolution images have a GSD larger than 300 m; medium resolution ranges from 300 to 30 m; high resolution from 30 to 5 m; and very high-resolution (VHR) images have a GSD below 5 m (ESA 2020). Another important measure is the Ground Resolved Distance (GRD) which represents the minimum distance between two separate targets on the ground, from a given altitude, that allows them to be detected as distinct entities by the satellite sensor. It is the smallest distance that can be resolved by the sensor (Valenzuela and Reyes 2019) and the spatial dimension of a pixel, which is reliant on the altitude and design of the sensor. Thus, a smaller GRD indicates a higher resolution. Unlike GSD, GRD is not a specific measurement but rather the smallest resolvable detail (JOUAV 2024). GRD is sensitive to factors such as GSD, atmospheric conditions, sensor resolution, and image quality (Valenzuela and Reyes 2019; JOUAV 2024). It is also influenced by the focal length of the camera in the sensor.

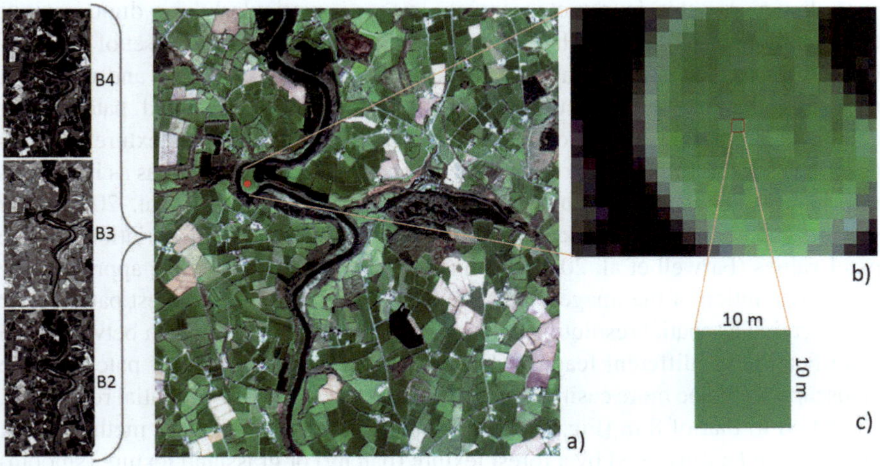

Fig. 6.5 A subset of Sentinel 2 imagery displayed with the true-colour composite bands red-blue-green (RGB) applied to the 4-3-2 spectral bands (**a**) and the respective pixel size (**b**, **c**). As a true RGB composite, all features in green indicate the presence of vegetation (e.g. forest, grass, crops)

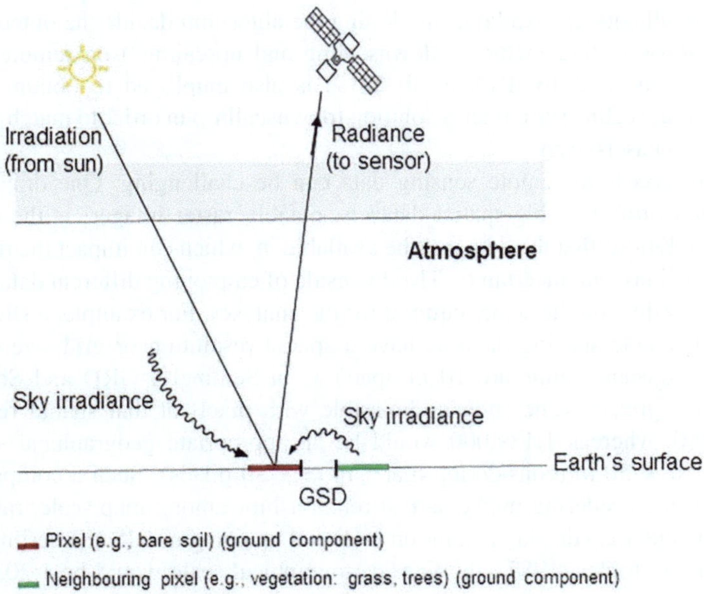

Fig. 6.6 The atmospheric effect in remote sensing data. The atmosphere affects and determines paths for energy or emitted radiance to illuminate an object on Earth's surface (i.e. equivalent ground pixel) and to reach the sensor. GSD ground sample distance

A key parameter in remote sensing imagery, particularly for digital image classification, is texture, which describes the structure or pattern of a set of elements or features in the image. Image texture can be smooth (i.e. simple and uniform), such as bare soil and agricultural fields, or rough (complex and patchy), for example, dense forests and coniferous habitats. Satellite image texture analysis describes aspects of habitat structure and composition over larger areas as it conveys information about spectral and spatial heterogeneity (Farwell et al. 2021). The satellite-image texture approach quantifies spectral and spatial variations in the pixel values (Farwell et al. 2021). To analyse habitats using such an approach, the spatial resolution of the imagery must be similar in size to the smallest patch under investigation as spatial resolution determines the ability to distinguish between, and identification of, different features. For example, a small feature (or patch) on the landscape would be more easily identified in a raster or grid of 1 m spatial resolution, compared to that of 8 m (Fig. 6.4). With the image texture analysis method, plant biomass can be illustrated by a forest texture (patchy) or grassland texture (smooth) while AGB biomass can be estimated using texture in tandem with a vegetation index (Gao et al. 2019; Tian et al. 2023).

Additionally, many processes are scale-dependent. Such an issue, in turn, poses a challenge for the researcher in deciding between selecting data as a function of spatial resolution (pixel size)—at the expense of disregarding important data—or employing all datasets available and letting the algorithm decide the outcome. The application of scaling methods (downscaling and upscaling[3]) on remote sensing images (Atkinson 2013; Park et al. 2019) is also employed to obtain a coarse resolution (upscaling) or a finer resolution (downscaling) in order to match the pixel size in the datasets used.

Having access to remote sensing data can be challenging. One drawback of dealing with different big-spatial datasets, notably raster images, is the different spatial resolutions that the data may be available in, which can impact the results by introducing bias and uncertainty. The downside of employing different datasets also leads to an effect on the scale required for the analyses. For example, a situation in which all remote sensing datasets have a spatial resolution or grid size of 10 m (i.e., two adjacent points are 10 m apart) as in Sentinel 1-GRD and Sentinel 2 images, the greater scale (finer) admissible with pixels at that spatial resolution is 1:37,800, whereas 1:189,000 would be an appropriate geographical scale for pixels at 50 × 50 m (considering that 1 m = 3780 pixels). Such a comparison is important in considering mathematical relationships among map scale, raster data resolution, and map display resolution in digital mapping or GIS (Kimerling 2009). According to Tobler (1987), the finer cartographical scale would be 1:20,000 and 1:100,000 respectively, since a line cannot be drawn much narrower than about 0.5 mm.[4] From this, on a 1:20,000 scale paper map, the minimum distance which

[3] Figures 1 and 2 in Park et al. (2019, p. 590).

[4] There are other Minimal Mappable Units (MMU) proposed, which go smaller, as demonstrated by Rossiter (1999) and Rossiter et al. (2022).

can be represented (resolution) is about 10 m; on a 1:250,000 scale paper map, the resolution is 125 m.[5] A rule of thumb is to divide the denominator of the map scale by 1000 to get the detectable size in metres; the approximate resolution is half of this amount (Tobler 1987). Map scale refers to the amount of reduction of the real world to its graphic representation (usually a paper map). However, a map represented in a geographic information system (GIS) can be zoomed in to a square metre or less and zoomed out until the screen displays the entire study area and beyond. Consequently, geographic data in a GIS is referred to as map scale (cartographical) and display scale (graphical), and a gridded output based on satellite-derived raster pixels should not be smaller or have finer spatial resolution than the original gridded dataset (input). In the remote sensing case, pixels from airborne or spaceborne-based platforms (air-photo or satellite image, respectively) should not be smaller than the resolution of the camera used to record it. In a GIS, analysis is done at the resolution of the data, not at the display scale.

Geographical scale plays a central role in sampling (e.g., soil sampling), geostatistical analysis, and statistical models (Nikparvar and Thill 2021; Singh and Sarma 2023) and, consequently, in deriving estimates or outputs (results). Furthermore, findings on one scale may not hold on another (Singh and Sarma 2023). As scale also defines the spatial extent of observations, separate from pixel resolution, it is important for the sampling strategy in order to cover the heterogeneity of the phenomena being analysed in the study area. Regarding spatial resolution, its relationship with geographical scale is a function of graphical representations of the data (e.g., polygon, line, point, regular grid) (Nikparvar and Thill 2021: 7). In remote sensing data, when imagery is acquired from satellites or drones, the information conveyed in the images means that continuous space is represented in a regular or irregular grid (pixels). Thus, the scale of analysis should be determined before fitting any spatial models.

4 Obtaining Information with Remote Sensing

How do sensors measure energy (from different objects or targets on Earth's surface) to obtain information about soil, vegetation, water, etc.? How can sensors monitor biodiversity across the landscape? Addressing this question is twofold. Firstly, in **optical remote sensing** and for isotropic radiation (i.e. radiation that theoretically has the same intensity in all directions), reflectance corresponds to the ratio between the amount of energy that leaves a unit area on the ground (outgoing radiance L_{up}) to the total amount of incident energy received by that unit area (irradiance or incoming radiance L_{down})—often expressed as exitance (E) and irradiance (I), respectively (Knudby 2021) (Eq. 6.1).

[5] Raster resolution = [scale/(1000 × 2)], and Map Scale = Raster resolution (m) × 2 × 1000.

$$R = \frac{L_{up}}{L_{down}} = \frac{E_{up}}{I_{down}} \quad (6.1)$$

where R is the reflectance; L_{up} and L_{down} refer to outgoing radiance and incoming radiance, respectively; E_{up} and I_{down} mean outgoing exitance and incoming irradiance, respectively.

In the optical remote sensing context, the electromagnetic radiation that has been emitted by the Sun interacts with both the Earth's atmosphere and the surface (Fig. 6.6). For example, the portion of the radiation emitted by the Sun that reaches the top of the Earth's atmosphere passes through the atmosphere and is reflected by the Earth's surface, passes through the atmosphere again before it arrives at the sensor, where it is measured (Knudby 2021). Solar radiation detected by a passive or optical sensor passes through the atmosphere twice: downwards and upwards. Electromagnetic radiation that is not absorbed and passes through the Earth's atmosphere, through atmospheric windows, varies by wavelength and undergoes atmospheric scattering and atmospheric absorption (e.g., by ozone, carbon dioxide, and water vapour) effects—radiation in shorter wavelengths is more readily scattered than longer wavelengths. The interactions between the electromagnetic radiation field and the atmosphere influence the intensity of energy measured by the sensor at different wavelengths, resulting in a source of noise (Knudby 2021), due to atmospheric particles.

Radiance and reflectance are two important concepts in remote sensing. Radiance refers to incident energy or incoming radiant energy that leaves a radiation source in a given direction. The incident radiant energy that reaches an object can be decomposed into the sum of the reflected energy, the absorbed energy and the transmitted energy,[6] following the principle of conservation of energy. Reflectance means reflectance of radiation and it is a measurement of the percentage of incoming or incident energy that is reflected by a surface,[7] being a fixed characteristic of an object (Walsh 2003). Differences in surface properties (e.g., vegetation, soil, water, urban areas) allow the distinction of one object from another by comparing the reflectance of different objects at each wavelength (Walsh 2003), also known as the spectral signature. On the Earth's surface, the intensity of reflectance depends on the degree of smoothness and roughness of a surface—for example, icebergs and arable lands represent smoother and rougher surfaces, respectively.

As the satellite sensors measure radiance, it is important to convert the measured top-of-atmosphere (TOA) radiance to TOA reflectance—which is the total reflectance of the Earth-Atmosphere system—and, from this, the bottom atmosphere reflectance (BOA reflectance) or surface reflectance. The latter is of direct relevance

[6] Incident energy (E_I) = absorbed energy (E_A) + transmitted energy (E_T) + reflected energy (E_R) (Walsh 2003).

[7] Reflectance = reflected energy/incident energy or E_R/E_I. Incident energy is the amount of incoming radiant energy; reflected energy is the amount of energy bouncing off the object (Walsh 2003).

to most remote sensing applications in environmental sciences. This conversion, known as atmospheric correction (i.e. converting measurements from the Earth-Atmosphere system to only the Earth's surface), is an important area of research in optical remote sensing since the state of the atmosphere and the Earth's surface both influence TOA reflectance (Knudby 2021). Such a procedure is crucial when researchers need to extract biophysical or geophysical parameters (e.g. of soil, water, or vegetation for biomass estimation, plant health assessment) from remote sensing data (Albrecht et al. 2024), since atmospheric correction is a quantitative-physical property of objects/elements/targets on the Earth's surface in the satellite image. Moreover, it accounts for the attenuation caused by scattering and absorption in the atmosphere and allows researchers to use images captured by different sensors, at different times, and directly compare derived reflectance values in each pixel (Knudby 2021), as well as the spectral signatures measured on Earth's surface for different terrestrial elements.

> What this means is that an area can be monitored and in the absence of change the estimated surface reflectance (by band, i.e. the spectral signature) should not change—which in turn means that if the spectral signature is indeed seen to change through time, actual change is happening on the surface. Proper atmospheric correction is thus (with a few exceptions) the basis for a large field of applied remote sensing—detection of change. (Knudby 2021, Chapter 4)

Such a procedure is important so that each cell or pixel element contains a value with physical meaning. In general, remote sensing data providers provide satellite imagery where the atmospheric correction has already been applied. In a situation where raw data may be required, the parameters, coefficients and formulas are available in the metadata and guidebooks provided by the data provider. However, the information and explanations outlined here are important for remote sensing data users to be aware of the availability of these two datasets (TOA and BOA) and the difference between them when optical remote sensing data are employed.

Common applications of remote sensing data and techniques within the visible spectrum/visible light region (optical remote sensing) include the Hemeroby analysis of landscape (e.g. degree of human impact) and spectral indices derived for estimating vegetation, water, or soil properties. The Hemeroby index can be used to identify and analyse the magnitude of human influence on landscapes (Steinhardt et al. 1999; Walz and Stein 2014), such that the lowest values (ahemerob) correspond to natural or non-disturbed landscapes and habitats. This method is useful for studying the degree to which landscapes have been altered and for landscape monitoring. Spectral indices employ band maths or spectral maths to calculate an index or ratio based on specific bands. For example, the normalized difference vegetation index (NDVI) (Rouse et al. 2023) (Eq. 6.2) provides the basis for remote sensing of phenology (USGS 2018). It is a widely used metric for quantifying greenness, health and density of vegetation using sensor data, and is also useful for analysing plant biomass (Prabhakara et al. 2015; Farias et al. 2023). It is calculated as a ratio between the red (R), which vegetation absorbs, and near-infrared (NIR), which vegetation strongly reflects, as follows:

$$NDVI = (NIR - R) / (NIR + R) \tag{6.2}$$

Radio detection and ranging (radar) or **microwave remote sensing**, of which the most common type is Synthetic Aperture Radar (SAR), measures the ratio between the power of the pulse transmitted and that of the echo received (the backscatter), that is, the signal that the object on the Earth's surface redirects directly back towards to radar antenna. The radar range equation (Eq. 6.3) is a fundamental equation that provides a measure of the proportion of emitted energy that is returned from a target. The normalized radar cross section (σ^0), otherwise known as the backscattering coefficient, is of significant importance as it is a measure of the physical and electrical properties of the target on the Earth (e.g., tree, soil, water, rock, buildings) in a form of polarisation intensity of an electromagnetic wave.

$$E_r = \frac{\left(E_s \cdot G^2 \cdot \lambda^2 \cdot \left(\sigma^0 \cdot A\right)\right)}{(4\pi)^3 \cdot (c \cdot t)^4} \tag{6.3}$$

where E_r is the energy returned from a target on Earth's surface; E_s is the emitted energy; G^2 is the antenna gain; λ^2 is the wavelength; σ^0 is the normalized radar cross-section, also known as radar backscattering; A refers to the spatial resolution; $(4\pi)^3$ is the description of a sphere; c refers to the speed of light; and t means time.

There are two important concepts when dealing with radar, namely the radar polarisation and electrical characteristics (or dielectric properties) of materials or targets on the Earth, expressed as the dielectric constant. Since these are complex processes and require detailed description and explanation in terms of the underlying theory, this chapter just highlights the importance they play in calculating backscattering values and obtaining information from remote sensing. Readers who are keen to explore further can find relevant information in most remote sensing books or in the specified satellite handbook provided by space agencies, such as in Zyl and Kim (2010), Fletcher (2012), and Varotsos and Krapivin (2020).

Radar polarisation refers to the mode in which the radar signals are transmitted and/or received and different modes are better suited to different applications. Polarisation refers to waves that are either vertically or horizontally orientated. Depending on how the sensor is designed in terms of polarisation configuration, it can have a Vertical Transmit—Horizontal Receive Polarisation (VH) or Vertical Transmit—Vertical Receive Polarisation (VV), for example. Hence, radar polarisation refers to like-polarised waves (HH or VV) and cross-or dual-polarised waves (HV or VH). The former is also named parallel polarisation or co-polarisation and means that the scatterers reflect the wave in the same polarisation. Conversely, in cross-polarised mode, the signal will be measured in a different plane as the transmitted and received polarisations are orthogonal (right angles) to one another. Radar systems can have one, two or all four of these transmit/receive polarisation combinations. Thus, radar polarisation carries the information on the backscattering values. Its importance relies on the fact that typically, the intensity of the backscattered signals in parallel polarisation is greater (usually stronger) than in cross-polarisation (Sano et al. 2020. Translated by Sandra Deodoro; Ibrahim et al. 2023), which is usually weak. However, cross-polarisation is useful when volume or multiple scattering occurs. Thus, it can allow the separation of different targets, for example, soil and vegetation, forest and non-forest, through a method called polarimetric decomposition. This approach is useful for classifying images or landscapes in terms of land use, and land cover, as well as for estimating surface physical properties, such as soil moisture, soil texture and biomass (Deodoro et al. 2024). Another useful application of radar polarisation is for calculating the radar vegetation index (RVI), the radar counterpart to NDVI. It was originally proposed for quad-polarisation by Kim and van Zyl (2009) and was later modified for dual-pol SAR data by Trudel et al. (2012) and specifically for using Sentinel-1 dual-pol data (VV-VH) by Nasirzadehdizaji et al. (2019), Gururaj et al. (2019) and Mandal et al. (2020).

The dielectric constant (ε), also called permittivity, is a measure of an object's electrical character, a parameter that indicates the reflectivity and conductivity of a material. It refers to a measure of the electrical potential energy of a substance under the influence of an electromagnetic field (Babaeian et al. 2019). Most natural materials have, in the microwave spectrum, a dielectric constant in the range of 3–8 when dry, whereas water has a dielectric constant of approximately 80 (ESA 2007). As a consequence, soil or vegetation with moisture will display greater reflectivity in SAR images.

Consideration of the factors that influence the radar backscattering coefficient is important when analysing radar data. These factors include (i) radar configuration—wave polarisation (i.e. horizontal, vertical), incidence angle and frequency or wavelength (bands); and, (ii) physical characteristics of the terrain or target surface—landforms/topography, roughness, vegetation cover, soil texture, moisture (vegetation and soil) and the dielectric constant of vegetation and soil (Das and Paul 2015; Liu et al. 2019). It is also important to consider the particular application and ensure that an appropriate wavelength is employed to address the stated research goals and that meaningful information can be obtained from the data (this is also valid when employing optical remote sensing). This is due to the fact that longer wavelengths have the ability for higher penetration within the vegetation volume.[8] For example, the P-band SAR sensor has a long wavelength (30–100 cm) (Table 6.2) and can penetrate tree cover to reach the underlying soil volume (Srivastava et al. 2006). While the penetration capability of SAR depends on the wavelength, plant parameters such as plant moisture, plant height and plant volume are also important factors (Srivastava et al. 2006).

While SAR requires a higher level of expertise, compared to optical data, it offers several advantages. It is relatively unaffected by illumination (i.e., it can image at night) or clouds and L-band sensors are minimally affected by vegetation (Beale et

[8] For a detailed illustration on this process, readers can refer to the Figure 2 in Schumann and Moller (2015). https://doi.org/10.1016/j.pce.2015.05.002.

Table 6.2 Approximate microwave measurement depths for bare soil and different land surface covers

Land surface cover	Microwave bands			
	X-band (2.5–3.75 cm) (8–12 GHz)	C-band (3.75–7.5 cm) (4–8 GHz)	L-band (15–30 cm) (1–2 GHz)	P-band (30–100 cm) (0.3–1 GHz)
Bare soil	1.25–1.87 cm	1.87–3.75 cm	7.5–15 cm	15–50 cm
Agriculture and rangeland	0.5–0.75 cm	0.75–1.5 cm	3–6 cm	6–20 cm
Forest	0.25–0.37 cm	0.37–0.75 cm	1.5–3 cm	3–10 cm

Source: Babaeian et al. (2019)

al. 2019). Such characteristics are explained by the fact that the wavelengths of the spectral bands of these sensors are much longer than the average diameter of the particles and molecules found in the atmosphere (Sano et al. 2020. Translated by Sandra Deodoro) and therefore not subject to the level of attenuation or scattering that optical sensors experience. The longer the wavelength (and the lower the frequency), the deeper the penetration into targets, particularly in forests or tall vegetation. Moreover, longer microwave wavelengths ($\lambda > 3.8$ cm) have some important advantages over optical remote sensing for agricultural applications (Moran et al. 2012). For example, SAR sensors have demonstrated their potential for the effective measurement and monitoring of surface soil characteristics at a high spatial resolution (Gorrab et al. 2015; Deodoro et al. 2023, 2024). Babaeian et al. (2019) provide a summary of the measurement depths of microwave bands (C-band, L-band, and P-band) for various land surface covers (Table 6.2), which shows the penetration capability of short- and long wavelengths SAR to reach the topsoil layer under different soil covering conditions—grass and forest.

It is important to consider the interaction between the geometry of the target and the wavelength in the microwave region of the electromagnetic spectrum (Liu et al. 2019). In other words, the size of the object influences the interaction between the object and the microwave radiation, that is, it indicates whether the object interacts with the incoming radiation. The wavelengths tend to interact more with objects that are similar in size to the wavelength, due to Rayleigh's Criterion (Bole et al. 2014; UIUC 2010; Harvard n.d.). This is a key aspect with regard to the size of the wavelength and the scatters, as longer wavelengths, such as L or P bands, interact less with smaller components (UIUC 2010). For example, leaves and small twigs have less impact on the microwaves and allow them to pass through the canopy.

Despite being less reliant on weather, atmospheric, and meteorological conditions for sensor operation and image acquisition compared to optical spectrum-based remote sensors, microwave remote sensing data is more complex and data interpretation and analysis is less straightforward. Factors that influence the return radar signal (i.e., echo or backscattering) need to be explicitly considered. For example, the backscattering intensity is dependent on interactions between the electrical characteristics of different surfaces, terrain features (e.g., roughness,

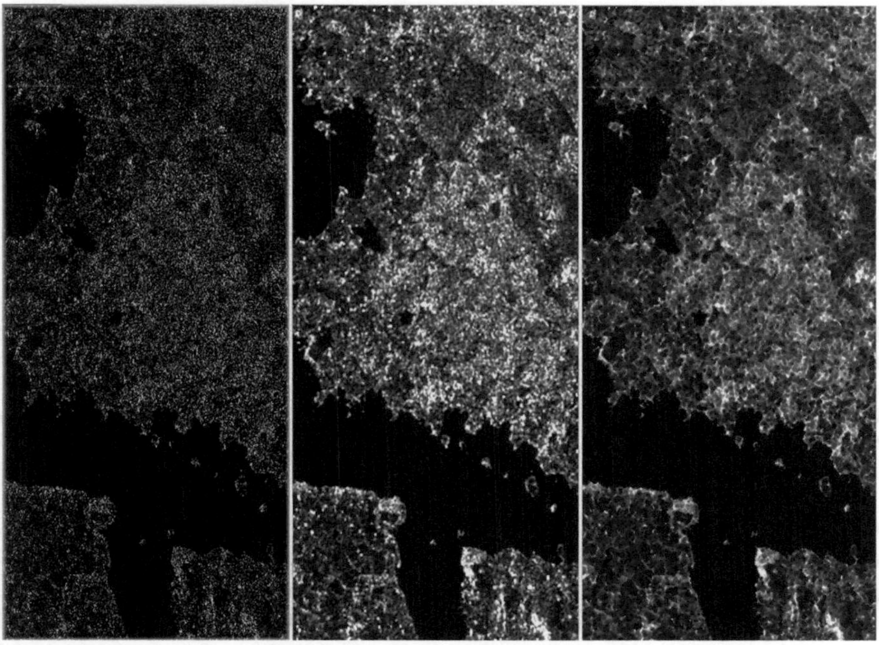

Fig. 6.7 An example of two different speckle filter applied using two different window sizes tested for an area in Ireland. Original image (on left), Box Car method at 5 × 5 (centre), Refined Lee at 7 × 7 (on right). (Source of satellite imagery: Sentinel 1)

slope) and geometric characteristics of targets (Liu et al. 2019; Beale et al. 2019), and thus additional factors need to be considered.

As with atmospheric correction in optical remote sensing, an important pre-processing step to obtain the radar backscattering is the correction of inherent speckle in radar images (i.e. a grainy "salt and pepper" texture in an image) using speckle filter methods and window sizes (i.e. to scan the pixels). Since they depend upon the area under investigation (e.g. size and physical characteristics), they are deemed as empirical methods and, as such, different filter methods and window sizes must be tested (Fig. 6.7) as they yield different image improvements and resulting values.

5 Earth Observation Platforms

Earth observation (EO) is a widely used collective term for platforms that measure reflected or emitted electromagnetic radiation, which comprises satellite or spaceborne platforms, aerial remote sensors (e. g., manned: aircrafts; unmanned: drones), and proximal sensors (proximal sensors such as Ground Penetrate Radar-GPR). However, in the broadest sense, it also includes any devices or instruments

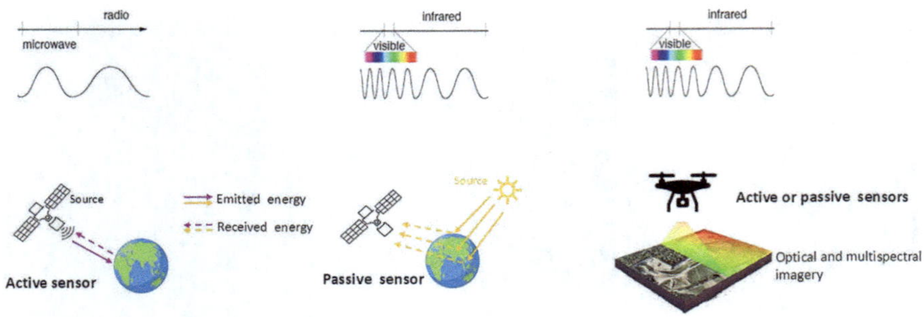

Fig. 6.8 A general overview of different types of remote sensing platforms commonly employed for biomass estimation, with an example of two remote sensing instruments (satellite and drone) in the context of the electromagnetic spectrum. (Image of Earth obtained from Freepik (https://www.freepik.com). Drone picture retrieved from Microsoft Windows Stock Images. LiDAR imagery (optical and multispectral imagery) retrieved from Wingtra (https://wingtra.com/))

that measure or monitor the Earth's surface, sub-surface or atmosphere (e.g. meteorological, hydrological etc). Reflecting the volume of EO data obtained by EO platforms, environmental scientists have been at the forefront of big data analysis, required to make sense of the data. Due to the variety of platforms, instruments and sensors, here we provide a very brief overview of satellite and unmanned aerial vehicle (UAV) or drone platforms that have applications in biomass estimation (Fig. 6.8).

5.1 Satellites Based Platforms

> If you really want a long history of understanding any process on Earth, Landsat is where you should go. (Tasso Azevedo, MapBiomas Initiative)

Scientists and researchers have been observing the Earth through the lens of satellites and collecting spectral information from its surface since the early 1970s, thanks in large part to the Landsat program. The program consists of a series of Earth-observing satellite missions, jointly managed by NASA and the United States Geological Survey (USGS) that has produced a unique satellite data archive enabling scientists to assess long-term changes on the Earth's surface. Landsat 8, launched in 2013, and Landsat 9, launched in 2021 continue to contribute to this program ensuring the continuity of this long-term satellite-based record. In addition to the Landsat Program, numerous countries have developed and deployed satellite platforms for EO. For example, the Japanese Aerospace Exploration Agency has undertaken a number of missions for monitoring the Earth and atmosphere. In Europe, the European Space Agency (ESA) has developed the Sentinel missions, of which Sentinel-2, an optical-based platform, is widely employed. The Sentinel program, which includes radar (e.g. S-1—C band) and optical platforms, was designed to support the needs of the Copernicus program, a core component of the European Union Space Programme. A comprehensive list of current and future satellite missions and their field of application is maintained by the World Meteorological Organisation (WMO) and is available at https://space.oscar.wmo.int/satelliteprogrammes.

5.2 *Unmanned Aerial Vehicle: Drones*

EO satellite orbits have a mean altitude between 700 and 800 km, whereas drones fly very close to the ground at a few dozen or at least a few hundred metres. As a result, they capture details that low-spatial resolution satellites (e.g. more than 10 m) cannot. Drones navigate using a global positioning system (GPS) and have a degree of autonomy governed by software-controlled flight plans and are remotely operated as robotic drones. Depending on payload, they can carry different types of sensors: optical cameras, optical gas imaging, multispectral, hyperspectral, thermal, and laser scanners/LiDAR. Typically, for environmental applications (e.g., forest and crop monitoring, and biomass estimation), drones can readily obtain a variety of high-resolution data, including red-blue-green (RGB), multi- or hyper-spectral (i.e., more detailed or narrowed wavelengths or continuous and contiguous ranges of wavelengths, for example, 400–2200 nm in steps of 1 nm), dense 3-dimensional data clouds using Light Detection and Ranging (LiDAR) or thermal data.

UAVs can be classified according to different characteristics, such as rotors, wings, power source, controllers, hardware, navigation (e.g., GPS/GNSS, IMU, and altimeter), mountings, and sensors. The most common drone is the Vertical Take-Off and Landing type (VTOL), that is, it can hover, take off and land vertically, and can include types of UAV such as fixed-wing, helicopters and other UAVs with powered rotors.

A drawback in the use of UAVs is that they are not permitted to fly anywhere as there may be local restrictions (e.g., near strategic geofences or military areas, airports) and their operation can be highly regulated, depending on the country of operation. Moreover, the operation of UAVs requires a license in many countries. Challenges related to flight time and avoiding obstacles—which require human intervention—may also emerge. Table 6.3 summarises the main differences between satellites and UAVs; nevertheless, they complement each other.

6 Above-Ground Biomass (Plant Biomass) Estimation from Remote Sensing Data

Aboveground biomass (AGB) refers to the total amount of plant-derived living and dead organic matter per unit of surface area (Zhang et al. 2023), thus, is an

Table 6.3 Comparative characteristics between satellites and drones

Characteristic	Satellite	Drone
Imagery	From the "space"	From the "sky"
Image resolution	Reliant upon spatial resolution	High
Width of field (scalability)	Large areas; global coverage	Small areas; local coverage
Angle	Fixed angle	Different angles
Temporal resolution	Frequent; as per a passing-over timeframe	On-demand
Time series archive	Possible	Possible, however, time-consuming. A repeatedly setting up of equipment is needed
Meteorological sensitivity (issue)	Clouds (optical remote sensing)	Rain and strong wind
Accessibility	No limited by terrain conditions	Some areas like mountains or forests are an issue as there is a risk of losing equipment[a]
Cost-effectiveness	Launch and maintenance are expensive; the cost per unit of data collected is relatively lower when compared to ground-based sensors[b]	The costs are higher if the fields are vast. The retrieved images need GIS specialists' interpretations[a]
Data availability	Free of charge by governmental space agencies	Data are not freely available
Autonomy	Operates on its own once reaches its orbit, transmitting data to the ground station[a]	Requires operators to fly
Legal restrictions	No subject to regulation due to territorial restrictions. "The probability of legal restrictions on field images on data-driven platforms is very low"[a]	Subject to regulation before to setting up a drone and during the fly

[a]EOS Data Analytics (https://eos.com/blog/drones-vs-satellites/)
[b]FlyPix AI-Geospatial AI Platform

important component of the terrestrial carbon cycle. It serves as useful information in a number of applications such as crop yield monitoring, grass management, pasture forage, and forest timber production (Zhang et al. 2023). Angerer et al. (2023) highlight that satellite imagery at a spatial resolution of 10 m and even 30 m, such as Sentinel and Landsat respectively, offers opportunities to estimate AGB on rangelands and pastures at scales practical for evaluating livestock stocking rates. In forest ecosystems, remote-sensing techniques have played an important role in AGB estimation at different scales, over the past five decades (Tian et al. 2023). Moreover, when AGB is modelled using remote sensing data and integrated with Geographic Information Systems (GIS), spatially comprehensive outputs can be produced (Angerer et al. 2023). That is, not only AGB estimates and associated accuracies (table outputs) but also their geographic location (map outputs).

Typically, AGB estimation using remote sensing data is calculated using a number of different approaches, including: (i) empirical models via statistical relationship (using linear or nonlinear regression) between ground measurements (locations, weather etc) and pixels from spectral bands (or some combination of bands) of remote sensing images; (ii) process models that use remote sensing variables as inputs to spatially and temporally predict the estimates or where remote sensing information is assimilated into the model to constrain the model estimated values; and (iii), machine learning models that combine field-collected, remote sensing and other explanatory data (e.g., terrain, climate, etc.) to estimate biomass (Angerer et al. 2023).

6.1 Application over Forested Landscapes

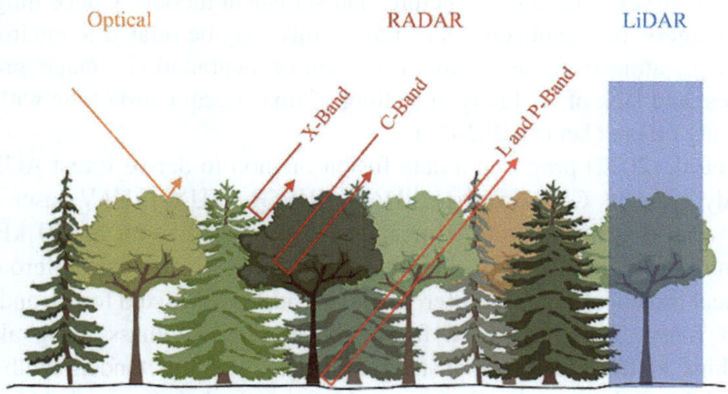

Diagram of multi-source remote sensing instrument to detect forest parameters such as AGB. (Source: Tian et al. (2023). CC BY 4.0. https://doi.org/10.3390/f14061086)

Forest aboveground biomass (AGB), as a component of total biomass, refers to the total mass of living organisms located above the ground, including trees, shrubs, herbaceous plants, and other biota (Ma et al. 2024). The measurement of AGB can be obtained with remote sensing techniques by employing ancillary parameters that serve as a proxy for AGB, such as sensor-derived surface reflectance and backscattering, remote sensing-derived texture information (e.g. Sarker and Nichol 2011; Kelsey and Neff 2014; Tian et al. 2023), and vegetation indices. These include NDVI, difference vegetation index (DVI), enhanced vegetation index (EVI), leaf area index (LAI), forest coverage, and forest structure parameters (e.g., tree height, diameter at breast height (DBH), canopy height.). These vegetation indices are commonly used as covariates considering that live green vegetation absorbs solar radiation in the red wavelengths for photosynthesis (Tian et al. 2023). In a sparse forest covers context, other vegetation indices (VI) are used to minimize interference from the atmosphere and soil background, such as the orthogonal transform-based perpendicular vegetation index (PVI), soil-adjusted VI (SAVI), and modified SAVI (MSAVI) (Fatehi et al. 2015; Sadeghi et al. 2018; Tian et al. 2023). Typically, optical remote sensors (multispectral and hyperspectral) are employed for forest AGB, however, microwave and LiDAR remote sensing have also been widely used for this application, since tree height can be also obtained from SAR (e.g. interferometry) and LiDAR data (Tian et al. 2023).

At large scales, estimating AGB using remote sensing combined with ground measurements is based on the relationship between the power law allometric equation between AGB and tree height, despite being derived at the plot scale, it remains applicable at large scale (e.g. Wu et al. 2015; Tian et al. 2023). While there is an advantage to using LAI-based estimation approaches, a preliminary evaluation of the relationship between leaf biomass and AGB is required so that AGB can be extrapolated (Zhang and Kondragunta 2006; Tian et al. 2023).

Tian et al. (2023) point out possible sources of uncertainty in forest AGB estimates (Table 6.4) which include those associated with remote sensing imagery, sample plot survey data, stand structure, and statistical models. Concerning remote sensing imagery, the main sources of uncertainty may be related to environmental factors (e.g., atmosphere and humidity), sensor degradation, image processing techniques, and lack of, or faulty, matching of image acquisition time with sample sites, among others (Tian et al. 2023).

Yang et al. (2022) proposed a data fusion method to derive forest AGB across nine study sites in China, using LiDAR (RIEGL VUX-1 UAV laser scanner, operated at between 200 to 300 m above ground level with a 400 kHz pulse repetition frequency) and Sentinel 2 images (level 1-C), covering coniferous forest, sub-tropical broadleaf forest, coniferous and broad-leaved mixed forest, and tropical broadleaf forest. They estimated forest aboveground biomass using allometric relationships at the plot level, instead of single trees, in tandem with LiDAR canopy height attributes and optical spectral indexes. In their approach,[9] four canopy height-related metrics and four spectral metrics were derived from the UAV and satellite data, and the LiDAR point clouds and optical images were integrated using allometric relationships (Eq. 6.4) (Yang et al. 2022):

$$\mathrm{AGB} = \left(a\ \mathrm{TH}^{b}\right)^{\mathrm{VI}} \tag{6.4}$$

where TH is a canopy height metric in each plot; VI is the vegetation index in each plot; and *a* and *b* are statistical parameters to be estimated from the data. The original formula of the proposed biomass is shown in Eq. (6.4). The authors employed a

[9] The workflow of the proposed methodology for estimating aboveground biomass (AGB) and the performance of the biomass models (i.e. measured and predicted AGB for all plots and different forest types) are found in Yang et al. (2022)—Figures 2 and 5, respectively.

Table 6.4 A summary of limitations and advantages of the main methods used for estimating forest aboveground biomass (AGB)

Methods	Descriptions	Limitations	Advantages
Empirical modelling	Parametric models Includes: linear regression (LR), multiple regression (MR), and non-linear regression models.	Assumptions about the distribution of the data must be met. Difficult to increase the scale and to apply to the whole area. Risk of over-fitting, which can be minimised with cross-validation.	Description of the process through equations and functions. Simple and intuitive empirical formulas. Easy to understand and interpret the results
	Non-parametric models Includes: k-nearest neighbours (KNN), artificial neural networks (ANN), support vector machines (SVM), random forests (RF), gradient boosting (GB), maximum entropy (ME), and deep learning (DL).	Requires highly accurate training data. Generally slow training process. Risk of over-fitting, which can be minimised with cross-validation. Complexity Low interpretability.	No empirically assumed distribution of the data Direct sample analyses High prediction accuracy.
Physical modelling	Includes: radiative transfer model and geometric optical model.	Complicated calculation process Only available for small areas	Clear physical meaning Good model stability and applicability.
Mechanistic modelling	Includes: climate-related models, physiological–ecological process models, and light-use efficiency models.	Extremely complex mechanisms Requires numerous parameters and are not easily accessible by the user.	Clear physical meaning High prediction accuracy.
Comprehensive modelling	Includes: FAREAST, LANDIS/LANDIS-II, FVS, and SORTIE-ND model	Requires numerous parameters and are not easily accessible Requires a large amount of information on tree species	Logical mechanism, flexible structure, and a variety of forms.

Source: Adapted from Tian et al. (2023)
Notes: *ANN* artificial neural networks, *GB* gradient boosting, *KNN* k-nearest neighbour, *LR* linear regression, *ME* maximum entropy, *MR* multiple regression, *RF* random forest, *SVM* support vector machine

modified approach by linearising the orginal formula and adopting the vegetation index ($NDVI_{mean}$) as the power of TH_{mean}, as they considered that canopy height is easier to obtain than the diameter at breast height (DBH) and the vegetation index can reflect the heterogeneity of the canopy and the metabolic rate of the species (Yang et al. 2022).

They collected 1370 field plots (20 × 20 m) measurements of DBH and tree height from 2012 to 2019. The AGB of each tree was estimated using a species-specific allometric equation according to Zhou et al. (2018) and Luo et al. (2020), and the total AGB of each plot was estimated by adding up the AGB of all individual trees within it and was used as the ground truth in the following analysis (Yang et al. 2022). Yang et al. (2022) found that the proposed biomass modelling method (i) achieved good prediction accuracy in all plots and different forest types; (ii) achieved a better performance than models solely based on tree height metrics, with an R^2 between predicted and measured AGB higher by about 10% and the RMSE decreased by about 22% in four forest types; (iii) remains robust even when using a small training sample size; and (iv), the proposed method also achieved a comparable or even better performance than advanced machine learning models such as RF, ANN, and SMR. The authors concluded that their method is expected to be examined and applied at larger scales for improved high-precision estimation of the forest biomass.

In a SAR-based approach, Mukhopadhyay et al. (2021) employed the integrated use of polarimetric (PolSAR) and interferometric (PolInSAR) imagery analysis for AGB mapping in India, using random forests (RF) and multiple linear regression (MLR) models. The datasets employed represented an image pair of fully polarimetric Radarsat-2 C-band and the related field data (systematic sampling technique). From their experimental results, they found that the RF algorithm provided a better prediction result of AGB (R^2 and RMSE around 0.65 and 24.33 Mg/ha, respectively) than the MLR model (0.54 and 33.05 Mg/ha, respectively).

Notwithstanding the success of remotely estimating forest AGB, variables such as canopy structure, tree species, and other environmental factors should be considered when developing biomass estimation algorithms or models (Tian et al. 2023) since electromagnetic energy is prone to atmospheric and environmental effects.

6.2 Application over Managed Landscapes: Grassland and Crop Land

Grasslands are the most widespread global land cover types. After forests, they are the second largest carbon sink and thus play a crucial role in regulating the global carbon cycle. Furthermore, the global demand for food derived from grassland and grassland-based agriculture is increasing (Ali et al. 2017), which places increasing pressure on food security and climate change.

Current approaches to estimating above-ground biomass (AGB) over grasslands can be divided into ground-based and remote sensing methods. The former can be either destructive (e.g. obtain grass cuttings that are then dried and weighed in the laboratory) or non-destructive (e.g. rising plate meter) (Bazzo et al. 2023); irrespective, they are designed for obtaining information from relatively small or defined geographical areas (e.g. at parcel, paddock or field scales) and represent point samples. Similar to forest AGB, grassland biomass can also be estimated using remote sensing data and techniques. For example, Liu et al. (2023) highlighted some common data derived from remote sensing techniques as inputs of machine learning models for aboveground biomass estimation of grassland and shrubland (e.g., spectral bands; SAR bands; topography variables such as aspect, elevation, slope; vegetation index; and environmental variables such as water vapour pressure, precipitation, and land surface temperature-LST). Amongst them, the vegetation index is important as it can capture plant phenology (USGS 2018). Moreover, Liu et al. (2023) observed that the vegetation indices (Table 6.5) were the most important for grassland AGB estimation, among all input variables in the random forest model. However, conventional ground-based methods are typically time-consuming and are only representative at small spatial scales (e.g., paddock/field and farm scales).

For example, in Ireland, grassland comprised almost 62% of the total cover (4.4 million hectares) of the national's terrestrial and transitional ecosystems in 2018 (CSO 2024) and is a major feed source for the pasture-based dairy farming and livestock industry; thus, grasslands are typically intensively managed systems (Ali et al. 2017). Grass pasture is typically grazed by cattle on a rotation basis and/or harvested 2–3 times per year to provide a source of overwinter feed and both management practices maintain a minimum grass height of 4 cm (O'Donovan and McEvoy n.d.). Ali et al. (2017) developed a variety of approaches including multiple linear regression (MLR), artificial neural network (ANN) and adaptive neuro-fuzzy inference system (ANFIS) models to estimate the grassland biomass (kg dry matter/ha/day) of two intensively managed grassland farms in Ireland. In situ measurements consisted of weekly measured biomass, which were used for model development. They employed a time series (46 images per year) of MODIS Terra surface reflectance 8-day composite (250 and 500 m) from which five vegetation indices and two raw spectral bands (RED = red band, NIR = Near Infrared band) were used as an input for all three models. Ali et al. (2017) observed that the ANFIS model produced an improved estimation of biomass as compared to the ANN and MLR, with R^2 values equal to 0.85 and 0.76, and RMSE to 11.07 and 15.35 for the two study sites, respectively.

In a different environmental scenario, with grass and shrub interspersed over complex topography and fragmented landscapes, Liu et al. (2023) integrated remote sensing data (Sentinel-1 SAR and multispectral Sentinel-2), environmental and topographical datasets to estimate AGB of mountain grassland in Southwest China. To account for terrain conditions, topography variables were also obtained from radar data—a digital elevation model (DEM) with 12.5 m spatial resolution derived from the ALOS L-band SAR. Elevation, slope and aspect were calculated for each pixel. Other ancillary variables included vegetation indices, environmental data (land surface temperature with 1 km spatial resolution; a downscaled precipitation product with 1 km spatial resolution, and water vapour pressure) and landcover types data with 10 m spatial resolution from which the grassland and shrubland were extracted. Field data, collected during the growing season of 2021 in order to match the ground samples with the satellite data, were based on five sample plots of 1 × 1 m at each sample point for grassland and one sample plot of 5 × 5 m for shrubland with an interval distance of over 1 km between the plots (Liu et al. 2023). They obtained a total of 519 samples including 298 samples for pure grassland and 221 samples for shrubland. They used traditional regression, machine learning (ML) and deep learning (DL) models (multiple stepwise regression-MSR, random forest-RF, support vector machine-SVM and convolutional neural network-CNN). Concerning the models, Liu et al. (2023) observed from the results that the RF model performed the best in the test dataset with an R^2 equal to 0.80 for shrubland and 0.75 for grassland. Regarding the input variables in the RF model, the vegetation indices (EVI, EVI2, NDVI, NIRv, MSR and DVI) (Table 6.5) played the most important role in grassland AGB estimation. They concluded that integrating SAR into grassland AGB estimation was useful to improve the accuracy of modelled mountain grassland AGB in their study area.

Remote sensing data and techniques have also been useful for monitoring cover crop biomass throughout crop cycles. These techniques are increasingly being employed in the context of precision agriculture, integrating multiple sensors to support farmers with informed decision-making.

Breunig et al. (2024) estimated the AGB of rye (*Secale cereale L.*) employing PlanetScope (PS) optical, Sentinel-1 SAR, and hybrid (optical plus SAR) datasets in southern Brazil (Fig. 6.9). They derived optical attributes (surface reflectance from PS's blue, green, red, and near-infrared-NIR bands, NDVI and EVI), radar attributes from Sentinel-1 (intensities from the ground range detected-GRD product; backscattering from VV and HH polarizations and both ratio and polarization indices, from the single look complex product-SLC). As in-situ AGB data for rye (*Secale cereal L.*), they collected 50 samples systematically distributed across every half-hectare in a farm at four dates, aligned with satellite image acquisition dates. Meteorological variables (temperature and rainfall) were used as ancillary data to provide analysis during the experiment, such as assessment of the relationships between AGB measurements and meteorological data, analysis of rye growing cycle by comparing the changes in measured cover crop biomass with local variations in precipitation and temperature (Breunig et al. 2024). Tri-dimensional perspectives of UAV RGB dense-cloud and the respective derived DEM were also employed for the early and late growing seasons to represent the landscape condition and complement the analysis. For AGB statistical modelling, multiple linear regression models were used and repeated 100 times for both individual (optical PS or Sentinel-1 SAR) and combined datasets (optical plus SAR) and variable importance was extracted. As a result, they observed that optical vegetation indices displayed higher correlation coefficients (r) for AGB estimation (r = +0.67 for both EVI and NDVI) compared to SAR attributes like VV, Ratio, and polarisation (r ranging from −0.52 to −0.58). With regards to model performance, they found that rye AGB estimates

Table 6.5 Summary of vegetation index equations found as most important for grassland AGB estimation

Vegetation index	Equation (general)	Equation (for Sentinel 2)	Parameters description
EVI	$G \times (NIR - Red)/(NIR + C1 \times Red - C2 \times Blue + L)$	$2.5 \times (NIR - Red)/(NIR + 6 \times Red - 7.5 \times Blue + 1)$	G = gain factor; L = value to adjust for the canopy background; C = coefficients for atmospheric resistance
EVI2	$G \times (NIR - Red)/(NIR + C1 \times Red + L)$	$2.5 \times (NIR - Red)/(NIR + 2.4 \times Red + 1)$	
NDVI	$(NIR - Red)/(NIR + Red)$	$(NIR - Red)/(NIR + Red)$	–
NIRv	$NDVI \times NIR$	$NDVI \times NIR$	–
MSR	$(\mathrm{NIR/Red} - 1)/(\sqrt{}(\mathrm{NIR/Red} + 1))$	$(NIR/Red - 1)/\left(\sqrt{NIR/Red + 1}\right)$	–
DVI	$NIR - Red$	$NIR - Red$	–

Source: Adapted from Liu et al. (2023)

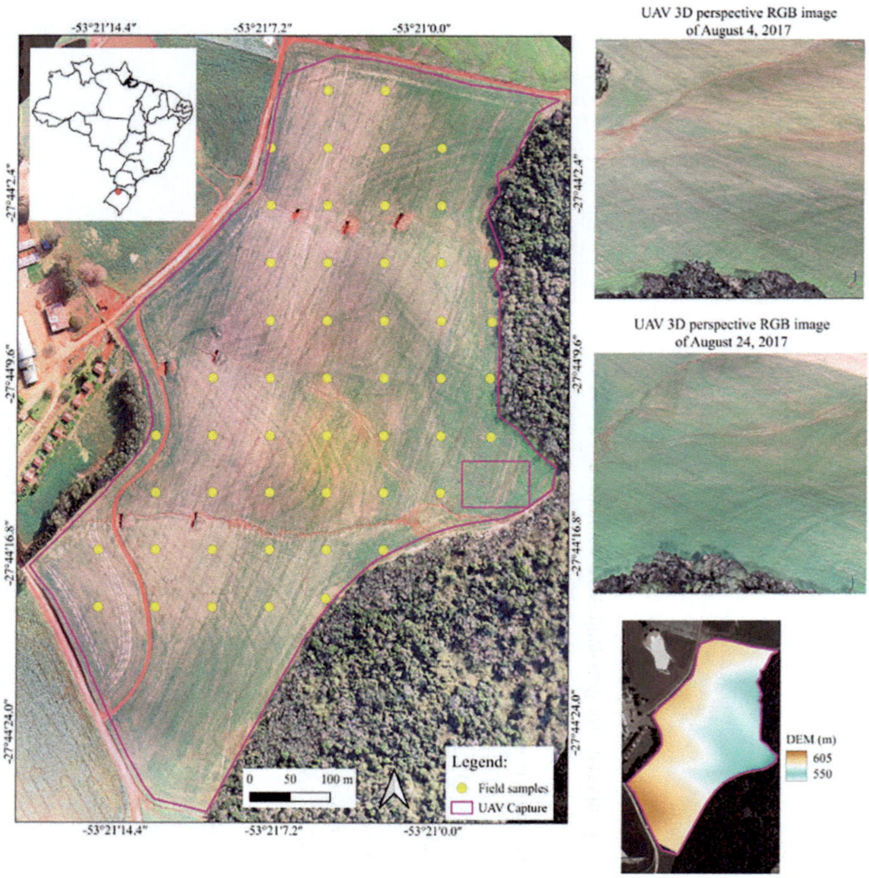

Fig. 6.9 Area cultivated with rye, in southern Brazil, where AGB biomass was estimated from remote sensing. (Source: Breunig et al. (2024). CC BY 4.0. https://doi.org/10.3390/rs16152686)

were improved with integrated optical and SAR data (hybrid model) ($R^2 = 0.62$, $p < 0.01$), reducing RMSE to 579 $kg{\cdot}ha^{-1}$, and the most important predictors were VV, NIR, blue, and EVI in this model approach. Regarding the VV predictor (i.e. SAR-based variable), Mercier et al. (2019) also found it as an important contributor in estimating rapeseed biomass. These findings could be linked to the vertical structure of rye, which enhances radar backscattering (vertical transmitted-vertical received or VV) as the growing cycle progresses and vegetation density increases (Breunig et al. 2024). These authors also point out that the prevalence of VV backscattering during the growth cycle could reduce optical saturation.

Breunig et al. (2024) also noted spatial variations in the predicted rye AGB identifying zones associated with varying biomass productivity during the crop cycle. They relate such findings to topography conditions and phenology captured by data time series, as greater amounts of rye AGB were observed at low-elevation areas of the farm compared to high-elevation terrains (Breunig et al. 2024). They also observed that just 65 days after planting, the biomass map already highlighted areas with lower biomass content in the central and western regions, ultimately yielding less than those in the eastern areas (Breunig et al. 2024). As a conclusion, they highlight that conducting such analyses immediately after plantation can guide farmers in land management decisions.

In a similar approach but employing Landsat-8 and Sentinel-2 data from the harmonized Landsat 8 and Sentinel-2 (HLS), Dong et al. (2020) investigated the potential of these datasets for crop biomass estimation for six crops in Manitoba, Canada. In their study, crop biomass was estimated using remotely sensed leaf area index (LAI) to reparametrize a simple crop growth model using a generic relationship between both vegetation and water indices (modified simple ratio red-edge, red-edge NDVI, Normalized Difference Water Index, NDWI1). Their findings indicated that the above-ground dry biomass of the crops analysed was more accurately estimated from the assimilation of LAI derived from the combined satellites ($R^2 = 0.81$, RMSE = 135.4 g/m^2), than that of LAI derived from Sentinel-2 data only ($R^2 = 0.80$, RMSE = 136.7 g/m^2) or Landsat-data only ($R^2 = 0.68$, RMSE = 191.0 g/m^2). Dong et al. (2020) also noted that the improved temporal coverage associated with Sentinel-2 and the availability of a red-edge band on this sensor may explain accurate estimates from that satellite for both crop growth stage and leaf dry biomass estimates.

7 Conclusion

This chapter presented a general background of the use of remote sensing data and techniques commonly employed for estimating above-ground biomass, in unmeasured sites, under different types of landscapes. When there is a large area to be monitored in a specific timespan (e.g. each month or year), spaceborne sensors are useful to capture and provide information in that spatial extent and period. If the impact of field logistics, such as unpredictable weather, suitable airstrip locations, and airspace restrictions (Neville 2024) is minimal, traditional airborne surveys can be a better choice, especially for local or regional areas (i.e. at a finer spatial resolution).

Using remote sensing enables the evaluation of vegetation change, and biomass over time, in order to map and monitor (altered) landscapes. This scientific capability is important for making data-informed policy recommendations and planning regarding the trade-off between land management and biodiversity, The value of using spaceborne platforms, which are being developed and deployed by commercial providers, offering increasingly finer spatial, spectral and temporal resolution can also be highlighted due to the oftentimes costlier and higher risks associated with ground-based studies, especially over large, remote areas.

In light of the current climate change projections—the increase in global air temperature, sea level rise, and extreme events (e.g., drought and flooding)—and its effects on biodiversity, given the direct relationship between biomass and carbon content, remote sensing-based approaches are playing, and will continue to do so, an increasingly important role in forest AGB estimation and carbon cycle studies (Tian et al. 2023).

References

Albrecht F, Walawender J, Belgiu M, Lang S (2024) Atmospheric correction. Living textbook. University of Twente. Available at https://ltb.itc.utwente.nl/page/671/concept/124371. Accessed 24 July 2024

Ali I, Cawkwell F, Dwyer E, Green S (2017) Modelling managed grassland biomass estimation by using multitemporal remote sensing data – a machine learning approach. IEEE J Select Topics Appl Earth Observ Remote Sens 10(7):3254–3264. https://doi.org/10.1109/JSTARS.2016.2561618

Angerer JP, Fox WF, Wolfe JE, Tolleson DR, Owen T (2023) Chapter 20 – Land degradation in rangeland ecosystems. In: Sivanpillai R, Shroder JF (eds) Hazards and disasters series, biological and environmental hazards, risks, and disasters, 2nd edn. Elsevier, pp 395–434. https://doi.org/10.1016/B978-0-12-820509-9.00007-1

Atkinson PM (2013) Downscaling in remote sensing. Int J Appl Earth Obs Geoinf 22:106–114. https://doi.org/10.1016/j.jag.2012.04.012

Babaeian E, Sadeghi M, Jones SB, Montzka C, Vereecken H, Tuller M (2019) Ground, proximal, and satellite remote sensing of soil moisture. Rev Geophys 57:530–616. https://doi.org/10.1029/2018RG000618

Bazzo COG, Kamali B, Hütt C, Bareth G, Gaiser T (2023) A review of estimation methods for aboveground biomass in grasslands using UAV. Remote Sens 15:639. https://doi.org/10.3390/rs15030639

Beale J, Snapir B, Waine T, Evans J, Corstanje R (2019) The significance of soil properties to the estimation of soil moisture from C-band synthetic aperture radar. Hydrol Earth Syst Sci. Preprint. https://doi.org/10.5194/hess-2019-294

Bole A, Wall A, Norris A (2014) Target detection—chapter 3. In: Radar and ARPA manual, 3rd edn. Butterworth-Heinemann, pp 139–213. https://doi.org/10.1016/B978-0-08-097752-2.00003-9. ISBN 9780080977522

Breunig FM, Dalagnol R, Galvão LS, Bispo PDC, Liu Q, Berra EF, Gaida W, Liesenberg V, Sampaio TVM (2024) Monitoring cover crop biomass in southern Brazil using combined PlanetScope and Sentinel-1 SAR data. Remote Sens 16:2686. https://doi.org/10.3390/rs16152686

Central Statistics Office—CSO (2024) Ecosystem accounts—grasslands and croplands 2000–2022. Key findings CSO statistical release 07 June 2024. Available at https://www.cso.ie/en/releasesandpublications/fp/fp-eagac/ecosystemaccountsgrasslandsandcroplands2000-2022/. Accessed 1 Dec 2024

Cracknell AP (2018) The development of remote sensing in the last 40 years. Int J Remote Sens 39(23):8387–8427. https://doi.org/10.1080/01431161.2018.1550919

Das K, Paul PK (2015) Present status of soil moisture estimation by microwave remote sensing. Cogent Geosci 1(1):1084669. https://doi.org/10.1080/23312041.2015.1084669

Deodoro SC, Moral RA, Fealy R, McCarthy T, Fealy R (2023) An assessment of Sentinel 1 SAR, geophysical and topographical covariates for estimating topsoil particle size fractions. Eur J Soil Sci 74(5):e13414. https://doi.org/10.1111/ejss.13414

Deodoro SC, Moral RA, Fealy R, McCarthy T, Fealy R (2024) Using the surface scattering mechanism from dual-pol SAR data to estimate topsoil particle size fractions. Int J Appl Earth Obs Geoinf 128:e10374. https://doi.org/10.1016/j.jag.2024.103742

Dong T et al (2020) Estimating crop biomass using leaf area index derived from Landsat 8 and Sentinel-2 data. ISPRS J Photogramm Remote Sens 168:236–250. https://doi.org/10.1016/j.isprsjprs.2020.08.003

Elachi C, Van Zyl JJ (2021) Introduction to the physics and techniques of remote sensing. Wiley, Hoboken, NJ

ESA—The European Space Agency (2007) EVISAT ASAR product handbook, Issue 22, 27 February 2007. Available at https://earth.esa.int/eogateway/documents/20142/37627/ASAR-Product-Handbook.pdf. Accessed 25 Oct 2023

ESA—The European Space Agency (2020) Newcomers earth observation guide, ver. 1.9. Available at https://business.esa.int/newcomers-earth-observation-guide. Accessed 17 Aug 2024

ESA—The European Space Agency (2023) Remote sensing data underpin research on soil. Available at https://earth.esa.int/eogateway/data-application/remote-sensing-data-underpin-research-on-soil. Accessed 24 July 2024

ESA—The European Space Agency (n.d.) Introduction to optical remote sensing for agriculture and water. Available at https://eo4society.esa.int/training_uploads/LTC2022/1_Monday/8_Introduction-to-Optical-Remote-Sensing-for-Agriculture-and-Water_Ramoino_Vekerdy.pdf. Accessed 27 Apr 2025

Farias GD et al (2023) Normalized Difference Vegetation Index (NDVI) for soybean biomass and nutrient uptake estimation in response to production systems and fertilization strategies. Front Sustain Food Syst 6:959681. https://doi.org/10.3389/fsufs.2022.959681

Farwell LS et al (2021) Satellite image texture captures vegetation heterogeneity and explains patterns of bird richness. Remote Sens Environ 253:112175. https://doi.org/10.1016/j.rse.2020.112175

Fatehi P, Damm A, Schaepman ME, Kneubuhler M (2015) Estimation of alpine forest structural variables from imaging spectrometer data. Remote Sens 7(12):16315–16338. https://doi.org/10.3390/rs71215830

Fletcher K (2012) Sentinel-1: ESA's radar observatory Mission for GMES operational services. European Space Agency-ESA Communications ESA SP-1322/1, Leiden

Gao Y, Liang Z, Wang B, Wu Y, Liu S (2019) UAV and satellite remote sensing images based aboveground biomass inversion in the meadows of Lake Shengjin. J Lake Sci 31(2): 517–528

Gorrab A, Zribi M, Baghdadi N, Mougenot B, Fanise P, Chabaane ZL (2015) Retrieval of both soil moisture and texture using TerraSAR-X images. Remote Sens 7(8):10098–10116. https://doi.org/10.3390/rs70810098

Gururaj P, Umesh P, Shetty A (2019) Assessment of spatial variation of soil moisture during maize growth cycle using SAR observations. In: Remote sensing for agriculture, ecosystems, and hydrology XXI, vol 11149. SPIE, Bellingham, WA, pp 372–379

Harvard (n.d.) Rayleigh's criterion. Harvard Natural Sciences-Lecture Demonstrations. Harvard University. https://sciencedemonstrations.fas.harvard.edu/presentations/rayleighs-criterion. Accessed 19 Jan 2024

Ibrahim GRF, Rasul A, Abdullah H (2023) Improving crop classification accuracy with integrated Sentinel-1 and Sentinel-2 data: a case study of barley and wheat. J Geovis Spat Anal 7:22. https://doi.org/10.1007/s41651-023-00152-2

Jensen JR (2009) Remote sensing of the environment: an earth resource perspective, 2nd edn. Pearson Education India, New Delhi, p 613

JOUAV—Unmanned Aircraft System (2024) Ground Sample Distance (GSD): definition, importance & calculation. Available at https://www.jouav.com/blog/ground-sample-distance.html. Accessed 17 Aug 2024

Kelsey KC, Neff JC (2014) Estimates of aboveground biomass from texture analysis of Landsat imagery. Remote Sens 6(7):6407–6422. https://doi.org/10.3390/rs6076407

Kim Y, Van Zyl JJ (2009) A time-series approach to estimate soil moisture using polarimetric radar data. IEEE Trans Geosci Remote Sens 47(8):2519–2527. https://doi.org/10.1109/TGRS.2009.2014944

Kimerling AJ (2009) Mathematical relationships among map scale, raster data resolution, and map display resolution. Imagery & Remote Sensing. Available at https://www.esri.com/arcgis-blog/products/product/imagery/mathematical-relationships-among-map-scale-raster-data-resolution-and-map-display-resolution/?rmedium=redirect&rsource=blogs.esri.com/esri/arcgis/2009/12/04/mathematical-relationships-among-map-scale-raster-data-resolution-and-map-display-resolution. Accessed 9 Feb 2024

Knudby A (2021) Remote sensing. Open Library. Available at https://ecampusontario.pressbooks.pub/remotesensing/front-matter/welcome-to-remote-sensing/. Accessed 26 July 2024

Liu P (2015) A survey of remote-sensing big data. Front Environ Sci 3:45. https://doi.org/10.3389/fenvs.2015.00045

Liu C-a, Chen Z-x, Shao Y, Chen J-s, Hasi T, Pan H-z (2019) Research advances of SAR remote sensing for agriculture applications: a review. J Integr Agric 18(3):506–525. https://doi.org/10.1016/S2095-3119(18)62016-7

Liu W et al (2023) Machine learning-based grassland aboveground biomass estimation and its response to climate variation in Southwest China. Front Ecol Evol 11:1146850. https://doi.org/10.3389/fevo.2023.1146850

Luo Y, Wang X, Ouyang Z, Lu F, Feng L, Tao J (2020) ChinAllome tree 1.0: China's normalized tree biomass equation dataset. Earth System Science Data Discuss 42:21–40. https://doi.org/10.5194/essd-2019-1

Ma T, Zhang C, Ji L, Zuo Z, Beckline M, Hu Y, Li X, Xiao X (2024) Development of forest aboveground biomass estimation, its problems and future solutions: a review. Ecol Indic 159:111653. https://doi.org/10.1016/j.ecolind.2024.111653

Mandal D et al (2020) Dual polarimetric radar vegetation index for crop growth monitoring using sentinel-1 SAR data. Remote Sens Environ 247:111954. https://doi.org/10.1016/j.rse.2020.111954

Mercier A, Betbeder J, Baudry J, Denize J, Leroux V, Roger J-L, Spicher F, Hubert-Moy L (2019) Evaluation of Sentinel-1 and -2 time series to derive crop phenology and biomass of wheat and rapeseed: Northen France and Brittany case studies. In: Neale CM, Maltese A (eds) Remote sensing for agriculture, ecosystems, and hydrology XXI. SPIE, Bellingham, WA, p 2

Moran MS, Alonso L, Moreno JF, Mateo MPC, De la Cruz DF, Montoro A (2012) A RADARSAT-2 quad-polarized time series for monitoring crop and soil conditions in Barrax, Spain. IEEE Trans Geosci Remote Sens 50(4):1057–1070. https://doi.org/10.1109/TGRS.2011.2166080

Mukhopadhyay R, Kumar S, Aghababaei H, Kulshrestha A (2021) Estimation of aboveground biomass from PolSAR and PolInSAR using regression-based modelling techniques. Geocarto Int 37:4181–4207. https://doi.org/10.1080/10106049.2021.1878289

Nasirzadehdizaji R, Sanli FB, Abdikan S, Cakir Z, Sekertekin A, Ustuner M (2019) Sensitivity analysis of multi-temporal sentinel-1 SAR parameters to crop height and canopy coverage. Appl Sci 9(4):655. https://doi.org/10.3390/app9040655

Neville L (2024) Ensuring sustainable urban development and protecting biodiversity with the Government of New South Wales. Available at https://www.planet.com/pulse/ensuring-sustainable-urban-development-and-protecting-biodiversity-with-the-government-of-new-south-wales. Accessed 6 Aug 2024

Nikparvar B, Thill J-C (2021) Machine learning of spatial data. ISPRS Int J Geo Inf 10:600. https://doi.org/10.3390/ijgi10090600

O'Donovan M, McEvoy M (n.d.) Managing your grass. Available at https://www.teagasc.ie/media/website/animals/dairy/managinggrass.pdf. Accessed 6 Jan 2025

Park N-W, Kim Y, Kwa G-H (2019) An overview of theoretical and practical issues in spatial downscaling of coarse resolution satellite-derived products. Korean J Remote Sens 35(4):589–607. https://doi.org/10.7780/kjrs.2019.35.4.8

Prabhakara K, Hively WD, McCarty GW (2015) Evaluating the relationship between biomass, percent groundcover and remote sensing indices across six winter cover crop fields in Maryland, United States. Int J Appl Earth Obs Geoinf 39:88–102. https://doi.org/10.1016/j.jag.2015.03.002

Rossiter DG (1999) Soil geographic databases SOL.SGDB. Lecture Notes. Soil Science Division International Institute for Aerospace Survey & Earth Sciences (ITC), Version 2, February 1999. Available at https://www.css.cornell.edu/faculty/dgr2/_static/files/pdf/SoilGeographicDataBases.pdf. Acessed 18 July 2024

Rossiter DG, Poggio L, Beaudette D, Libohova Z (2022) How well does digital soil mapping represent soil geography? An investigation from the USA. Soil 8:559–586. https://doi.org/10.5194/soil-8-559-2022

Rouse JW Jr, Haas RH, Schell JA, Deering DW (2023) Monitoring the vernal advancement and retrogradation (green wave effect) of natural vegetation. Prog. Rep. RSC 1978-1. Remote Sensing Center, Texas A&M University, College Station, 93p

Sadeghi Y, St-Onge B, Leblon B, Prieur JF, Simard M (2018) Mapping boreal forest biomass from a SRTM and TanDEM-X based on canopy height model and Landsat spectral indices. Int J Appl Earth Obs Geoinf 68:202–213. https://doi.org/10.1016/j.jag.2017.12.004

Sano EE, Matricardi EAT, Camargo FF (2020) State-of-the-art of the radar remote sensing: fundamentals, sensors, image processing, and applications. Rev Bras Cartogr 72:1458–1483. https://doi.org/10.14393/rbcv72nespecial50anos-56568

Sarker LR, Nichol JE (2011) Improved forest biomass estimates using ALOS AVNIR-2 texture indices. Remote Sens Environ 115(4):968–977. https://doi.org/10.1016/j.rse.2010.11.010

Schumann G, Moller D (2015) Microwave remote sensing of flood inundation. Phys Chem Earth, Parts A/B/C 83:84–95. https://doi.org/10.1016/j.pce.2015.05.002

Singh S, Sarma K (2023) Exploring soil spatial variability with GIS, remote sensing, and geostatistical approach. J Soil, Plant Environ 2(1):79–99. https://doi.org/10.56946/jspae.v2i1.186

Srivastava HS, Patel P, Navalgund RR (2006) How far SAR has fulfilled its expectation for soil moisture retrieval? SPIE Digital Library, 64100. In: Microwave remote sensing of atmosphere and environment-II, AE107, pp 1–12. https://doi.org/10.1117/12.693946

Steinhardt U, Herzog F, Lausch A, Müller E, Lehmann S (1999) Hemeroby index for landscape monitoring and evaluation. In: Pykh YA, Hyatt DE, Lenz RJ (eds) Environmental indices—system analysis approach. EOLSS Publ, Oxford, pp 237–254

Tian L, Wu X, Tao Y, Li M, Qian C, Liao L, Fu W (2023) Review of remote sensing-based methods for forest aboveground biomass estimation: progress, challenges, and prospects. Forests 14(6):1086. https://doi.org/10.3390/f14061086

Tobler WR (1987) Measuring spatial resolution. In: Proceedings, land resources information systems conference, Beijing, pp 12–16

Trudel M, Charbonneau F, Leconte R (2012) Using RADARSAT-2 polarimetric and ENVISAT-ASAR dual-polarization data for estimating soil moisture over agricultural fields. Can J Remote Sens 38(4):514–527

UIUC (2010) Effects of wavelength on the ability to detect an object. University of Illinois at Urbana-Champaign. http://ww2010.atmos.uiuc.edu/(Gh)/guides/rs/rad/basics/wvl.rxml. Accessed 19 Jan 2024

USGS (2018) United States Geological Survey. NDVI, the Foundation for Remote Sensing Phenology. Available at https://www.usgs.gov/special-topics/remote-sensing-phenology/science/ndvi-foundation-remote-sensing-phenology#overview. Accessed 29 July 2024

Valenzuela ÁQ, Reyes JCG (2019) Basic spatial resolution metrics for satellite imagers. IEEE Sensors J 19(13):4914–4922. https://doi.org/10.1109/JSEN.2019.2902512

Varotsos CA, Krapivin VF (2020) Microwave remote sensing tools in environmental science, 1st edn. Springer Nature Switzerland AG, p XXIII-468. https://doi.org/10.1007/978-3-030-45767-9

Walsh M (2003) Remote sensing. Engineering and design. Engineer manual. Manual no. 1110-2-2907, Department of the Army. US Army Corps of Engineers, Washington, DC

Walz U, Stein C (2014) Indicators of hemeroby for the monitoring of landscapes in Germany. J Nat Conserv 222:79–289. https://doi.org/10.1016/j.jnc.2014.01.007

Wu X, Wang X, Wu Y, Xia X, Fang J (2015) Forest biomass is strongly shaped by forest height across boreal to tropical forests in China. J Plant Ecol 8(6):559–567. https://doi.org/10.1093/jpe/rtv001

Yang Q et al (2022) Allometry-based estimation of forest aboveground biomass combining LiDAR canopy height attributes and optical spectral indexes. Forest Ecosyst 9:100059. https://doi.org/10.1016/j.fecs.2022.100059

Zhang XY, Kondragunta S (2006) Estimating forest biomass in the USA using generalized allometric models and MODIS land products. Geophys Res Lett 33(9):L09402. https://doi.org/10.1029/2006GL025879

Zhang Z, Lu L, Zhao Y, Wang Y, Wei D, Wu X, Ma X (2023) Recent advances in using Chinese earth observation satellites for remote sensing of vegetation. ISPRS J Photogramm Remote Sens 195:393–407. https://doi.org/10.1016/j.isprsjprs.2022.12.006

Zhou G, Yin G, Tang X (2018) Carbon storage of Forest ecosystems in China: biomass equation. Longmen Bookstore of Science Press, Beijing

Zyl JV, Kim Y (2010) Synthetic aperture radar polarimetry. JPL space science and technology series. In: Yuen JH (ed) National Aeronautics and Space Administration-NASA. Jet Propulsion Laboratory, California Institute of Technology, La Cañada Flintridge

Chapter 7
The Relationship Between Biomass of Native Vegetation and Diversity of Small Mammals in Agro-Silvicultural Landscapes of Southeast Brazil

Lina C. Vásquez-Uribe, Renato M. Toledo, Rafael de Andrade Moral, Vânia R. Pivello, and Luciano Martins Verdade

Abstract Agroecosystems show spatio-temporal variation that generates "pulses" in biomass production throughout the landscape, likely affecting the use of space and the trophic relationships of wild species. Plant biomass is considered a primary source of energy; thus, it can influence species composition and abundance. This study aimed to evaluate the relationship between above-ground biomass and patterns of diversity, distribution, and abundance of non-volant small mammals in the main agro-silvicultural landscapes of São Paulo state in southeast Brazil: sugarcane and *Eucalyptus* plantations, with fragments of native vegetation. From October 2021 to September 2022, we carried out monthly sampling campaigns over a grid of 300 traps in a *Eucalyptus*-dominated landscape and 240 traps in a sugarcane-dominated landscape, adapted from RAPELD methodology. Remote images taken by unmanned aerial vehicles (UAV) of native vegetation and abandoned pasture were used to estimate above-ground biomass. Within agrosilvicultural landscapes, an increase in biomass prompts a shift in the abundance of small mammals, transitioning from open area species to forest species. Particularly, *Akodon montensis*, *Calomys tener*, *Didelphis albiventris*, *Nectomys squamipes* and *Oligoryzomys nigripes* show potential as ecological indicators of processes within agro-silvicultural landscapes. This study supports the use of native vegetation biomass as a proxy for assessing small mammals' abundance in agrosilvicultural landscapes. Moreover, native vegetation recovery emerges as one of the most effective strategies for enhancing the conservation value of agrosilvicultural landscapes.

L. C. Vásquez-Uribe · R. M. Toledo · V. R. Pivello
Universidade de São Paulo/IB/Departamento de Ecologia, São Paulo, Brazil

R. A. Moral
Department of Mathematics and Statistics, Maynooth University, Maynooth, Co. Kildare, Ireland

L. Martins Verdade (✉)
Wildlife Monitoring Consultancy, Campinas, São Paulo, Brazil
e-mail: luciano.verdade@wildlifemgmtco.com

L. Martins Verdade, R. A. Moral (eds.), *Wildlife Monitoring*,
https://doi.org/10.1007/978-3-032-05821-8_7

Keywords Above-ground biomass · Agroecosystems · Biodiversity conservation · Rodents · Marsupials · Remote sensing

1 Introduction

Agricultural landscapes have a matrix of agricultural fields (i.e., agroecosystems), with distinct nutrient cycles and species compositions, usually artificially maintained. Agroecosystems show spatio-temporal variation that generates "pulses" in biomass production throughout the landscape, which probably affect the use of space and the trophic relationships of wild species (Verdade et al. 2016). For example, the biomass variation of sugarcane fields (0–100 $t.ha^{-1}.year$, Goldemberg et al. 2008) and *Eucalyptus* plantations (0–300 $m^3.ha^{-1}.year$ of wood in 6–7-year commercial cycle, Stape et al. 2004) can prompt resident species to deal with the oscillation of food resources (Wang et al. 2006), with unknown consequences in terms of demography and evolution (Verdade et al. 2014). *Eucalyptus* trees, due to their rapid growth, high productivity, great adaptability, and wide range of uses, have been managed for commercial exploration in 6–7-year harvest cycles (Mora and Garcia 2000). In sugarcane plantations, the harvest occurs every year (between April and November), usually for four consecutive years before it is replanted (de Oliveira Florentino et al. 2018).

Consequently, the relatively high temporal heterogeneity in agroecosystems can affect habitat quality (Catling et al. 2001; Fox et al. 2003; Monamy and Fox 2000; Pinotti et al. 2015) and resource availability for wildlife (DeWalt et al. 2003; Pinotti et al. 2011). As plant biomass is considered a primary source of energy, it can modulate wild animal species composition and abundance (Korthals et al. 2001; Wardle et al. 2003). Mammals may act as prey and predators, and their direct and indirect interactions with both plants and other animals have a strong influence on the ecosystem dynamics (Gonçalves et al. 2012; Sinclair 2003). In addition, seasonal changes in plant phenology and in the growth stage of crops determine qualitative and quantitative variations in resources of both crop fields and their borders. These fluctuations subsequently impact the patterns of habitat use by rodents (Fraschina et al. 2012).

Likewise, the vegetation structure is considered an important predictor for small mammal assemblages (Benedek and Sîrbu 2018b), as it provides food resources and refuge against predators (Moro and Gadal 2007). Thus, the conservation and management of small remnants of native vegetation can be of major impact in maintaining biodiversity in fragmented landscapes (Archibald et al. 2011; Ranta et al. 1998; Ribeiro et al. 2009; Watson 2003). Several studies have shown a correlation between changes in small mammal assemblages as vegetation undergoes alterations in composition and structure due to succession (Atkeson and Johnson 1978; Briani et al. 2004; Churchfield 1997; Fox 1982, 1990; Hirth 1959; Pearson 1959). Furthermore, the species responses are related to their respective life histories (Friend 1982; Kelly et al. 2013). Still, plant biomass can be an appropriate indicator of small mammal species richness, since plant production (i.e., biomass) directly implies the availability of consumable resources for species at the base of the food chain (Wang et al. 2001). For example, intensive biomass production maintains great biodiversity in forests, that can influence nutrient cycling (Verschuyl et al. 2011). In this sense, the evaluation of characteristics related to the structural complexity of the vegetation, such as species richness, height, diameter at breast height (DBH), and the number of trees can be good indicators of small mammals' abundance (Catling et al. 2001; Clarke and White 2008). Also, these structural characteristics are used in allometric relationships for estimating biomass and carbon in plants (Chave et al. 2005, 2014; Montagu et al. 2005; Vieilledent et al. 2012). In sugarcane plantations, for example, the mean stem diameter and height of individuals have the potential to provide accurate estimates of plant biomass (De Carvalho et al. 2019); in *Eucalyptus* plantations, the species type, wood specific gravity, mean individual DBH and total tree height are frequently used as predictors (Chave et al. 2005; De Carvalho et al. 2019).

Specifically, several authors have conducted research on the influence of agro-silvicultural environments on population dynamics and on the structure of the small mammal assemblage, considering the vegetation characteristics. According to Benedek and Sîrbu (2018a), the use of agricultural land can influence the structure of the small mammal community mainly due to the habitat characteristics. For instance, hedgerow networks in agricultural landscapes of Canada and France provide habitat for specific rodent species and support a high biomass of prey (small mammals) for predators (Michel et al. 2006; Sullivan et al. 2012).

In Costa Rican coffee plantations, varying habitats — sunny and shady — offer distinct conditions for small mammals. Abundance tends to rise with soil covered by herbaceous plants and the presence of mature trees. Meanwhile, species richness shows a positive correlation with the diversity of tree species within these habitats (Caudill et al. 2015; Daily et al. 2003). Also, mature coffee shrubs can provide cover and facilitate movement, allowing greater connectivity of mammal populations (Daily et al. 2003). In the savanna-agricultural landscape of Eswatini, southern Africa, fields with tall sugarcane (>2 m) showed diversity and species composition akin to neighboring native savanna, differently from relatively empty fields with recently planted sugarcane (Mamba et al. 2019). The Brazilian cacao-growing region (locally called *cabrucas*), in the northeast, harbors mammal species of conservation concern, with similar richness and composition in both the forest remnants (Atlantic Forest) and the agroforests (Cassano et al. 2012). Also, in southeastern Brazil, Martin et al. (2012) found a significant variation in the composition of resident small mammal species throughout the commercial cycle of *Eucalyptus* production. As the vegetation structure evolves with the development of the trees, resident species are replaced by others. *Eucalyptus* plantations serve as habitat for part of the small mammal assemblage, and the diversity of taxonomic groups increases over the course of the commercial cycle (Verdade et al. 2020; Vásquez et al. 2021).

Monitoring for biodiversity conservation should include four critical areas of assessment: changes in habitat extent and landscape structure, habitat degradation, alterations in biodiversity, and tracking of pressures and threats within and outside protected areas (Nagendra et al. 2013). In this evaluation process, landscape ecology provides the scientific basis for the study and management of landscapes, using different metrics defined as measurable biotic and abiotic characteristics of the environment to obtain quantitative data concerning ecological resources and functional characteristics (McAlpine and Eyre 2002). These metrics contribute to understanding the relationship between patterns and processes, and they must perform well in highly fragmented landscapes, where habitat alteration and biodiversity loss have been more pronounced (Corry 2005). Moreover, landscape metrics rely on remote sensing data, enabling the acquisition of temporal information on land use (Chmielewski et al. 2014).

Remote sensing can deliver data on habitat quantity (e.g., amount, configuration) and quality (e.g., structure, distribution of species, habitat types and/or communities) (He et al. 2011) at diverse spatial scales and temporal frequencies (Wulder et al. 2004). Hence, these data are increasingly required in conservation management (Mairota et al. 2015). Additionally, a wide range of remote sensing data sources (e.g., hyperspatial, hyperspectral) and products (e.g., vegetation indices such as the Normalized Difference Vegetation Index- NDVI) can be used as a key tool for ecological monitoring, particularly when coupled with field data (Nagendra 2001; Deodoro et al. 2026, this volume). Such approaches hold great promise for adaptive conservation management, where the integration of remote sensing analyses with field datasets generates impacts on land cover and habitat change (Nagendra and Rocchini 2008) or fragmentation (Mairota et al. 2013).

The present study aimed to evaluate the relationship between biomass and patterns of diversity, distribution, and abundance of non-volant small mammals in the main agro-silvicultural landscapes of São Paulo state: sugarcane and *Eucalyptus* plantations, containing fragments of natural vegetation. To do so, we tested two working hypotheses: H_1: There is a spatial correlation between the plant biomass of native vegetation in landscapes dominated by Eucalyptus and sugarcane plantations and the species richness and abundance of small mammals and H_2: The abundance of small mammals can be estimated based on biomass quantification.

2 Study Area

This study was carried out at Fazenda Agrícola Rio Claro-ARC (7320 ha), in Lençóis Paulista municipality, and Fazendas Arca (1123 ha) and Três Lagoas (3242 ha), both located in the Angatuba municipality, São Paulo state, Brazil (Fig. 7.1). These sites have a seasonal climate, where the dry season spans from April to September (Santana et al. 2008). Local climate is classified as humid subtropical (Cfa), according to the Köppen's classification (Kottek et al. 2006; Rubel et al. 2017), with average temperatures ranging from 17.0 to 30.0 °C throughout the year (Agritempo 2022). The annual precipitation averages between 1200 and 1300 mm (CEPAGRI 2021; Leda et al. 2014; Santana et al. 2008). The prevalent

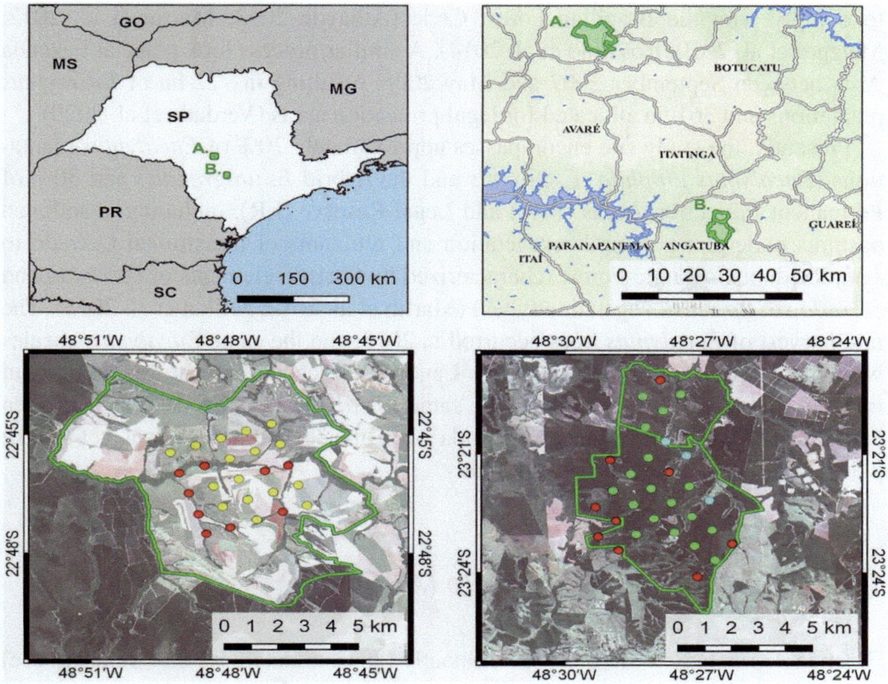

Fig. 7.1 Study areas: (**a**) Agrícola Rio Claro-ARC (Lençóis Paulista, São Paulo state, Brazil); (**b**) Fazendas Arca and Três Lagoas-FATL (Angatuba, São Paulo state, Brazil). Bold dots (•) represent sampling units: yellow (sugarcane), red (native vegetation), green (*Eucalyptus*) and blue (abandoned pasture)

soil type is Dystrophic Red Latosol (LVd25) (Gonçalves et al. 2011). Notably, both municipalities lie in a transition zone between the Atlantic Forest and Cerrado biomes.

2.1 *Fazendas Arca and Três Lagoas (FATL)*

Between 1870 and 1930, native vegetation in the Angatuba region was replaced by cotton and coffee plantations. The economic crisis of 1929 led to the collapse of agriculture, facilitating the natural recovery of native vegetation. In the 1970s, a new extensive deforestation occurred to create pastures planted with African grasses (*Urochloa* spp., Syn. *Brachiaria* spp.) (Lisboa and Prado 2019). During the 2000s, there was a shift from pastures to silviculture for cellulose pulp production, based on *Eucalyptus* plantations. Between August 2006 and November 2007, 2224 hectares of exotic pastures at Fazenda Três Lagoas were converted into *Eucalyptus* spp. plantations, while the remaining area (896 ha) was left for natural revegetation to comply with the Brazilian Forest Code (Athayde 2013; Martin et al. 2012; Metzger et al. 2010; Rosalino et al. 2014). A similar process took place at Fazenda Arca between September 2007 and May 2008, resulting in 722 ha of *Eucalyptus* plantations and 361 ha allocated for legal protection areas (Verdade et al. 2020).

Presently, the study site encompasses approximately 70% of *Eucalyptus* plantations (*Eucalyptus urophila*, *E. grandis* and the hybrid *E. urograndis*) and 30% of Permanent Protection Areas (PPA) and Legal Reserve (LR), including abandoned pastures undergoing natural revegetation and remnants of transitional Cerrado to semideciduous Atlantic Forest (characterized by floristic elements of *cerradão* and *cerrado stricto sensu* vegetation types) (Martin et al. 2012; Millan et al. 2015). The first harvest of *Eucalyptus* trees occurred in 2013, and the same *Eucalyptus* species were planted in 2014. At Fazenda Três Lagoas, the second harvest was carried out in 2021, and a new plantation with the same *Eucalyptus* species occurred between late 2021 and early 2022. As for Fazenda Arca, the second harvest was yet to take place during data collection for this study.

2.2 *Fazenda Agrícola Rio Claro (ARC)*

Fazenda Agrícola Rio Claro is predominantly (approximately 90% of this land use) planted with sugarcane (*Saccharum* sp., comprising), that represents the primary agricultural activity in the region (Donatelli et al. 2004; Leda et al. 2014). Sugarcane harvesting takes place annually between April and November. The remaining 10% of the farmland is occupied by native vegetation: fragments of *cerradão* and *cerrado sensu stricto* (phytophysiognomies according to Coutinho (1978)), *várzeas* (floodplains) and *capoeiras* (secondary forests) of the semideciduous Atlantic Forest (Duraflora 1999; Martins 1997).

A study conducted in the region (Leda et al. 2014) reported transformations from 1984 to 2014, with substantial reduction in pasture areas (439.53 ha), forestry (443.86 ha) and water bodies (312.31 ha). Conversely, areas for sugarcane plantations, urbanization, and the regeneration of native vegetation increased by, respectively, 229.04, 350.09, and 312.31 ha (Leda et al. 2014).

3 Methodology

3.1 *Sampling Methodology*

3.1.1 Small Mammals

The sampling design was adapted from RAPELD (Rapid Assessments of Biodiversity and Long-Term Ecological Research Projects), a protocol previously used in the Biodiversity Research Program (PPBio) (Magnusson et al. 2014). This method is designed to enhance the probability of accurately sampling biological communities and estimating the distribution, abundance, and biomass of species at each site. Additionally, it facilitates biogeographic comparisons between sites by using a spatial grid design, where nodes are located 1 km apart (Magnusson et al. 2005). Recognized as a suitable system for addressing specific research questions within a long-term monitoring framework, the RAPELD method enables rapid and comparable investigations of biotic factors (Magnusson et al. 2005).

Field data collection was carried out from October 2021 to September 2022, comprising a total of 12 sampling campaigns for each study site. Specifically, for the *Eucalyptus* landscape, the field data collection in FATL took place at the end of the second cycle and early third cycle. In the case of ARC, data collection occurred at the end of the second cycle of the *Eucalyptus* plantation. The field campaigns were conducted monthly, with five consecutive days allocated for each landscape (*Eucalyptus* and sugarcane). Traps were kept open for four consecutive nights and checked every morning.

A grid comprising 54 sampling units (SU) was established, with units positioned 1 km apart from each other. Of these, 30 SU were located at the *Eucalyptus*-dominated landscape, and 24 at the sugarcane-dominated landscape (Fig. 7.1). Each SU comprised 10 Sherman traps of two different sizes: five measuring 41 × 12.5 × 14.5 cm and five measuring 25 × 8 × 9 cm. These traps were arranged in pairs (one of each size), positioned on the ground along a line, being 20 m apart. This configuration resulted in 300 traps per sampling grid in the *Eucalyptus* landscape and 240 in the sugarcane landscape. The total sampling effort was 25,920 trap.nights, with 14,400 trap.nights at FATL (i.e., 10 traps/SU × 30 SU × 4 nights per monthly campaign × 12 monthly campaigns) and 11,520 trap.nights at ARC (i.e., 10 traps/SU × 24 SU × 4 nights per monthly campaign × 12 monthly campaigns).

This spatial grid upholds the assumption of spatial independence between SU, considering the home range of Neotropical small mammal species and their maximum movement distances (Cáceres and Monteiro-Filho 2001; Martin et al. 2012; Pires et al. 2010). The traps were baited with a mixture of banana, peanut, cornmeal, oat, and cod liver oil, as commonly employed to attract a variety of animal species (Barros et al. 2015; Jones et al. 1996; Kok et al. 2013). These baits are recognized for their effectiveness in achieving a high capture rate (Kok et al. 2013). We chose to use Sherman traps due to the agricultural management practices in sugarcane plantations, which restrict the use of other types of traps, such as pitfalls.

Species identification was based on morphometric characteristics (Gardner 2007; Paglia et al. 2012; Patton et al. 2015). Every individual was physically immobilized, weighted, measured, sexed, marked by subcutaneous microchips (Vera-Tag®) and ear punching (Martin et al. 2012), and had a hair sample taken from its back. The following information was recorded: individual body mass (g), body length (mm) and sex; list and number of species detected per sampling unit and study area. In addition, up to three individuals were collected as vouchers for species reference and deposited at the Mammals Laboratory collection at the University of São Paulo under Biodiversity Authorization and Information System (SISBIO) collection license No. 71789 and Ethics Committee on the Use of Animals (CEUA) license No. 003-2019.

3.1.2 AGB of Native Vegetation

We quantified aboveground biomass (AGB) of native vegetation which excludes cultivated plants but includes spontaneous stems of indigenous, exotic, and naturalized species, by combining remote sensing and field sampling.

Remote Sensing

A single-winged UAV model XMOBOTS ARATOR 5B equipped with Sony RGB 5100 and X1M5 sensors was employed to capture 3D multispectral data from all sampling units, including the vegetation plots. Flights were conducted in February 2020. The 3D multispectral data was generated at a spatial resolution of 8 cm (including R, G, B, red-edge, near-infrared bands, and elevation) using the software Metashape 1.6.2. Additionally, a topographic survey collected ground elevation data in the image areas to enhance the accuracy of vegetation height measurements.

Field Sampling

Vegetation plots of 40 × 40 m were established in the 21 sampling units having a natural/seminatural fragment in their center (comprising native vegetation and abandoned pasture), the central coordinate of each of these land units was also the center of the vegetation plots. Inside the vegetation plots, single-stemmed and multi-stemmed trees with a combined basal area greater than 7 cm^2 (equivalent to the standard DBH >3 cm) were identified and measured. The nomenclature followed the "Flora do Brasil 2020" database (https://floradobrasil.jbrj.gov.br), according to the APG IV system (APG 2016). Tree heights were recorded in the field using metric precision and a telescopic pole as a reference. Direct measurements were possible for trees up to 8 m in height.

Each tree sampled in the field had its dry-matter aboveground biomass estimated by an improved allometric equation for pantropical tree communities (Chave et al. 2014), using the diameter, height, and species identity as follows:

(a) $AGB = 0.063(\rho D^2 H)^{0.976}$; where AGB is the dry weight of aboveground biomass (kg), ρ is the specific density (g/cm^3), D is the diameter at breast height (cm), and H is the height (m);
(b) The specific wood density of each species was obtained from Chave et al. (2009), Zanne et al. (2009) and Sullivan et al. (2025). Species without references were associated with the average specific density of their respective genus.

Likewise, individuals identified only at the genus or family levels were related to the mean specific density for the respective taxonomic level. Unidentified individuals were excluded from the analysis. Then, individual biomasses were summed by plot and expressed in $t.ha^{-1}$; and,

(c) We sampled an additional 14 plots without trees and shrubs with DBH >3 cm, to be used as control references of open fields being spared from recent agricultural activities and lacking AGB of DBH >3 cm.

Native Vegetation AGB Mapping and Monitoring Buffers

We combined the 35-plot level AGB measurements with the UAV dataset to select the linear model of greater parsimony and simplicity that combined the Normalized Difference Vegetation Index (NDVI) and vegetation height as predictors of AGB through linear and non-linear models in R (R Core Team 2024). The response variable was log-transformed to satisfy the model's assumptions. Then, the selected model (i.e., $AGB = 0.842 \times (\text{vegetation height}^{(NDVI+1)})$; adjusted-$R^2 = 0.882$) was applied to estimate biomass in the semi-natural fragments inside all sampling units with a spatial resolution of 20 m. Using photointerpretation, we delimited sampling units' agricultural and silvicultural areas. With resulting native vegetation AGB maps we calculated the totals of non-cultivated AGB within buffers around monitoring points considering diameters of 50, 100, 200, 300 and 400 m.

3.2 *Analytical Methodology*

The following statistical analyses were performed to test hypotheses H_1 and H_2 defined previously, where species richness and abundance of small mammals corresponded to dependent variables and above-ground biomass (AGB) at buffers of 50, 100, 200, 300, and 400 m corresponded to independent variables as follows:

(a) Indicator species analysis: this method tests the relationship between the abundances (or the presence and absence) of small mammals' species and a categorical variable, for which we seek indicator species (e.g., high biomass and low biomass for each habitat type). The technique combines two indices: fidelity (high occurrence or density under certain conditions) and sensitivity (low occurrence or density under other conditions). We used the "indicspecies" package (De Cáceres et al. 2010) for R software (R Core Team 2024);
(b) Bray-Curtis dissimilarity analysis: to evaluate compositional relationships of small mammals we used the "vegan" package for R software (Anderson et al. 2017);
(c) Richness data: We fitted Poisson generalized linear models to the richness data, specifying different intercepts and slopes over biomass for each group (small mammals of open vs. forested areas). We assessed the significance of

the difference between the two slopes via a likelihood-ratio (LR) test between nested models;

(d) Abundance data: We fitted negative binomial generalized linear models to the abundance data, specifying different intercepts and slopes over biomass for each group (small mammals of open vs. forested areas). We assessed the significance of the difference between the two slopes via F-tests between nested models; and,

(e) Goodness-of-fit: We assessed goodness-of-fit of the fitted models via half-normal plots with a simulated envelope (Moral et al. 2017).

4 Results

4.1 Small Mammals' Diversity

A total of 355 individuals (229 at FATL and 126 at ARC from 14 species (three Didelphimorphia and 11 Rodentia) were captured (Table 7.1). At FATL the small mammal assemblage was dominated by *Calomys tener* (90 individuals), followed by *Oligoryzomys nigripes* (52 ind.), and *Didelphis albiventris* (35 ind.). However, *O. nigripes* prevailed in native vegetation. At ARC the small mammal assemblage was also dominated by *C. tener* (78 individuals), followed by *D. albiventris* (14 ind.) and *Mus musculus* (14 ind.). However, *D. albiventris* prevailed in native vegetation (Table 7.1). The general capture success rate (i.e., 100 × No. of captures/sampling effort, expressed in %) was 1.36% (1.59% for FATL and 1.09% for ARC). More specifically, at FATL, the sampling success for each habitat type was 1.45% in abandoned pasture, 1.33% in *Eucalyptus* plantation and 2.15% in native vegetation). At ARC, the sampling success was 0.62% in native vegetation and 1.37% in sugarcane.

4.2 Indicator Species and Dissimilarity Among Habitat Types

C. tener was the indicator species of agrosilvicultural landscapes according to the indicator species analysis (ISA) (ISA stat: 0.92, $p < 0.001$). At FATL, *A. montensis*, *D. albiventris*, *C. subflavus* and *N. squamipes* were identified as indicator species for native vegetation (ISA stat: 0.5770, $p = 0.002$; ISA stat: 0.81, $p = 0.001$; ISA stat: 0.37, $p = 0.045$ and ISA stat: 0.43, $p = 0.027$ respectively). Additionally, in native vegetation at ARC, *Rhipidomys* sp. was found to be the indicator species (ISA stat: 0.43, $p = 0.010$).

When analyzing the dissimilarity of species in terms of capture frequency at different sampling units in FATL and ARC, *A. montensis* exhibited a distribution pattern closely related to *D. albiventris*. This pair of species showed less dis-

Table 7.1 Diversity of small mammals in agricultural and silvicultural landscapes (respectively, Agrícola Rio Claro at Lençóis Paulista municipality and *Fazendas* Arca and Três Lagoas at Angatuba municipality) in, São Paulo state, Brazil

				Angatuba				Lençóis Paulista			
Order	Family	Species	Common names	NV	AP	Eu	Total	NV	Sc	Total	Grand total
Didelphimorphia	Didelphidae	*Didelphis albiventris*	White-eared opossum	21	2	12	35	14	0	14	49
		Gracilinanus microtarsus	Brazilian gracile opossum	1	0	0	1	1	0	1	2
		Marmosa paraguayana	Tate's wooly mouse opossum	7	0	0	7	0	0	0	7
Rodentia	Cricetidae	*Akodon montensis*	Montane grass mouse	18	0	0	18	1	0	1	19
		Calomys tener	Delicate vesper mouse	0	4	86	90	0	78	78	168
		Cerradomys subflavus	Flavescent cerradomys	0	4	0	4	0	0	0	4
		Holochilus brasiliensis	Brazilian marsh rat	2	0	0	2	0	0	0	2
		Necromys lasiurus	Hairy-tailed akodont	1	1	0	2	0	6	6	8
		Nectomys squamipes	Atlantic water rat	10	0	0	10	6	0	6	16
		Oligoryzomys flavescens	Flavescent colilargo	6	1	1	8	0	1	1	9
		Oligoryzomys nigripes	Black-footed colilargo	27	9	16	52	1	0	1	53
		Rhipidomys sp.	Long-tailed climbing mouse	0	0	0	0	3	0	3	3
	Muridae	*Mus musculus*	House mouse	0	0	0	0	0	14	14	14
		Rattus rattus	Black rat	0	0	0	0	1	0	1	1
No. of captures				93	21	115	229	27	99	126	355
Species richness				9	6	5	11	7	5	10	14
Sampling effort[a]				4320	1440	8640	14,400	4320	7200	11,520	25,920
Sampling success[b]				2.2	1.5	1.3	1.6	0.6	1.4	1.1	1.4

AP abandoned pastures, *Eu Eucalyptus* plantation, *NV* native vegetation, *Sc* sugarcane plantations

[a]Trap.nights

[b]No. of captures/sampling effort (%)

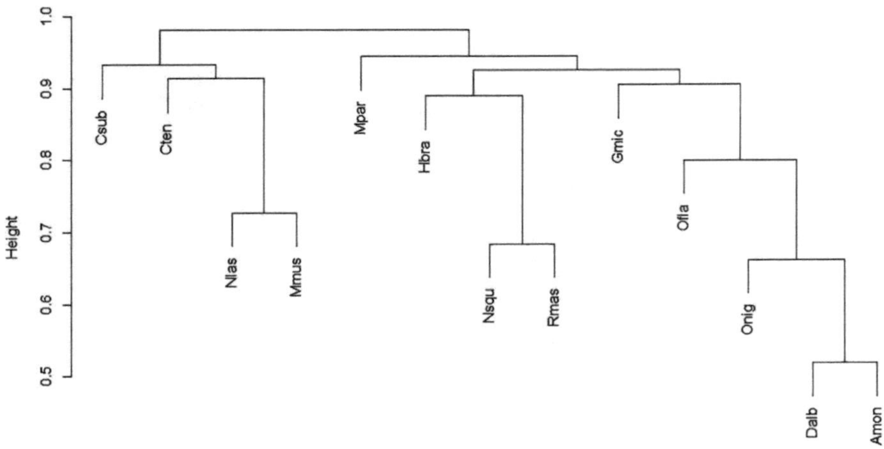

Fig. 7.2 Bray-Curtis dissimilarity analyses in agro-silvicultural landscapes (respectively, Rio Claro Agrícola at Lençóis Paulista municipality and *Fazendas* Arca and Três Lagoas at Angatuba municipality) in, São Paulo state, Brazil. (*Amon Akodon montensis*, *Cten Calomys tener*, *Csub Cerradomys subflavus*, *Dalb Didelphis albiventris*, *Gmic Gracilinanus microtarsus*, *Hbra Holochilus brasiliensis*, *Mpar Marmosa paraguayana*, *Mmus Mus musculus*, *Nlas Necromys lasiurus*, *Nsqu Nectomys squamipes*, *Ofla Oligoryzomys flavescens*, *Onig Oligoryzomys nigripes*, *Rmas Rhipidomys* sp.)

similarity than between *O. nigripes* and *O. flavescens*. These four species were associated with forests at FATL. A similar pattern was observed with the pairs *N. squamipes* and *Rhipidomys* sp., as well as *H. brasiliensis* and *M. paraguayana*, completing the group of species composing the branch represented on the right side of the dendrogram (Fig. 7.2). This branch brings together species indicating their association with natural forests and regeneration. At the other extreme of the dendrogram (left side), that reflects a preference for anthropic sites, the following species were observed: *C. subflavus*, *C. tener*, *N. lasiurus*, and *M. musculus* (exotic species) (Fig. 7.2).

4.3 The Relationship Between AGB and Small Mammals' Diversity

According to the fitted generalized linear models, the species richness of open areas (i.e., *C. tener*, *C. subflavus*, *N. lasiurus*, and *M. musculus*) decreased as AGB increased in all buffers. Conversely, species richness of forest areas (i.e., A. *montensis*, *D. albiventris*, *G. microtarsus*, *H. brasiliensis*, *M. paraguayana*, *N. squamipes*, *O. flavescens*, *O. nigripes*, and *Rhipidomys* sp.) exhibited an opposite trend (buffer 50 m: $LR = 52.30$, d.f. = 1, $p < 0.001$; buffer 100 m: $LR = 56.57$, d.f. = 1, $p < 0.001$; buffer 200 m: $LR = 45.35$, d.f. = 1, $p < 0.001$; buffer 300 m:

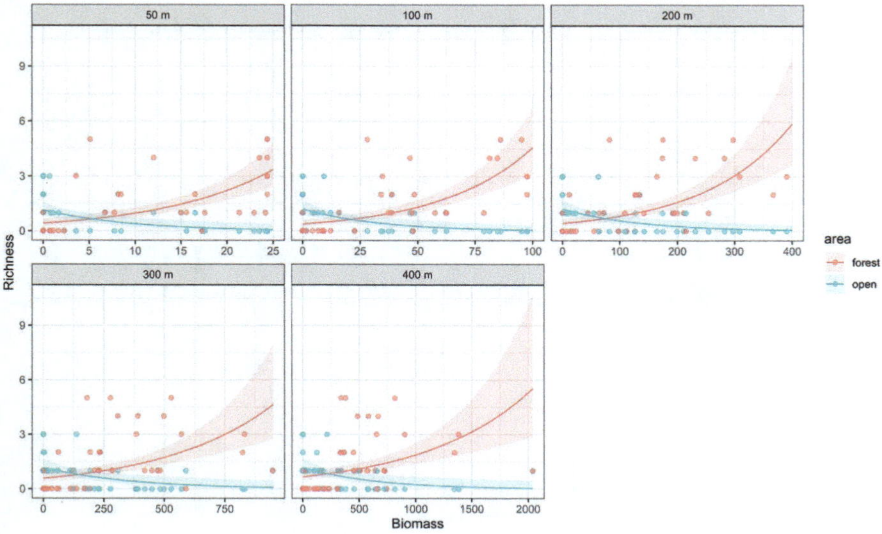

Fig. 7.3 Estimated mean species richness (solid lines) and 95% confidence intervals for the true mean richness (shaded ribbons) from the Poisson generalized linear model fitted to species richness of functional groups (open areas species and forest species) using above-ground biomass (AGB) as the predictor, at buffers of 50, 100, 200, 300 and 400 m in agro-silvicultural landscapes (respectively, Rio Claro Agrícola at Lençóis Paulista municipality and *Fazendas* Arca and Três Lagoas at Angatuba municipality) in, São Paulo state, Brazil

LR = 29.08, d.f. = 1, $p < 0.001$; buffer 400 m: LR = 22.34, d.f. = 1, $p < 0.001$) (Figs. 7.3 and 7.4). A similar pattern occurred with relative abundance: the number of individuals in species of open areas decreased as biomass increased in all buffers, whereas the abundance of forest species increased (buffer 50 m: $F_{1,104} = 55.79$, $p < 0.001$; buffer 100 m: $F_{1,104} = 66.11$, $p < 0.001$; buffer 200 m: $F_{1,104} = 48.26$, $p < 0.001$; buffer 300 m: $F_{1,104} = 22.73$, $p < 0.001$; buffer 400 m: $F_{1,104} = 46.27$, $p < 0.001$) (Figs. 7.3 and 7.4). These patterns corroborate our first hypothesis (H_1).

The relative abundance of each species significantly varied in relation to AGB for *A. montensis*, *C. tener*, *D. albiventris* and *N. squamipes* at 50, 100, 200, 300 and 400 m, and for *O. nigripes* at 100, 200, 300 and 400 m (Table 7.2; Fig. 7.5) This result partially corroborates our second hypothesis (H_2).

5 Discussion

In this study we identified a range of small mammals commonly found in agro-silvicultural areas within São Paulo State, Brazil, as documented by several other studies (Gheler-Costa et al. 2012; Lyra-Jorge and Pivello 2001; Martin et al. 2012; Rosalino et al. 2014; Umetsu and Pardini 2007; Vásquez et al. 2021) which comprise

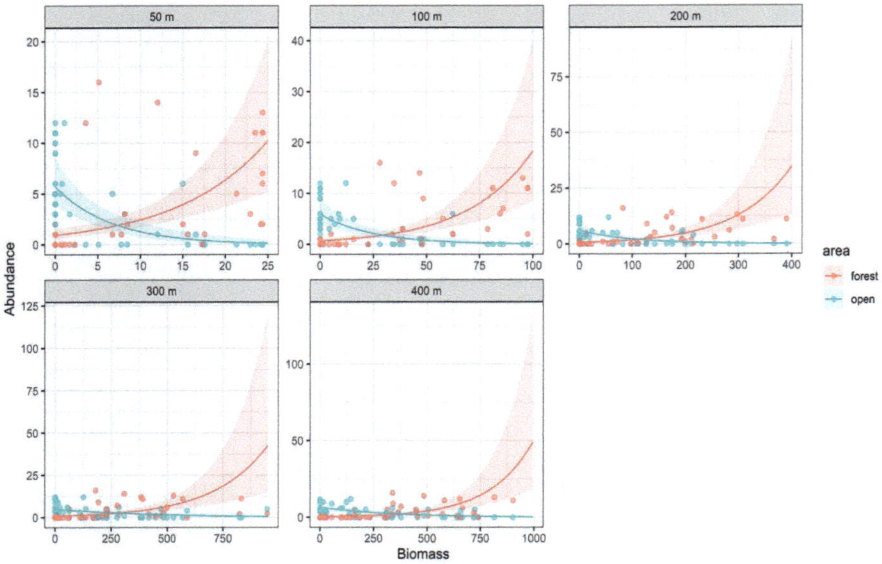

Fig. 7.4 Estimated mean species richness (solid lines) and 95% confidence intervals for the true mean richness (shaded ribbons) from the negative binomial generalized linear model fitted to the abundance data of species functional groups (open area species and forest species) using above-ground biomass (AGB) as the predictor, at buffers of 50, 100, 200, 300 and 400 m in agro-silvicultural landscapes (respectively, Rio Claro Agrícola at Lençóis Paulista municipality and *Fazendas* Arca and Três Lagoas at Angatuba municipality) in, São Paulo state, Brazil

10% of the 133 native small mammals species found in the Cerrado (Alho 2005; Carmignotto et al. 2014; Ernest and Mares 1986; Gardner 2007; Patton et al. 2015). However, we obtained relatively low capture success, which can be related to the type of trap used. Sherman traps vary in effectiveness depending on their size, location, bait, and how they interact with the morphological or ecological characteristics of the animals and the composition of the local community (De Cáceres et al. 2010; Rezende et al. 2012). Sherman traps are adequate for population studies as they reduce the mortality of captured animals, increase the likelihood of recaptures, and offer a high probability of capturing larger animals which can escape pitfall traps by climbing or leaping (White et al. 1982; De Barros et al. 2015; Stephens and Anderson 2014).

The species distribution observed in the study area reflects their habitat adaptations. For example, studies in the Cerrado have suggested a patchy distribution for the nonvolant small mammals (e.g., Ernest and Mares 1986; Lacher and Alho 2001; Santos-Filho et al. 2012) with distinct species assemblages found in different habitats (Carmignotto et al. 2012). Also, the relative abundance of species varies among localities, and some species can be found in high numbers at a given site but may be rare or even absent in others (e.g., Marinho-Filho et al. 1994). Usually, most species tend to be locally rare, and few are very abundant within all or most assemblages (Vieira and Palma 2005). Usually, agro-silvicultural

Table 7.2 F-test statistics and associated p-values for the slopes obtained from the negative binomial generalized linear model fitted to the abundance data of small mammal species using above-ground biomass (AGB) as the predictor, at buffers of 50, 100, 200, 300, and 400 m in agricultural and silvicultural landscapes (respectively, Agrícola Rio Claro-ARC at Lençóis Paulista municipality and Fazendas Arca and Três Lagoas-FATL at Angatuba municipality) in, São Paulo state, Brazil

	50 m		100 m		200 m		300 m		400 m	
Species	F statistic	p value	F statistic	p value	F statistic	p value	F statistic	p value	F statistic	p value
A. montensis	25.14	<0.001	35.88	<0.001	26.06	<0.001	15.23	<0.001	12.12	<0.001
C. tener	44.19	<0.001	42.28	<0.001	24.20	<0.001	6.81	0.012	4.38	0.041
D. albiventris	42.15	<0.001	49.46	<0.001	40.91	<0.001	24.38	<0.001	19.26	<0.001
N. squamipes	33.28	<0.001	39.57	<0.001	39.21	<0.001	18.06	<0.001	12.96	<0.001
O. nigripes	2.88	0.096	5.01	0.030	4.79	0.033	3.51	0.067	6.03	0.017

All F-test statistics are associated with 1 and 52 degrees of freedom

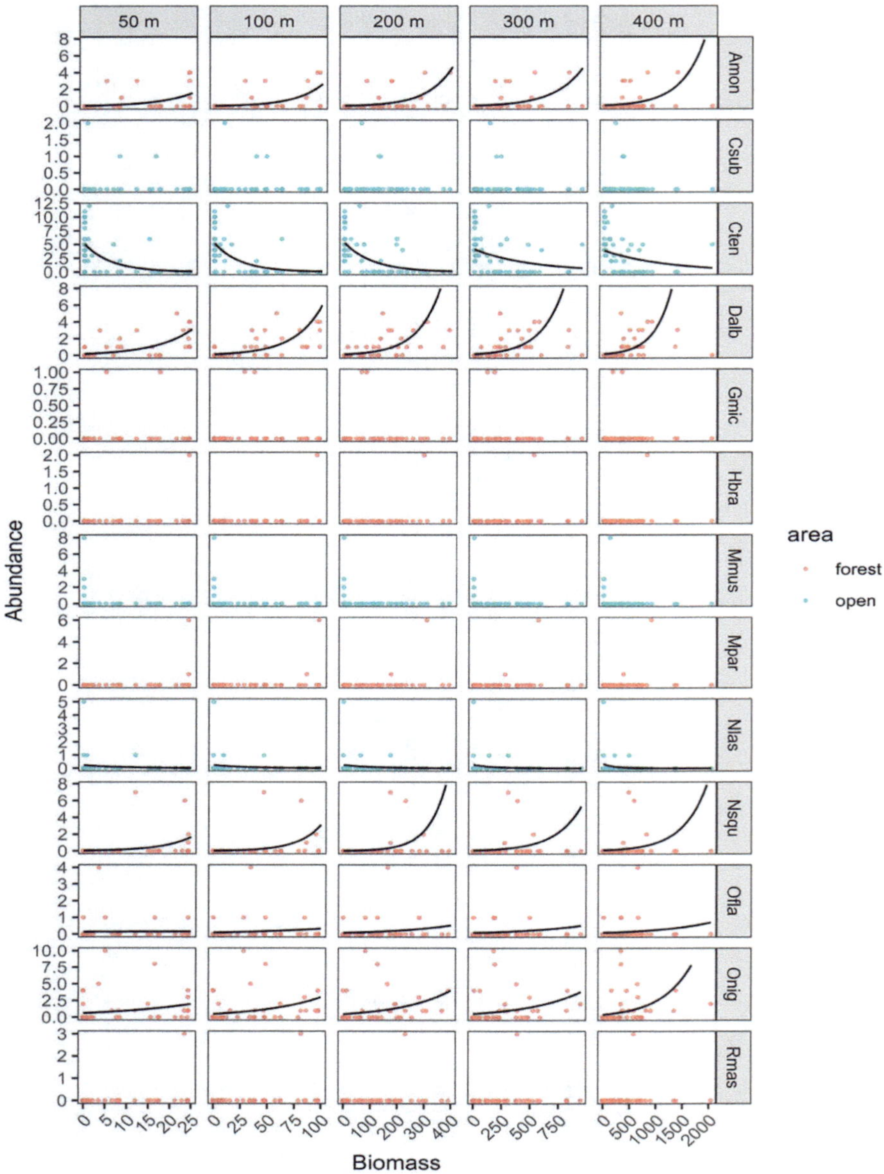

Fig. 7.5 Estimated mean abundances from the negative binomial generalized linear models considering all species (*Amon Akodon montensis*, *Cten Calomys tener*, *Csub Cerradomys subflavus*, *Dalb Didelphis albiventris*, *Gmic Gracilinanus microtarsus*, *Hbra Holochilus brasiliensis*, *Mpar Marmosa paraguayana*, *Mmus Mus musculus*, *Nlas Necromys lasiurus*, *Nsqu Nectomys squamipes*, *Ofla Oligoryzomys flavescens*, *Onig Oligoryzomys nigripes*, *Rmas Rhipidomys* sp.), using above-ground biomass (AGB) as the predictor, at buffers of 50, 100, 200, 300 and 400 m in agro-silvicultural landscapes (respectively, Rio Claro Agrícola at Lençóis Paulista municipality and Fazendas Arca and Três Lagoas at Angatuba municipality) in, São Paulo state, Brazil

landscapes that maintain both agriculture and native patches have their habitat spatial heterogeneity increased, which is beneficial for generalist and opportunistic mammals (MacDonald et al. 2007; Mills et al. 1991). The generalist species tend to occur in almost all habitats within the agricultural landscape, whereas specialist species tend to occur in habitats with high vegetation cover or with specific habitat requirements (Coda et al. 2015; Gomez et al. 2017; Martínez et al. 2014).

Our findings revealed that *C. tener*, *N. lasiurus* and *C. subflavus* occur specifically in open areas (De Vivo and Carmignotto 2015; Percequillo et al. 2008) of the Cerrado (e.g., grasslands, gallery forests, woodland savanna) (Percequillo 2015; Salazar-Bravo 2015; Ulyses et al. 2015), but also in forest formations of the Cerrado (cerradão) and Atlantic Forest (secondary vegetation) adjacent to Cerrado (Patton et al. 2015). In fact, *C. tener* was the most abundant rodent in both study sites, being the most widespread genus of rodent fauna, found in most habitats (Salazar-Bravo 2015). *N. lasiurus* was primarily captured in sugarcane habitats and *C. subflavus* was only captured in abandoned pasture habitat, which aligns with findings from other studies conducted in open vegetation areas in Brazil (Alho 2005; Becker et al. 2007; Bonvicino et al. 2014; Pires et al. 2010).

On the other hand, A. *montensis*, *D. albiventris*, *M. paraguayana*, *G. microtarsus*, *H. brasiliensis*, *N. squamipes*, *O. flavescens*, *O. nigripes* and *Rhipidomys* sp. are species more associated with forest habitats (Gardner 2007; Patton et al. 2015). The marsupial *D. albiventris* occurs in the Caatinga and Cerrado habitats, as well as enclaves (Gardner 2007) with a greater abundance in forest habitats, although absent in sugarcane plantations, according to this study. *Oligoryzomys nigripes* is the most habitat-generalist rodent among all Brazilian *Oligoryzomys* species. It occurs in both primary and secondary vegetation in the Atlantic Forest and Cerrado (Bonvicino and Weksler 1998), and this is related to its high abundance in native vegetation at FATL. The species *H. brasiliensis* and *N. squamipes* were only detected in native vegetation, in the proximity of water bodies, as these species are associated with riparian habitats (Gardner 2007). *Oligoryzomys flavescens* occurs in the Atlantic Forest, as well as in Cerrado gallery forests (Patton et al. 2015) and agroecosystems; this species was found to be associated with high vegetation cover and graminoid richness (Gorosito et al. 2018), however in this study its capture rate was relatively low. The marsupials (*M. paraguayana* and *G. microtarsus*) are associated with Cerrado and forest physiognomies (Alho 2005; Gardner 2007), with *G. microtarsus* mainly located in tree crowns, showing a preference for forest environments (Paglia et al. 2020; Patton et al. 2015), although its capture rate in this study was low. *Rhipidomys* sp. are arboreal rodents that occurs in remnants of the eastern Atlantic Forest and is considered a generalist habitat species with a high tolerance for fragmented environments (Grelle 2003; Pardini 2004; Pardini and Umetsu 2006; Patton et al. 2015).

Furthermore, our findings reveal that as biomass increases, there is a shift in species abundance, characterized by an exchange from open-area species to forest species. This supports the hypothesis that the population dynamics of small mammals may be influenced by their interaction with biomass in agro-silvicultural landscapes. Thus, species respond to ecological scales based on their individual ecological needs, spatial behaviors, mobility, and reliance on habitat (Dorigo et al. 2021; Kozakiewicz et al. 1999; Serafini et al. 2019). For instance, *C. tener* was dominant in the *Eucalyptus* plantation at FATL where *Eucalyptus* trees were still relatively small (early third cycle), giving the plantation an open habitat characteristic. In contrast, *O. nigripes* and *D. albiventris* replace it as the dominant species in native vegetation habitats and in the *Eucalyptus* plantation at ARC, where *Eucalyptus* trees reached a considerable height (final second cycle). In the same way, *C. tener* dominates in open areas such as sugarcane fields, where homogeneity and uniformity in vegetation structure may exclude specialized species (Hurst et al. 2014), while *D. albiventris* dominates in forested areas such as native vegetation habitats. Thus, landscape heterogeneity plays a crucial role in biodiversity conservation and promotes the persistence of species within agrosilvicultural landscapes (Ecke et al. 2002; Scott et al. 2008; Serafini et al. 2019).

Specifically, the relationship between the abundance of *A. montensis*, *C. tener*, *D. albiventris*, *N. lasiurus* and *N. squamipes* with AGB in different buffers is attributed to ecological preferences and use of space for each species. *N. lasiurus* and *C. tener* are usually the most abundant rodents in studies conducted in open vegetation. Home ranges of *N. lasiurus* vary from 0.02 to 0.52 ha (e.g., Pires et al. 2010) and of *C. tener*, the ranges are up to 0.32 for males and up to 0.24 ha for females. The daily distance traveled by *C. tener* and *N. lasiurus* is around 50 m (Wilson et al. 2017), showing that they are short-moving species.

The association between *N. squamipes* and biomass is linked to aquatic environments, with the shape of its home range tending to follow the shape of water bodies (Blundell et al. 2001; Lima et al. 2015). *A. montensis* is found in forest patches within highly deforested landscapes and agricultural areas across all landscapes (Püttker et al. 2013; Umetsu and Pardini 2007). The home range of male individuals of *A. montensis* is between 1460 and 1092 m^2 for females, with linear distances moved between trapping sessions ranging from 100 to 370 m (Wilson et al. 2017). *Didelphis albiventris* shows notable adaptability to fragmented habitats, using both canopy and ground levels (Vieira and Camargo 2012). It inhabits a large home range averaging 2.33 ha (Chemisquy et al. 2021; Sanches et al. 2012). Accordingly, these last two species show the ability to cover long distances.

Furthermore, regarding the relationship between forest species abundance AGB, higher abundances were generally recorded in the 50 and 100 m buffers. This finding supports the arguments of several authors, who suggest that demographic variability in small mammal populations occurs over relatively short distances (approximately 100 m) (Bowman et al. 2000, 2001; Krohne and Burgin 1990). In this way, movement patterns of animals tend to mirror the structure of the environment. In heterogeneous forests, an effective strategy for facilitating movement seems to involve a combination of extensive long-distance movements interspersed with concentrated shorter moves within selected patches (Wells et al. 2006).

The present results emphasize the importance of patches of native vegetation in the agricultural matrix for the conservation of small mammals' diversity, and therefore, to increase the agricultural landscapes conservation value (Dorigo et al. 2021; Verdade et al. 2014). Small mammals are impacted by alterations in plant communities, which lead to changes in habitat structure and food availability (Fraschina et al. 2012). According to Biz et al. (2017), quantifying the effects of the matrix type provides information on the species movement behavior and its persistence in fragmented landscapes. This is the case of highly permeable *Eucalyptus* plantations with remnants of isolated native vegetation. In this way, relatively simple indices can improve our understanding of species distribution, and could be applied in modeling, monitoring, and managing complex tropical landscapes (Umetsu et al. 2008).

Changes in biomass directly influence primary consumers such as small mammals, serving as an indicator of food availability and vegetation cover, thereby impacting their abundance and species richness (Reed et al. 1994; Rossi and Leiner 2022; Smith et al. 2017; Szymañski et al. 2020). Verschuyl et al. (2011) suggests that intensive biomass production in forests supports a large proportion of biodiversity, and can change the species composition, nutrient cycles and, ultimately, habitat biodiversity. Therefore, small mammals can serve as valuable indicators of sustainable forest management, offering insights into biodiversity structure and dynamics (Teder et al. 2007; Trivellone et al. 2014), being able to assist decision-making processes for public policies, local managers, and stakeholders (Balmford 2005; Rüdisser et al. 2012). For instance, this approach can contribute to the management of agricultural areas in compliance with the Brazilian Forest Code.

6 Conclusions

The following conclusions can be taken from this study:

(a) In agrosilvicultural landscapes, an increase in biomass leads to a tendency of an increase in abundance of small mammals, when transitioning from open area species to forest species;
(b) Small mammals, particularly *Akodon montensis*, *Calomys tener*, *Didelphis albiventris*, *Nectomys squamipes* and *Oligoryzomys nigripes*, show potential as ecological indicators of the processes occurring within agrosilvicultural landscapes;
(c) The findings of this study support the use of native vegetation AGB as a proxy for assessing small mammal abundance in agricultural landscapes. Moreover, native vegetation recovery emerges as one of the most effective strategies for enhancing the conservation value of agricultural landscapes;
(d) Future studies should prioritize the influence of AGB of small mammals' diversity in other biomes, like the Amazon Forest, Caatinga, Pampas, and Pantanal; and,
(e) Long-term ecological monitoring is necessary to assess evolutionary and ecological patterns, including spatial-temporal dimensions, within anthropic

environments. The methodology employed in this study might be valuable in such circumstances.

Acknowledgements This study was supported by Fundação de Amparo à Pesquisa do Estado de São Paulo (Proc. No. 2017/01304-4) and PhD fellowship was obtained from Conselho Nacional de Desenvolvimento Científico e Tecnológico (Cnpq) (Proc. No. 142456/2019-5) and Coordenação de Aperfeiçoamento de Pessoal de Nível Superior (Capes) (Proc. No. 33002037020P1). Appropriate licenses were taken from the Instituto Chico Mendes de Conservação da Biodiversidade—ICMBio (Proc. No. 71789) and "Comissão de Ética no Uso dos Animais" (CEUA) of University of São Paulo/CENA (Proc. No. 003-2019). Alexandre Percequillo and Edson Fiedler de Abreu helped with the identification of animal species and Erico Silva assisted with surveying the vegetation. Fazendas Três Lagoas and Arca and Agrícola Rio Claro provided operational and logistic support.

References

Agritempo (2022) Sistema de Monitoreo Agrometeorológico. http://www.agritempo.gov.br/agritempo/index.jsp

Alho CJR (2005) Intergradation of habitats of non-volant small mammals in the patchy Cerrado landscape. Arquivos Do Museu Nacional, Rio de Janeiro 63(1):41–48

Anderson MJ, Walsh DCI, Robert Clarke K, Gorley RN, Guerra-Castro E (2017) Some solutions to the multivariate Behrens–fisher problem for dissimilarity-based analyses. Aust N Z J Stat 59(1):57–79. https://doi.org/10.1111/anzs.12176

APG – Angiosperm Phylogeny Group, Chase MW, Christenhusz MJM, Fay MF, Byng JW, Judd WS, Soltis DE et al (2016) An update of the Angiosperm Phylogeny Group classification for the orders and families of flowering plants: APG IV. Bot J Linn Soc 181(1):1–20

Archibald RD, Craig MD, Bialkowski K, Howe C, Burgess TI, Hardy GESJ (2011) Managing small remnants of native forest to increase biodiversity within plantation landscapes in the south west of Western Australia. For Ecol Manag 261(7):1254–1264. https://doi.org/10.1016/j.foreco.2011.01.004

Athayde E (2013) Importância da heterogeneidade natural na conservação da diversidade de árvores em paisagem agrosilvicultural. Doctoral thesis, UNESP

Atkeson TD, Johnson AS (1978) Succession of small mammals on pine plantations in the Georgia Piedmont. Am Midl Nat 101:2

Balmford A (2005) Ecology: the convention on biological diversity's 2010 target. Science 307(5707):212–213. https://doi.org/10.1126/science.1106281

Barros CS, Püttker T, Pardini R (2015) Timing and environmental cues associated with triggering of reproductive activity in Atlantic forest marsupials. Mamm Biol 80(2):141–147. https://doi.org/10.1016/j.mambio.2014.07.008

Becker RG, Paise G, Baumgarten LC, Vieira EM (2007) Small mammals' community structure and density of Necromys lasiurus (Rodentia, Sigmodontinae) in open areas of the Cerrado in central Brazil/Estrutura de comunidades de pequenos mamiferos e densidade de Necromys lasiurus (Rodentia, Sigmodontinae) em areas abertas de cerrado no Brasil Central. Mastozoologia Neotropical 2(14):157–169

Benedek AM, Sîrbu I (2018a) Responses of small mammal communities to environment and agriculture in a rural mosaic landscape. Mamm Biol 90:55–65. https://doi.org/10.1016/j.mambio.2018.02.008

Benedek AM, Sîrbu I (2018b) Responses of small mammal communities to environment and agriculture in a rural mosaic landscape. Mamm Biol 90:55–65

Biz M, Cornelius C, Metzger JPW (2017) Matrix type affects movement behavior of a neotropical understory forest bird. Perspect Ecol Conserv 15(1):10–17. https://doi.org/10.1016/j.pecon.2017.03.001

Blundell G, Maier J, Debevec E (2001) Linear home ranges: effects of smoothing, sample size, and autocorrelation on kernel estimates. Ecol Monogr 71:469. https://doi.org/10.2307/3100069

Bonvicino CR, Weksler M (1998) A new species of Oligoryzomys (Rodentia, Sigmodontinae) from Northeastern and Central Brazil. Zeitschrift Für Säugetierkunde 63:90–103

Bonvicino CR, Casado F, Weksler M (2014) A new species of Cerradomys (Mammalia: Rodentia: Cricetidae) from Central Brazil, with remarks on the taxonomy of the genus. Zoologia (Curitiba) 31:525–540. https://doi.org/10.1590/S1984-46702014000600002

Bowman J, Forbes G, Dilworth T (2000) The spatial scale of variability in small-mammal populations. Ecography 23(3):328–334. http://www.jstor.org/stable/3683103

Bowman J, Forbes GJ, Dilworth TG (2001) The spatial component of variation in small-mammal abundance measured at three scales. Can J Zool 79(1):137–144. https://doi.org/10.1139/z00-188

Briani DC, Palma ART, Vieira EM, Henriques RPB (2004) Post-fire succession of small mammals in the Cerrado of central Brazil. Biodiversity Conserv 13:1023–1037

Cáceres NC, Monteiro-Filho ELA (2001) Food habits, home range and activity of Didelphis aurita (Mammalia, Marsupialia) in a Forest fragment of southern Brazil. Stud Neotropical Fauna Environ 36(2):85–92. https://doi.org/10.1076/snfe.36.2.85.2138

Carmignotto AP, de Vivo M, Langguth A (2012) Mammals of the Cerrado and Caatinga. In: Bones, clones, and biomes. University of Chicago Press, Chicago, IL, pp 307–350. https://doi.org/10.7208/chicago/9780226649214.003.0014

Carmignotto AP, Bezerra AMR, Rodrigues FHG (2014) Nonvolant small mammals from a southwestern area of Brazilian Cerrado: diversity, habitat use, seasonality, and biogeography. Therya 5(2):535–558. https://doi.org/10.12933/therya-14-197

Cassano CR, Barlow J, Pardini R (2012) Large mammals in an agroforestry Mosaic in the Brazilian Atlantic Forest. Biotropica 44(6):818–825. https://doi.org/10.1111/j.1744-7429.2012.00870.x

Catling P, Coops N, Burt R (2001) The distribution and abundance of ground-dwelling mammals in relation to time since wildfire and vegetation structure in South-Eastern Australia. Wildl Res 28:555–565. https://doi.org/10.1071/WR00041

Caudill SA, DeClerck FJA, Husband TP (2015) Connecting sustainable agriculture and wildlife conservation: does shade coffee provide habitat for mammals? Agric Ecosyst Environ 199:85–93. https://doi.org/10.1016/j.agee.2014.08.023

CEPAGRI (2021) Centro de Pesquisas Meteorológicas e Climáticas Aplicadas a Agricultura. http://www.cpa.unicamp.br

Chave J, Andalo C, Brown S, Cairns MA, Chambers JQ, Eamus D, Fölster H, Fromard F, Higuchi N, Kira T, Lescure J-P, Nelson BW, Ogawa H, Puig H, Riéra B, Yamakura T (2005) Tree allometry and improved estimation of carbon stocks and balance in tropical forests. Oecologia 145(1):87–99. https://doi.org/10.1007/s00442-005-0100-x

Chave J, David C, Jansen S, Lewis SL, Swenson NG, Zanne AE (2009) Towards a worldwide wood economics spectrum. Ecol Lett 12(4):351–366

Chave J, Réjou-Méchain M, Búrquez A, Chidumayo E, Colgan MS, Delitti WBC, Duque A, Eid T, Fearnside PM, Goodman RC, Henry M, Martínez-Yrízar A, Mugasha WA, Muller-Landau HC, Mencuccini M, Nelson BW, Ngomanda A, Nogueira EM, Ortiz-Malavassi E et al (2014) Improved allometric models to estimate the aboveground biomass of tropical trees. Glob Chang Biol 20(10):3177–3190. https://doi.org/10.1111/gcb.12629

Chemisquy MA, Morinigo FM, Fameli A, González-Ittig RE (2021) Genetic diversity of the white- eared opossum Didelphis albiventris (Didelphimorphia: Didelphidae) in Argentina. Mastozoología Neotropical 28(1):1–11. https://doi.org/10.31687/saremMN.21.28.1.0.20

Chmielewski S, Chmielewski TJ, Tompalski P (2014) Land cover and landscape diversity analysis in the west Polesie biosphere reserve. Int Agrophys 28(2):153–162. https://doi.org/10.2478/intag-2014-0003

Churchfield S (1997) Community structure and habitat use of small mammals in grasslands of different successional age. J Zool 242:519–530

Clarke DJ, White JG (2008) Recolonisation of powerline corridor vegetation by small mammals: timing and the influence of vegetation management. Landsc Urban Plan 87(2):108–116. https://doi.org/10.1016/J.LANDURBPLAN.2008.04.009

Coda J, Gomez D, Steinmann AR, Priotto J (2015) Small mammals in farmlands of Argentina: responses to organic and conventional farming. Agric Ecosyst Environ 211:17–23. https://doi.org/10.1016/j.agee.2015.05.007

Corry RC (2005) Characterizing fine-scale patterns of alternative agricultural landscapes with landscape pattern indices. Landsc Ecol 20(5):591–608. https://doi.org/10.1007/s10980-004-5036-8

Coutinho LM (1978) O conceito do cerrado. Rev Bras Bot 1:17–23

Daily G, Ceballos G, Pacheco J, Suzan G, Anchez-Azofeifa A (2003) Countryside biogeography of neotropical mammals: conservation opportunities in agricultural landscapes of Costa Rica. Conserv Biol 17:1473–1884. https://doi.org/10.1111/j.1523-1739.2003.00298.x

De Barros C, Püttker T, Trevizan Pinotti B, Pardini R (2015) Determinants of capture-recapture success: an evaluation of trapping methods to estimate population and community parameters for Atlantic forest small mammals. Zoologia (Curitiba) 32:334–344. https://doi.org/10.1590/S1984-46702015000500002

De Cáceres M, Legendre P, Moretti M (2010) Improving indicator species analysis by combining groups of sites. Oikos 119(10):1674–1684. https://doi.org/10.1111/j.1600-0706.2010.18334.x

De Carvalho EX, Menezes RSC, de Sá Barreto Sampaio EV, DES N, Tabosa JN, de Oliveira LR, Simões AL, Sales AT (2019) Allometric equations to estimate sugarcane aboveground biomass. Sugar Tech 21:1039–1044. https://doi.org/10.1007/s12355-019-00741-w

De Vivo M, Carmignotto AP (2015) Family Sciuridae G. Fischer, 1817. In: Patton JL (ed) Mammals of South America, Volume 2: Rodents. University of Chicago Press, Chicago, IL, pp 1–48

Deodoro S et al (2026) Chapter 6: The use of remote sensing data for above-ground biomass estimation. In: Verdade LM, Moral RA (eds) Wildlife monitoring. Springer, Cham

DeWalt SJ, Maliakal SK, Denslow JS (2003) Changes in vegetation structure and composition along a tropical forest chronosequence: implications for wildlife. For Ecol Manag 182(1):139–151. https://doi.org/10.1016/S0378-1127(03)00029-X

Donatelli RJ, de Costa TVV, Ferreira CD (2004) Dinâmica da avifauna em fragmento de mata na Fazenda Rio Claro, Lençóis Paulista, São Paulo Brasil. Rev Bras Zool 21(1):97–114. https://doi.org/10.1590/S0101-81752004000100017

Dorigo L, Boscutti F, Sigura M (2021) Landscape and microhabitat features determine small mammal abundance in forest patches in agricultural landscapes. Peer J 9:e12306. https://doi.org/10.7717/peerj.12306

Duraflora (1999) Identificação, mapeamento e quantificação de vegetal natural: fazenda Rio Claro. Lençóis Paulista - Águas de Santa Bárbara, municípios

Ecke F, Löfgren O, Sörlin D (2002) Population dynamics of small mammals in relation to forest age and structural habitat factors in northern Sweden. J Appl Ecol 39:781–792. https://doi.org/10.1046/j.1365-2664.2002.00759.x

Ernest KA, Mares MA (1986) Ecology of Nectomys squamipes, the neotropical water rat, in Central Brazil: home range, habitat selection, reproduction and behaviour. J Zool 210(4):599–612. https://doi.org/10.1111/j.1469-7998.1986.tb03658.x

Fox BJ (1982) Fire and mammalian secondary succession in an Australian coastal heath. Ecology 83:1332–1341

Fox BJ (1990) Changes in the structure of mammal communities over successional time scales. Oikos 59:321–329

Fox BJ, Taylor JE, Thompson PT (2003) Experimental manipulation of habitat structure: a retrogression of the small mammal succession. J Anim Ecol 72(6):927–940. https://doi.org/10.1046/j.1365-2656.2003.00765.x

Fraschina J, León VA, Busch M (2012) Long-term variations in rodent abundance in a rural landscape of the pampas. Argentina Ecol Res 27(1):191–202. https://doi.org/10.1007/s11284-011-0888-2

Friend GR (1982) Bird populations in exotic pine plantations and indigenous eucalypt forests in gippsland, Victoria. Emu 82(2):80–91. https://doi.org/10.1071/MU9820080

Gardner AL (2007) Mammals of South America. Volume 1: marsupials, xenarthrans, shrews, and bats. J Mammalogy 90(2):521–523. https://doi.org/10.1644/08-MAMM-R-296.1

Gheler-Costa C, Vettorazzi CA, Pardini R, Verdade LM (2012) The distribution and abundance of small mammals in agroecosystems of southeastern Brazil. Mammalia 76(2):185–191. https://doi.org/10.1515/mammalia-2011-0109

Gomez MD, Coda JA, Serafini VN, Steinmann AR, Priotto JW, Gomez MD (2017) Small mammals in agroecosystems: responses to land use intensity and farming management. Mastozoología Neotropical 24(2):289–300. http://www.sarem.org.ar, http://www.sbmz.com.br

Gonçalves H, Waldir S, Júnior C, De R, Dart O, Luiz M, Áglio D, Silva De Sousa J, Guedes J, Fontana, PA, Leonor A, Martins S, Pacobahyba De Oliveira A, Atualizada L (2011) Documentos 130 Empresa Brasileira de Pesquisa Agropecuária Embrapa Solos Ministério da Agricultura, Pecuária e Abastecimento. http://www.cnps.embrapa.br/solosbr/publicacao.html

Gonçalves P, Alcobia S, Simões L, Santos-Reis M (2012) Effects of management options on mammal richness in a Mediterranean agro-silvo-pastoral system. Agrofor Syst 85(3):383–395. https://doi.org/10.1007/s10457-011-9439-7

Goldemberg J, Coelho ST, Guardabassi P (2008) The sustainability of ethanol production from sugarcane. Energy Policy 36:2086–2097

Gorosito IL, Marziali Bermúdez M, Busch M (2018) Advantages of combining generalized linear models and occupancy models to find indicators of habitat selection: small mammals in agroecosystems as a case study. Ecol Indic 85:1–10. https://doi.org/10.1016/j.ecolind.2017.10.003

Grelle C (2003) Forest structure and vertical stratification of small mammals in a secondary Atlantic Forest, Southeastern Brazil. Stud Neotropical Fauna Environ 38:81–85. https://doi.org/10.1076/snfe.38.2.81.15926

He KS, Rocchini D, Neteler M, Nagendra H (2011) Benefits of hyperspectral remote sensing for tracking plant invasions. Divers Distrib 17(3):381–392. https://doi.org/10.1111/j.1472-4642.2011.00761.x

Hirth HF (1959) Small mammals in old field succession. Ecology 40(3):417–425. https://doi.org/10.2307/1929758

Hurst ZM, McCleery RA, Collier BA, Silvy NJ, Taylor PJ, Monadjem A (2014) Linking changes in small mammal communities to ecosystem functions in an agricultural landscape. Mamm Biol 79(1):17–23. https://doi.org/10.1016/j.mambio.2013.08.008

Jones C, McShea MJ, Conroy WJ, Kunz TH (1996) Capturing mammals. In: Measuring and monitoring biological diversity. Smithsonian Institution Press, Washington, DC, pp 115–122

Kelly LT, Dayman R, Nimmo DG, Clarke MF, Bennett AF (2013) Spatial and temporal drivers of small mammal distributions in a semi-arid environment: the role of rainfall, vegetation and life-history. Austral Ecol 38(7):786–797. https://doi.org/10.1111/aec.12018

Kok AD, Parker DM, Barker NP (2013) Rules of attraction: the role of bait in small mammal sampling at high altitude in South Africa. Afr Zool 48(1):84–95. https://doi.org/10.3377/004.048.0109

Kottek M, Grieser J, Beck C, Rudolf B, Rubel F (2006) World Map of the Köppen-Geiger Climate Classification Updated. Meteorologische Zeitschrift 15:259–263

Korthals GW, Smilauer P, Van Dijk C, van der Putten WH (2001) Linking above- and below-ground biodiversity: abundance and trophic complexity in soil as a response to experimental plant communities on abandoned arable land. Funct Ecol 15(4):506514. https://doi.org/10.1046/j.0269-8463.2001.00551.x

Kozakiewicz M, Gortat T, Kozakiewicz A, Barkowska M (1999) Effects of habitat fragmentation on four rodent species in a polish farm landscape. Landsc Ecol 14(4):391–400. https://doi.org/10.1023/A:1008070610187

Krohne DT, Burgin AB (1990) The scale of demographic heterogeneity in a population of Peromyscus leucopus. Oecologia 82(1):97–101. http://www.jstor.org/stable/4219207

Lacher TE, Alho CJR (2001) Terrestrial small mammal richness and habitat associations in an Amazon Forest — Cerrado contact zone. Biotropica 33(1):171–181. https://doi.org/10.1646/0006-3606(2001)033

Leda VC, Nicolete DAP, Pezzoni Filho JC, Sartori AAC, Zimback CRL (2014) Uso do Solo na Sub-Bacia do Ribeirão da Prata. Lençóis Paulista – SP Nativa 2(3):170–174. https://doi.org/10.14583/2318-7670.v02n03a08

Lima D, Medeiros de Pinho G, Fernandez F (2015) Spatial patterns of the semi-aquatic rodent Nectomys squamipes in Atlantic forest streams. J Nat Hist 50:1–15. https://doi.org/10.1080/00222933.2015.1059516

Lisboa MAM, Prado BHS (2019) Fazenda das Areias e Fazenda da Comquista: história, memória e cultura. s.n., São Paulo

Lyra-Jorge MC, Pivello VR (2001) Combining live trap and pitfall to survey terrestrial small mammals in savanna and forest habitats. Mammalia 65:524–530

MacDonald DW, Tattersall FH, Service KM, Firbank LG, Feber RE (2007) Mammals, agri-environment schemes and set-aside – what are the putative benefits? Mammal Rev 37(4):259–277. https://doi.org/10.1111/j.1365-2907.2007.00100.x

Magnusson WE, Lima AP, Luizão R, Luizão F, Costa FRC, de Castilho CV, Kinupp VF (2005) RAPELD: a modification of the gentry method for biodiversity surveys in long-term ecological research sites. Biota Neotropica 5(2):19–24. https://doi.org/10.1590/S1676-06032005000300002

Magnusson WE, Lawson B, Baccaro F, de Castilho CV, Guy Castley J, Costa F, Drucker DP, Franklin E, Lima AP, Luizão R, Mendonça F, Pezzini F, Schietti J, Toledo JJ, Tourinho A, Verdade LM, Hero J-M (2014) Multi-taxa surveys: integrating ecosystem processes and user demands. In: Verdade LM, Lyra-Jorge MC, Piña CI (eds) Applied ecology and human dimensions in biological conservation. Springer, Berlin Heidelberg, pp 177–187. https://doi.org/10.1007/978-3-642-54751-5_12

Mairota P, Cafarelli B, Boccaccio L, Leronni V, Labadessa R, Kosmidou V, Nagendra H (2013) Using landscape structure to develop quantitative baselines for protected area monitoring. Ecol Indic 33:82–95. https://doi.org/10.1016/j.ecolind.2012.08.017

Mairota P, Cafarelli B, Didham RK, Lovergine FP, Lucas RM, Nagendra H, Rocchini D, Tarantino C (2015) Challenges and opportunities in harnessing satellite remote-sensing for biodiversity monitoring. Eco Inform 30:207–214. https://doi.org/10.1016/j.ecoinf.2015.08.006

Mamba M, Fasel NJ, Mahlaba TAM, Austin JD, McCleery RA, Monadjem A (2019) Influence of sugarcane plantations on the population dynamics and community structure of small mammals in a savanna-agricultural landscape. Global Ecol Conserv 20:e00752. https://doi.org/10.1016/j.gecco.2019.e00752

Marinho-Filho J, Reis ML, Oliveira PS, Vieira EM, Paes MN (1994) Density and small mammal numbers: conservation of the Cerrado biodiversity. Anais Academia Brasileira de Ciências 66:149–157

Martin PS, Gheler-Costa C, Lopes PC, Rosalino LM, Verdade LM (2012) Terrestrial non-volant small mammals in agro-silvicultural landscapes of Southeastern Brazil. For Ecol Manag 282:185–195. https://doi.org/10.1016/J.FORECO.2012.07.002

Martínez JJ, Millien V, Simone I, Priotto JW (2014) Ecological preference between generalist and specialist rodents: spatial and environmental correlates of phenotypic variation. Biol J Linn Soc 112(1):180–203. https://doi.org/10.1111/bij.12268

Martins CS (1997) Uso de habitat pelo bugio, Alouatta fusca Clamitans, em um fragmento florestal em Lençois Paulista-SP. [Unicamp]. http://repositorio.unicamp.br/jspui/handle/REPOSIP/315850

McAlpine CA, Eyre TJ (2002) Testing landscape metrics as indicators of habitat loss and fragmentation in continuous eucalypt forests (Queensland, Australia). Landsc Ecol 17(8):711–728. https://doi.org/10.1023/A:1022902907827

Metzger JP, Lewinsohn TM, Joly CA, Verdade LM, Martinelli LA, Rodrigues RR (2010) Brazilian law: full speed in reverse? Science 329(5989):276–277. https://doi.org/10.1126/science.329.5989.276-b

Michel N, Burel F, Butet A (2006) How does landscape use influence small mammal diversity, abundance and biomass in hedgerow networks of farming landscapes? Acta Oecol 30(1):11–20. https://doi.org/10.1016/j.actao.2005.12.006

Millan CH, Develey PF, Verdade LM (2015) Stand-level management practices increase occupancy by birds in exotic eucalyptus plantations. For Ecol Manag 336:174–182. https://doi.org/10.1016/j.foreco.2014.10.005

Mills J, Ellis B, McKee K, Maiztegui J, Childs J (1991) Habitat associations and relative densities of rodent populations in cultivated areas of Central Argentina. J Mammal 72(3):470–479. https://doi.org/10.2307/1382129

Monamy V, Fox BJ (2000) Small mammal succession is determined by vegetation density rather than time elapsed since disturbance. Austral Ecol 25(6):580–587. https://doi.org/10.1111/j.1442-9993.2000.tb00063.x

Montagu KD, Düttmer K, Barton CVM, Cowie AL (2005) Developing general allometric relationships for regional estimates of carbon sequestration—an example using Eucalyptus pilularis from seven contrasting sites. For Ecol Manag 204(1):115–129. https://doi.org/10.1016/j.foreco.2004.09.003

Mora AL, Garcia CH (2000) A cultura do eucalipto no Brasil. Sociedade Brasileira de Silvicultura (SBS)

Moral RA, Hinde J, Demétrio CGB (2017) Half-normal plots and overdispersed models in R: the hnp package. J Stat Softw 81(10):1–23. https://doi.org/10.18637/jss.v081.i10

Moro D, Gadal S (2007) Benefits of habitat restoration to small mammal diversity and abundance in a pastoral agricultural landscape in mid-Wales. Biodivers Conserv 16:301–315. https://doi.org/10.1007/978-1-4020-6865-2_21

Nagendra H (2001) Using remote sensing to assess biodiversity. Int J Remote Sens 22(12):2377–2400. https://doi.org/10.1080/01431160117096

Nagendra H, Rocchini D (2008) High resolution satellite imagery for tropical biodiversity studies: the devil is in the detail. Biodivers Conserv 17(14):3431–3442. https://doi.org/10.1007/s10531-008-9479-0

Nagendra H, Lucas R, Honrado JP, Jongman RHG, Tarantino C, Adamo M, Mairota P (2013) Remote sensing for conservation monitoring: assessing protected areas, habitat extent, habitat condition, species diversity, and threats. Ecol Indic 33:45–59. https://doi.org/10.1016/j.ecolind.2012.09.014

de Oliveira Florentino H, Irawan C, Aliano AF, Jones DF, Cantane DR, Nervis JJ (2018) A multiple objective methodology for sugarcane harvest management with varying maturation periods. Ann Oper Res 267(1–2):153–177. https://doi.org/10.1007/s10479-017-2568-2

Paglia AP, da Fonseca GAB, Rylands AB, Herrmann G, Aguiar LMS, Chiarello AG, Leite Y, Paise G, Vieira EM, Prado PI (2020) Small mammals respond to extreme habitat fragmentation in the Brazilian Atlantic Forest according to the landscape continuum model. Mammal Res 65(2):309–322. https://doi.org/10.1007/s13364-019-00464-z

Paglia AP, Fonseca GAB da, Rylands AB, Herrmann G, Aguiar LMS, Chiarello AG, Leite YLR, Costa LP, Siciliano S, Kierulff MCM, Mendes SL, Tavares V da C, Mittermeier RA, Patton JL (2012) Lista anotada dos mamíferos do Brasil. In: Occasional papers in conservation biology. Conservação Internacional

Pardini R (2004) Effects of forest fragmentation on small mammals in an Atlantic Forest landscape. Biodivers Conserv 13(13):2567–2586. https://doi.org/10.1023/B:BIOC.0000048452.18878.2d

Pardini R, Umetsu F (2006) Pequenos mamíferos não-voadores da Reserva Florestal do Morro Grande: distribuição das espécies e da diversidade em uma área de Mata Atlântica. Biota Neotropica 6(2). https://doi.org/10.1590/S1676-06032006000200007

Patton JL, Pardiñas UFJ, D'Elía G (2015) Mammales of South America, Volume 2: Rodents. University of Chicago Press, Chicago, IL. (Vol. 2, Issue September 2013. https://doi.org/10.7208/chicago/9780226169606.001.0001

Pearson PG (1959) Small mammals and old field succession on the piedmont of New Jersey. Ecology 40:249–255

Percequillo A (2015) Genus Cerradomys Weksler, Percequillo, and Voss, 2006. In: Mammals of South America, Volume 2: Rodents. University of Chicago Press, Chicago, IL, pp 300–307. https://doi.org/10.7208/chicago/9780226169606.001.0001

Percequillo A, Hingst-Zaher E, Bonvicino CR (2008) Systematic review of genus Cerradomys Weksler, Percequillo and Voss, 2006 (Rodentia: Cricetidae: Sigmodontinae: Oryzomyini), with description of two new species from eastern Brazil. Am Mus Novit 3622:1–46. https://doi.org/10.1206/495.1

Pinotti BT, Naxara L, Pardini R (2011) Diet and food selection by small mammals in an old-growth Atlantic forest of South-Eastern Brazil. Stud Neotropical Fauna Environ 46:1–9. https://doi.org/10.1080/01650521.2010.535250

Pinotti BT, Pagotto C, Pardini R (2015) Wildlife recovery during tropical Forest succession: assessing ecological drivers of community change. Biotropica 47:765–774. https://doi.org/10.1111/btp.12255

Pires A, Fernandez FA, Feliciano BR, de Freitas D (2010) Use of space by Necromys lasiurus (Rodentia, Sigmodontinae) in a grassland among Atlantic Forest fragments. Mamm Biol 75(3):270–276. https://doi.org/10.1016/J.MAMBIO.2009.04.002

Püttker T, Bueno AA, De Barros CDS, Sommer S, Pardini R (2013) Habitat specialization interacts with habitat amount to determine dispersal success of rodents in fragmented landscapes. J Mammal 94(3):714–726. https://doi.org/10.1644/12-MAMM-A-119.1

R Core Team (2024) R: a language and environment for statistical computing. The R Foundation for Statistical Computing. http://www.R-project.org

Ranta P, Blom T, Niemela J, Joensuu E, Siitonen M (1998) The fragmented Atlantic rain forest of Brazil: size, shape and distribution of forest fragments. Biodivers Conserv 7(3):385–403. https://doi.org/10.1023/A:1008885813543

Reed BC, Brown JF, VanderZee D, Loveland TR, Merchant JW, Ohlen DO (1994) Measuring phenological variability from satellite imagery. J Veg Sci 5(5):703–714. https://doi.org/10.2307/3235884

Rezende C, Zimbres B, Marinho-Filho J (2012) Ecologia espacial de mamíferos: principais perguntas e métodos de estudo. In: Freitas T, Vieira E (eds) Mamíferos do Brasil: genética, sistemática, ecologia e conservação, vol 2. Sociedade Brasileira de Mastozoologia, Brazil, pp 79–89

Ribeiro MC, Metzger JP, Martensen AC, Ponzoni FJ, Hirota MM (2009) The Brazilian Atlantic Forest: how much is left, and how is the remaining forest distributed? Implications for conservation. Biol Conserv 142:1141–1153

Rosalino LM, Martin PS, Gheler-Costa C, Lopes PC, Verdade LM (2014) Neotropical small mammals' diversity in the early cycle of commercial Eucalyptus plantations. Agrofor Syst 88:427–436. https://doi.org/10.1007/s10457-014-9702-9

Rossi RC, Leiner NO (2022) Weather, fire, and density drive population dynamics of small mammals in the Brazilian Cerrado. J Mammal 103(5):1127–1140. https://doi.org/10.1093/jmammal/gyac053

Rüdisser J, Tasser E, Tappeiner U (2012) Distance to nature—a new biodiversity relevant environmental indicator set at the landscape level. Ecol Indic 15(1):208–216. https://doi.org/10.1016/j.ecolind.2011.09.027

Rubel F, Brugger K, Haslinger K, Auer I (2017) The climate of the European Alps: Shift of very high resolution Köppen-Geiger climate zones 1800–2100. Meteorologische Zeitschrift 26:115–125

Salazar-Bravo J (2015) Genus Calomys Water house, 1837. In: Mammals of South America, Volume 2: Rodents. University of Chicago Press, Chicago, IL, pp 481–507. https://doi.org/10.7208/chicago/9780226169606.001.0001

Sanches VQA, Gomes M, Passos F, Graciolli G, de Aquino Ribas AC (2012) Home- range and space use by Didelphis albiventris (Lund 1840) (Marsupialia, Didelphidae) in Mutum

Island, Paraná river. Brazil Biota Neotropica 12(4):50–55. https://doi.org/10.1590/S1676-06032012000400004

Santana R, Felix De Barros N, Leite H, Comerford NB, Ferreira De Novais R (2008) Estimativa de biomassa de plantios de eucalipto no Brasil. Revista Árvore 32(4):697–706

Santos-Filho M, Friero-Costa F, Ignácio A, Silva M (2012) Use of habitats by non-volant small mammals in Cerrado in Central Brazil. Braz J Biol 72(4):893–902

Scott DM, Joyce CB, Burnside NG (2008) The influence of habitat and landscape on small mammals in Estonian coastal wetlands. Estonian J Ecol 57(4):279. https://doi.org/10.3176/eco.2008.4.05

Serafini VN, Priotto JW, Gomez MD (2019) Effects of agroecosystem landscape complexity on small mammals: a multi-species approach at different spatial scales. Landsc Ecol 34(5):1117–1129. https://doi.org/10.1007/s10980-019-00825-8

Sinclair ARE (2003) Mammal population regulation, keystone processes and ecosystem dynamics. Philos Trans R Soc Lond B Biol Sci 358(1438):1729–1740. https://doi.org/10.1098/rstb.2003.1359

Smith SH, Steenhof K, McClure CJW, Heath JA (2017) Earlier nesting by generalist predatory bird is associated with human responses to climate change. J Anim Ecol 86(1):98–107. https://doi.org/10.1111/1365-2656.12604

Stape JL, Binkley D, Ryan MG (2004) Eucalyptus production and the supply, use and efficiency of use of water, light and nitrogen across a geographic gradient in Brazil. For Ecol Manag 193(1):17–31. https://doi.org/10.1016/j.foreco.2004.01.020

Stephens R, Anderson E (2014) Effects of trap type on small mammal richness, diversity, and mortality. Wildl Soc Bull 38:619–627. https://doi.org/10.1002/wsb.418

Sullivan TP, Sullivan DS, Thistlewood HMA (2012) Abundance and diversity of small mammals in response to various linear habitats in semi-arid agricultural landscapes. J Arid Environ 83:54–61. https://doi.org/10.1016/j.jaridenv.2012.03.003

Sullivan MJ, Phillips OL, Galbraith D, Almeida E, de Oliveira EA, Almeida J et al (2025) Variation in wood density across south American tropical forests. Nat Commun 16(1):2351

Szymañski CR, Alvarez JA, Campos CM, Tabeni S (2020) A first assessment of the land management effect on the ecological role of large trees as habitat refuges for desert small mammals. Basic Appl Ecol 48:136–145. https://doi.org/10.1016/j.baae.2020.09.005

Teder T, Moora M, Roosaluste E, Zobel K, Pärtel M, Kõljalg U, Zobel M (2007) Monitoring of biological diversity: a common-ground approach. Conserv Biol 21(2):313–317. https://doi.org/10.1111/j.1523-1739.2006.00575.x

Trivellone V, Schoenenberger N, Bellosi B, Jermini M, de Bello F, Mitchell EAD, Moretti M (2014) Indicators for taxonomic and functional aspects of biodiversity in the vineyard agroecosystem of southern Switzerland. Biol Conserv 170:103–109. https://doi.org/10.1016/j.biocon.2013.12.008

Ulyses FJ, Pardiñas P, Ortiz PE, Jayat JP, Salazar-Bravo J (2015) Genus Calomys Water house, 1837. In: Mammals of South America, Volume 2: Rodents. University of Chicago Press, Chicago, IL, pp 481–507. https://doi.org/10.7208/chicago/9780226169606.001.0001

Umetsu F, Pardini R (2007) Small mammals in a mosaic of forest remnants and anthropogenic habitats— evaluating matrix quality in an Atlantic forest landscape. Landsc Ecol 22(4):517–530. https://doi.org/10.1007/s10980-006-9041-y

Umetsu F, Paul Metzger J, Pardini R (2008) Importance of estimating matrix quality for modeling species distribution in complex tropical landscapes: a test with Atlantic forest small mammals. Ecography 31:359–370. https://doi.org/10.1111/j.2008.0906-7590.05302.x

Vásquez LC, Marques TS, de Abreu EF, Cioci R, Piña CI, Verdade LM (2021) Diversity of small mammals on the early second commercial cycle of eucalyptus plantations in Southeast Brazil. For Ecol Manag 489:119052. https://doi.org/10.1016/j.foreco.2021.119052

Verdade L, Piña C, Lyra-Jorge MC (2014) Redirections in conservation biology. In: Verdade LM, Lyra-Jorge MC, Piña CI (eds) Applied ecology and human dimensions in biological conservation. Springer, Berlin Heidelberg, pp 3–17. https://doi.org/10.1007/978-3-642-54751-5_1

Verdade L, Gheler-Costa C, Lyra-Jorge MC (2016) The multiple facets of agricultural landscapes. In: Biodiversity in agricultural landscapes of Southeastern Brazil. Gruyter Open Ltd., Berlin, pp 2–13

Verdade L, Moral RA, Calaboni A, do Amaral MVSG, Martin PS, Amorim LS, Gheler-Costa C, Piña CI (2020) Temporal dynamics of small mammals in eucalyptus plantations in Southeast Brazil. Global Ecol Conserv 24:e01217. https://doi.org/10.1016/j.gecco.2020.e01217

Verschuyl J, Riffell S, Miller D, Wigley TB (2011) Biodiversity response to intensive biomass production from forest thinning in north American forests—a meta-analysis. For Ecol Manag 261(2):221–232. https://doi.org/10.1016/j.foreco.2010.10.010

Vieilledent G, Vaudry R, Andriamanohisoa SFD, Rakotonarivo OS, Randrianasolo HZ, Razafindrabe HN, Rakotoarivony CB, Ebeling J, Rasamoelina M (2012) A universal approach to estimate biomass and carbon stock in tropical forests using generic allometric models. Ecol Appl 22(2):572–583. https://doi.org/10.1890/11-0039.1

Vieira E, Camargo N (2012) Uso do espaço vertical por marsupiais brasileiros. In: Cáceres NC (ed) Os marsupiais do Brasil: biologia, ecologia e conservação, 2nd edn. Editora da UFMS, Campo Grande, pp 347–363

Vieira EM, Palma ART (2005) Pequenos mamíferos de Cerrado: distribuição dos gêneros, estrutura das comunidades nos diferentes habitats. In: Scariot A, Felfili JM, Sousa-Silva JC (eds) Cerrado: ecologia, biodiversidade e conservação. Ministério do Meio Ambiente, pp 265–282

Wang X, Feng Z, Ouyang Z (2001) Vegetation carbon storage and density of forest ecosystems in China. Ying Yong Sheng Tai Xue Bao = The Journal of Applied Ecology/Zhongguo Sheng Tai Xue Xue Hui, Zhongguo Ke Xue Yuan Shenyang Ying Yong Sheng Tai Yan Jiu Suo Zhu Ban 12:13–16

Wang T, Hung C, Randall D (2006) The comparative physiology of food deprivation: from feast to famine. Annu Rev Physiol 68:223–251. https://doi.org/10.1146/annurev.physiol.68.040104.105739

Wardle DA, Yeates GW, Williamson W, Bonner KI (2003) The response of a three trophic level soil food web to the identity and diversity of plant species and functional groups. Oikos 102(1):45–56. https://doi.org/10.1034/j.1600-0706.2003.12481.x

Watson DM (2003) Long-term consequences of habitat fragmentation—highland birds in Oaxaca, Mexico. Biol Conserv 111:283–303

Wells K, Pfeiffer M, Lakim M, Kalko E (2006) Movement trajectories and habitat partitioning of small mammals in logged and unlogged rain forests on Borneo. J Anim Ecol 75(5):1212–1223. https://doi.org/10.1111/j.1365-2656.2006.01144.x

White GC, Anderson DR, Burnham KP, Otis DL, Laboratory LAN (1982) Capture-recapture and removal methods for sampling closed populations. Los Alamos National Laboratory, Los Alamos, NM. https://pubs.usgs.gov/publication/70120390

Wilson DE, Mittermeier RA, Lacher TE (eds) (2017) Handbook of the mammals of the world: rodents II, vol 7. Lynx Edicions, Barcelona, ES

Wulder M, Hall RJ, Coops NC, Franklin SE (2004) High spatial resolution remotely sensed data for ecosystem characterization. Bioscience 54(6):511. https://doi.org/10.1641/0006-3568(2004)054[0511:HSRRSD]2.0.CO;2

Zanne AE, Lopez-Gonzalez G, Coomes DA, Ilic J, Jansen S, Lewis SL, Miller RB, Swenson NG, Wiemann MC, Chave J (2009) Global wood density database. Dryad Identifier

Part III
Technological Innovations II: Analytical Methodology

Chapter 8
An Introduction to Bayesian Hierarchical Modelling Applied to Wildlife Monitoring

Niamh Mimnagh and Estevão Prado

Abstract This chapter provides a comprehensive introduction to Bayesian modelling, with a focus on its application to wildlife monitoring and ecological inference. It outlines the advantages of Bayesian hierarchical models for managing uncertainty, incorporating prior knowledge and integrating diverse data sources. It then offers practical guidance on implementing Bayesian models, diagnosing convergence, handling missing data, and choosing appropriate software tools. Advanced topics in spatial and temporal modelling are explored, culminating in a case study that uses hierarchical Poisson regression to examine how prey species influence red fox abundance across Japanese landscapes. The chapter aims to equip ecologists and applied statisticians with the tools and understanding needed to apply Bayesian methods in conservation and wildlife research.

Keywords Bayesian statistics · Markov Chain Monte Carlo · Hierarchical models · Wildlife monitoring · Ecological modelling

1 Introduction

This chapter explores the power and flexibility of Bayesian modelling in the context of wildlife monitoring, an area where uncertainty, complexity, and data limitations are the norm rather than the exception. Through this lens, we aim to show how Bayesian approaches not only accommodate but also leverage these challenges to yield more robust ecological insights.

Section 2 begins by examining the role of Bayesian hierarchical models in modern conservation, showcasing their strengths in handling multi-level variability and integrating diverse data sources. To ground readers in the core concepts of

N. Mimnagh (✉)
Department of Mathematics and Statistics, Maynooth University, Co. Kildare, Ireland

E. Prado
Department of Mathematical Sciences, Lancaster University, Lancaster, UK

L. Martins Verdade, R. A. Moral (eds.), *Wildlife Monitoring*,
https://doi.org/10.1007/978-3-032-05821-8_8

likelihoods, priors and posterior inference, Sect. 3 provides a foundational overview of Bayesian statistics, with a particular focus on regression models.

Section 4 presents N-mixture models for estimating wildlife abundance under imperfect detection. Here, we introduce Bayesian parameter estimation techniques, such as Markov Chain Monte Carlo (MCMC), and demonstrate their implementation in ecologically meaningful scenarios. A practical guide to implementing these models follows in Sect. 5, covering essential steps such as data preparation, handling missing values, structuring hierarchical components and assessing model fit. We review common software tools, highlighting their strengths for different modelling contexts.

Finally, Sect. 6 explores advanced topics in spatial and temporal modelling, discusses methods for managing missing data and illustrates the application of these techniques through a detailed case study on red fox abundance in Japan. This case study exemplifies how Bayesian modelling can reveal nuanced ecological relationships and inform conservation strategies across complex landscapes.

2 Application of Bayesian Hierarchical Modelling to Wildlife Monitoring

The monitoring of wildlife plays a crucial role in animal conservation and management, helping ecologists and policy-makers track the health, behaviour and population dynamics of species. It facilitates more informed decisions on issues such as habitat preservation and management of endangered species (Morrison 2013). Bayesian hierarchical modelling provides a powerful framework for wildlife monitoring, as it accommodates the complexity and uncertainty inherent in ecological data while allowing for the integration of multiple sources of information (Hobbs and Hooten 2015). In contrast, traditional methods often rely on incomplete or noisy data, which can limit the effectiveness in guiding conservation efforts.

Bayesian hierarchical models are particularly suited for wildlife monitoring because they can account for different levels of variability (both within and across populations) and can incorporate spatial and temporal heterogeneity. A key advantage of Bayesian models is their ability to quantify uncertainty (Wikle 2003; Gelman et al. 2013). Rather than providing point estimates, these models produce a range of plausible values—posterior distributions—for parameters of interest, such as population size, survival rates or habitat use probabilities. This approach enables the assessment of not only the most likely outcomes but also the risks and uncertainties associated with various scenarios.

For instance, when using Bayesian models to predict the future population size of an endangered species, the model might produce a 95% credible interval that indicates a range of possible population sizes while accounting for variability in covariates such as mortality and detection probability. This probabilistic framework provides decision-makers with more robust information, enabling them to weigh the risks of alternative management strategies.

Moreover, the hierarchical nature of these models allows for the inclusion of multiple sources of variation, such as individual animal behaviour, population-level dynamics and broader ecological patterns (Royle and Young 2008). This layered approach facilitates nuanced interpretations of the data and more targeted conservation interventions. For example, a Bayesian hierarchical model might reveal that, while a population is declining overall, certain sub-populations are faring better due to differences in habitat quality. This insight could inform localised conservation efforts, enabling resources to be directed more effectively.

2.1 Related Works

Bayesian modelling has been widely applied to analyse multiple data sources for estimating abundance patterns across wildlife communities. For example, Gilbert et al. (2024) integrate data from distance sampling surveys and single-visit surveys to estimate abundances of 11 herbivore species in the Masai Mara National Reserve, Kenya. Similarly, Sigourney et al. (2022) applies Bayesian hierarchical modelling to aerial survey data to estimate trends in the abundance of the Atlantic harbour seal in Maine, USA, while Jiménez et al. (2016) uses occupancy data and Bayesian modelling to monitor wolf populations in Spain. Mimnagh et al. (2022) use Bayesian hierarchical models to estimate the abundance of bald eagles in the Alexander Archipelago (Alaska, USA), accounting for inter-species correlations and environmental covariates. In a related application, Mimnagh et al. (2026) develop a Bayesian hierarchical framework to estimate abundance from scarce animal vestige data, highlighting the adaptability of the approach across challenging ecological settings. The hierarchical structures of these models provide more accurate estimates for species that are scarce by combining information across communities and surveys (Kéry and Royle 2020). Moreover, Bayesian approaches provide a way to visualise the degree of uncertainty around population parameter estimates, which is crucial for ecological decision-making.

Bayesian hierarchical modelling has also been utilised to understand habitat preferences and movement patterns based on telemetry data. For instance, Hooten et al. (2016) uses telemetry data and Bayesian models to investigate movement patterns of Canada lynx in Colorado, USA. Similarly, Adam et al. (2019) uses hierarchical hidden Markov models to estimate the movements of the Atlantic cod through the English Channel and the horn shark off the coast of California, USA. Additionally, Torney et al. (2021) applies Bayesian modelling to estimate movement patterns of sheep in Patagonia, Argentina. By modelling spatial distributions in relation to environmental covariates, the researchers are able to identify habitat locations and pathways for conservation efforts. The hierarchical structure of these methods enables the inference of the latent behaviours underlying movement processes, such as resting, foraging and migrating.

Furthermore, Bayesian hierarchical modelling has been applied to monitor disease prevalence in wildlife populations. For example, Hoyt et al. (2020) use these models to assess the spread of white-nose syndrome in bat populations at 101 sites across the globe, while Song and Lawson (2009), Bondo et al. (2024) examine chronic wasting disease in deer in Wisconsin, USA, and Pennsylvania, USA, respectively. These models incorporate hierarchical structures by nesting individuals within populations and populations within regions, enabling the estimation of both local and broader-scale disease dynamics. Bayesian inference further provides credible intervals around disease spread estimates, which enables better-informed management decisions, such as targeted interventions in high-risk areas.

3 Basics of Bayesian Statistics

This section introduces the concept of Bayesian statistics in the context of regression models, which we find to be more intuitive for applied scientists, as these models are widely used to tackle real-world problems. Where necessary, we also highlight the differences between Bayesian and non-Bayesian (i.e., frequentist) inference paradigms.

In both Frequentist and Bayesian statistics, we generally aim to model a variable of interest, y_i, based on a set explanatory variables, $x_i = (x_{i1}, x_{i2}, \ldots, x_{id})$, where $i = 1, \ldots, n$. The contribution of the explanatory variables is measured by a vector of unknown parameters, $\theta = (\theta_1, \ldots, \theta_d)^\top$. Assuming that the observations are independent, under both inference approaches, we have the likelihood function

$$p(y|x, \theta) = \prod_{i=1}^{n} p(y_i|x_i, \theta). \tag{8.1}$$

The likelihood terms $p(y_i|x_i, \theta)$ are provided by a probability density function (pdf) or a probability mass function (pmf), chosen according to the characteristics of y.

In Bayesian statistics, the parameter vector θ is assumed to be both *unknown* and *random*, whereas, under its Frequentist counterpart, it is treated as *unknown* but *fixed*. The unknown part arises from not knowing exactly how the explanatory variables contribute to predicting the response variable, y. For instance, if $x_{i,j}$, for $i = 1, \ldots, n$, has a positive marginal effect on y (i.e., the greater $x_{i,j}$, the greater y), then θ_j should be positive. In practice, however, optimisation algorithms are necessary to precisely estimate θ_j.

Viewing θ as a random quantity in Bayesian statistics implies that point estimates alone are insufficient to fully describe it. Instead, θ is fully characterised by its *posterior distribution*:

$$p(\theta|x, y) = \frac{p(y|x, \theta)p(\theta)}{\int p(y|x, \theta)p(\theta)d\theta}. \tag{8.2}$$

Using the Bayes' rule, the expression above combines information from the data, via the product of the likelihood terms in (8.1), and prior information, $p(\theta)$, which reflects the scientist's prior beliefs or knowledge about θ.

3.1 Example: Bayesian Linear Regression

At first glance, the concepts of likelihood terms, prior and posterior distributions might seem abstract. To help the reader better understand these ideas, we illustrate them using a simple Bayesian regression model to show what each of these elements might look like in practice.

Let's assume we are interested in predicting a univariate response variable for n independent observations based on d explanatory variables, as follows:

$$y_i = x_i\theta + \epsilon_i, \tag{8.3}$$

where $\epsilon_i \sim \mathsf{N}(0, \sigma^2)$;[1] for simplicity, we assume that σ^2 is known. In the model above, y_i represents a response variable and x_i is a row vector containing the values of d explanatory variables for the i-th observation. The linear model assumes that the explanatory variables are linearly related to the response only, although more complex relationships can be captured using polynomial terms or splines (Wood 2017). Here, $p(y_i|x_i, \theta) = \mathsf{N}(y_i|x_i\theta, \sigma^2)$, where the last two terms denote the mean and the variance of the normal distribution. From Eq. (8.1), we have:

$$\begin{aligned} p(y|x, \theta) &= \prod_{i=1}^{n} p(y_i|x_i, \theta) \\ &= \prod_{i=1}^{n} \mathsf{N}(y_i|x_i\theta, \sigma^2) \\ &= \prod_{i=1}^{n} (2\pi\sigma^2)^{-1/2} \exp\left(-\frac{1}{2\sigma^2}(y_i - x_i\theta)^2\right) \\ &= (2\pi\sigma^2)^{-n/2} \exp\left(-\frac{1}{2\sigma^2}\sum_{i=1}^{n}(y_i - x_i\theta)^2\right). \end{aligned} \tag{8.4}$$

The prior distribution, $p(\theta)$, should reflect the available knowledge about θ. For example, if the marginal effects of the explanatory variables can take any value on the real line, then $p(\theta)$ should account for this and the uncertainty associated with each element of θ. In practice, $p(\theta)$ is a probability distribution that represents

[1] This notation reads: 'Epsilon follows a normal distribution with mean zero and variance σ^2'.

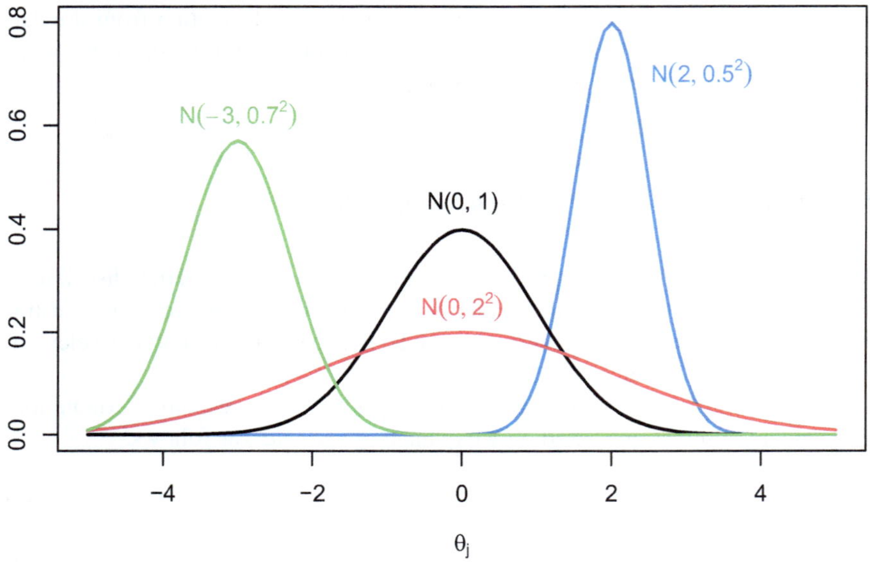

Fig. 8.1 Shape of the normal distribution with different means and variances

any prior information about θ before fitting the model. In this simple example, we might assume the prior knowledge about each entry of the parameter vector θ is well-represented by the assumption:

$$\theta_j \sim \mathsf{N}(\mu, \sigma_\mu^2), \text{ for } j = 1, \ldots, d, \tag{8.5}$$

where μ and σ_μ^2 are pre-specified hyper-parameters that control the mean and variance, respectively. This choice implies that we believe a priori that θ_j is centered around μ with an uncertainty of σ_μ^2. The normal distribution allows us to represent varying degrees of prior information and uncertainty by adjusting its mean and variance. Figure 8.1 displays a univariate normal distribution with different mean and variance parameters. Each combination of μ and σ_μ^2 represents a possible prior belief about θ_j. The parameterisations with negative and positive means suggest that, a priori, θ_j is believed to differ from zero, as there is little probability mass near zero. On the other hand, prior distributions centered around zero imply that, a priori, θ_j is expected to be close to zero, but with varying degrees of uncertainty. A prior with mean zero and a variance of 1 suggests that θ_j is believed to be around zero, likely falling within the range of -2 and 2 with high probability. Conversely, a prior with larger variance indicates greater uncertainty, implying that θ_j could lie within a broader range, such as -4 and 4.

Since we have the likelihood function, $p(y|x, \theta)$, and the prior distributions, $p(\theta_j)$, we can find the posterior distribution of θ_j using Eq. (8.2):

$$
\begin{aligned}
p(\theta_j|x, y) &= \frac{p(y|x, \theta)p(\theta_j)}{\int p(y|x, \theta) \prod_{j=1}^{d} p(\theta_j)d\theta_j} \\
&= \frac{\prod_{i=1}^{n} p(y_i|x_i, \theta)p(\theta_j)}{\int \prod_{i=1}^{n} p(y_i|x_i, \theta)p(\theta_j)d\theta_j} \\
&= \frac{\prod_{i=1}^{n} \mathsf{N}(y_i|x_i\theta, \sigma^2)\mathsf{N}(\theta_j|\mu, \sigma_\mu^2)}{\int \prod_{i=1}^{n} \mathsf{N}(y_i|x_i\theta, \sigma^2)\mathsf{N}(\theta_j|\mu, \sigma_\mu^2)d\theta} \\
&= \mathsf{N}(\theta_j^\star, \sigma_j^{2\star}),
\end{aligned}
$$

where

$$
\theta_j^\star = \sigma_j^{-2\star} \left(\mu/\sigma_\mu^2 + \sigma^{-2} \sum_{i=1}^{n} y_i x_{i,j} \right)
$$

and

$$
\sigma_j^{2\star} = \left(\sigma_\mu^{-2} + \sigma^{-2} \sum_{i=1}^{n} x_{i,j}^2 \right)^{-1} .
$$

Each element of $\theta = (\theta_1, \ldots, \theta_d)$ has a normal distribution as posterior distribution. This fact comes with no surprise due to the conjugacy of the normal distribution; a limited class of pdfs and pmfs have this property. In practice, it means that under the Bayesian perspective, θ_j is characterised by a probability distribution that nicely blends the prior information about θ_j, namely μ and σ_μ^2, and the data of the problem under investigation, namely $x_{i,j}$ and y_i.

Notice that the variance of the posterior distribution of θ_j is inversely proportional to the sum of squares of $x_{.,j}$. Thus, the greater the number of observations n, the narrower the posterior distribution. For this particular linear model, if we compare the Bayesian posterior distribution with the asymptotic distribution of the maximum likelihood estimator (MLE), they both lead to the same distribution, except for the lack of prior information in the distribution of the MLE.

In this example, we used a Bayesian linear regression model to exemplify how prior and posterior distributions work in practice. The likelihood function is a product of univariate normal distributions as well as the prior distribution on θ. This choice facilitates mathematical tractability, making the posterior distribution an analytical expression. Were the likelihood function or the prior on θ other distributions, we most likely would end up getting a posterior distribution without closed-form (because it would not be possible to integrate θ in the denominator of (8.2)), which in turn would require computational tools to allow inference.

4 Bayesian Modelling in Wildlife Monitoring

As mentioned in Sect. 2.1, the Bayesian paradigm can also be applied to perform inference in statistical models used for wildlife monitoring. In this section, we present from a Bayesian perspective the N-mixture models (Royle 2004), which are a class of models used to estimate population sizes from animal count surveys, as an example of Bayesian hierarchical modelling in practice using non-trivial, non-conjugate models. Although these models were originally proposed in a non-Bayesian framework, developing Bayesian N-mixture models is relatively straightforward.

Let y_{it} denote the number of distinct animals observed at location $i = 1, \ldots, I$ and sampling occasions $t = 1, \ldots, T$. In an ideal scenario, y_{it} would represent the number of *all* animals at site i and time t, allowing for simpler models, such as generalised linear models (Nelder and Wedderburn 1972; McCullagh and Nelder 1989), to predict animal abundance. However, animal count surveys are typically imperfect, as they often fail to represent the entire population since animal observation depends on multiple factors, including the observer's experience, vegetation density, time of day and season.

Assuming that the population under investigation does not vary over time (i.e., it is closed to births, deaths, migration and immigration), the imperfect counts can be modelled as

$$y_{it} | N_i, \pi \sim \text{Binomial}(N_i, \pi), \tag{8.6}$$

where N_i denotes the true, unobserved population size and π is the detection probability. This model accounts for the fact that y_{it} are discrete, imperfect counts and introduces unknown latent variables, N_i, which denote the true animal abundance at each site.

In this framework, the detection probability π is the primary parameter of interest, while N_i are latent variables (or random effects) that are typically predicted rather than estimated in the classical sense. Under a Frequentist approach, one might attempt to fit the model by maximising the conditional likelihood, which treats N_i as fixed effects.

$$\begin{aligned} \prod_{i=1}^{I} \prod_{t=1}^{T} p(y_{it} | N_i, \pi) &= \prod_{i=1}^{I} \prod_{t=1}^{T} \mathsf{Bin}(y_{it} | N_i, \pi), \\ &= \prod_{i=1}^{I} \prod_{t=1}^{T} \binom{N_i}{y_{it}} p^{y_{it}} (1 - \pi)^{N_i - y_{it}}. \end{aligned} \tag{8.7}$$

However, as noted by Royle (2004), maximising (8.7) can be challenging and unstable, particularly for large number of sites and sparse counts. To estimate N_i and p, Royle (2004) instead uses the integrated likelihood function:

$$\ell(\{N_i\}, \pi \mid \{y_{it}\}) = \prod_{i=1}^{I} \left(\sum_{N_i = \max_t\{y_{it}\}} \left(\prod_{t=1}^{T} \binom{N_i}{y_{it}} \pi^{y_{it}} (1-\pi)^{N_i - y_{it}} \right) p(N_i) \right),$$

where $\{y_{it}\} = (y_{11}, y_{12}, \ldots, y_{IT})$ and $\{N_i\} = (N_1, N_2, \ldots, N_I)$. Before maximising the above expression using numerical optimisation algorithms, $p(N_i)$ must be specified. From a non-Bayesian perspective, Royle (2004) considers N_i as 'nuisance parameters' (or random effects) with an associated density $p(N_i)$. From a Bayesian point of view, $p(N_i)$ represents the prior distribution on N_i, incorporating prior knowledge about the N_i. Since the N_i are discrete random variables, a prior distribution that has support on the natural numbers, such as the Poisson and negative Binomial distributions, is appropriate.

4.1 Prior Distributions on N_i and π

In practice, when analysing, for example, survey data of endangered animal species, we generally expect the latent population per site, N_i, to be low, and $p(N_i)$ should reflect this expectation. Figure 8.2 displays two Poisson distributions with different means as potential candidates for $p(N_i)$. These distributions could represent the prior beliefs of two experts on the population size under investigation. The prior with a mean of 3 has most of its probability mass concentrated between 0 and 10, whereas the Poisson distribution with a mean of 12 has a greater variance and considers a priori that the population size at each site is likely in the range of 5 to 20. Although Fig. 8.2 only illustrates two configurations of a Poisson distribution, other discrete distributions could be explored to gain flexibility, since the Poisson distribution assumes that the mean is equal to the variance.

Ideally, the prior for the detection probability should account for the constraint $\pi \in (0, 1)$. The most natural choices are the Beta distribution or any probability density function truncated to the unit interval. Figure 8.3 presents four potential prior distributions for the detection probability. For example, the Beta distribution with both shape parameters set to 1 is equivalent to the uniform(0,1) distribution, and considers that, a priori, any value between 0 and 1 is equally likely for the detection probability. This lack of specific information on possible values for π yields a *flat*, or *non-informative* prior.[2] In contrast, a Beta distribution with shape parameters of 0.5 implies that the true detection probability is likely to be near 0 or 1. In practice, an observer with less/more experience may have a lower/higher detection probability, and the prior should reflect this.

[2] Although this terminology can be questioned, since the non-belief on any particular value for a parameter constitutes prior information, and therefore the flat prior is not completely non-informative.

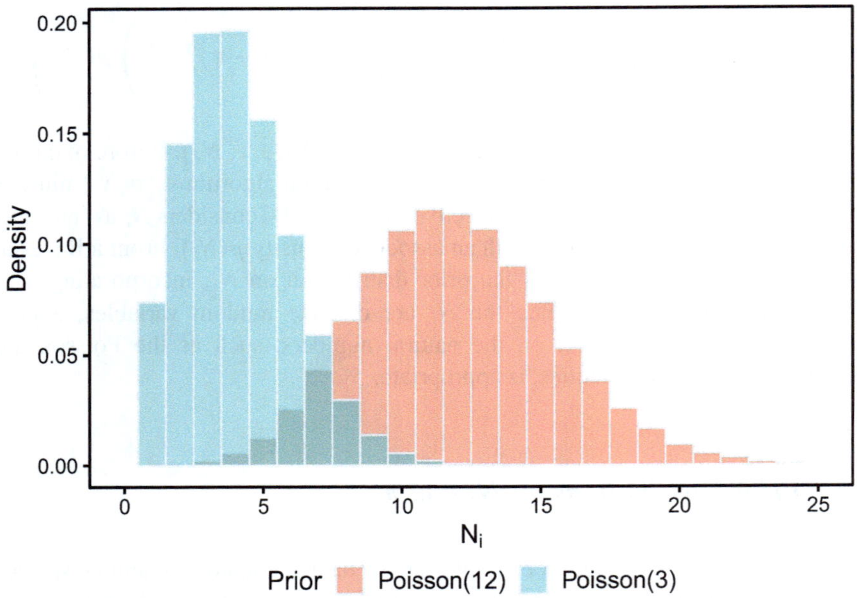

Fig. 8.2 Examples of possible prior distributions $p(N_i)$. Shape of the Poisson distribution with means 3 and 12

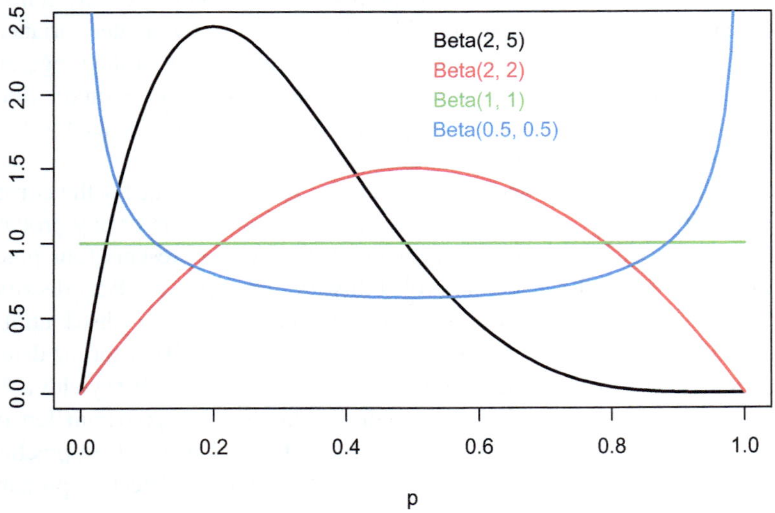

Fig. 8.3 Examples of possible prior distributions for the detection probability, π

In the Bayesian community, justifying the choice of priors and their hyperparameters is standard practice, as these choices can influence model results.

So far in this section, we (1) stated our model, (2) identified the parameters of interest and (3) presented how prior distributions could be placed on these parameters. Of course, the choices of prior distributions for N_i and π are not restricted to the Poisson and Beta distributions shown here; infinitely many other options exist. In general, choosing a prior depends on the specific problem, with no universal rule. However, priors should always reflect prior knowledge or the lack thereof about the phenomenon under investigation.

4.2 Posterior Distributions of N_i and π

Equipped with the likelihood function (8.7) and the prior distributions on the parameters of interest, we can find the joint posterior distribution of N_i and π. For mathematical convenience, we will assume a Poisson and Beta distribution as priors for N_i and p.

We first use the definition of the posterior distribution in (8.2) and substitute in the Binomial likelihood function and the Poisson and Beta priors. The result in the third line of equation (8) follows as the denominator in expression (2) is a constant that does not involve N_i and p. The last line explicitly presents all terms from the penultimate line.

$$\begin{aligned}
p(\{N_i\}, \pi|x, y) &= \frac{\prod_{i=1}^{I}\prod_{t=1}^{T} p(y_{it}|N_i, \pi)p(N_i)p(\pi)}{\int\int\prod_{i=1}^{I}\prod_{t=1}^{T} p(y_{it}|N_i, \pi)p(N_i)p(\pi)dN_i d\pi} \\
&= \frac{\prod_{i=1}^{I}\prod_{t=1}^{T} \text{Bin}(y_{it}|N_i, \pi)\text{Pois}(N_i|\lambda)\text{Beta}(\pi|\alpha, \beta)}{\int\int\prod_{i=1}^{I}\prod_{t=1}^{T} \text{Bin}(y_{it}|N_i, \pi)\text{Pois}(N_i|\lambda)\text{Beta}(\pi|\alpha, \beta)dN_i d\pi} \\
&\propto \prod_{i=1}^{I}\prod_{t=1}^{T} \text{Bin}(y_{it}|N_i, \pi)\text{Pois}(N_i|\lambda)\text{Beta}(\pi|\alpha, \beta) \\
&\propto \left[\prod_{i=1}^{I}\left[\prod_{t=1}^{T}\binom{N_i}{y_{it}}\pi^{y_{it}}(1-\pi)^{N_i-y_{it}}\right]\lambda^{N_i}e^{-\lambda}(N_i!)^{-1}\right]\times \\
&\quad \times \frac{\Gamma(\alpha+\beta)}{\Gamma(\alpha)\Gamma(\beta)}\pi^{\alpha-1}(1-\pi)^{\beta-1}.
\end{aligned} \tag{8.8}$$

At first glance, the posterior distribution in (8.8) seems more complicated than that of the Bayesian linear model. The main difference is that the choices of prior distribution for the parameters of interest in Sect. 3.1 led to a closed-form posterior distribution. In contrast, the joint posterior distribution of a Bayesian N-mixture model does not present the same mathematical simplicity since the Binomial, Poisson and Beta distributions are not conjugate.

4.3 Inference via Markov Chain Monte Carlo

Since the posterior distribution (8.8) does not present a closed-form expression, the alternative is to draw enough samples from this distribution to obtain metrics of interest, such as the posterior mean. To obtain such samples for N_i and p, computational methods are required, as direct sampling from (8.8) is not feasible. Markov chain Monte Carlo (MCMC) (Brooks et al. 2011) methods are a class of simulation algorithms used in Bayesian statistics to indirectly draw non-independent samples from posterior distributions defined up to a constant of proportionality.

By construction, MCMC methods generate a Markov chain where the target posterior distribution serves as the limiting distribution. A central method in this area, the Metropolis-Hastings (MH) algorithm (Metropolis et al. 1953; Hastings 1970) proposes, at each iteration, new posterior samples and decides whether to accept them or not as parameter values. The acceptance step requires evaluation of the posterior distribution at both the proposed and current parameter values.

In this section, we introduce the MH algorithm and explain how it can be employed to draw samples from the posterior distributions of N_i and π. To simplify the inference process, we focus on the posterior distribution of the individual parameters, conditional on all the other parameters—known as full conditional distributions—rather than on their joint posterior distribution in (8.8). Notably, the use of full conditionals facilitates the configuration of the MH algorithm, particularly in setting the proposal step.

Before presenting the full conditionals of N_i and π, we define the Bayesian N-mixture model using a more conventional notation commonly used to represent Bayesian hierarchical models. For site $i = 1, \ldots, I$ and time $t = 1, \ldots, T$,

$$\begin{aligned} y_{it}|N_i, \pi &\sim \mathsf{Bin}(N_i, \pi), \\ N_i &\sim \mathsf{Pois}(\lambda), \\ \pi &\sim \mathsf{Beta}(\alpha, \beta). \end{aligned} \tag{8.9}$$

Note that the first 'layer' of our model is the likelihood term, where y_{it} denotes the number of observed animals at site i and time t, N_i is the true, unobserved animal population at site i and π is the detection probability. Here, N_i and π are unknown. The second and third layers define prior distributions for these parameters of interest, where λ, α and β are known hyperparameters. Finally, the full conditional posterior distributions of N_i and p are given by:

$$\begin{aligned} p(N_i|y, \pi) &= \frac{\prod_{t=1}^{T} p(y_{it}|N_i, \pi) p(N_i)}{\int \prod_{t=1}^{T} p(y_{it}|N_i, \pi) p(N_i) dN_i} \\ &= \frac{\prod_{t=1}^{T} \mathsf{Bin}(y_{it}|N_i, \pi)\mathsf{Pois}(N_i|\lambda)}{\int \prod_{t=1}^{T} \mathsf{Bin}(y_{it}|N_i, \pi)\mathsf{Pois}(N_i|\lambda) dN_i} \end{aligned}$$

$$\propto \prod_{t=1}^{T} \text{Bin}(y_{it}|N_i, \pi)\text{Pois}(N_i|\lambda)$$

$$\propto \left[\prod_{t=1}^{T} \binom{N_i}{y_{it}} \pi^{y_{it}}(1-\pi)^{N_i - y_{it}}\right] \lambda^{N_i} e^{-\lambda}(N_i!)^{-1} \tag{8.10}$$

$$p(\pi|y, \{N_i\}) = \frac{\prod_{i=1}^{I}\prod_{t=1}^{T} p(y_{it}|N_i, \pi)p(\pi)}{\int \prod_{i=1}^{I}\prod_{t=1}^{T} p(y_{it}|N_i, \pi)p(\pi)d\pi}$$

$$= \frac{\prod_{i=1}^{I}\prod_{t=1}^{T} \text{Bin}(y_{it}|N_i, \pi)\text{Beta}(\pi|\alpha, \beta)}{\int \prod_{i=1}^{I}\prod_{t=1}^{T} \text{Bin}(y_{it}|N_i, \pi)\text{Beta}(\pi|\alpha, \beta)dp}$$

$$\propto \prod_{i=1}^{I}\prod_{t=1}^{T} \text{Bin}(y_{it}|N_i, \pi)\text{Beta}(\pi|\alpha, \beta)$$

$$\propto \left[\prod_{i=1}^{I}\prod_{t=1}^{T} \pi^{y_{it}}(1-\pi)^{N_i - y_{it}}\right] \pi^{\alpha-1}(1-\pi)^{\beta-1}. \tag{8.11}$$

As per the joint posterior distribution in (8.8), these full conditionals are also known only up to a constant of proportionality, thus requiring MCMC or approximation methods for inference. The advantage of MCMC methods over approximate methods, such as variational inference (Blei et al. 2017), is that the former provide asymptotically exact samples from the target distribution, whereas the approximate methods typically assume that the target distribution is well-behaved and approximate it using a more tractable distribution (e.g., normal).

Algorithm 8.1 illustrates how the MH algorithm can be used to generate samples from (8.10) and (8.11). First, the observed animal counts, y_{it}, are provided and initial values are set for the latent counts, N_i, and detection probability, π, and hyperparameters of the prior distributions. Next, for each of the M iterations, a candidate parameter value π' is drawn from a proposal distribution $q_\pi(\cdot|\pi)$ and accepted as a sample from the posterior distribution of π with probability $a\left(\pi, \pi'\right)$. The same rationale applies to sampling the N_i, though they are updated individually, one at a time.

Due to the Markovian structure of the proposal distribution, posterior samples from MH algorithms are inherently dependent. The literature on optimal scaling for the random-walk Metropolis algorithm suggests that as the dimension of the parameter vector goes to infinity, the MH algorithm is most efficient when the acceptance rate is approximately 0.234, assuming both the proposal and target distributions are normal and the entries of the parameter vector are independent and identically distributed (i.i.d.) (Gelman et al. 1997; Roberts and Rosenthal 2001), meaning they are all drawn from the same distribution and do not depend on one another. However, empirical and other theoretical works (Bédard 2007; Neal and

Roberts 2008; Sherlock 2013) indicate that this acceptance rate maximises the MH's efficiency (measured in terms of effective sample size) in settings where both proposal and target distribution are non-normal or the entries of the parameter vector are not i.i.d.

To clarify the method, we elaborate on the choice of the proposal distributions to fully specify the MH algorithm and facilitate straightforward inference. One of the simplest and most intuitive proposal distribution for N_i is $q(\cdot|N_i) = \mathsf{Pois}(N_i)$. Since N_i is a latent discrete random variable, it is logical to choose a proposal distribution that respects this discreteness. A normal distribution, for instance, would be inappropriate as it takes values over the entire real line, including the negative values, which are not allowed for N_i.

For the detection probability $0 < \pi < 1$, several proposal distribution options are available. For example, with $q_\pi(\cdot|\pi) = \mathsf{Beta}(\alpha, \beta)$, one can set $\mathbb{E}(\pi') = \alpha/(\alpha + \beta) \approx \pi$, though caution is required as $\text{Var}(\pi') \to 0$ when $\beta \to 0$. Other alternatives, such as the Simplex distribution (Barndorff-Nielsen and Jørgensen 1991) and the unit-inverse normal distribution (Ghitany et al. 2019), are similarly bounded within the unit interval but provide greater flexibility than the Beta distribution, incorporating both location and scale parameters. For a comprehensive overview of distributions on the unit interval, see Ghitany et al. (2019) and references therein.

Algorithm 8.1 Metropolis-Hastings algorithm

Input: $\{y_{it}\}$.
Initialise: $\{N_i\}$, π and set all hyperparameters of the prior distributions.
for $(m = 1 \text{ to } M)$ **do**
  Propose a candidate parameter value $\pi' \sim q_\pi(\cdot|\pi)$.
  Compare the current (π) and proposed (π') via a MH step with

$$a\left(\pi, \pi'\right) = \min\left(1, \frac{p(\pi'|y, \{N_i\})q_\pi(\pi|\pi')}{p(\pi|y, \{N_i\})q_\pi(\pi'|\pi)}\right) \tag{8.12}$$

  Sample $u \sim \mathsf{Unif}\,(0, 1)$: If $a\left(\pi, \pi'\right) > u$, set $\pi = \pi'$, otherwise set $\pi = \pi$.
  for $(i = 1 \text{ to } I)$ **do**
    Propose a candidate parameter values $N_i' \sim q(\cdot|N_i)$.
    Compare the current (N_i) and proposed (N_i') via a MH step with

$$a\left(N_i, N_i'\right) = \min\left(1, \frac{p(N_i'|y, p)q(N_i|N_i')}{p(N_i|y, p)q(N_i'|N_i)}\right) \tag{8.13}$$

    Sample $u \sim \mathsf{Unif}\,(0, 1)$: If $a\left(N_i, N_i'\right) > u$, set $N_i = N_i'$, otherwise set $N_i = N_i$.
  end for
end for
Output: Samples from the posterior distributions of N_i and π.

5 Guide to Practical Implementation

The practical implementation of Bayesian hierarchical models for wildlife monitoring involves several key steps, from data preparation and model specification to diagnosing convergence and assessing model fit. Additionally, various software and tools may be used to implement these models, each offering specific advantages for ecologists and wildlife biologists. This section provides a guide on how to approach each step.

5.1 Data Preparation

Data preparation is a foundational step in the development of Bayesian hierarchical models, with significant implications for model performance, convergence, and interpretability. In ecological and wildlife monitoring contexts, raw datasets often originate from diverse sources and exhibit heterogeneity in structure, resolution, and quality. Before these data can be used in formal modelling, they must be processed and cleaned in a way that preserves ecological meaning and supports robust inference.

Population counts may be overdispersed, zero-inflated, or affected by imperfect detection, requiring transformation or modelling techniques that can accommodate these features (Royle 2004; Mimnagh et al. 2022).

Environmental covariates may be collected at varying spatial or temporal scales. These data must be aligned with the biological processes of interest, which may involve temporal aggregation, or normalisation. When covariate values are missing for certain sites or time periods, imputation methods can be used to retain the integrity of the dataset (described in Sect. 6.3.2).

Spatial data require careful handling of coordinate systems, resolution matching, and extraction of covariate values at relevant ecological locations (e.g., animal home ranges, survey plots). Consistency across spatial layers is crucial to avoid introducing bias.

In a Bayesian context, preparing data also involves ensuring that priors, likelihoods, and hierarchical structures are appropriately aligned with the characteristics of the input data. Thorough exploratory data analysis, summary statistics, and visualisation are essential during this stage to uncover anomalies, identify missingness patterns, and validate ecological assumptions.

Ultimately, well-prepared data support the reliability of the entire modelling workflow. They facilitate more accurate parameter estimation, allow for better convergence diagnostics (e.g., trace plots, Gelman-Rubin statistic (Gelman and Rubin 1992)), and lead to models that are both computationally efficient and ecologically interpretable.

5.2 Handling Missing Data

Wildlife monitoring datasets frequently contain missing values due to detection issues, logistical constraints, or equipment failure. Bayesian methods offer robust solutions for handling missing data by allowing for the incorporation of prior distributions that account for uncertainty. Several techniques can be employed, including:

- **Hierarchical modelling of missing data:** In a fully specified Bayesian hierarchical model, missing data can be treated as latent variables and estimated jointly with model parameters. This is especially powerful when the missingness is not completely random, but can be explained by other observed variables in the model. By sharing information across individuals, sites, or time periods, hierarchical models borrow strength from the non-missing data to inform estimates of the missing values. This approach improves estimation precision and reduces bias, particularly in ecological studies with sparse data, such as site occupancy or count surveys.
- **State-space models:** In time-series or spatial movement data—such as animal telemetry or population monitoring—state-space models (Jonsen et al. 2005; Hooten et al. 2017) offer a framework for handling missing observations. These models separate the true underlying ecological process (e.g., animal location or abundance) from the observation process (e.g., detection or measurement error), allowing for inference even when some observations are missing or partially observed. By modelling the latent states as a Markovian process and incorporating known observation error, state-space models can estimate missing values while accounting for both process variability and observational uncertainty.

5.3 Model Specification

Model specification involves defining the structure of the Bayesian hierarchical model, including the levels of hierarchy and the prior distributions for the parameters. This process ensures that the model accurately represents the ecological or wildlife system being studied, while incorporating appropriate assumptions about variability and uncertainty.

5.3.1 Hierarchical Structure

Wildlife monitoring often involves data that can be structured hierarchically. For example, individual survival or movement behaviour might be modelled at the first level, nested within population-level processes (such as overall population growth) at a second level, and region-level environmental factors (such as climate or habitat) at a third level. Each level can have its own set of parameters and uncertainty distributions, allowing the model to account for variation at multiple scales.

A general hierarchical model can be fully expressed via:

- Likelihood function for individual i in group j. For example, $y_{ij} \sim f(\theta_{ij})$.
- First level: Individual parameters depend on group-level parameters (e.g., $\theta_{ij} \sim g(\beta_j)$).
- Second level: group-level parameters depend on higher-level parameters (e.g., $\beta_j \sim h(\gamma)$).

In this representation, $f(\cdot)$ denotes the likelihood function, $g(\cdot)$ and $h(\cdot)$ represent conditional distributions at different hierarchical levels. This structure is particularly useful for modelling species distributions, abundance, survival probabilities, and habitat preferences while accounting for variation across individuals, populations, and environments.

5.3.2 Likelihood Function

The likelihood function links the data to the model, and is chosen based on the nature of the response variable. The choice of likelihood function is of great importance, as it directly impacts the accuracy of parameter estimation and model fit.

- Count data is often collected as part of species abundance or detection surveys, and is modelled using discrete distributions such as the Poisson or negative binomial distributions.
- Binary data is often collected during presence-absence surveys, and can be used to determine site occupancy. It is generally modelled using the Bernoulli distribution.
- Continuous data may be collected for variables such as body size, movement distances, or habitat characteristics, and can be modelled using continuous distributions, such as the normal distribution.

5.3.3 Choice of Prior Distributions

One of the strengths of Bayesian modelling is the ability to specify prior distributions for model parameters. Prior distributions can be uninformative (vague) or informative, depending on the data and the research question.

- Uninformative priors can be used when little prior knowledge is available. These priors aim to exert minimal influence on the posterior distribution. Examples of uninformative priors include uniform distributions or weakly informative normal distributions with large variance.
- Informative priors can be used when prior knowledge from previous studies, expert opinion, or biological constraints is available. Informative priors can

improve inference, particularly in cases of sparse data. For example, survival probabilities may be informed by prior mark-recapture studies.

The selection of priors should be assessed for their impact on the posterior distribution, ensuring that they do not exert undue influence over model inference, unless this can be justified by strong prior knowledge.

5.4 Diagnosing Convergence and Model Fit

After a model has been run, it is critical to ensure that it has converged and that the fit is appropriate for the data. This involves various diagnostic techniques to assess the quality of parameter estimates and the model's predictive performance.

5.4.1 Markov Chain Monte Carlo Diagnostics

As pointed out in Sect. 4.3, Markov chain Monte Carlo (MCMC) sampling is typically used to estimate the posterior distributions of model parameters. Several diagnostic tools are available to assess convergence.

- Trace plots offer a visual inspection of MCMC samples over iterations, and can reveal mixing behaviour. Well-mixed chains should resemble a stationary process without strong autocorrelation.
- The Gelman-Rubin statistic ($\hat{R}$) (Gelman and Rubin 1992) measures between-chain and within-chain variance. Values close to 1 indicate convergence, while values significantly greater than 1 suggest lack of convergence.
- Effective Sample Size (ESS) quantifies the number of independent samples obtained. A sufficient ESS ensures that the chains are long enough to give reliable parameter estimates. A low ESS suggests high autocorrelation, requiring longer chains or reparameterisation.

Poor convergence can often be addressed by increasing the number of iterations, using more informative initial values, or a reparametrisation of the model.

5.4.2 Posterior Predictive Checks

To assess model fit, posterior predictive checks can be performed. This involves generating new data from the posterior predictive distribution, comparing summary statistics (e.g., means, variances) of the simulated and observed data, and using graphical methods, such as histograms or density plots, to assess how well the model reproduces key features of the observed data. If the model is well-specified, the predicted data should resemble the observed data. Discrepancies between simulated and observed data may suggest model misspecification, and may require adjustments to the likelihood function or inclusion of additional covariates.

By combining MCMC diagnostics and posterior predictive checks, researchers can ensure that their Bayesian hierarchical models are robust, well-calibrated, and capable of making reliable ecological inferences.

5.5 Software for Bayesian Hierarchical Modelling

Several software packages can be used to implement Bayesian hierarchical models, each offering varying levels of flexibility, efficiency, and ease of use. The choice of software often depends on factors such as model complexity, computational efficiency, and user experience with probabilistic programming languages. Below, we provide an overview of three commonly used tools for Bayesian hierarchical modelling in wildlife monitoring; see Appendix 1 for an example of an occupancy model implementation.

5.5.1 JAGS (Just Another Gibbs Sampler)

`JAGS` (Plummer 2003) is a widely used software for Bayesian analysis that employs Gibbs sampling—a type of Markov Chain Monte Carlo (MCMC) method that draws samples from the posterior distribution by iteratively sampling each parameter from its conditional distribution. It is particularly useful for hierarchical models with count data, as it allows direct specification of discrete latent parameters, such as occupancy states or individual animal presence in capture-recapture models. `JAGS` may be integrated into R through the `rjags` and `R2jags` (Su and Yajima 2015) packages, which enable users to specify models using a syntax derived from the Bayesian inference Using Gibbs Sampling (BUGS) language (Spiegelhalter et al. 1996)—a language for describing Bayesian hierarchical models—and execute MCMC sampling within the R environment.

One of the strengths of `JAGS` is its ease of use, especially for those familiar with the BUGS language. It is also relatively lightweight in terms of memory-usage, making it suitable for analyses with limited computational resources. However, its reliance on Gibbs sampling can be a drawback, as this method can mix slowly in high-dimensional parameter spaces, leading to longer runtimes and higher autocorrelation between samples.

5.5.2 Stan

`Stan` (Carpenter et al. 2017) is a high-performance probabilistic programming language that uses Metropolis-Hastings methods (Hamiltonian Monte Carlo (HMC) and its variant, the No-U-Turn Sampler (NUTS)), to perform Bayesian inference.

Stan may be accessed in R via the rstan package (Stan Development Team 2024), which provides an interface for compiling and running models written in the Stan modelling language.

Compared to JAGS, Stan is generally more efficient for complex hierarchical models with many continuous parameters. HMC enables faster mixing and reduces autocorrelation in posterior samples, often leading to improved convergence diagnostics. However, Stan does not directly support discrete parameters, meaning they must be marginalised out, which can complicate model specification for problems that naturally involve latent discrete states.

A downside of Stan is that it tends to be more memory-intensive than JAGS, particularly for large datasets. This can be a limiting factor for users with constrained computational resources. However, for models with many continuous parameters, Stan is often the preferred choice due to its sampling efficiency and robust diagnostic tools.

5.5.3 Nimble

Nimble (de Valpine et al. 2017) is an R-based framework that extends JAGS by allowing users to write hierarchical Bayesian models using a BUGS-like syntax, while providing additional flexibility for customising MCMC algorithms. Unlike JAGS and Stan, Nimble enables users to define and modify their own samplers, which can lead to improved sampling efficiency for specific model structures.

One major advantage of Nimble is its ability to work with discrete latent variables, like JAGS, while also providing options for adaptive MCMC methods that improve mixing. Additionally, Nimble allows for model compilation into C++ which can significantly speed up execution times compared to standard JAGS implementations.

The added flexibility of Nimble makes it particularly useful for wildlife monitoring applications that require custom model components, such as state-space models or multi-species occupancy models. However, the increased complexity in defining and tuning samplers may present a steeper learning curve for new users compared to JAGS or Stan.

6 Advanced Topics in Bayesian Hierarchical Modelling

As Bayesian hierarchical modelling becomes more widely used in wildlife monitoring, researchers increasingly incorporate advanced techniques to capture the spatial and temporal complexity of ecological data. These advanced topics include spatial and temporal models to account for geographic and time-dependent variability, as well as methods for handling missing data, which are common in wildlife studies. This section outlines the principles and applications of these advanced approaches. By using advanced Bayesian methods, researchers can gain deeper insights into the drivers of wildlife population dynamics and improve conservation and management strategies. These advanced models, when implemented correctly, allow for more accurate predictions and a better understanding of complex ecological processes.

6.1 Spatial Models

Spatial models are critical for understanding how wildlife populations are distributed across landscapes and how they interact with environmental variables. Wildlife often exhibit spatial dependence, meaning that observations closer to each other in space are more likely to bear similarities to one another than those further apart. Ignoring spatial autocorrelation can lead to biased estimates and misleading conclusions. Bayesian hierarchical models offer a flexible way to incorporate spatial structure into wildlife monitoring. Several key methods exist for incorporating spatial structure into a Bayesian hierarchical model.

6.1.1 Spatial Autocorrelation

In many wildlife datasets, individuals or populations that are geographically close tend to exhibit similar ecological characteristics, such as habitat preferences, movement patterns or survival rates. This phenomenon—spatial autocorrelation—occurs when nearby observations are more similar than those farther apart. Ignoring this spatial dependence can lead to bias in parameter estimates and incorrect inferences about ecological processes.

To address spatial autocorrelation, Bayesian hierarchical models may incorporate spatially structured random effects that account for spatial dependence between observations. These effects help mitigate bias, improve parameter estimation, and enhance predictive accuracy. Spatial dependence may be modelled using various approaches. Gaussian Processes assume a continuous spatial correlation structure, using covariance functions to define dependencies between locations. These models provide flexibility in capturing spatial gradients and transitions in environmental factors. Conditional Autoregressive (CAR) (Besag 1974) and Intrinsic Conditional Autoregressive (ICAR) (Besag et al. 1991) models define spatial relationships in a discrete framework, where the value at one location is influenced by the values at its neighbouring locations, enforcing a local smoothing effect.

Geostatistical models (Cressie 1989) are designed to analyse and predict spatially continuous phenomena using point data. In ecological studies, these models are particularly useful for understanding patterns in species distribution, environmental gradients, or resource availability across a landscape.

These models handle spatial autocorrelation using a covariance function (or variogram), which defines how the similarity between observations decays with distance. One of the most commonly used covariance structures is the Matérn function, which provides flexible control over both the range (i.e., the distance beyond which spatial correlation becomes negligible) and the smoothness of the spatial field.

In a Bayesian hierarchical framework, geostatistical models are often implemented as Gaussian Process (GP) models, where the spatial random effect is modelled using a multivariate normal distribution.

One of the key advantages of geostatistical models is their ability to separate spatial structure from the effects of measured covariates, enabling researchers to more accurately identify and quantify the drivers of ecological patterns.

By integrating spatial modelling techniques into Bayesian hierarchical frameworks, researchers can obtain more reliable estimates of population parameters, better quantify spatial heterogeneity, and improve ecological inference, ultimately leading to more informed conservation and management decisions.

6.2 Temporal Models

Temporal models account for changes in wildlife populations and behaviours over time. They are particularly important for tracking trends in population size, migration patterns, or responses to environmental changes like climate variability. Temporal dependence, much like spatial dependence, needs to be incorporated to avoid biased estimates and to accurately model time-varying processes.

6.2.1 Time-Series Models

Time-series models allow researchers to analyse and predict the temporal dynamics of wildlife populations by incorporating dependencies between observations at different time points. As wildlife populations are often influenced by past conditions, these models help capture trends, cycles, and fluctuations that may arise due to ecological or environmental factors.

Autoregressive (AR) models assume that the current population size is a function of previous population sizes, making them particularly useful when population changes are driven by intrinsic biological processes, such as reproduction or survival rates. Moving Average (MA) models, on the other hand, account for short-term fluctuations by modelling dependence on past error terms, which can arise due to sudden environmental changes or sampling variability.

A more comprehensive approach, the Autoregressive Integrated Moving Average (ARIMA) (Box et al. 2015) model combines autoregression, differencing and moving averages. ARIMA models can capture both short-term and long-term patterns in population trends by adjusting for non-stationarity, ensuring that population estimates are not biased by trends or shifts in mean levels over time. These models are widely used in wildlife monitoring to forecast population changes, assess the impact of external stressors, and inform conservation strategies.

6.2.2 State-Space Models

State-space models (Aoki 2013) are a type of hierarchical model used for modelling time-series data in ecological studies, particularly when direct observations of wildlife populations are noisy or incomplete. These models consist of two components: the observation process and the latent biological process. The observation process accounts for the way the data are collected, including errors and inconsistencies in measurements, while the biological process represents the true underlying population dynamics or movement patterns. By separating these two processes, state-space models allow researchers to infer hidden states, such as true population size or behavioural changes, even when direct observations are imperfect.

A key advantage of state-space models is their ability to incorporate latent states, which explicitly capture both measurement error and process variability. This makes them particularly useful for wildlife monitoring studies, where data collection is often challenging due to factors like weather conditions, observer biases, and the elusive nature of certain species.

For linear state-space models, Kalman filters (Welch and Bishop 1995) provide an efficient way to estimate hidden states over time, making them useful for structured population models with relatively simple dynamics. However, many ecological systems are non-linear and exhibit complex, non-Gaussian dynamics. In such cases, particle filters (Djuric et al. 2003) can be employed to approximate the hidden states, allowing for flexible inference in models with more intricate interactions between environmental factors and wildlife populations.

State-space models are widely applied in wildlife monitoring, including estimating survival rates (Tan and Ke 2003), tracking migration patterns (Patterson et al. 2008), and improving population estimates (Hostetler and Chandler 2015) in datasets with high levels of uncertainty. By incorporating hierarchical Bayesian frameworks, these models enhance predictive accuracy and provide a robust approach to understanding the dynamics of species in changing environments.

6.2.3 Dynamic Models

Dynamic models provide a framework for capturing population changes over time, allowing parameters to evolve in response to environmental variability, habitat modification, or conservation efforts. These models enable researchers to analyse complex temporal dynamics by integrating factors such as colonisation, extinction and behavioural shifts. One prominent example is the Dynamic Occupancy Model (Royle and Kéry 2007), which estimates the probability of species presence over multiple time periods, while accounting for changes in habitat suitability or species dispersal. By explicitly modelling colonisation and extinction processes, these models help identify factors that influence species persistence.

Another approach, Hidden Markov Models (HMMs) (Rabiner and Juang 1986) are widely used to describe transitions between discrete states over time. For instance, species may shift between breeding and non-breeding states, or alternate between migratory and resident behaviours based on environmental factors. HMMs incorporate latent states that account for uncertainty in observed transitions, improving inference on species movement and life-history strategies.

An extension of HMMs, Dynamic Bayesian Networks (DBNs) (Murphy 2002) provide an even more flexible framework by allowing dependencies among multiple time-varying variables. DBNs integrate information from various ecological and environmental sources, making them useful for understanding how interconnected factors influence population dynamics. These models can incorporate habitat changes, climate variability, and management interventions, providing insights into species' responses to conservation strategies or long-term environmental trends. Overall, dynamic models are invaluable tools for assessing how ecological processes unfold over time, and for guiding adaptative wildlife management and conservation planning.

6.3 *Missing Data*

Missing data are a common challenge in wildlife monitoring due to various factors, such as equipment failure, difficult terrain, and the elusive nature of certain species. In such cases, Bayesian hierarchical models offer a powerful approach for handling missing data, as they provide a framework for incorporating uncertainty and inferring missing values using the observed data and prior information. Using these models, researchers can ensure that incomplete datasets do not lead to biased or unreliable estimates of ecological parameters.

6.3.1 Data Augmentation

One common approach to handling missing data in a Bayesian framework is data augmentation. In this approach, missing values are treated as latent (unknown) variables, and the model estimates these values as part of the overall inference process (Royle and Dorazio 2012). This method can be particularly useful in capture-recapture studies, where individual detection histories are often incomplete due to animals being missing during surveys. By augmenting these histories, researchers can obtain more accurate estimates of true population size and survival probabilities.

In the spatially explicit capture-recapture (SECR) model (Borchers and Efford 2008; Royle and Young 2008), missing detections present an additional challenge because movement data are inherently linked to spatial coordinates. Data augmentation can help fill in gaps by using observed detections to infer likely movement patterns, leading to more reliable estimates of home range size and density. Similarly, in the case of incomplete survey data, Bayesian models can use observed patterns to infer missing observations. For example, if a species is detected in a given area during some time periods but not others, a hierarchical model can estimate the likelihood that it was present but undetected, rather than assuming a true absence.

6.3.2 Missing Covariate Data

When dealing with missing environmental covariate data, Bayesian hierarchical models provide a flexible framework that allows uncertainty to be accounted for during parameter estimation. Environmental covariates, such as temperature, precipitation, habitat quality or resource availability, often play a crucial role in explaining patterns of species distribution, survival and reproduction. However, missing values in these covariates can arise due to equipment failures, incomplete field surveys, or logistical constraints in data collection. Ignoring or improperly handling missing covariate data can introduce bias into statistical models and lead to misleading ecological inferences.

One common Bayesian approach to addressing missing covariate data is multiple imputation (Rubin 1996), which involves generating plausible values for missing observations by drawing from the posterior distribution of the missing data. Instead of assuming a single best estimate for each missing covariate, multiple imputation creates a range of possible values based on observed data patterns and prior information. This results in a more realistic representation of missing data, and is particularly useful when missing values occur at random, or when they are only weakly related to other observed variables.

Another powerful approach involves estimating missing covariates alongside the primary response variables within a hierarchical Bayesian framework (Schafer and Yucel 2002). This method is particularly advantageous when the missingness in covariate data is not random but is related to the underlying ecological process being studied. For example, in a study of animal movement, habitat quality may be missing more frequently in remote or difficult-to-access areas. Instead of treating these missing values as independent of the observed data, a model can explicitly estimate habitat quality based on spatial correlations or ecological constrains, improving the accuracy of inferences about species-habitat relationships.

Beyond these techniques, Bayesian hierarchical models can incorporate informative prior distributions for missing covariates, using external knowledge to guide the imputation process (Gelman 2002). For example, if long-term climate records suggest a certain range of temperatures in a given region, these data can be used as priors to inform the estimation of missing temperature values in wildlife studies. Similarly, when habitat quality indices are missing for certain locations, priors derived from satellite imagery or historical land-use data can provide additional constraints, ensuring that imputed values remain ecologically realistic.

By integrating missing data directly into the modelling process, Bayesian approaches enable ecologists to extract meaningful insights even from incomplete datasets. Rather than discarding observations with missing covariates—an approach that can significantly reduce statistical power and introduce biases—these models allow for a more comprehensive analysis that retains all available information while properly accounting for uncertainty. This leads to more reliable predictions and deeper ecological understanding, ultimately informing more effective conservation and management strategies. As ecological studies increasingly rely on large-scale and heterogeneous data sources, advanced Bayesian techniques for handling missing covariate data will continue to play a critical role in improving the robustness and applicability of wildlife population models.

7 Case Study

The red fox (*Vulpes vulpes*) is one of the most widely distributed carnivores in the world, found throughout much of the Northern Hemisphere (Larivière and Pasitschniak-Arts 1996). It is a known predator of the Japanese squirrel (*Sciurus lis*) (Thorington Jr et al. 2012), which inhabits the islands of Honshu, Shikoku, and Kyushu, as well as the Japanese hare (*Lepus brachyurus*) (Enari and Sakamaki 2012). Additionally, the Japanese badger (*Meles anakuma*) is found throughout Honshu, Kyushu, Shikoku, and Shōdoshima, and there is evidence of red foxes utilising burrows created by the Japanese badger (Kondo 2018). Given these interactions, it is plausible that the abundance of the red squirrel, Japanese hare and Japanese badger may have an impact on the abundance of the red fox at shared sites.

This study aims to estimate the impact of the abundances of the Japanese squirrel, Japanese hare, and Japanese badger on red fox populations using Bayesian hierarchical count models. The analysis is motivated by data collected in the Satochi-satoyama regions in Japan (Ministry of the Environment (Japan) 2014). These are areas classified as 'socio-ecological production landscapes', including farmlands, plantation forests and grasslands that were historically managed in a sustainable manner (Takeuchi 2001). The areas of Satochi-satoyama are recognised as important habitats for biodiversity, with almost half of Japan's regions of high concentrations of endangered species located within these landscapes (Ministry of the Environment (Japan) 2010). The dataset consists of observations collected at 37 sites throughout the Satochi-satoyama areas of Japan, between 2005 and 2012. These sites provide an opportunity to assess temporal and spatial variation in red fox abundance, while considering the influences of other species' populations. To analyse the relationship between red fox abundance and the presence of the Japanese squirrel, Japanese hare and Japanese badger, we introduce a Bayesian Poisson regression model; see Appendix 2 for the model implementation in `JAGS`.

For each observation i, the count of foxes is modelled as a Poisson-distributed random variable:

$$\text{Fox}_{ij} \sim \text{Poisson}(\lambda_{ij}),$$

where $i = 1, \ldots, N$, $j = 1, \ldots, S$ and

$$\log \lambda_{ij} = \beta_0 + \beta_{j,1} \cdot \text{Hare}_i + \beta_{j,2} \cdot \text{Squirrel}_i + \beta_{j,3} \cdot \text{Badger}_i + \beta_2 \cdot \text{Year}_i.$$

In this model, β_0 is the global intercept term and β_2 is a fixed-effect capturing temporal trends across years, whereas $\boldsymbol{\beta}_j = (\beta_{j,1}, \beta_{j,2}, \beta_{j,3})^\top$ are site-specific random effects that capture the impact on fox abundance of hare, squirrel and badger abundances, respectively. The intercept β_0 and the year effect β_2 are assigned weakly informative normal prior distributions (i.e., $\beta_0, \beta_2 \sim \mathsf{N}(0, 100)$). We assume that the parameter vector $\boldsymbol{\beta}_j \sim \mathsf{MVN}(\boldsymbol{\mu}_\beta, \boldsymbol{\Sigma})$, where

- $\boldsymbol{\mu}_\beta = (\boldsymbol{\mu}_{\beta_1}, \boldsymbol{\mu}_{\beta_2}, \boldsymbol{\mu}_{\beta_3})^\top$ represents the global mean effects across all sites, with priors:

$$\mu_{\beta_k} \sim \mathsf{N}(0, 100).$$

- $\boldsymbol{\Sigma}$ is the covariance matrix capturing dependencies among the site-specific effects, derived from the precision matrix $\boldsymbol{\Omega}$ as:

$$\boldsymbol{\Omega} \sim \mathsf{Wishart}(\boldsymbol{R}, 4),$$
$$\boldsymbol{\Sigma} = \boldsymbol{\Omega}^{-1}.$$

The Wishart prior distribution on $\boldsymbol{\Omega}$ ensures a positive-definite covariance structure, and controls the structure of variability and correlation among the site-specific effects.

This hierarchical approach allows site-specific variation in the way that hare, squirrel and badger abundance influences red fox populations. The multivariate normal prior distribution on $\boldsymbol{\beta}_j$ accounts for correlations between the effects of the various prey species. This model can be used to understand spatial and temporal drivers of fox abundance, and to predict fox populations at unobserved sites.

Figure 8.4 presents the impact of hare, badger and squirrel abundance on red fox abundance across sites in Japan. The effect estimates ($\boldsymbol{\beta}_j$) are mapped spatially, with red indicating a positive effect, blue indicating a negative effect, and white indicating no significant relationship (i.e., the 99% credible interval contains zero). Figure 8.4a presents the effect of hare abundance on fox abundance, which varies across sites. In several locations, hare abundance exhibits a negative effect on fox abundance, suggesting that higher hare populations are associated with fewer foxes. In other locations, sites show a weak positive effect, implying that fox abundance may increase with hare abundance in certain locations. However, many sites remain white, which indicates that in these locations, hare abundance does not significantly affect fox abundance. This pattern for alternating positive and negative impacts is also present when examining squirrel abundance (panel (b)) and badger abundance (panel (c)).

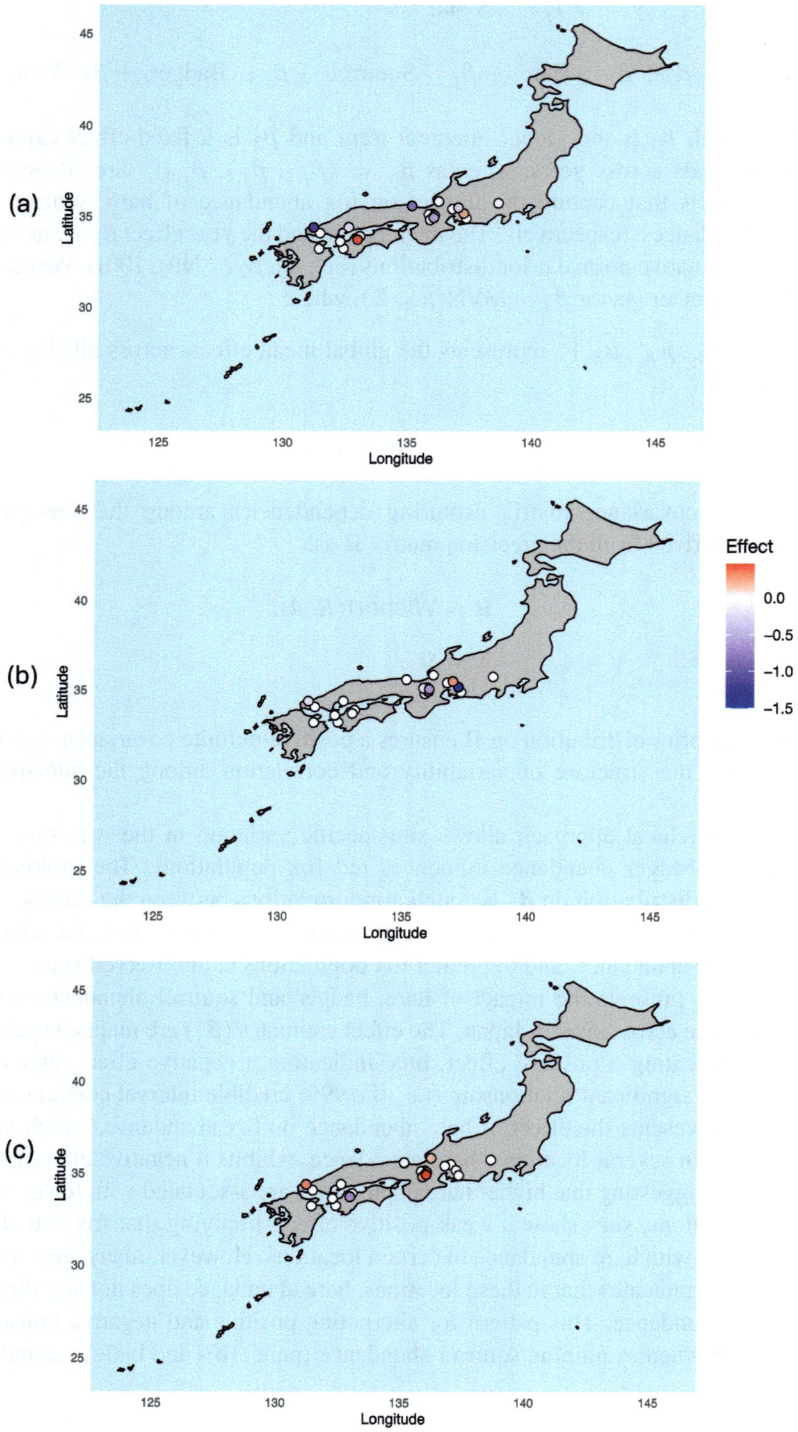

Fig. 8.4 Site-level effect of (**a**) Japanese hare abundance, (**b**) Japanese squirrel abundance and (**c**) Japanese badger abundance on fox abundance

These spatial differences suggest that the abundances of hares, badgers and squirrels may play a role in shaping local fox populations, possibly due to competition for resources, habitat differences or other ecological interactions, but the effect is inconsistent across the landscape. The spatial variability in impact highlights the complexity of ecological interactions between foxes and their prey or competitors. In some regions, the presence of hares, badgers or squirrels influences fox abundances, whereas in others, no strong relationship is observed. Understanding these spatial variations is crucial for predicting predator-prey relationships and ecosystem dynamics across different landscapes.

Appendix 1: Example of an Occupancy Model Implementation in JAGS, Nimble and Stan

To illustrate how Bayesian hierarchical models are implemented, consider an occupancy model where temporal variation is introduced via an intercept that varies with time. In this case, we could use the following representation:

$$y_i \sim \text{Bernoulli}(p_{it}),$$

where

$$\text{logit}(p_{it}) = \alpha_t + \beta_1 x_1 + \beta_2 x_2,$$
$$\alpha_t \sim \mathsf{N}(\mu_\alpha, 1/\tau_\alpha),$$
$$\beta_j \sim \mathsf{N}(0, 100),$$

and $i = 1, \ldots, n, t = 1, \ldots, T$ and $j = 1, 2$. We would write JAGS or Nimble code to implement this model as follows:

```
model {
 # Likelihood
  for (i in 1:N) {
   y[i] ~ dbern(p[i])
   logit(p[i])<- alpha[year[i]] + beta1*x1[i] + beta2*x2[i]
   }

 # Priors for varying intercept over time
  for (t in 1:T) {
   alpha[t] ~ dnorm(mu_alpha, tau_alpha)
   }

 # Priors for fixed effects
 beta1 ~ dnorm(0, 0.01)
 beta2 ~ dnorm(0, 0.01)
```

```
 # Hyperpriors for the random intercept over time
 mu_alpha ~ dnorm(0, 0.01)
 tau_alpha <- pow(sigma_alpha, -2)
 sigma_alpha ~ dunif(0, 10)
}
```

In the above model, a binary outcome (`y[i]`) for `N` individuals at `T` time points is predicted. The probability of occurrence for each individual, `p[i]`, is modelled using a logistic regression model. The logit link function makes sure that the values from the linear predictor are in the unit interval.

- `alpha[year[i]]`: A year-specific random intercept, accounting for year-to-year variation. It follows a normal distribution centred at μ_α with precision τ_α. This structure introduces a hierarchical model, allowing the model to share information across years.
- `beta1 * x1[i] + beta2 * x2[i]`: The effects of two continuous covariates.

The prior distributions for the fixed effects (`beta1` and `beta2`) are weakly-informative normal distributions. `JAGS` defines the normal distribution using the mean and the precision (1/variance). In this case, the prior distributions have a mean of 0 and a variance of 10, which allows for a wide range of possible parameter values. The large variance reflects uncertainty about the true parameter values, thus allowing the data to shape the posterior distribution.

The same model may be implemented in `Stan` as follows:

```
data {
 int<lower=0> N;                      // Number of observations
 int<lower=0> T;                     // Number of time periods
 int<lower=0, upper=1> y[N];  // Occupancy data
 int<lower=1, upper=T> year[N]; // Year
 vector[N] x1;
 vector[N] x2;
}

parameters {
 vector[T] alpha;           // Varying intercept for each year
 real beta1;                 // Fixed effect for covariate 1
 real beta2;                 // Fixed effect for covariate 2
 real mu_alpha;              // Mean hyperprior
 real<lower=0> sigma_alpha; // Standard deviation hyperprior
}

model {
 // Priors
 mu_alpha ~ normal(0, 10);
```

```
 sigma_alpha ~ normal(0, 10);
 alpha ~ normal(mu_alpha, sigma_alpha);
 beta1 ~ normal(0, 10);
 beta2 ~ normal(0, 10);

 // Likelihood
 for (i in 1:N) {
  y[i] ~ bernoulli_logit(alpha[year[i]] + beta1 * x1[i]
                  + beta2 * x2[i]);
 }
}
```

In the code above, Stan requires definition of the data, parameters and model block separately. In the data block, we define the structure and types of input data. This includes specifying the number of observations (N), the number of time periods (T) the binary outcome (y), the time period associated with each observation (year) and the covariates (x1 and x2). In the parameter block, we then declare all unknown parameters (alpha, beta1, beta2, mu_alpha, sigma_alpha). The likelihood and priors are then provided in the model block. The normal prior distribution is now defined by its mean and standard deviation, and Stan does not require the separate specification of the logit-linked probability of occupancy, as this can all be performed in one step using the bernoulli_logit function.

The choice between JAGS, Stan and Nimble depends on the specific needs of the analysis. Each of these probabilistic programming languages provides functions for diagnostic checks, posterior predictive checks, and residual analysis, allowing the user to assess convergence and model fit directly within the software environment. If the model includes discrete latent variables and computational efficiency is a concern, JAGS or Nimble may be preferable. For models with complex continuous parameters where sampling efficiency is important, Stan offers significant advantages. Users that require customised MCMC samplers or compiled code for faster execution may find Nimble to be the most flexible solution.

In practice, wildlife researchers may benefit from experimenting with different tools to identify which best balances computational efficiency, model complexity, and ease of implementation for their specific applications.

Appendix 2: Case Study Code

```
library(R2jags)
fox_data<-readr::read_csv(fox_data.csv)

cat('
model {
```

```
  # Likelihood
  for(i in 1:N){
    fox[i] ~ dpois(lambda[i])
    lambda[i] <- exp(beta0 +
                     beta[site[i],1] * hare[i] +
                     beta[site[i],2] * squirrel[i] +
                     beta[site[i],3] * badger[i] +
                     beta2 * year[i])
  }

  # Priors for global intercept and effect of year
  beta0 ~ dnorm(0, 0.01)
  beta2 ~ dnorm(0, 0.01)

 # Site-level random effects (Multivariate Normal)
  for(j in 1:S){
    beta[j, 1:3] ~ dmnorm(mu_beta[], Omega[,])
  }

  # Priors for covariance matrix
  Omega[1:3,1:3] ~ dwish(R[,], 4)
  Sigma[1:3,1:3] <- inverse(Omega[,])

  # Hyperpriors for mean slopes
  for(k in 1:3){
    mu_beta[k] ~ dnorm(0, 0.01)
  }
}', file={model_code <- tempfile()})


 data_jags <- list(
  N = nrow(fox_data),                  # Number of
                                       observations (rows)
  S = length(unique(fox_data$site)), # Number of unique sites
  R=diag(3),                           # diagonal matrix for
                                     inter-species correlations
  fox = fox_data$'Vulpes vulpes',
  hare = fox_data$'Lepus brachyurus',
  squirrel = fox_data$'Sciurus lis',
  badger = fox_data$'Meles anakuma',
  year = fox_data$year,
  site = fox_data$site)


# Parameters to monitor
```

```
params <- c("beta0", "beta","beta2", "Sigma")
model_run=NULL
list2env(data_jags , envir=globalenv())
model_run = jags.parallel(data = names(data_jags),
                          parameters.to.save = params,
                          model.file = model_code,
                          n.iter = 80000,
                          n.burnin = 10000,
                          n.thin = 5, n.chains=3)

summary<-model_run$BUGSoutput$summary
```

References

Adam T, Griffiths CA, Leos-Barajas V, Meese EN, Lowe CG, Blackwell PG, Righton D, Langrock R (2019) Joint modelling of multi-scale animal movement data using hierarchical hidden Markov models. Methods Ecol Evol 10(9):1536–1550

Aoki M (2013) State Space Modeling of Time Series. Springer Science & Business Media

Barndorff-Nielsen OE, Jørgensen B (1991) Some parametric models on the simplex. J Multivariate Anal 39(1):106–116

Bédard M (2007) Weak convergence of Metropolis algorithms for non-i.i.d. target distributions. Ann Appl Probab 17(4):1222–1244

Besag J (1974) Spatial interaction and the statistical analysis of lattice systems. J R Stat Soc Ser B (Methodol) 36(2):192–225

Besag J, York J, Mollié A (1991) Bayesian image restoration, with two applications in spatial statistics. Ann Inst Stat Math 43:1–20

Blei DM, Kucukelbir A, McAuliffe JD (2017) Variational inference: a review for statisticians. J Am Stat Assoc 112(518):859–877

Bondo KJ, Rosenberry CS, Stainbrook D, Walter WD (2024) Comparing risk of chronic wasting disease occurrence using bayesian hierarchical spatial models and different surveillance types. Ecol Model 493:110756

Borchers DL, Efford MG (2008) Spatially explicit maximum likelihood methods for capture–recapture studies. Biometrics 64(2):377–385

Box GE, Jenkins GM, Reinsel GC, Ljung GM (2015) Time series analysis: forecasting and control. Wiley

Brooks S, Gelman A, Jones G, Meng X-L (2011) Handbook of Markov chain Monte Carlo. CRC Press.

Carpenter B, Gelman A, Hoffman MD, Lee D, Goodrich B, Betancourt M, Brubaker M, Guo J, Li P, Riddell A (2017) Stan: a probabilistic programming language. J Stat Softw 76:1–32

Cressie N (1989) Geostatistics. Am Stat 43(4):197–202

de Valpine P, Turek D, Paciorek CJ, Anderson-Bergman C, Lang DT, Bodik R (2017) Programming with models: writing statistical algorithms for general model structures with nimble. J Comput Graph Stat 26(2):403–413

Djuric PM, Kotecha JH, Zhang J, Huang Y, Ghirmai T, Bugallo MF, Miguez J (2003) Particle filtering. IEEE Signal Process Mag 20(5):19–38

Enari H, Sakamaki H (2012) Landscape-scale evaluation of habitat uses by sympatric mammals foraging for bark and buds in a heavy snowfall area of northern Japan. Acta Theriol 57:173–183

Gelman A (2002) Prior distribution. Encyclopedia Environ 3(4):1634–1637

Gelman A, Rubin DB (1992) Inference from iterative simulation using multiple sequences. Stat Sci 7(4):457–472

Gelman A, Gilks WR, Roberts GO (1997) Weak convergence and optimal scaling of random walk metropolis algorithms. Ann Appl Probab 7(1):110–120

Gelman A, Carlin JB, Stern HS, Dunson DB, Vehtari A, Rubin DB (2013) Bayesian data analysis, 3rd ed. Chapman and Hall/CRC, Boca Raton

Ghitany M, Mazucheli J, Menezes A, Alqallaf F (2019) The unit-inverse gaussian distribution: a new alternative to two-parameter distributions on the unit interval. Commun Stat Theory Methods 48(14):3423–3438

Gilbert NA, Blommel CM, Farr MT, Green DS, Holekamp KE, Zipkin EF (2024) A multispecies hierarchical model to integrate count and distance-sampling data. Ecology 105:e4326

Hastings WK (1970) Monte Carlo sampling methods using Markov chains and their applications. Biometrika 57(1):97–109

Hobbs NT, Hooten MB (2015) Bayesian models: a statistical primer for ecologists. Princeton University Press

Hooten MB, Buderman FE, Brost BM, Hanks EM, Ivan JS (2016) Hierarchical animal movement models for population-level inference. Environmetrics 27(6):322–333

Hooten MB, Johnson DS, McClintock BT, Morales JM (2017) Animal movement: statistical models for telemetry data. CRC Press.

Hostetler JA, Chandler RB (2015) Improved state-space models for inference about spatial and temporal variation in abundance from count data. Ecology 96(6):1713–1723

Hoyt JR, Langwig KE, Sun K, Parise KL, Li A, Wang Y, Huang X, Worledge L, Miller H, White JP, et al (2020) Environmental reservoir dynamics predict global infection patterns and population impacts for the fungal disease white-nose syndrome. Proc Natl Acad Sci 117(13):7255–7262

Jiménez J, García, EJ, Llaneza L, Palacios V, González, LM, García-Domínguez F, Múñoz-Igualada J, López-Bao JV (2016) Multimethod, multistate bayesian hierarchical modeling approach for use in regional monitoring of wolves. Conserv Biol 30(4):883–893

Jonsen ID, Flemming JM, Myers RA (2005) Robust state–space modeling of animal movement data. Ecology 86(11):2874–2880

Kéry M, Royle JA (2020) Applied hierarchical modeling in ecology: analysis of distribution, abundance and species richness in R and BUGS: Volume 2: Dynamic and advanced models. Academic Press.

Kondo A (2018) Interspecific burrow sharing between mammals in countryside in Japan. Mammal Study 43(3):213–218

Larivière S, Pasitschniak-Arts M (1996) Vulpes vulpes. Mammalian Species (537):1–11

McCullagh P, Nelder J (1989) Generalized linear models, 2nd edn. Chapman and Hall/CRC monographs on statistics and applied probability series. Chapman & Hall

Metropolis N, Rosenbluth AW, Rosenbluth MN, Teller AH, Teller E (1953) Equation of state calculations by fast computing machines. J Chem Phys 21(6):1087–1092

Mimnagh N, Parnell A, Prado E, Moral RA (2022) Bayesian multi-species n-mixture models for unmarked animal communities. Environ Ecol Stat 29(4):755–778

Mimnagh N, Ferreira I, Verdade LM, Moral RA (2026) Counting animals we can't see: a triple count model for scarce vestige data. In: Verdade LM, Moral RA (eds) Wildlife management: integrating conservation and innovation in human-altered landscapes. Springer

Ministry of the Environment (Japan) (2010) Living in harmony with nature, building vibrant communities: action plan for the conservation and sustainable use of socio-ecological production landscapes (Satochi-Satoyama). Technical report, Government of Japan

Ministry of the Environment (Japan) (2014) Monitoring site 1000 village survey - medium and large mammal survey data (2006–2012) Technical report, Government of Japan

Morrison ML (2013) Wildlife restoration: techniques for habitat analysis and animal monitoring, volume 1. Island Press

Murphy KP (2002) Dynamic Bayesian networks: representation, inference and learning. University of California, Berkeley

Neal P, Roberts G (2008) Optimal scaling for random walk metropolis on spherically constrained target densities. Methodol Comput Appl Probab 10:277–297
Nelder JA, Wedderburn RW (1972) Generalized linear models. J R Stat Soc Ser A Stat Soc 135(3):370–384
Patterson TA, Thomas L, Wilcox C, Ovaskainen O, Matthiopoulos J (2008) State–space models of individual animal movement. Trends Ecol Evol 23(2):87–94
Plummer M (2003) Jags: a program for analysis of bayesian graphical models using gibbs sampling. In Proceedings of the 3rd International Workshop on Distributed Statistical Computing, vol 124. Vienna, Austria
Rabiner L, Juang B (1986) An introduction to hidden Markov models. IEEE ASSP Mag 3(1):4–16
Roberts GO, Rosenthal JS (2001) Optimal scaling for various metropolis-hastings algorithms. Stat Sci 16(4):351–367
Royle JA (2004) N-mixture models for estimating population size from spatially replicated counts. Biometrics 60(1):108–115
Royle JA, Dorazio RM (2012) Parameter-expanded data augmentation for bayesian analysis of capture–recapture models. J Ornithol 152(Suppl 2):521–537
Royle JA, Kéry M (2007) A bayesian state-space formulation of dynamic occupancy models. Ecology 88(7):1813–1823
Royle JA, Young KV (2008) A hierarchical model for spatial capture–recapture data. Ecology 89(8):2281–2289
Rubin DB (1996) Multiple imputation after 18+ years. J Am Stat Assoc 91(434):473–489
Schafer JL, Yucel RM (2002) Computational strategies for multivariate linear mixed-effects models with missing values. J Comput Graph Stat 11(2):437–457
Sherlock C (2013) Optimal scaling of the random walk metropolis: general criteria for the 0.234 acceptance rule. J Appl Probab 50(1):1–15
Sigourney DB, Murray KT, Gilbert JR, Ver Hoef JM, Josephson E, DiGiovanni Jr RA (2022) Application of a Bayesian hierarchical model to estimate trends in Atlantic harbor seal (Phoca vitulina vitulina) abundance in Maine, USA, 1993–2018. Marine Mammal Sci 38(2):500–516
Song H-R, Lawson A (2009) Space-time Bayesian survival modeling of chronic wasting disease in deer. Prev Veterinary Med 91(1):46–54
Spiegelhalter D, Thomas A, Best N, Gilks W (1996) Bugs 0.5: Bayesian inference using gibbs sampling manual (version II). MRC Biostatistics Unit, Institute of Public Health, Cambridge, UK, 1–59
Stan Development Team (2024) RStan: the R interface to Stan. R package version 2.32.6
Su Y-S, Yajima M (2015) R2jags: Using r to run 'jags'. R package version 0.5-7, 34
Takeuchi K. (2001) Nature conservation strategies for the 'satoyama' and 'satochi', habitats for secondary nature in Japan. Glob Environ Res 5(2):193–198
Tan WY, Ke W (2003) State space models for survival analysis. Handbook Stat 23:519–535
Thorington Jr RW, Koprowski JL, Steele MA, Whatton JF (2012) Squirrels of the world. JHU Press
Torney CJ, Morales JM, Husmeier D (2021) A hierarchical machine learning framework for the analysis of large scale animal movement data. Movement Ecol 9:1–11
Welch G, Bishop G (1995) An introduction fo the Kalman filter. University of North Carolina
Wikle CK (2003) Hierarchical Bayesian models for predicting the spread of ecological processes. Ecology 84(6):1382–1394
Wood SN (2017) Generalized additive models: an introduction with R. Chapman and Hall/CRC

Chapter 9
Distributional Regression Techniques for Wildlife Management

Mikis Dimitrios Stasinopoulos, Fernanda De Bastiani, Kevin Burke, and Julian Merder

Abstract Distributional regression models are powerful "input-output" frameworks in which the response variable is assumed to follow a well-defined theoretical distribution. Unlike traditional regression models that focus solely on the mean, distributional regression models offer a more comprehensive perspective by capturing multiple aspects of the response distribution. This includes not only location parameters but also variability, skewness, kurtosis, quantiles, and exceedance probabilities. Such flexibility is especially valuable in wildlife research, where understanding the full shape of a distribution can provide critical insights into species abundance, habitat suitability, and the likelihood of extreme ecological events. As such, they serve as valuable tools for translating complex ecological data into actionable insights for wildlife management and conservation planning.

Keywords Exceedance probability prediction · Extreme events · Ecological risk assessment · GAMLSS · Generalized additive models · Non-linear regression

M. D. Stasinopoulos
University of Greenwich, London, UK
e-mail: d.stasinopoulos@gre.ac.uk

F. De Bastiani (✉)
Federal University of Pernambuco, Recife, Brazil
e-mail: fernanda.bastiani@ufpe.br

K. Burke
University of Limerick, Limerick, Ireland
e-mail: kevin.Burke@ul.ie

J. Merder
University of Canterbury, Christchurch, New Zealand
e-mail: jmerder@carnegiescience.edu

L. Martins Verdade, R. A. Moral (eds.), *Wildlife Monitoring*,
https://doi.org/10.1007/978-3-032-05821-8_9

1 Introduction

Distributional regression models are regression models where all the characteristics of the distribution of the response can be modelled as functions of the explanatory variables. The emergence of distributional regression modelling can be traced back to the least squares models when the normality assumption was explicitly added to justify theoretical results related to the standard errors of the coefficients of the model (Fox 2002). Many models were developed during the nineteenth century, including polynomial regression, log-linear models, and the famous logistic function introduced by Verhulst to model population growth (Cramer 2002). Fisher introduced nonlinear regression in the 1920s (Dodge 2008), although detailed studies of it did not take place until the 1970s.

The next big step came in the 1970s with the introduction of generalized linear models (GLM) by Nelder and Wedderburn (1972). GLMs assume that the response variable belongs to an exponential family of distributions. The exponential family includes a few important but rather limited distributions in terms of flexibility: the normal, the gamma, the inverse Gaussian, the Poisson, one version of the beta-binomial, the binomial and the Tweedie,[1] among other distributions. GLMs allow the response to be continuous or discrete, depending on the data. The exponential family is theoretically an important class of distributions, especially for modelling the mean of the response, but it is very restrictive in other aspects. The reason is simple, the exponential family includes distributions with up to two parameters. While two parameters can model the location and scale of the data, this is not sufficient for higher order distributional features such as skewness and kurtosis.

GAMLSS was introduced by Rigby and Stasinopoulos (2005) to overcome some of the problems associated with GLM, such as heterogeneity of variance, over-dispersion (by allowing modelling of scale parameters), excess of zeros (by modelling the probability of zero), and skewness and kurtosis (by allowing modelling of relevant parameters). Within the GAMLSS framework, the response distribution is generalized to include any possible theoretical parametric distribution, and all parameters of the assumed distribution can be modelled as a function of the explanatory variables using linear or additive smoothing effects. The GAMLSS models are the first distributional models to move away from the idea that we need to model only the mean model (see Kneib (2013) and Kneib et al. (2023)). Note that related topics like quantile regression models introduced by Koenker and Bassett (1978) were the first regression type models to move away from the idea of modelling only the mean. Quantile regressions have the advantage of not requiring an assumed distribution for the response. That is, they are nonparametric as far as the distribution is concerned, but unfortunately they model only one specific feature of the distribution of the response, its quantiles. Because they model each quantile separately, they run the risk of quantiles crossing each other, results that can violate

[1] The Tweedie family has three parameters a location, a scale and a power parameter, describing the mean and variance relationship (which at early stages of modelling was fixed).

the required monotonic order. Also, at very low or very high quantiles, the values are very unstable because there is not enough information in the data to support the fit.

The amount of data collected, that is, the number of observations in the data, n, is usually proportional to the information we can extract from the data. More data means more information to extract. Thus, the first consideration when deciding whether to use a distributional regression model should be n, the number of observations in the data. For small data sets, i.e. $n < 100$, our experience is that using a distributional regression model is hardly worth the effort. There is not enough information to justify the extra complexity. For larger data sets, the user will find that the effort pays off.

In this chapter, we will argue that the use of distributional regression models in wildlife monitoring could improve our understanding of the complex relationships that occur in the natural world. Section 2 gives different scenarios where we think distributional regression models could be useful. Section 3 gives a general introduction to distributional regression models and its justification while Sect. 4 demonstrates the use of the GAMLSS package in **R**. Conclusions are drawn in Sect. 5.

2 Scenarios

This section highlights several possible wildlife management scenarios where we believe the use of distributional regression models could benefit research.

2.1 Heterogeneity

Ignoring the variation in the data could affect the conclusions. For example, consider an invasive species study where we have counts for two habitats (natural, anthropologically altered). The species could be assumed to have a negative binomial distribution within the habitat cells. In both habitat types, the species has the same mean across cells, but three times the variance in the anthropologically altered habitat type. This means that in such anthropologically altered habitats, the species are much more likely to cause damage because extremes are much more frequent and disruptive, despite the same mean. This behaviour is missed by a standard GLM model.

In the case of climate change, changes in variability may occur before changes in the mean. Consider a herd of animals foraging for food—suitable patches of shelter/food can become more spatially concentrated under climate change. This, in turn, may lead to the animals becoming more spatially concentrated in these same areas, thus causing increased spatial variability in animal counts. Therefore, while a response in the mean abundance of animals may take some time, a change in the variance of abundance may occur much sooner, indicating that the species is under threat.

Animals may group together due to various factors, such as flocking behaviour or aggregation in response to environmental variability. The degree of clustering can be a function of environmental variables or the abundance of another species. A negative binomial distribution can be assumed to model abundance per area, allowing the dispersion parameter σ to vary as a function of the environment or the abundance of other species, providing new insights beyond the mean.

In a coral reef or pelagic ecosystem, schooling fish may respond to predation pressure by clustering more closely together, so that the dispersion parameter σ becomes a function of the abundance of predators, while the expected mean abundance of fish (parameter μ), remains relatively constant.

2.2 Overdispersion

Note that heterogeneity is a term used in statistics for unequal variance in responses. Over-dispersion is usually associated with discrete response variables and has to do with the fact that if we assume a Poisson distribution for the response, its variance is often larger than we would expect for the Poisson distribution. In a distributional regression model, both terms refer to the fact that the variation in the response could be a function of the explanatory variables and can be modelled. Within the GAMLSS framework, there are distributions to deal with overdispersion, such as the two-parameter negative binomial (NBI) distribution (where the variance is a quadratic function of the mean), which is a good first candidate, or for example mixture distributions. Also, there are distributions that can deal with both over-dispersion and under-dispersion, such as the double Poisson and the discrete Burr XII distributions. Moreover GAMLSS allows for the discretization of continuous distributions that may be able to handle over-dispersion or under-dispersion. See Rigby et al. (2019) for more details.

2.3 Excess or Shortage of Zeros

Consider a biomass example where we are not interested only in the mean but in how likely it is to have zero yield in a year. We could do some wildlife management where we increase the mean over time, but we also have a risky practice such that the probability of getting a zero count for the year also increases, e.g., zero biomass or zero yield. This could have much more devastating effects (economically or for ecosystem health) than the advantage of doubling the expected mean in a year. Understanding the sources of zero inflation is crucial in wildlife management, as it allows for more precise modelling of species abundance and distribution, ultimately informing more effective conservation strategies. GAMLSS can deal with an excess of zeros and also has options to adjust a distribution in zero, such as the zero adjusted gamma, see Rigby et al. (2019).

2.4 *Tail Probabilities*

In many studies, the research interest is not in the center of the response distribution, but in the tails. Typical examples are environmental toxicity studies. For example, Merder et al. (2023) modeled concentrations of microcystin, a potent toxin produced by cyanobacteria. The primary interest in such studies lies in predicting values above health-based thresholds, rather than average concentrations, which provide limited insight into environmental or public health risks. This highlights the importance of modelling the full distribution of outcomes, especially its upper tail. Another example is Fendrich et al. (2024) who modelled arsenic levels in European topsoil. Again the interest is in the right tail of the response distribution, rather than the mean.

2.5 *High Skewness and/or Kurtosis*

Changes in the size distribution of fish populations, particularly in terms of skewness and kurtosis, can signal sustainability problems. A right-skewed distribution, where small fish dominate and large individuals are rare, often results from selective fishing that targets larger fish. This demographic shift lowers reproductive output and may erode genetic diversity, as larger fish disproportionately contribute to both recruitment and the maintenance of genetic variation. Over time, the population may struggle to recover and risk stock collapse. Distributional regression, such as GAMLSS, could be used to see if the size distribution changes over time, indicating where the shape parameters respond much stronger and earlier than the mean.

In this section, we have mentioned cases where distributional regression models could have an advantage over more classical regression models. In the next section, we highlight their differences.

3 Distributional Regression Models

To analyse data, we use *models*. This Chapter concentrates on a specific statistical model which falls in the category of *input-output* models. That is, we have information on variable(s) X, the *input*, and we want to use this information to say something about the variable(s) Y, the *output*. We believe that X has information about Y but we do not know what or how. Borrowing a diagram introduced by Breiman (2003) to describe the relationship between X and Y we have;

$$X \rightarrow \boxed{\text{NATURE}} \rightarrow Y.$$

The input X affects the output Y through a filter called, **NATURE**, representing reality. We observe X and Y but we only guess what reality can be. In order to investigate the relationship we use a model

$$X \rightarrow \boxed{\text{Model}} \rightarrow Y.$$

Unlike nature, which is a real phenomenon (albeit complex and unknown), a *model* is an imaginary creation. It could be a mathematical model, an algorithmic model or anything that if appropriately constructed, could help to unscramble the complexity of natural phenomena. Any conceptual idea can be a model, but in order to accept them scientifically, we would need the extra requirements that there should be an objective measure to check how good the model is. That is, we need a *goodness of fit* measure. This measure would tell us how well a model represents reality and to provide a way of comparing alternative models. *Stochastic* models (that is, classical regression models), differential equation models, agent-based models and *algorithmic* models (random forest, neural network) satisfy those requirements and therefore are all suitable candidates models for scientific inquiry.

The most commonly used measure of goodness of fit is the mean squared error $\sum (y_i - \hat{y}_i)^2/n$ where y_i is the observed value of the response and $\hat{y}_i$ is the predicted value from the model. The values of y_i and $\hat{y}_i$ could be evaluated either at the same data set used to fit the model, the *training* data, or at the values of an independent *test* data set. Creating a test data set requires a sufficiently large number of observational points in the first place. This is not always possible in wildlife research. Test data sets are useful to detect *over-fitting*. Over-fitting occurs when the fitted values, $\hat{y}_i$, are too close to the observed data, y_i, in such a way that the model does not generalise well when used for prediction of new observations outside of the training data.

In this Chapter we are dealing with statistical models that contain *stochastic* (*probabilistic*) components. The reason we need a stochastic component in a model is simple. Let us imagine that we can repeat the process $X \rightarrow \boxed{\text{NATURE}} \rightarrow Y$ several times. We would usually expect the response Y (even if the X is held constant) to be different each time. The output (response) Y will typically have natural variation which is not taken into account by the basic input-output model. We call this variation the *natural error variation* of Y. The reason why this natural error variation occurs, could have different explanations, but they are not important for what will follow. The important thing is that natural variation in the response exists, it is observable in practice and we need to take it into account when we build up a model for the response Y. This natural error variation introduces a specific *stochastic* component into the model, the *conditional distribution* of Y given X. We will add this extra component in the diagram of the input output model as;

$$X \rightarrow \boxed{\text{Model}} \rightarrow \mathcal{D}(Y|X) \tag{9.1}$$

where $\mathcal{D}(Y|X)$ represents the natural error distribution of Y conditional on the X. The distribution $\mathcal{D}(Y|X)$ is unknown but when we build a statistical model we assume that it takes the form of a mathematical distribution $D(Y|X;\boldsymbol{\theta})$. The distribution $D(Y|X;\boldsymbol{\theta})$ is an assumption of the model. Fortunately, it is an assumption which can be easily checked using residuals diagnostic plots. Mathematical theoretical distributions have *parameters*. Those parameters are associated with different properties of the distributions like *location*, *scale*, *skewness*, *kurtosis*. Let us denote the distributional parameters as $\boldsymbol{\theta} = (\theta_1, \ldots, \theta_k)^\top$. Then we can rewrite Eq. (9.1) as;

$$X \rightarrow \boxed{\theta = g(X)} \rightarrow D(Y|X, \theta) \tag{9.2}$$

Notice the difference between Eqs. (9.2) and (9.1). Firstly, the unknown distribution $\mathcal{D}(Y|X)$ is substituted with the theoretical distribution $D(Y|X,\boldsymbol{\theta})$. Secondly, the model in Eq. (9.1) is replaced with $\boldsymbol{\theta} = g(X)$ which explicitly states that the distributional parameters $\boldsymbol{\theta}$ are a function, $g()$, of the explanatory (input) variables. The input-output model of Eq. (9.2) represents a **distributional regression model**. We say 'regression' because X affect Y like the ordinary regression does and 'distributional' because the X affect the whole distribution of Y through the whole set of parameters $\boldsymbol{\theta}$ (rather than just the mean parameter).

The classical mean regression model is a subclass of the model in Eq. (9.2) where the location parameter $\boldsymbol{\theta}_1 = \boldsymbol{\mu}$, the mean of the distribution of $D(Y|X,\boldsymbol{\theta})$, is modelled linearly,

$$\boldsymbol{\mu} = \alpha + \beta_1 \mathrm{x}_1 + \cdots + \beta_p \mathrm{x}_p \tag{9.3}$$

and where the assumed distribution $D(Y|X,\boldsymbol{\theta})$ is the normal distribution, a distribution with two parameters $(\boldsymbol{\theta}_1, \boldsymbol{\theta}_2) = (\boldsymbol{\mu}, \sigma^2)$, the mean and variance. Note that in Eq. (9.3) we assumed that we have p explanatory variables. Also note in the classical mean regression model the mean $\boldsymbol{\mu}$ is modelled as a linear function of the explanatory variables while the variance σ^2 is assumed to be constant. That makes the linear model incapable of modelling heterogeneity in variance. Aitkin (1987) proposed a linear model for the log-variance

$$\log \boldsymbol{\sigma}^2 = \alpha_\sigma + \beta_{1\sigma} \mathrm{x}_{1\sigma} + \cdots + \beta_{p\sigma} \mathrm{x}_{p\sigma} \tag{9.4}$$

to account for heterogeneity. There are three features worth noticing in Eq. (9.4). The logarithm in $\boldsymbol{\sigma}^2$ was introduced to make sure that the variance is always positive. In the terminology of the generalized linear model (GLM) the σ has a log *link* function. Also note that $\log\sigma^2 = 2 \log(\sigma)$ so the model for the variance σ^2 is proportional to a model for the standard deviation σ; for numerical reasons (and also because σ is in the same units as the response) it is easier to model σ rather than σ^2. The third point is that the explanatory variables in σ could be completely different from the explanatory variables used to model μ Eq. (9.3). That is, different explanatory variables could affect the mean and the standard deviation.

In a distributional regression model, the behaviour of Y can be studied through its conditional distribution, $\mathcal{D}(Y|X, \boldsymbol{\theta})$. For example, the explanatory variables X can affect the location i.e. the mean or median, the scale (variation, volatility) or the skewness and kurtosis of the distribution. In fact, they can affect any other characteristic of the distribution of the response Y of interest to the practitioner. It should be apparently clear that the mean squared error as a measure of goodness of fit is not appropriate for a distributional regression model. It can only measure how far the observation, y_i, is from its predicted mean, $\hat{y}_i$. We need a measure of how good the distribution $f(y_i \mid (\theta_{1i}, \ldots, \theta_{ki})$ is for each of the response values y_i for $i = 1, 2, \ldots, n$. A common measure for this is the log-likelihood defined as

$$\ell = \sum_{i}^{n} \log f\Big(y_i \mid (\theta_{1i}, \ldots, \theta_{ki})\Big),$$

but there are also distributionally-focused residuals as described in Sect. 3.3.

The aim of a distributional regression model is twofold, firstly, to find $f()$ and more specifically to find **which** and **how** the X's affect the specific part of interest in the distribution of Y; and secondly, to investigate the **shape** of the natural error distribution of Y. The latter is discussed briefly in Sect. 3.1. The former involves selection of the variables and is explained briefly in Sect. 3.2.

3.1 Different Types of Distributions

The theoretical distributions used in distributional regression models can generally be classified into three categories: continuous, discrete, and mixed distributions.

The *continuous* distributions have three commonly-used subclasses; continuous distributions which take values from minus infinity to plus infinity; continuous distributions in the positive line with range of values from 0 to infinity; and continuous distributions which take values in the finite interval zero to one.

The *discrete* distributions take integer values. They can be divided into two commonly-used categories; one takes values from 0 to infinity (often referred to as *Poisson-type* distribution) and the other which takes finite values from 0 up to a positive integer number. The latter are often referred to as *binomial* type distributions while the maximum possible value is referred as the *binomial denominator*.

A *mixed* distribution is a combination of a continuous distribution having one or more discrete points added to it. We will use the following terminology to

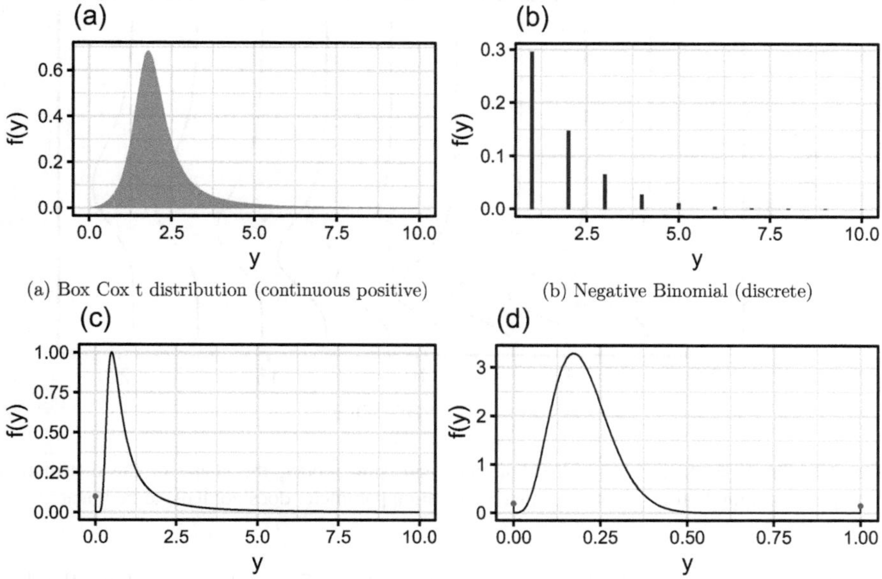

(a) Box Cox t distribution (continuous positive)

(b) Negative Binomial (discrete)

(c) Zero adjusted Box Cox t (mixed distribution)

(d) Inflated beta (mixed distribution)

Fig. 9.1 The three types of theoretical distributions: (**a**) Box Cox t distribution (continuous positive), (**b**) negative binomial (discrete), (**c**) zero adjusted Box Cox t (mixed distribution), and (**d**) inflated beta (mixed distribution)

distinguish between whether the continuous part is defined on the positive real line, $(0, \infty)$ or in the finite interval $(0, 1)$. For example, a *zero-adjusted* distribution is a continuous distribution on the positive real line with an extra zero point mass added. An *inflated distribution* is a continuous distribution defined on the interval from 0 to 1, with two added discrete point masses at zero and one.

Figure 9.1 demonstrates the three types of theoretical distributions. Figure 9.1a shows the Box Cox *t* distribution defined in the positive real line. Figure 9.1b shows a discrete distribution, the negative binomial. Figure 9.1c, d show the two different types of mixed distributions. Panel (c) shows a zero adjusted distribution, in this case, a combination of a continuous in the positive real line distribution (BCT) with an added point at zero. Panel (d) shows an inflated distribution, a distribution on the interval 0 to 1 (the beta distributions) with added discrete points at 0 and 1.

Notice that the practitioner has to make sure that the assumed theoretical distribution $D(Y|X, \boldsymbol{\theta})$ in Eq. (9.2) has the same range of values as the response variable Y.

There is an important point we can make about the theoretical distributions, $D(Y|X, \boldsymbol{\theta})$, which has to do with the number of parameters:

Distributions with more parameters tend to be more flexible.

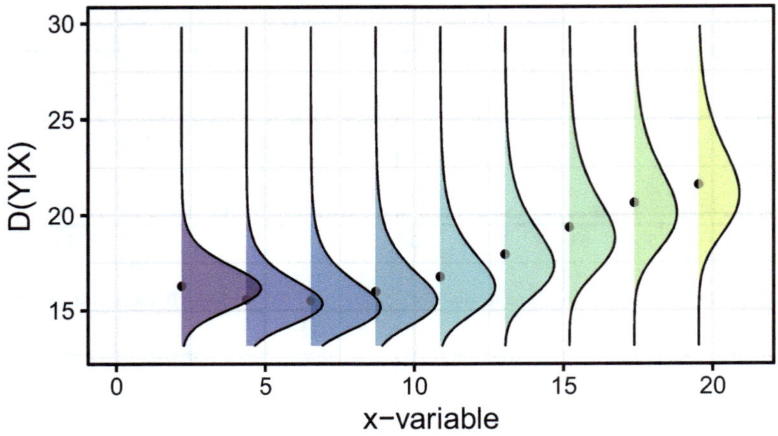

Fig. 9.2 The basic feature of a distributional regression model, the shape of the distribution changes according to the *X*. Note also that the location parameter does not have to be linear

One-parameter distributions are very restrictive. The problem is that all other characteristics of the one-parameter distribution are fixed. As an example, consider the Poison distribution, from the exponential family: it has only one parameter, the mean, which is also the variance of the distribution. The problem is that most data sets with count responses have variance greater that their mean, creating the problem known as *over-dispersion*, see Sect. 2. More generally, one-parameter distributions have their variation, skewness and kurtosis fixed. Two-parameter distributions allow different variation but their skewness and kurtosis are fixed. Three-parameter distributions allow modelling the location (mean or median) the scale (variance or dispersion) but only one of the skewness or kurtosis (not both). In order to model both skewness and kurtosis a four-parameter distribution is needed. Concluding, the more parameters the more flexible the distribution is and modelling all the parameters allows for changing the shape of the distribution of the response according to the values of the explanatory variables, as Fig. 9.2 demonstrates (with one explanatory variable).

3.2 Modelling Relationships

Equation (9.2) explicitly stated that all distribution parameters are functions of the explanatory variables $\boldsymbol{\theta}(\mathrm{x})$. To make the interpretation of the model easier, but also to conform with the specific algorithm used to fit the GAMLSS model (which is based on modelling each parameter separately), it is easier to split this relationship into K separate regression equations where each parameter of the distribution is modelled separately as a function of explanatory variables X_k taken from the list of all available explanatory variables X. We can generally write the dependence of the parameters on the explanatory variables as;

$$\eta_k = g_k(\boldsymbol{\theta}_k) = \mathcal{ML}_k(\mathrm{X}_k) \tag{9.5}$$

for $k = 1, \ldots, K$. Where the functions $g_k()$ are *link* functions ensuring the vector of the kth parameter, $\boldsymbol{\theta}_k$, takes values in the correct range (as we did in Eq. (9.4) for ensuring that σ remain positive). We use the notation $\mathcal{ML}()$ to represent any *machine learning regression* technique suitable for exploring the relationships between the explanatory variables and the distribution parameters. The choice could a linear model like Eq. (9.3);

$$g_k(\boldsymbol{\theta}_k) = \alpha + \beta_1 \mathrm{x}_{1k} + \cdots + \beta_p \mathrm{x}_{pk}$$

an additive smoothing model of the type;

$$g_k(\boldsymbol{\theta}_k) = \alpha + s_1(\mathrm{x}_{1k}) + \cdots + s_k\left(\mathrm{x}_{pk}\right)$$

where $s()$ denotes non-parametric smooth functions or any available algorithmic regression model such as a random forest

$$g_k(\boldsymbol{\theta}_k) = \mathcal{RF}_k(\mathrm{X}_k).$$

The choice of the machine learning regression technique $\mathcal{ML}()$ is the practitioner's preference often determined by the type and the size of data. Key factors to consider include:

(i) *interpretation* of the model; mathematical models such as additive smooth models are more interpretable than algorithmic models such as neural networks or random forests, see Molnar (2020).
(ii) the *number of coefficients* in the model; the number of coefficients is compared relative to the number of observations. If the number of coefficients is close to or greater than the number of observations, dimension-reduction techniques such as LASSO and principal components regression are appropriate. In all other cases the interpretation is crucial.

3.3 *Goodness of Fit and Residuals*

The most common metric used to check the goodness of fit in a distributional regression model is the *log-likelihood*. Maximizing the log-likelihood using the training data is known as the *maximum likelihood estimation* of the coefficients in the model. The likelihood, L, is defined as the probability of obtaining the sample (seen as a function of the parameters rather that as a function of the observations), which for independent samples is the product of the individual probabilities.[2]

It is common in statistical modelling instead of the log-likelihood to use the *deviance*, D, which is defined as minus twice the value of the log-likelihood i.e. $D = -2 \log L = -2\ell$. The deviance, like the log-likelihood, is a measure of goodness of fit but it is only useful relative to other model deviances obtained from the same data set. The smaller the deviance the better the model fits to the data. The deviance like the log-likelihood can be used in both the training data set as a measure of goodness of fit or in the test data a set as a measure the model's predictive power. Again the comparison is relative to other models using the exact same data. If no test data exist it is common to compare models, especially with different number of coefficients, using the *generalised Akaike information criterion* (Akaike 1983) which is defined as:

$$GAIC = \mathrm{D} + k \times df$$

where D is the deviance and *df* are the *degrees of freedom* used to fit the model. That is the number of independent fitted coefficients in the model and k is a number which is chosen by the user leading to different well known measures of fit. For example, $k = 2$ corresponds to the Akaike information criterion (AIC) (Akaike 1973), while $k = \log(n)$, where n is the total number of observations, corresponds to the Bayesian Information Criterion (BIC) (Schwarz 1978). Another value of interest is $k = 3.8414$, which for all practical purposes can be set to 4. This value corresponds to a $\chi^2(1)$ test statistic used in a likelihood ratio test to check the significance of adding one extra term to the model.

The residuals of any distributional regression model are useful for checking the assumptions of the model and especially how well the distribution ($\mathcal{D}$) fits to the data. Residuals within a distributional regression model are defined differently from the residuals of the classical regression models. In classical models the residuals are defined as the difference between the response and its estimated mean value $(y_i - \hat{y}_i)$, which only relate to the mean model. Residuals in a distributional regression model are the *normalised quantile residuals*, NQR, due to Dunn and Smyth (1996). Another popular name for NQR frequently used is *z-scores*. We need two transformations, to get the z-scores from a fitted distributional model. Both

[2] For discrete responses the probability of observing y_i is given as $f(y_i|\theta_{1i}, \ldots, \theta_{k,i})$ where $f()$ is the mass function of the response. For independent samples the probability of observing the sample as a function of the parameters is given as $L(\theta_{1i}, \ldots, \theta_{k,i}) = \prod_i^n f(y_i|\theta_{1i}, \ldots, \theta_{k,i})$ and the log-likelihood is defined as $\ell = \sum_i^n \log f(y_i|\theta_{1i}, \ldots, \theta_{k,i})$. For continuous variables the probability of observing y_i is given as $[F(y_i + \Delta_i|\theta_{1i}, \ldots, \theta_{k,i}) - F(y_i - \Delta_i)|\theta_{1i}, \ldots, \theta_{k,i})]$ where $F()$ is the cumulative distribution function of the response and Δ_i is a small number around y_i. This probability can be approximated by $2\Delta_i[f(y_i|\theta_{1i}, \ldots, \theta_{k,i})]$. Therefore, for independent samples and ignoring the constant $2\Delta_i$ which does not depend on the parameters, the likelihood of a continuous response is defined similarly to the discrete case as $L(\theta_{1i}, \ldots, \theta_{k,i}) = \prod_i^n f(y_i|\theta_{1i}, \ldots, \theta_{k,i})$ and the log-likelihood as $\ell = \sum_i^n \log f(y_i|\theta_{1i}, \ldots, \theta_{k,i})$.

transformations are easy to obtain via the cumulative distribution function (cdf) of the assumed distribution for the response and the q-function (inverse cdf) of the normal distribution $\Phi^{-1}(\cdot)$. Let y_i and $F\left(y_i, \hat{\boldsymbol{\theta}}_i\right)$ be the *ith* observed response and its fitted cumulative distribution function (cdf), respectively. The first transformation gets the probability integral transformed (PIT) residuals defined as,

$$\hat{u}_i = F\left(y_i, \hat{\boldsymbol{\theta}}_i\right). \tag{9.6}$$

If the distribution of y_i is specified "correctly" then the PIT residuals are expected to behave as independent and identically distributed (i.i.d) uniform random variables, i.e. $u_i \sim U(0, 1)$. It is rather difficult to check departures from the uniform distribution, so the PIT residuals are transformed to z-scores using

$$\hat{z}_i = \Phi^{-1}\left(\hat{u}_i\right).$$

If the assumed distribution of the response is approximately "correct" the z-scores are i.i.d. standard normally distributed random variables i.e. $z_i \sim N(0, 1)$. All the diagnostics tools developed for linear models can be then used for the z-scores obtained by fitting different distributions to the data.

The two transformations needed to create the z-score residuals are shown in Fig. 9.3. In Fig. 9.3a the response is transformed to a PIT using the fitted cdf function $F\left(y_i, \hat{\boldsymbol{\theta}}_i\right)$ while in Fig. 9.3b the PIT are transformed to z-scores using the inverse cdf of the normal distribution (otherwise known as the q-function of the normal distribution). While the above residuals work perfectly for continuous distributions, they are not ideal for discrete counts especially if the range of the response is small. A randomisation of the PIT residuals between the values y_i and $y_i + 1$ usually solves the problem; see for example Chap. 10 of Stasinopoulos et al. (2017).

In conclusion, if the assumption of the distribution for the response is adequate for the given data then we expect

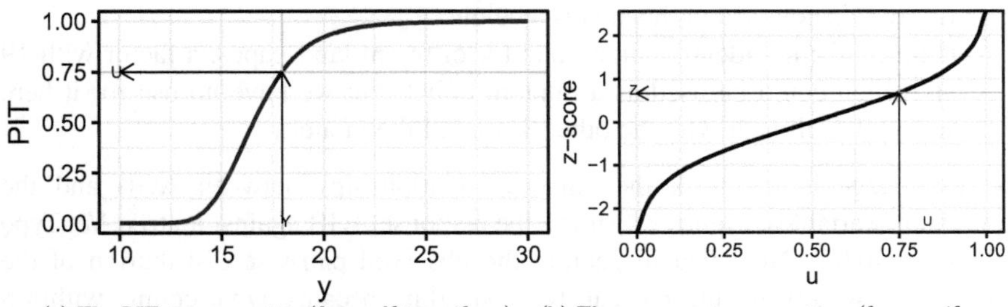

(a) The PIT transormation (from cdf to uniform) (b) The z-scores transormation (from uniform to normal)

Fig. 9.3 The two transformations needed to obtain z-score residuals from a distributional regression model. (**a**) The PIT transformation (from cdf to uniform). (**b**) The z-scores transformation (from uniform to normal)

$$\hat{z}_i = \Phi^{-1}\left(F_o\left(y_i, \hat{\boldsymbol{\theta}}_i\right)\right) \tag{9.7}$$

for $i = 1, 2, \ldots, n$ to behave like i.i.d normally distributed variables.

Note that in the examples that will follow we adopt the convention that the distribution parameters $(\theta_1, \theta_2, \theta_3, \theta_4) = (\mu, \sigma, \nu, \tau)$. When we refer to μ we generally talk about the location parameter, σ refers to the scale parameter. Both ν and τ are shape parameters. Occasionally ν is a proper skewness parameter and τ is a proper kurtosis parameter, see for example Rigby et al. (2019).

4 Biomass Example

This section should be treated as a practical introduction to distributional regression models rather than a full demonstration of their capabilities. We consider data for aboveground biomass (AGB) previously analysed by de Toledo et al. (2024) and presented in the supplemental material for their paper.

Aboveground Biomass refers to the total amount of plant-derived living and dead organic matter per unit of surface area. The data set contain 6149 observations on the following variables:

- **AGB**: (response) the above ground biomass in tonnes per hectare. There are several ways to calculated AGB but here it is calculated using the formula $\text{AGB} = 0.842 \times (\text{VegHeight}^{(\text{NDVI} + 1)})$;
- **VegType**: a factor indicating the type of vegetation with three levels: regenerating fields (RF), pioneer woods (PW) and dense forests (DF);
- **matrix**: a factor with two levels, "cana", for sugarcane and "euca" for eucalyptus, describing the surrounding land use;
- **NDVI**: Normalized Difference Vegetation Index, a remote sensing index for assessing vegetation health, density, and productivity; it is used to calculate AGB and therefore it is not used in the analysis below;
- **VegHeight**: average vegetation height in meters; it is used to calculate AGB and therefore it is not used in the analysis below;
- **RefLandUnit**: the identification code for different landscapes; a factor with 19 levels which could be used as a random effect (but we have no use for it here because it is collinear with the other factors of the model).

The first stage is to check the pairwise relationships between AGB and the explanatory variables. Figure 9.4 shows the response AGB against matrix, VegType and RefLandUnit. Note that inspecting the observed pairwise distribution of the response, as we do in Fig. 9.4, can be misleading because what counts within a distributional regression model is the conditional distribution of the response given **all** explanatory variables in the model. Nevertheless plots such as Fig. 9.4 can be used, with caution, as evidence for high skewness or kurtosis. The response in the pairwise relationships seems highly skewed, therefore it probably needs a skewed

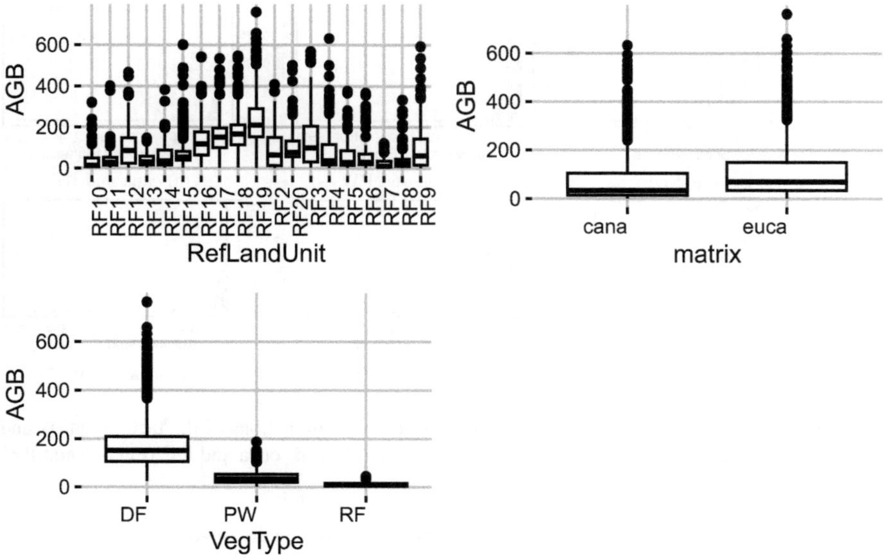

Fig. 9.4 The Aboveground biomass (the response) against the three data factors RefLandUnit, matrix and VegType

distribution. Also there is evidence of differences in behaviour of the response against the explanatory factors. More specifically, for the factor VegType, the level DF appears to behave differently from levels PW and RF.

It is a rather fortunate feature of our data that the explanatory variables matrix and VegType are factors. This gives us the opportunity to partition the data into six distinct sets according to the levels of the factors matrix and VegType; i.e. [(cana, DF), (cana, PW), (cana, RF), (euca, DF), (euca, PW), and (euca, RF)]. Figure 9.5 displays a histogram and a density plot of the response AGB at the six different partitions of the data. We can confirm from the plots in Fig. 9.5 that the distribution for AGB is skewed to the right and varies considerably in shape at different locations.

The assumption of an appropriate distribution for the response variable is a fundamental requirement in any distributional regression model. At a minimum, the assumed distribution must align with the range of the response variable. In our case, the response AGB takes values on the positive real line. Therefore, suitable candidates include continuous distributions defined on $(0, \infty)$ such as the two-parameter Gamma distribution. However since the plots in Fig. 9.4 also indicate a highly skewed distribution, we will consider distributions with greater flexibility, including three- and four-parameter positive real-line distributions. de Toledo et al. (2024) used log(AGB) as the response to fit a linear additive model with main effects for matrix and VegType. Using distributional regression we do not need to transform the response, because fitting a normal model to log(AGB) with an identity link is mathematically equivalent to fitting a log-normal distribution to the untransformed

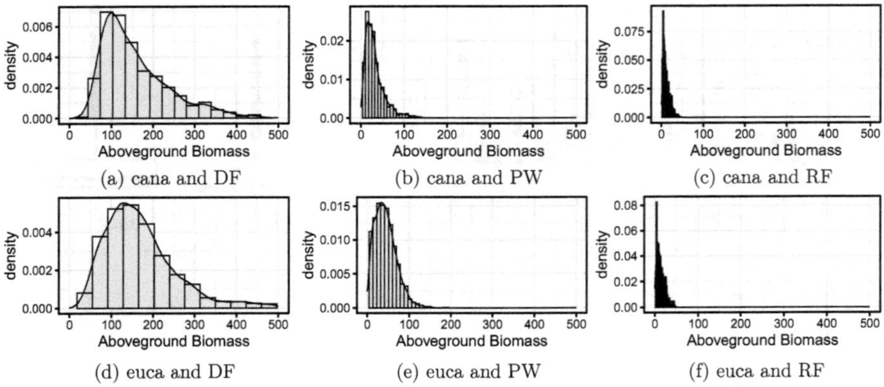

Fig. 9.5 Histograms of aboveground biomass at different combinations of the factors matrix and VegType. (**a**) cana and DF. (**b**) cana and PW. (**c**) cana and RF. (**d**) euca and DF. (**e**) euca and PW. (**f**) euca and RF

Table 9.1 The GAIC table for the biomass data analysis, including distribution name, GAMLSS-package distribution name, and associated numbers of degrees of freedom (df)

Distribution	GAMLSS	AIC	BIC	df
Gamma	GA	60195.55	60276.23	12
Inverse Gaussian	IG	60374.08	60454.76	12
log Normal	LOGNO	60239.90	60320.59	12
Box-Cox-Cole-Green	BCCGo	61221.28	61221.28	18
Box-Cox *t*	BCTo	59997.90	60159.27	24
Box-Cox Power exp.	BCPEo	59889.95	60051.33	24
log SHASH	logSHASH	59860.13	60021.52	24

AGB (using a log link for the median of the distribution). This flexibility makes transformations unnecessary in many cases, provided the underlying distribution is appropriately specified.

The different distributions we tried in our example together with their equivalent gamlss.dist package names are shown in the first and the second columns of Table 9.1, respectively. Note that the package gamlss.dist is downloaded automatically when the package GAMLSS is downloaded in **R**. For more details about the distributions see Rigby et al. (2019).

Because the number of observations in the data is high, 6149, we can afford to fit models for all the parameters of the distributions. Here we upgrade the model fitted by de Toledo et al. (2024) and instead of fitting the main effects model for matrix and VegType, i.e. ~Matrix + VegType, we fit the first order interactions between the two factors i.e. ~Matrix * VegType. Note that the interaction models for the variables matrix and VegType was fitted for all the parameters the distributions used. Columns three and four of Table 9.1 shows the AIC, and the BIC, respectively, of the fitted distributions while column 5 shows the fitted degrees of freedom. The gamma (GA), the inverse Gaussian (IG) and the log-normal distributions (LOGNO), in Table 9.1, have two-parameters. The Box-Cox Cole and Green (BCCGo) has three-parameters and the Box-Cox t (BCTo), Box-Cox power exponential (BCPEo) and log-SHASH (logSHASH) have four-parameters. The distribution BCCGo is introduced by Cole and Green (1992) and it assumes a normal distribution after a power transformation of Y. Rigby and Stasinopoulos (2006) created the BCTo distribution as a power transformation of Y^{ν} having a shifted and scaled (truncated) t distribution with degrees of freedom parameter τ, and it is presented as a model for a dependent variable Y exhibiting both skewness and leptokurtosis. The BCPEo distribution is defined by a power transformation Y^{ν} having a shifted and scaled (truncated) standard power exponential distribution with parameter τ, and it provides a model for a dependent variable Y exhibiting both skewness and kurtosis (leptokustosis or platykurtosis), see Rigby and Stasinopoulos (2004) for more details. The logSHASH distribution is a logarithmic version of the Sinh-arcinh distribution of Jones (2005). From Table 9.1 we can see that the logSHASH distribution with four parameters is best for both AIC and BIC. The log-Normal distribution often used in the literature is not a particular good fit at least according to AIC or BIC.

The effect for the parameters of the logSHASH distribution i.e. μ, σ, ν and τ, are shown in Fig. 9.6. The effects shown are interaction plots in the linear predictors scales, i.e. η_{μ}, η_{σ}, η_{ν} and η_{τ}. Note that the interpretation of the parameters of the original SHASH distribution is; μ is the median [that implies that for logSHASH the location parameters μ is still the median], σ is a scale parameter, ν is the left tail heaviness parameter and τ is the right tail heaviness parameter. Figure 9.6 μ shows that the median is higher in the dense forest, DF, than the pioneer woods, PW, and it is even lower in regenerating fields, RF. The direction of the relationship is reversed as far the rest of the parameters are concerned, see Fig. 9.6 σ, ν, τ, respectively. The regenerating fields, RF, have higher scale and tail parameters than the dense forest, DF, and the pioneer woods, PW. This means that the distribution fitted for RF has different scale and tail parameters when compared to DF, and the PW. It is clear from Fig. 9.6 that VegType does not play a large role in this model, albeit, for σ and ν, the values for DF are increased slightly for cana when compared to euca.

Figure 9.7 shows diagnostic plots for the fitted logSHASH distribution. Figure 9.7a shows a worm plot of the residuals (z-scores) (Buuren 2007). A worm plot is a de-trended QQ-plot. That is, a QQ-plot turned 45° to the right so the QQ-plot is now on a horizontal line. This highlights any discrepancy from normality. [Remember, that, if the model is adequate, the z-score should look like a normally distributed variable.] Most of the worm plot points lie within the 95% piecewise interval indicating that we have no evidence to question the logSHASH distribution assumption. Figure 9.7b shows a bucket plot of the z-score (De Bastiani et al. 2022). Bucket plots are designed to detect skewness and/or kurtosis. The idea here is to see whether the point of skewness and excess kurtosis of the fitted residuals lies away from the middle cycle (which represent the normal distribution). In our case the

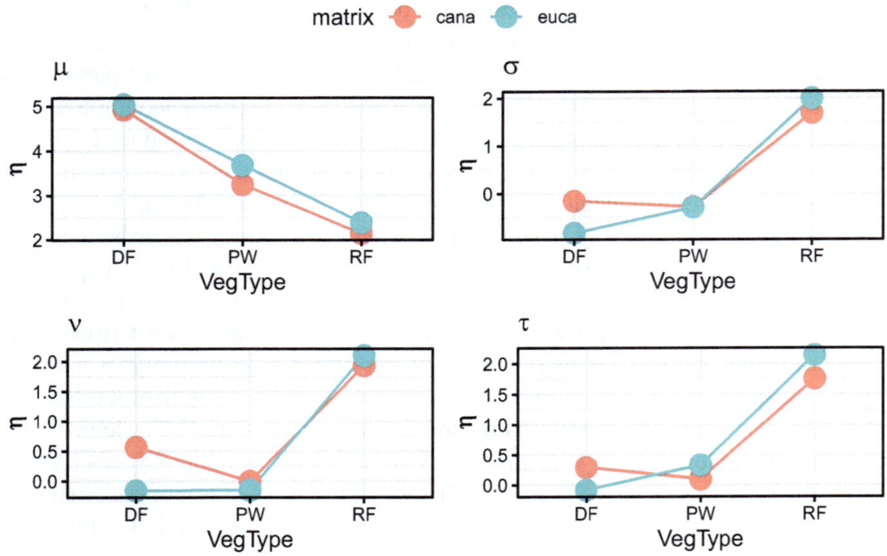

Fig. 9.6 Interaction terms in the linear predictors scale for VegType against matrix for the parameters of the logSHASH distribution

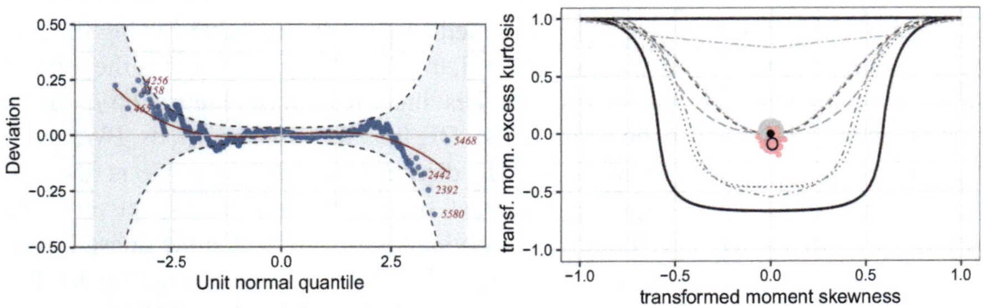

(a) Worm plot of the residuals.

(b) Bucket plot of the residuals.

Fig. 9.7 Diagnostic plots from the regression model using the logSHASH distribution. (**a**) A worm plot of the residuals (detrended qqplot). (**b**) A bucket plot of the residuals (transformed moment skewness against transform moment excess kurtosis)

point, annotated as O, is close to the middle of the plot indicating that there is no reason to worry about skewness or kurtosis in the residuals, which in turn means that the logSHASH distribution has accounted for skewness and kurtosis in the data.

Figure 9.8 demonstrates the essence of GAMLSS and the distributional regression model. Each value of the response value is associated with a fitted distribution rather than a fitted mean value like a standard regression model. In our example, we have fitted interactions for two factors, matrix with two levels, and VegType with three levels. That is, we have fitted six different distributions altogether. Figure 9.8

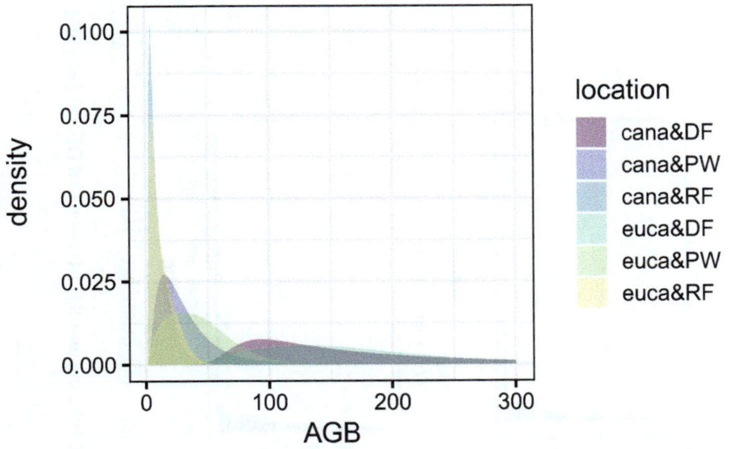

Fig. 9.8 The shape of the six different fitted logSHASH distributions for AGB

plots those six fitted distributions. Given the fitted distributions one could make proper probabilistic statements about the behavior of the response.

Figure 9.9 shows the same information as Fig. 9.8 but each combination of matrix and VegType is shown separately. The plots also show the histograms of the data (similar to Fig. 9.5). The fitted distributions match their equivalent histograms well.

5 Conclusions

The concept of a fully distributional regression specification, GAMLSS, is only 20 years old. Within this period many practitioners have embraced it to explain different aspects of their data. We feel that distributional regression has a great future within the field of wildlife preservation. However, larger and richer datasets will be needed to exploit its full potential to gain a deeper understanding of natural phenomena. The analysis performed here is done using the GAMLSS and related packages in **R**, which are under constant development. More about the available functions can be found in Stasinopoulos et al. (2017). More about the available distributions in Rigby et al. (2019).

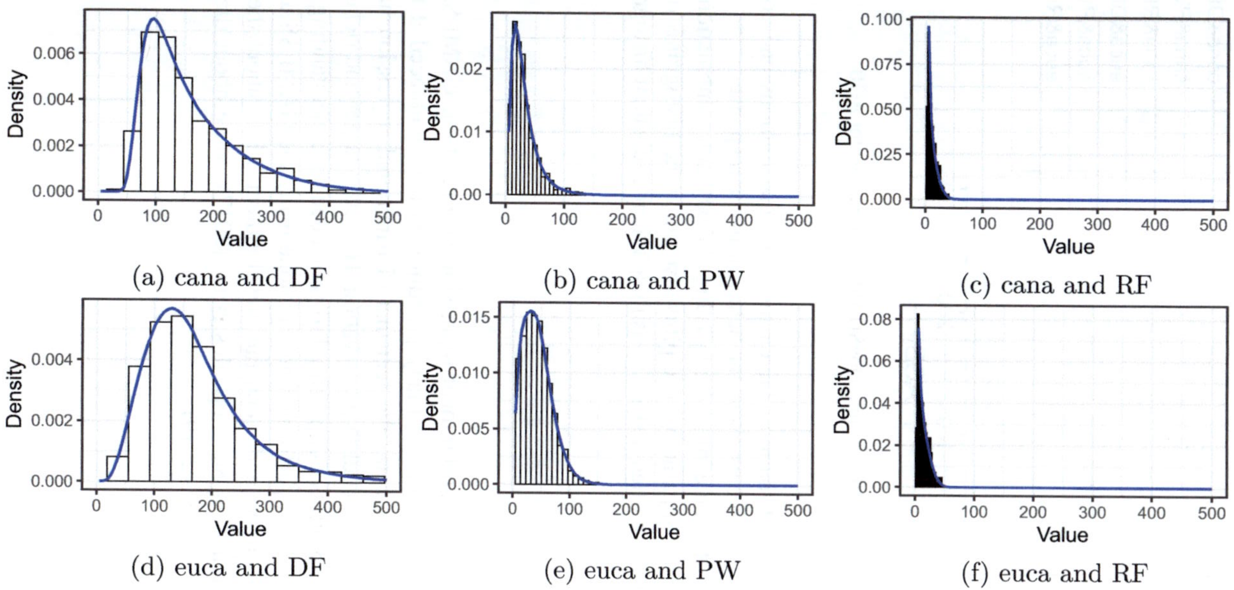

Fig. 9.9 The AGB histograms and the fitted pdf plots at different locations. (**a**) cana and DF. (**b**) cana and PW. (**c**) cana and RF. (**d**) euca and DF. (**e**) euca and PW. (**f**) euca and RF

References

Aitkin M (1987) Modelling variance heterogeneity in normal regression using GLIM. Appl Stat 36:332–339

Akaike H (1973) Information theory and an extension of the maximum likelihood principal. In: Petrov BN, Gaski F (eds) Second international symposium on information theory. Akademiai Kaido, Budapest, pp 267–281

Akaike H (1983) Information measures and model selection. Bull Int Stat Inst 50(1):277–290

Breiman L (2003) Statistical modeling: the two cultures. Qual Control Appl Stat 48(1):81–82

Buuren S (2007) Worm plot to diagnose fit in quantile regression. Stat Modell 7(4):363

Cole TJ, Green PJ (1992) Smoothing reference centile curves: the LMS method and penalized likelihood. Stat Med 11:1305–1319

Cramer JS (2002) The origins of logistic regression (Technical report), vol 119. Tinbergen Institute, pp 167–178

De Bastiani F, Stasinopoulos DM, Rigby RA, Heller GZ, Silva LA (2022) Bucket plot: a visual tool for Skewness and Kurtosis comparisons. Braz J Probab Stat 36(3):421–440

de Toledo RM, Pivello VR, Perring MP, Verdade LM (2024) Natural vegetation biomass and the dimension of forest quality in tropical agricultural landscapes. Ecol Appl 34(3):e2950

Dodge Y (2008) Nonlinear regression. In: The concise encyclopedia of statistics. Springer, New York

Dunn PK, Smyth GK (1996) Randomized quantile residuals. J Comput Graph Stat 5:236–244

Fendrich AN, Van Eynde E, Stasinopoulos DM, Rigby RA, Mezquita FY, Panagos P (2024) Modeling arsenic in European topsoils with a coupled semiparametric (GAMLSS-RF) model for censored data. Environ Int 185:108544

Fox J (2002) An R and S-Plus companion to applied regression. Sage Publications, Thousand Oaks. http://www.socsci.mcmaster.ca/jfox/Books/Companion/

Jones MC (2005) In discussion of Rigby, R. A. and Stasinopoulos, D. M. (2005) Generalized additive models for location, scale and shape. Appl Stat 54:507–554

Kneib T (2013) Beyond mean regression (with discussion and rejoinder). Stat Modell 13(4):275–303

Kneib T, Silbersdorff A, Säfken B (2023) Rage against the mean–a review of distributional regression approaches. Econ Stat 26:99–123

Koenker R, Bassett G (1978) Regression quantiles. Econometrica 46:33–50

Merder J, Harris T, Zhao G, Stasinopoulos DM, Rigby RA, Michalak AM (2023) Geographic redistribution of microcystin hotspots in response to climate warming. Nat Water 1(10):844–854

Molnar C (2020) Interpretable machine learning. Lulu com. https://christophm.github.io/interpretable-ml-book/

Nelder JA, Wedderburn RWM (1972) Generalized linear models. J R Stat Soc Ser A 135:370–384

Rigby RA, Stasinopoulos DM (2004) Smooth centile curves for skew and kurtotic data modelled using the Box-Cox power exponential distribution. Stat Med 23:3053–3076

Rigby RA, Stasinopoulos DM (2005) Generalized additive models for location, scale and shape (with discussion). Appl Stat 54:507–554

Rigby RA, Stasinopoulos DM (2006) Using the Box-Cox t distribution in GAMLSS to model skewness and kurtosis. Stat Modell 6(3):209–229

Rigby RA, Stasinopoulos DM, Heller GZ, De Bastiani F (2019) Distributions for modeling location, scale, and shape: using GAMLSS in r. Chapman & Hall/CRC, Boca Raton

Schwarz GE (1978) Estimating the dimension of a model. Ann Stat 6(2):461–464

Stasinopoulos DM, Rigby RA, Heller GZ, Voudouris V, De Bastiani F (2017) Flexible regression and smoothing: using GAMLSS in r. Chapman & Hall/CRC, Boca Raton

Chapter 10
A Tutorial on Observation-Driven Count Time Series Models Applied to Ecology

Konstantinos Fokianos

Abstract We study simple properties of observation-driven models for the analysis of count time series. Recent literature developed stochastic properties of such models together with associated inference. In addition the availability of software makes their fitting easily accessible. The tutorial offers an introduction to this research area by discussing well established models, their properties and associated inference for both univariate and multivariate count time series. Several other topics are also discussed and real data examples illustrate the methodology for analysis of ecological time series.

Keywords Autocorrelation · Discrete data · Estimation · Interventions · Prediction

1 Introduction

This chapter is devoted to the analysis of count time series, that is a stream of integer-valued observations. Such data are found in ecology in many setups, like counting the number of plant species in a forest or counting bird abundance, for example. The purpose of this chapter is to provide readers with some recent developments of methodology which is directly applicable to such data. I will be using some datasets to illustrate the analysis and point to irregularities, possible extensions and special aspects of count time series data. This work is based on recent statistical developments related to modeling and inference for count time series. The main goal of this tutorial is to familiarize the readers with these developments and illustrate the methodology of observation-driven models; this is explained in more detail in the next sections. There are numerous methods that can be applied—depending always on the problem at hand. For example Auger-Méthé et al. (2021)

K. Fokianos (✉)
University of Cyprus, Nicosia, Cyprus
e-mail: fokianos@ucy.ac.cy

L. Martins Verdade, R. A. Moral (eds.), *Wildlife Monitoring*,
https://doi.org/10.1007/978-3-032-05821-8_10

study the large class of state-space models; see also West and Harrison (1997) and Shumway and Stoffer (2011) among others. Broadly speaking, models for time series analysis can be classified either as parameter driven or observation-driven; the classification is due to Cox (1981, p. 101). Given a stream of observations, observation-driven models are specified by means of the past values of the process (possibly infinite) and an error term. Parameter driven models are defined by assuming a latent structure; for example, the state space models, as advanced by West and Harrison (1997) among others, constitute an important class of examples falling under this framework. Another interesting class of count time series models is the so called integer autoregressive processes (see Alzaid and Al-Osh (1988), Weiß (2008) and more recently Joe (2016)) which are defined by means of the thinning operator and their interpretation is based on branching processes theory. Integer autoregressive models initiated this line of research on observation-driven count time series models since their properties are nicely studied but maximum likelihood inference is challenging. Due to space limitations, this tutorial focuses explicitly on observation-driven models. A recent survey of several results related to the analysis of count time series models is given by Davis et al. (2016), Weiß (2018) and Davis et al. (2021) where the interested reader will engage more thoroughly in this topic. Application of observation-driven models for ecological time series have been studied by Ainley et al. (2017), McGrane-Corrigan et al. (2024), Franchi and Truquet (2025), among others.

Since we are dealing with count data, it is essential to have an appropriate distribution in mind. The most elementary distribution employed for count data analysis is the Poisson distribution. It is well known that it is commonly used to model rates of random events that occur (arrive) in some fixed time interval. Suppose that $\lambda > 0$ denotes the rate of arrivals and let Y be the random variable that represents the number of arrivals over a fixed time interval. Then, we say that Y follows the Poisson distribution with probability mass function (pmf)

$$\mathrm{P}[Y = y] = \frac{\exp(-\lambda)\lambda^y}{y!}, \quad y = 0, 1, 2, \ldots. \tag{10.1}$$

It is easy to show that the mean and variance of Y are both equal to λ, that is $\mathrm{E}[Y] = \mathrm{Var}[Y] = \lambda$. This is a characterization property of the Poisson distribution which limits its applicability; see Hinde and Demétrio (1998) among others. Another important data generating process for the analysis of count data is the negative Binomial distribution whose pmf is

$$\mathrm{P}[Y = y] = \frac{\Gamma(\phi + y)}{\Gamma(y+1)\Gamma(v)}\Big[\frac{\phi}{\phi+\lambda}\Big]^{\phi}\Big[\frac{\lambda}{\phi+\lambda}\Big]^{y}, \quad y = 0, 1, 2, \ldots, \tag{10.2}$$

where $\Gamma(.)$ denotes the Gamma function and both λ and ϕ are positive parameters. This particular parametrization of the negative Binomial distribution is important in applications because the expectation of Y is again λ but its variance is quadratic in λ, vis. $\mathrm{Var}[Y] = \lambda + \lambda^2/\phi$. Hence the negative Binomial is overdispersed, that is its variance is greater than its mean. This phenomenon emerges often when count data analysis is implemented and should be taken into account properly. When ϕ assumes large values then Eq. (10.2) converges to the pmf of Poisson. The negative Binomial distribution is a Gamma-Poisson mixture and as such it belongs to the large class of mixed Poisson distributions; see Karlis and Xekalaki (2005) for a review.

These two distributions are fundamental in the development of univariate count time series models. They have been used extensively in applications, together with their modifications, and provide inference and predictions with good statistical guarantees. These distributions in combination with generalized linear models (McCullagh and Nelder 1989) provide an appropriate framework for the analysis of count time series models. This class of models extends the usual autoregressive models (see Shumway and Stoffer 2011, for example) to count time series. In addition, estimation and inference are based on the maximum likelihood theory (or quasi maximum likelihood). However, the generalization of these distributions to higher dimensions in difficult. Though several appropriate multivariate pmf have been suggested in the literature (see Cameron and Trivedi 2013 and Inouye et al. 2017, for example), their applicability is limited in the sense that their functional form does not allow straightforward inference. Some further discussion about these issues and a new data generating process for multivariate integer valued vectors is given by Fokianos (2024).

This chapter consists of the following topics. Section 2 develops examples of linear and non-linear models for count time series. The class of linear models is useful to get a deep insight of properties of such processes. Other models though, notably log-linear based models, accommodate covariates and bypass constrained optimization for model estimation. Section 3 lists additional distributional assumptions that have been used for modeling count time series models. In this case the linear model serves as a prototype to discover properties of such process. Section 4 discusses covariates and the associated modelling of interventions in the context of the linear and log-linear models of Sect. 2. Section 5 contains some real examples of ecological time series. These data will be employed for illustrating the relevance of the methodology. Their corresponding statistical analysis, is based on the models that are examined and the inference developed in Sect. 6, and it is given in Sect. 7. There is also a growing body of literature on the topic of modeling multivariate count time series. This is a challenging problem and a new research direction is outlined in Sect. 8. The chapter concludes with some additional topics related to modeling ecological time series data.

Notation For a stationary time series say (Y_t), we denote by $\mathrm{E}[Y_t]$ its expectation and by $c(h)$ (respectively $\rho(h)$) its autocovariance (respectively autocorrelation) function, i.e. $c(h) = \mathrm{Cov}(Y_t, Y_{t+h})$ and $\rho(h) = c(h)/c(0)$, for $h = 0, \pm 1, \pm 2, \ldots$. If $h = 0$ then $c(0) = \mathrm{Var}[Y_t]$, that is the variance of (Y_t).

2 Examples of Count Time Series Models

2.1 Linear Poisson Autoregressive Models

Following the conventional case of developing and studying continuous valued time series, it is sensible to formalize the description of observation-driven count time series models by considering initially the popular autoregressive models. These models postulate that a process depends only on its immediate past plus an error term. The simplest example is an autoregressive model of order 1 which will be called Poisson AR(1) model. To grasp the main concepts some notation should be introduced.

Suppose that $\{Y_t\}$ denotes a univariate count time series which is observed in equidistant time points $t = 1, 2, \ldots, n$. It can represent the number of species observed over a fixed period of time. We define the Poisson AR(1) model by assuming that

$$Y_t \mid Y_{t-1} \sim \text{Poisson}(\lambda_t), \quad \lambda_t = d + b_1 Y_{t-1}, \tag{10.3}$$

with d, b_1 non-negative parameters and $\{\lambda_t\}$ denoting the mean of Y_t given Y_{t-1}. Accordingly, the left hand-side of (10.3) implies that

$$\text{P}[Y_t = y \mid Y_{t-1}] = \frac{\exp(-\lambda_t)\lambda_t^y}{y!}, \quad y = 0, 1, 2, \ldots,$$

where λ_t is given by the right hand side of (10.3). The non-negativity condition imposed on both parameters d and b_1 imply that $\lambda_t \geq 0$. If $b_1 = 0$, then we have the case of independent and identically distributed Poisson data with mean d, which is not interesting.

To give some intuition about (10.3), introduce a Poisson process, say $N_t(.)$, of unit intensity. Then the left-hand side of Eq. (10.3) implies that Y_t is equal to the number of events, say $N_t(\lambda_t)$ of $N_t(.)$ in the time interval $[0, \lambda_t]$. Hence, (10.3) simply counts the number of occurrences in the time interval $[0, \lambda_t]$ and depends on its previous value Y_{t-1}. In addition, the autocorrelation of the observed series Y_t is described in terms of the parameter b_1. It can be shown that the condition $0 < b_1 < 1$ implies that model (10.3) is stationary and its autocorrelation function is always positive and decays fast as the distance between observation increases. In fact (see Ferland et al. 2006, Fokianos et al. 2009, among others) the following are true, provided that $0 < b_1 < 1$, and the process Y_t is stationary.

- The mean of the process $\{Y_t\}$ is given by

$$\text{E}[Y_t] = \frac{d}{1 - b_1}.$$

- The autocovariance function of the process $\{Y_t\}$ is given by

$$c(h) = \mathrm{E}[Y_t]\frac{b_1^{|h|}}{1-b_1^2}, \quad h = 0, \pm 1, \ldots.$$

- The autocorrelation function process $\{Y_t\}$ is given by

$$\rho(h) = b_1^{|h|}, \quad h = 0, \pm 1, \ldots.$$

These formulas prove that standard properties of the Poisson AR(1) model mimic the properties of ordinary autoregressive model; see for instance Shumway and Stoffer (2011, Sec. 3). In addition, the second bullet point shows that $c(0) = \mathrm{Var}[Y_t] > \mathrm{E}[Y_t]$ because $0 < b_1 < 1$. Hence, the marginal distribution of (Y_t) is overdispersed under a Poisson AR(1) model. This is a typical phenomenon that occurs in count data, i.e. the variance is larger than the mean. It was just shown that under a conditional linear Poisson distributed autoregressive model this stylistic fact is taken properly into account. Even though the Poisson assumption adapts to overdispersion other distributional assumptions (like the Negative Binomial distribution (10.2)) might be more appropriate for some applications—see Sect. 3.1. Furthermore, the autocovariance function of the Poisson AR(1) model is always positive. This is an additional stylistic fact observed in count time series data which is accommodated by model (10.3). Finally, the form of the autocorrelation function $\rho(.)$ proves the claim that Poisson AR(1) processes are short-memory models.

Model (10.3) is easily generalized to include $p > 1$ past observations as in the case of usual autoregressive models for continuous data. Denote by $\underline{Y}_t = (Y_t, Y_{t-1}, \ldots, Y_{t-(p-1)})$ and consider the model

$$Y_t \mid \underline{Y}_{t-1} \sim \text{Poisson}(\lambda_t), \quad \lambda_t = d + \sum_{i=1}^{p} b_i Y_{t-i}, \quad t \geq p+1, \tag{10.4}$$

with d, $\{b_i, i = 1, 2\ldots, p\}$ non-negative parameters, and $\lambda_t = \mathrm{E}[Y_t \mid \underline{Y}_{t-1}]$. Again this means that

$$\mathrm{P}[Y_t = y \mid \underline{Y}_{t-1}] = \frac{\exp(-\lambda_t)\lambda_t^y}{y!}, \quad y = 0, 1, 2, \ldots,$$

where λ_t is given by the right hand side of (10.4). Intuitive interpretation of (10.4) in terms of a Poisson process is identical to the analysis of (10.3). To avoid complicated notation, explicit equations for the autocovariance and autocorrelation function of a Poisson AR(p) model will not be given. It is mentioned though that their form resembles that of ordinary autoregressive processes. The condition $0 < \sum_{i=1}^{p} b_i < 1$ guarantees stationarity of (10.4). Under this condition, the autocorrelation function of model (10.4) is always positive but decays fast as the distance between observation increases. In addition, a linear Poisson AR(p) model introduces overdispersion as discussed in the case of model (10.3).

The assumption of linearity, in general, yields interesting results and offers deep insights of the properties of count processes. Unfortunately, it is not a realistic assumption and a model such as (10.4) is restrictive in applications in the sense that all autoregressive coefficients are positive thus making the autocorrelation function to be always positive. Conceptually, negative autocorrelation might be observed from real data and thus there exists necessity to develop appropriate methodology. Furthermore, covariates can be included in (10.4) but by taking care that the mean of the process remains positive. Both of these restrictions limit the application of linear models to real data. The next section introduces log-linear, and more generally non-linear models, that can accommodate more complex data structures. We will be discussing two important examples.

2.2 Non-linear Poisson Autoregressive Models

The first class of models that relaxes the constraint of positivity of the autocorrelation function and easily accommodates covariates is the log-linear model which is a canonical link model for count data; see McCullagh and Nelder (1989). Let the time-varying mean of the count process $\{Y_t\}$ be denoted again by λ_t and define the new process $\nu_t = \log \lambda_t$. Then the Poisson autoregressive log-linear model for count time series of order p is defined by (see Zeger and Liang 1986; Fokianos and Tjøstheim 2011)

$$Y_t \mid \underline{Y}_{t-1} \sim \text{Poisson}(\exp(\nu_t)), \quad \nu_t = d + \sum_{i=1}^{p} b_i \log(Y_{t-i}+1) \; t \geq p+1, \quad (10.5)$$

where the notation is as before. The parameters d and $\{b_i, i = 1, 2, \ldots, p\}$ belong, in general, to $\mathbb{R}$ but restrictions on the parameter space are necessary to assure stationarity of the process. Such conditions have been studied by Douc et al. (2013) and Sim et al. (2021).

Model (10.5) is analogous to (10.4). Instead of modeling the mean process λ_t, we model the log-mean process by regressing ν_t to past values of a transformed response. This is a one-to-one transformation of Y_{t-i} which is quite standard in coping with zero data values and it maps zeroes of Y_{t-i} into zeroes of $\log(Y_{t-i} + 1)$. Moreover, both λ_t and Y_t are transformed into the same scale and this ensures stability of the process.This point is better understood by considering the following simple model

$$Y_t \mid Y_{t-1} \sim \text{Poisson}(\exp(\nu_t)), \quad \nu_t = d + b_1 Y_{t-1}.$$

The stability of this process is guaranteed only when $b_1 < 0$, otherwise $\{Y_t\}$ will explode fast.

Certainly other specifications are possible—for instance suppose that we consider a model for the log-mean process by introducing $\log(Y_{t-1} + c)$, where c is an arbitrary constant different than one. But this modeling approach does not affect estimation; see Fokianos and Tjøstheim (2011) for a small-scale simulation study which illustrates this claim.

It is a challenging problem to get an explicit expression for the autocorrelation function of model (10.5) because of the non-linearity introduced by the logarithmic function. However, by simulating a very long path of the series, we get an accurate indication of the range of possible values of the autocorrelation function of process (10.5). Empirically it has been shown (see Fokianos and Tjøstheim 2011) that (10.5) can explain both positive and negative correlation. As a final remark, we observe that the log-linear model easily includes covariates without any concerns about the sign of the corresponding regression coefficients.

There are additional non-linear models that can be applied to count time series (see Neumann 2011; Tjøstheim 2012; Fokianos 2012; Douc et al. 2013; Wang et al. 2014; Weiß and Jahn 2024, among others) but we will be concerned with the so called threshold model whose study has been motivated by the success of the self-excited threshold autoregressive moving average process (see Tong 1990). A simple Poisson threshold model of order 1 is defined by assuming that conditionally on Y_{t-1}, Y_t is Poisson distributed with time-varying mean λ_t given by

$$\lambda_t = \begin{cases} d_1 + b_1 Y_{t-1}, & \text{if } Y_{t-1} \leq r, \\ d_2 + b_2 Y_{t-1}, & \text{if } Y_{t-1} > r, \end{cases} \tag{10.6}$$

with $d_i > 0$, $a_i > 0$, $b_i > 0$, $i = 1, 2$ and $r \in \mathbb{N}$. The parameter r is called the threshold parameter and if $Y_{t-1} \leq r$, then we say that Y_t belongs to the lower regime; otherwise Y_t belongs to the upper regime. Apparently model (10.6) introduces a non-smooth transition between regimes for the observed process $\{Y_t\}$ which yields either positive or negative correlation among successive observations. The threshold model can be easily generalized to include multiple regimes and/or higher lagged values of Y_t. In addition, a log-linear threshold model can be studied but such topic has not been addressed in the literature so far, to the best of our knowledge.

3 Other Distributional Assumptions

The Poisson assumption is an important starting point for developing appropriate regression models for count time series. However, because of its characteristic property that the mean is equal to variance, the Poisson assumption has been criticized since it cannot accommodate overdispersion. In the context of observation-driven count time series models this claim does not hold—recall the discussion in page 2.1. However, statistical inference cannot be based alone on Poisson assumption and some alternative distributions should be considered. For instance, the negative binomial pmf (10.2) takes into account properly a quadratic mean-variance relationship. In addition, there are several applications where an excess of zero values is seen and therefore the Poisson assumption might be questionable. In this section we give some alternative distributional assumptions that are useful for data analysis. We will illustrate the methodology based on the linear model (10.4) but other models, like the log-linear model, can be considered instead. We follow the notation of Sect. 2.

3.1 Negative Binomial

The negative binomial distribution was discussed in Sect. 1; see Eq. (10.2). This particular parametrization allows $\mathrm{E}[Y_t \mid \underline{Y}_{t-1}] = \lambda_t$ so model (10.4) is defined as before; the only difference being that the conditional distribution of Y_t given $\underline{Y}_{t-1}$ is negative Binomial with parameters λ_t and ϕ; see Zhu (2011) and Barreto-Souza and Ombao (2022) also. The fact that the negative Binomial pmf (10.2) is parameterized in such a way that its mean is equal to the Poisson mean allows for consistent estimation of all regression parameters by using the Poisson log-likelihood function as a quasi-likelihood; see Christou and Fokianos (2014) who study negative Binomial processes for modeling count time series in the general framework of mixed Poisson time series. Mixed Poisson models have been reviewed by Karlis and Xekalaki (2005) and alternative estimation methodology in this context has been studied by Aknouche et al. (2018).

With the same notation for model (10.3) and assuming that $0 < b_1 < 1$, we obtain for the negative Binomial conditional distribution the following properties:

- The mean of the process $\{Y_t\}$ is given by

$$\mathrm{E}[Y_t] = \frac{d}{1 - b_1}.$$

- The autocovariance function of the process $\{Y_t\}$ is given by

$$c(h) = \left(\mathrm{E}[Y_t] + \frac{\mathrm{E}^2[Y_t]}{\phi}\right) \frac{b_1^{|h|}}{1 - b_1^2(1 + \phi^{-1})}, \quad h = 0, \pm 1, \ldots.$$

- The autocorrelation function process $\{Y_t\}$ is given by

$$\rho(h) = b_1^{|h|}, \quad h = 0, \pm 1, \ldots.$$

Similar comments made for the case of the Poisson AR(1) model hold in this case as well. The negative binomial linear model introduces an overdispersed marginal distribution for the observed process Y_t whose autocovariance function is always positive. The form of the autocorrelation function of Y_t shows that this is a short memory process. Finally, it is worth mentioning that the autocorrelation function of both the Poisson and negative Binomial AR(1) models is identical.

3.2 *Mixture Models*

Mixture models are generalizations of count processes discussed earlier; see Zhu et al. (2010) and Doukhan et al. (2021). Their interpretation is simple and intuitive; the observed process switches between a finite number of regimes according to an unobservable stochastic mechanism. These models are called Hidden Markov Models (see Zucchini and MacDonald (2009), among others, for some general concepts, and McClintock et al. (2020) for a tutorial related to ecological applications) if the unobserved stochastic mechanism is a Markov chain or simple mixture models when the unobservable process consists of independent and identically distributed (iid) random variables.

We advance modeling by considering the first order mixture linear model but the approach can be extended following the earlier arguments. To be specific suppose that (Z_t) is an iid sequence of random variables with $\mathrm{P}[Z_t = k] = p_k$ for $k = 1, \ldots, K$, that is the process can switch in K different regimes. Assume further that (Y_t) is a count time series and recall the notation of Sect. 2. Then a first order autoregressive Poisson mixture model is defined by

$$\mathrm{P}[Y_t = y \mid Y_{t-1}] = \sum_{k=1}^{K} p_k \frac{\exp(-\lambda_{k,t})\lambda_{k,t}^{y}}{y!}, \quad y = 0, 1, 2, \ldots. \tag{10.7}$$

where $\lambda_{k,t} = d_k + b_{1,k} Y_{t-1}$, $k = 1, 2 \ldots, K$. Simple calculations show that

$$\mathrm{E}[Y_t \mid Y_{t-1}] = \sum_{k=1}^{K} p_k \lambda_{k,t},$$

$$\mathrm{Var}[Y_t \mid Y_{t-1}] = \sum_{k=1}^{K} p_k \lambda_{k,t}(1 + \lambda_{k,t}) - \Big(\sum_{k=1}^{K} p_k \lambda_{k,t}\Big)^2.$$

Since $\sum_{k=1}^{K} p_k \lambda_{k,t}^2 - \Big(\sum_{k=1}^{K} p_k \lambda_{k,t}\Big)^2 = \sum_{\substack{k=1 \\ k \neq l}}^{K} \sum_{l=1}^{K} p_k p_l (\lambda_{k,t} - \lambda_{l,t})^2 > 0$ we have shown that simple a Poisson mixture model takes into account overdispersion. Furthermore, if $p_1 = 1$ and $p_k = 0$, for $k \neq 1$, then (10.7) reduces to model (10.3). First order stationarity of (10.7) is implied by the condition $\sum_{k=1}^{K} p_k b_{1,k} < 1$. When $p_1 = 1$ and $p_k = 0$ for $k > 2$, then the above condition reduces to $0 < b_{1,1} < 1$ which is standard condition for the Poisson AR(1) model.

In this setup, observe that the mean processes $\lambda_{k,t}$ can be modeled by an autoregressive model of order p or a log-linear model—recall (10.5)—or by some suitable non-linear model. Furthermore, the Poisson assumption can be replaced by some other appropriate distributional assumption, like the negative Binomial distribution as discussed before.

3.3 Zero-Inflated Models

In some cases, we observe count time series with a large number of zero values. To take into account the presence of zero values, several authors (see Lambert 1992; Zhu 2012; Blasco-Moreno et al. 2019 among others) have studied some of the models discussed earlier but the Poisson assumption is replaced by the zero-inflated Poisson distribution. We say that a random variable Y follows the zero-inflated Poisson model with parameters $\omega \in (0, 1)$ and $\lambda > 0$, if its probability mass function is given by

$$\mathrm{P}[Y = y] = \begin{cases} \omega + (1-\omega)\exp(-\lambda), & y = 0, \\ (1-\omega)\dfrac{\lambda^y \exp(-\lambda)}{y!}, & y > 0. \end{cases}$$

Denote by $Y \sim \mathrm{ZIP}(\lambda, \omega)$. When $\omega = 0$ then the zero-inflated Poisson distribution reduces to the standard Poisson model. It is easy to see that

$$\mathrm{Var}[Y] = (1-\omega)\lambda(1+\omega\lambda) \geq (1-\omega)\lambda = \mathrm{E}[Y].$$

Therefore, the zero-inflated Poisson distribution accommodates overdispersion which is a consequence of the splitting of data into two processes because of the excess of zeros.

With this notation, the first order autoregressive ZIP model for the analysis of a count time series is defined by

$$Y_t \mid Y_{t-1} \sim \mathrm{ZIP}(\lambda_t, \omega), \quad \lambda_t = d + b_1 Y_{t-1}, \quad t \geq 1, \tag{10.8}$$

where the notation follows again that of Sect. 2. The main properties of model (10.8) are the following; see Zhu (2012) who shows that the condition $0 < (1-\omega)b_1 < 1$ implies stationarity of the process.

- The mean and variance of Y_t are given by

$$\mathrm{E}[Y_t] = \frac{(1-\omega)d}{1-b_1}; \quad \mathrm{Var}[Y_t] = \frac{(1-\omega)d[1+\omega d - (1-\omega)b_1]}{[1-(1-\omega)b_1^2][1-(1-\omega)b_1]^2}.$$

- The autocorrelation function of Y_t is

$$\rho(h) = [(1-\omega)b_1]^{|h|}, \quad |h| \geq 1.$$

Note again that (10.8) introduces a short memory process and it is positive for all lags h. In addition a log-linear model can be employed instead for modeling zero-inflated time series data.

4 Covariates and Interventions

In several applications we observe a stream of count time series together with an m-dimensional covariate vector process, say $\{X_t\}$. In this case, the aim is to discover whether the covariate process affects the response. In principle all models discussed earlier include covariates but the log-linear model (10.5) facilitates covariate modeling in a straightforward way. Indeed, (10.5) becomes

$$\nu_t = d + \sum_{i=1}^{p} b_i \log(Y_{t-i} + 1) + \gamma^T X_t,$$

where γ is an m-dimensional vector of parameters. This specification allows for the components of the vector γ to take either positive or negative values. Comparing with a linear model specification (see Agosto et al. 2016)

$$\lambda_t = d + \sum_{i=1}^{p} b_i Y_{t-i} + \gamma^T X_t,$$

the log-linear model has an advantage as it does not call for $\gamma^T X_t > 0$. However, an appropriate transformation, such as the exponential, resolves this issue, i.e.

$$\lambda_t = d + \sum_{i=1}^{p} b_i Y_{t-i} + \exp(\gamma^T X_t),$$

but interpretation of results should be done carefully.

An important type of covariate is the deterministic process given by

$$X_t = \delta^{t-\tau} \mathbb{1}(t \geq \tau) \tag{10.9}$$

where τ is any time point between the first time and last time of observation, δ is a known constant which acts as a decay rate and $\mathbb{1}(.)$ denotes the indicator function. This type of covariate has been studied by Fokianos and Fried (2010, 2012) and implemented by Liboschik et al. (2016, 2017) to model retrospectively interventions in count time series. This is motivated by the fact that knowing the effect of known interventions helps implementing new policies or to search for unknown intervention effects and find explanations for them a posteriori.

The specific form of covariate process covers various types of interventions (Box and Tiao 1975; Chen and Liu 1993) for different choices of the constant δ: a singular effect for $\delta = 0$ (spiky outlier) at $t = \tau$, an exponentially decaying change in location for $\delta \in (0, 1)$ (transient shift) and a permanent change of location for $\delta = 1$ (level shift). Similar to the case of covariates, the effect of an intervention is essentially additive for the linear model and multiplicative for the log-linear model. Estimation, testing and the case of multiple intervention has been examined thoroughly in the work by Liboschik et al. (2017) where we refer the reader for further details.

4.1 Discussion

Note that all type of models discussed can be extended to include a feedback process—that is a hidden process—as follows. Consider again the simple linear Poisson autoregressive model (10.3). Then, a recent stream of literature has focused on the properties of its generalization given by (see Streett 2000; Ferland et al. 2006; Fokianos et al. 2009 among others)

$$Y_t \mid \mathscr{F}_{t-1} \sim \text{Poisson}(\lambda_t), \quad \lambda_t = d + a_1\lambda_{t-1} + b_1 Y_{t-1}, \tag{10.10}$$

where $\mathscr{F}_t$ denotes the history of the process up to time t, that is all observation up to time t, and λ_t is a Poisson intensity process, as before. The parameters d, a_1, b_1 are assumed to be positive and they need to satisfy the condition $0 < a_1 + b_1 < 1$ so that (10.10) is stationary. When $a_1 = 0$, then model (10.3) is recovered.

Model (10.10) provides a parsimonious way of modeling count time series. In principle, when count time series are available and their autocorrelation function assumes relatively high values for large lags, then we should expect a model of the form (10.4) to accommodate this fact by entertaining a large number of lagged variables. However such an approach can be avoided when employing model (10.10). Several results for model (10.10) have been reported in the literature, see Rydberg and Shephard (2000), Streett (2000), Heinen (2003), Ferland et al. (2006), Fokianos et al. (2009), Neumann (2011) and Doukhan et al. (2012), among others and for a recent application which is related to the number of cases in epidemic models for studying the spread of infection diseases in a population see Bracher and Held (2022). Model (10.10) can be defined in terms of the log-linear model (10.5) and in the case of the threshold model (10.6); see previous references. In addition, (10.10) can be extended the case of the negative Binomial conditional distribution, i.e.

$$Y_t \mid \mathscr{F}_{t-1} \sim \text{NegBin}(\lambda_t, \phi), \quad \lambda_t = d + a_1\lambda_{t-1} + b_1 Y_{t-1}, \tag{10.11}$$

with the same notation as before and NegBin denotes the pmf (10.2). In terms of mean evaluation both processes (10.10) and (10.11) have the same mean structure; this is important property in terms of model fitting. However, the additional parameter ϕ generates a richer class of autocovariance functions. Some details of model fitting will be given in Sect. 6 and a recent extension of model (10.11) that accommodates time-varying dispersion has been studied by Barreto-Souza et al. (2025).

5 Case Studies

BioTime (see Dornelas et al. 2018) is a large database of biodiversity time series that contains many species abundance data. These measurements have been obtained from all over the world by a multitude of environmental studies. BioTime time series will be employed for illustrating the methodology outlined in this chapter. These time series are not equidistant but we proceed with the analysis treating them as if they were observed sequentially in time.

5.1 Lizard Data

The first data example we consider is the lizard pitfall trap data which is available at https://biotime.st-andrews.ac.uk/selectStudy.php?study=316. More specifically we study the species *Cnemidophorus inornatus*. A simple time series plot (together with the autocorrelation and partial autocorrelation functions) is shown in Fig. 10.1. The plot reveals two unusual measurements around the observation times 60 and 68. In addition, we note that the sample autocorrelation function is positive for most of the lags; this stylistic fact was discussed earlier where the theoretical autocovariance function of the linear model (10.3) was shown to be positive. In addition, the sample mean of the series is 2.11 and the corresponding sample variance is equal to 3.62; hence these data are overdispersed. Based on this short analysis, we will fit a simple autoregressive model which takes into account possible intervention effects.

5.2 Birds Data

We study now weekly records of bird species observed at Konza Prairie—for more see https://biotime.st-andrews.ac.uk/selectStudy.php?study=333. This dataset contains 9261 records of 132 distinct species and we will study the species *Molothrus ater*. The first step is to make again a standard time series plot (together with the autocorrelation and partial autocorrelation functions). This is shown in Fig. 10.2. The plot shows that the process does not exhibit any unusual effects. We

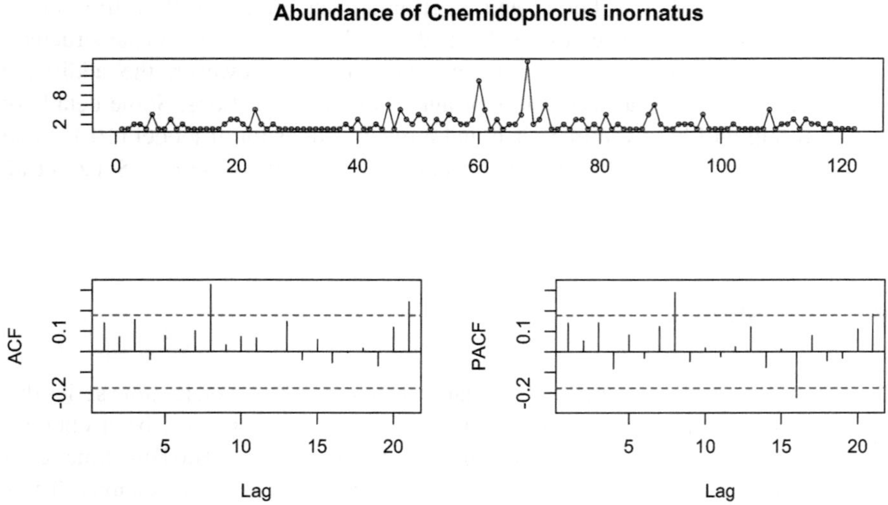

Fig. 10.1 Time series plot, acf (lower left plot) and partial acf (lower right plot) of *Cnemidophorus inornatus* abundance data. There are 122 observations during the period 21/7/1989 to 3/10/2006

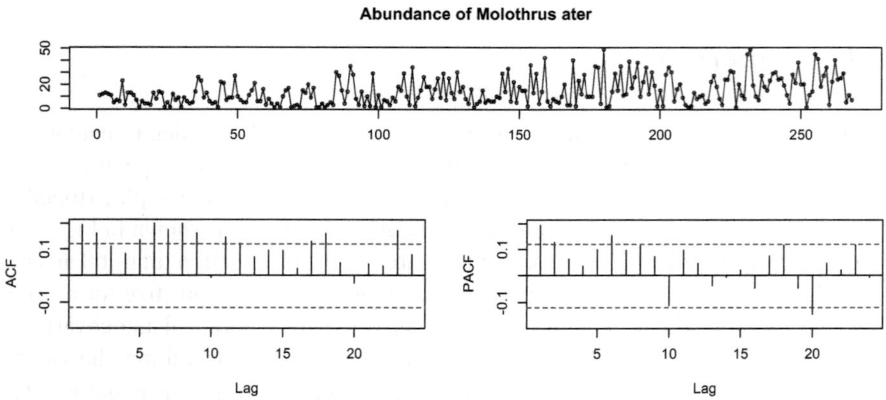

Fig. 10.2 Time series plot, acf (lower left plot) and partial acf (lower right plot) of *Molothrus ater* abundance data. There are 268 observations during the period 27/05/1981 to 15/06/2008

observe that the autocorrelation function is positive for all lags shown and does not decay to zero. However it does decay to zero but for lager lag values. The sample mean is 14.05 and the sample variance around 120.86. These data provide all standard characteristics for application of model (10.10) as we will see.

6 Inference

In this section, we will be discussing estimation and prediction for observation-driven count time series models. We will be concerned with maximum likelihood (under a Poisson assumption) and quasi-maximum likelihood estimations (under a negative Binomial assumption). Other distributions can be fitted in this framework but the discussion focuses on these topics for ease of presentation. Relevant references are given throughout the chapter.

6.1 Maximum Likelihood Estimation

We discuss (conditional) maximum likelihood inference—MLE—for the Poisson linear model (10.4) to illustrate the main ideas. Fitting of more complex models, like (10.10), is implemented in the same way but for ease of presentation we focus on pure autoregressive models. Similar results hold for the case of log-linear model (10.5) but those will be illustrated by means of a working example. The threshold model (10.6) is fitted in the same vein provided that the threshold parameter r is known. The cases of linear and log-linear models has been examined in detail by Liboschik et al. (2017) and all methodology is implemented by the R package `tscount`.

The Poisson based likelihood can be regarded as a "working" likelihood which yields quasi-maximum likelihood estimators (QMLE) if other distributional assumptions are used instead of the Poisson. It turns out that this framework can accommodate various data generating processes and can be generalized in numerous ways following the ideas of Wedderburn (1974); see also Christou and Fokianos (2014) and Ahmad and Franq (2016) in the context of time series. In addition, it provides consistent and asymptotically normally distributed estimator of the autoregressive coefficients provided that the mean model process has been correctly specified. Hence, asymptotic confidence intervals and testing is implemented in a straightforward way.

Assume that $\{Y_t, t = 1, 2, \ldots, n\}$ denotes an observed sample from a count time series and consider fitting model (10.4) to those data. Let θ be the $(p+1)$ dimensional vector of unknown parameters, that is $\theta = (d, a_1, b_1, \ldots, b_p)'$. Suppose further that the true value of the parameter is denoted by $\theta_0 = (d_0, b_{1,0}, \ldots, b_{p;0})'$. The conditional likelihood function for θ is given by

$$L(\theta) = \prod_{t=1}^{n} \mathrm{P}\Big[Y_t = y_t \mid \underline{Y}_{t-1}\Big] = \prod_{t=1}^{n} \frac{\exp(-\lambda_t(\theta))\lambda_t^{y_t}(\theta)}{y_t!},$$

because of the Poisson assumption. Note that $\lambda_t(\theta) = d + \sum_j b_j Y_{t-j}$ by (10.4). Hence, the log-likelihood function is given up to a constant, by

$$l(\theta) = \sum_{t=1}^{n} l_t(\theta) = \sum_{t=1}^{n} \Big(y_t \log \lambda_t(\theta) - \lambda_t(\theta) \Big).$$

The MLE $\widehat{\theta}$ is defined as the maximizer of the log-likelihood function; that is $\widehat{\theta} = \arg\max_{\theta \in \Theta} l(\theta)$. The score function is

$$S_n(\theta) = \frac{\partial l(\theta)}{\partial \theta} = \sum_{t=1}^{n} \frac{\partial l_t(\theta)}{\partial \theta} = \sum_{t=1}^{n} \left(\frac{y_t}{\lambda_t(\theta)} - 1 \right) \frac{\partial \lambda_t(\theta)}{\partial \theta},$$

where $\partial \lambda_t(\theta)/\partial \theta$ is a $(p+1)$-dimensional vector whose components are given by

$$\frac{\partial \lambda_t}{\partial d} = 1, \quad \frac{\partial \lambda_t}{\partial b_k} = Y_{t-k}, \quad k = 1, 2, \ldots, p.$$

Furthermore, the Hessian matrix is obtained by further differentiation of the score equations, i.e.

$$\begin{aligned} H_n(\theta) &= -\sum_{t=1}^{n} \frac{\partial^2 l_t(\theta)}{\partial \theta \partial \theta'} \\ &= \sum_{t=1}^{n} \frac{y_t}{\lambda_t^2(\theta)} \left(\frac{\partial \lambda_t(\theta)}{\partial \theta} \right) \left(\frac{\partial \lambda_t(\theta)}{\partial \theta} \right)' - \sum_{t=1}^{n} \left(\frac{y_t}{\lambda_t(\theta)} - 1 \right) \frac{\partial^2 \lambda_t(\theta)}{\partial \theta \partial \theta'}. \end{aligned}$$

Then the conditional information matrix is given by

$$G_n(\theta) = \sum_{t=1}^{n} \text{Var}\left[\frac{\partial l_t(\theta)}{\partial \theta} \mid \underline{Y}_{t-1} \right] = \sum_{t=1}^{n} \frac{1}{\lambda_t(\theta)} \left(\frac{\partial \lambda_t(\theta)}{\partial \theta} \right) \left(\frac{\partial \lambda_t(\theta)}{\partial \theta} \right)',$$

and it is an important quantity for the asymptotic distribution of the MLE $\widehat{\theta}$. Under certain regularity conditions, it can be proved that $\widehat{\theta}$ is consistent and asymptotically normal (see Fokianos et al. 2009 for details)

$$\widehat{\theta} \xrightarrow{p} \theta_0$$

and

$$\sqrt{n}\left(\widehat{\theta} - \theta_0\right) \xrightarrow{D} \mathcal{N}(0, G^{-1}),$$

where the matrix G is defined by

$$G(\theta) = \text{E}\left(\frac{1}{\lambda_t} \left(\frac{\partial \lambda_t}{\partial \theta} \right) \left(\frac{\partial \lambda_t}{\partial \theta} \right)' \right),$$

with E[·] is taken with respect to the stationary distribution and is evaluated at $\theta = \theta_0$. All the above quantities can be computed, and they are employed for standard inference. For instance, an estimator of the conditional information matrix G is given by

$$\hat{G} = \frac{1}{n}\sum_{t=1}^{n} \frac{1}{\lambda_t(\widehat{\theta})}\left(\frac{\partial \lambda_t(\widehat{\theta})}{\partial \theta}\right)\left(\frac{\partial \lambda_t(\widehat{\theta})}{\partial \theta}\right)'$$

The estimate $\hat{G}$ can be calculated by simply substituting $\widehat{\theta}$ to the above quantities.

Remark 1 In the case of mixed Poisson models, like the negative Binomial model (10.11) it is hard, in general, to analytically maximize available likelihood function. We resort to Quasi-MLE (QMLE) methodology as discussed earlier by treating the Poisson log-likelihood function as if it was the true log-likelihood. The same calculations persist regardless the existence of a hidden process, i.e. $a_1 = 0$ in (10.11). For the specific case of negative Binomial model, by recalling (10.2), the conditional information matrix is

$$\begin{aligned} G_{1n}(\theta) &= \sum_{t=1}^{n} \mathrm{Var}\Big[\frac{\partial l_t(\theta)}{\partial \theta} | \mathscr{F}_{t-1}\Big] \\ &= \sum_{t=1}^{n}\Big(\frac{1}{\lambda_t(\theta)} + \frac{1}{\phi}\Big)\left(\frac{\partial \lambda_t(\theta)}{\partial \theta}\right)\left(\frac{\partial \lambda_t(\theta)}{\partial \theta}\right)'. \end{aligned}$$

It turns out that consistency and asymptotic normality of $\hat{\theta}$ continue to hold; see Christou and Fokianos (2014). In fact, it can be shown that the QMLE is consistent and asymptotically normally distributed, i.e.

$$\sqrt{n}(\hat{\theta} - \theta_0) \xrightarrow{D} \mathscr{N}(0, G^{-1}(\theta_0)G_1(\theta_0)G^{-1}(\theta_0)),$$

where the matrix G is as before and the matrix G_1 is given by

$$G_1(\theta) = E\left(\left(\frac{1}{\lambda_t(\theta)} + \frac{1}{\phi}\right)\left(\frac{\partial \lambda_t}{\partial \theta}\right)\left(\frac{\partial \lambda_t}{\partial \theta}\right)'\right). \tag{10.12}$$

Obviously, the asymptotic variance matrix is a sandwich type matrix because of the issue of mispecification; see White (1982). For the case of Poisson distribution (that is ϕ tending to infinity) we discover again the earlier results, However, in general, the limiting information matrix depends on the parameter ϕ and we need to estimate consistently. An estimate is found by solving the equation

$$\sum_{t=1}^{n} \frac{(Y_t - \hat{\lambda}_t)^2}{\hat{\lambda}_t(1 + \hat{\lambda}_t \phi^{-1})} = n - p - q - 1,$$

where $\hat{\lambda}_t = \lambda_t(\hat{\theta})$, p is the order of observations and q is the order of the hidden process. For example, when fitting model (10.3) but with the negative Binomial as response distribution, then $p = 1$ and $q = 0$. Similarly for model (10.11), $p = q = 1$. If the solution of this equations exists, then it is consistent and can be calculated by existing software (cf. Dean et al. 1995; Breslow 1990; Cameron and Trivedi 1998, Ch. 3; Lawless 1987; Liboschik et al. 2017).

6.2 Residuals

An integral part of any model fitting process is the development of diagnostic tools for examining adequacy of the fit; this section provides, additionally, some tools for examining these issues. A detailed discussion concerning diagnostic methods for count time series models has been given by Kedem and Fokianos (2002, Sec. 1.6), among others. To examine the adequacy of the fit, we consider the so called Pearson residuals

$$e_t \equiv \frac{Y_t - E(Y_t \mid \underline{Y}_{t-1})}{\sqrt{Var(Y_t \mid \underline{Y}_{t-1})}}.$$

For the case of Poisson linear model (10.4) (or the corresponding log-linear model (10.5)), the Pearson residuals are given by

$$e_t = \frac{Y_t - \lambda_t}{\sqrt{\lambda_t}}, \quad t \geq 1. \tag{10.13}$$

For the same models with the assumption of negative Binomial distribution we obtain

$$e_t = \frac{Y_t - \lambda_t}{\sqrt{\lambda_t + \dfrac{\lambda_t^2}{\phi}}}, \quad t \geq 1. \tag{10.14}$$

Under the true model, the random variable $\{e_t, t \geq 1\}$ is a white noise sequence with $\mathrm{Var}[e_t] = 1$. To compute the Pearson residuals in either case, substitute λ_t by $\hat{\lambda}_t \equiv \lambda_t(\hat{\theta})$. Plots of the autocorrelation function and the cumulative periodogram (see Brockwell and Davis 1991, Sec. 10.2) give some clue about the whiteness of the sequence $\{e_t, t \geq 1\}$.

6.3 Information Criteria

Model selection can be based on standard information criteria, like Akaike's information criterion (AIC) introduced by Akaike (1973), and used widely in applications. The AIC is defined as a function of the number of independent model parameters,

$$\text{AIC}(p) = -2\log l(\theta) + 2\text{dim}(\theta), \tag{10.15}$$

where dim(.) denotes the dimension of a vector. We choose the model corresponding to p that minimizes (10.15). Evidently, the AIC penalizes against models with unrealistically large p. To improve the large sample performance of the AIC in the sense of obtaining consistent estimates of the order p, several modifications of the AIC have been suggested in the literature. In Rissanen (1978), Schwarz (1978) the $2p$ term in (10.15) is replaced by $p \log n$, i.e.

$$\text{BIC}(p) = -2\log l(\theta) + \text{dim}(\theta)\log n \tag{10.16}$$

where n is the number of observations. The BIC estimates the model order consistently in most cases. There are several information criteria that have been suggested over the year but we will be using Quasi (QIC) information criteria which is a special case of the AIC which takes into account that estimation is done by quasi-likelihood methods. See Pan (2001) for more details.

6.4 Prediction

Following Gneiting et al. (2007), we take the point of view that predictions should be probabilistic in nature. In addition, they should strive to maximize the sharpness of the predictive distribution subject to calibration. Calibration refers to the statistical consistency between the predictive distribution and the observations. The notion of sharpness refers to the concentration of the predictive distribution and is a property of the forecasts only. It follows that if the predictive distribution is more concentrated, then the forecasts obtained are sharper. Therefore, this will yield better predictions subject to calibration. In this section we provide only a tool to assess probabilistic calibration of a model. Assessing marginal calibration and sharpness has been discussed and implemented byJung and Tremayne (2011), Christou and Fokianos (2015) and Liboschik et al. (2017), among others. Calculation of all these measures requires an assumption on the conditional distribution of the process.

6.4.1 Assessment of Probabilistic Calibration

To ascertain whether or not a certain distributional assumption fits better the data than the Poisson distribution, consider the Probability Integral Transformation (PIT) histogram, as explained below. This tool is used for checking the statistical consistency between the predictive distribution and the distribution of the observations. If the observation is drawn from the predictive distribution, then the PIT has a standard uniform distribution. In the case of count data, the predictive distribution is discrete and therefore the PIT is no longer uniform. To remedy this, several authors have suggested a randomized PIT. However, Czado et al. (2009) introduced a non-randomized uniform version of the PIT.

In our context, we fit any model discussed earlier, such as (10.4) or (10.5) under the Poisson or any other assumption. After obtaining consistent estimators for the regression parameters, we estimate the mean process λ_t by $\hat{\lambda}_t = \lambda_t(\hat{\theta})$ and any other additional parameters. Then, the PIT is based on the conditional cumulative distribution

$$F^t(u|Y_t = y) = \begin{cases} 0, & u \leq P_{y-1}, \\ (u - P_{y-1})/(P_y - P_{y-1}), & P_{y-1} \leq u \leq P_y, \\ 1, & u \geq P_y, \end{cases} \quad (10.17)$$

where P_y is equal to the assumed conditional distribution. For instance, in the case of model (10.4) P_y is the Poisson distribution. Subsequently, we form the mean PIT by

$$\bar{F}(u) = \frac{1}{n} \sum_{t=1}^{n} F^{(t)}(u|y), \quad 0 \leq u \leq 1.$$

The mean PIT is compared to the cumulative distribution function of the standard uniform distribution. The comparison is performed by plotting a non randomized PIT histogram, which can be used as a diagnostic tool. After selecting the number of bins, say J, we compute

$$f_j = \bar{F}\Big(\frac{j}{J}\Big) - \bar{F}\Big(\frac{j-1}{J}\Big)$$

for equally spaced bins $j = 1, \ldots, J$, plot the histogram with height f_j for bin j, and check for uniformity. Deviations from uniformity hint at reasons for forecasting failures and model deficiencies. U-shaped histograms point at under-dispersed predictive distributions, while hump or inverse-U shaped histograms indicate overdispersion.

7 Case Studies—Continued

This section is a continuation of Sect. 5. Some models are fitted to real data and their performance is examined. All univariate models are fitted with the help of the R package `tscount`; see Liboschik et al. (2017) for more. Details are available by the author upon request.

7.1 Lizard Data

Based on Fig. 10.1, we apply a linear Poisson AR model which includes only Y_{t-8} as the lagged value. Fitting with $p = 1, 2, \ldots, 8$ yields that an appropriate model for the lizard data is given by

$$Y_t \mid Y_{t-8} \sim \text{Poisson}(\lambda_t), \quad \lambda_t = d + b_8 Y_{t-8},$$

when considering model selection criteria like AIC, BIC and QIC. The MLE are given by $\hat{d} = 1.540$ and $\hat{b}_8 = 0.262$ with corresponding 95% confidence intervals (based on normal approximation) (1.12, 1.95) and (0.08, 0.43), respectively. The sign of $\hat{b}_8$ confirms the positivity of the autocorrelation function. Further modeling by using the negative binomial distribution and/or the log-linear model did not improve the fit. To take into account the unusual observations observed at times $t = 60$ and $t = 68$ we model various type of intervention effects by following the methodology outlined in Sect. 4 and implemented by Liboschik et al. (2017). The end result is the discovery of a spiky outlier at $t = 60$-recall (10.9).

The final model selected for these data is given by

$$Y_t \mid Y_{t-8} \sim \text{Poisson}(\lambda_t), \quad \lambda_t = 1.744 + 0.137 Y_{t-8} + 8.838 \mathbb{1}(t = 60),$$

with corresponding standard errors 0.202, 0.080 and 3.319 for each regression parameter. Furthermore, for the model without intervention we obtain that AIC=427.414 and BIC=433.022 while for the model with intervention AIC=421.934 and BIC=430.346.

7.2 Birds Data

Recall Fig. 10.2. The sample autocorrelation plot shows that this is not a short memory process since the autocorrelation function decays for large lag values. To illustrate the general methodology, we will apply model (10.11) using the QMLE method—recall the discussion of Remark 1 from Sect. 6.1. Fitting various first order models to these data yields the results summarized in Table 10.1. We use the QIC to select a model as inference is based on QMLE. Table 10.1 shows that a suitable model for modeling these data is given by (10.11). The estimator $\hat{\phi}$ is computed along the lines of Remark 1 and its standard error is calculated using simple simulation. For instance, consider the linear model with $p = q = 1$. Then, for the given estimates of d, a_1, b_1 and ϕ, we simulate time series data of the same sample size using (10.11). For each simulated time series, we fit (10.11) to obtain $\hat{\phi}$. This is repeated 500 times and the standard error of the 500 realizations of $\hat{\phi}$ is computed to get an estimate of the desired standard error.

Table 10.1 Estimators and their standard errors (in parentheses) after fitting linear and log-linear models to *Molothrus ater* abundance data under the negative Binomial model

Model	$\hat{d}$	$\hat{a}_1$	$\hat{b}_1$	$\hat{\phi}$	QIC
Linear $p = 1$, $q = 0$	11.194 (1.008)	–	0.202 (0.068)	1.872 (0.246)	3305.058
Linear $p = 1$, $q = 1$	0.081 (0.199)	0.943 (0.035)	0.050 (0.025)	1.768(0.252)	3131.399
Log-linear $p = 1, q = 0$	2.172 (0.147)	–	0.190 (0.057)	1.873 (0.235)	3298.789
Log-linear $p = 1, q = 1$	0.264 (0.162)	0.800 (0.088)	0.107 (0.041)	1.941 (0.225)	3209.519

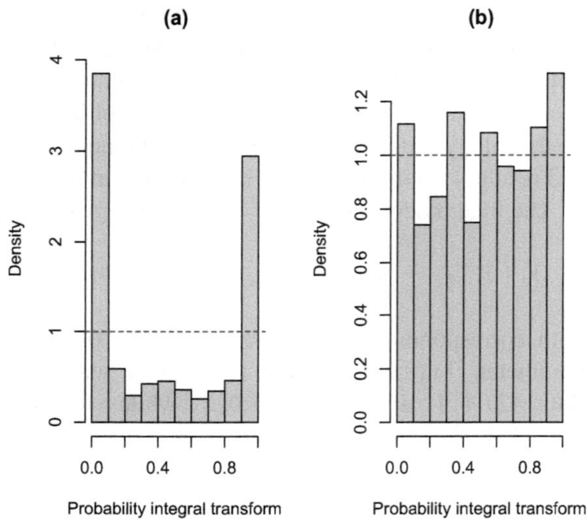

Fig. 10.3 PIT plots for the linear model (10.10), when $p = q = 1$, applied to *Molothrus ater* abundance data. (a) Poisson assumption. (b) Negative Binomial assumption. Fitting is based on MLE for the case of Poisson and on QMLE for the case of negative Binomial as discussed in remark 1

Some further support in favor of the linear model (10.11) is given by Figs. 10.3 and 10.4. The left-hand side plot of Fig. 10.3 shows the PIT histogram when the linear model with feedback (10.10) is fitted to the data. The right-hand side plot shows the same information but under the negative Binomial distribution. Obviously, the latter fits better the data. Furthermore, Fig. 10.4 shows a time series plot of the Pearson residuals (recall (10.14)). All plots do not reveal any gross departures from a white noise assumption.

8 Multivariate Models

In this section, we discuss multivariate observation-driven models for count processes. The main issue in this framework is the choice of the distribution function. Unfortunately, multivariate count distributions have complicated forms and do not

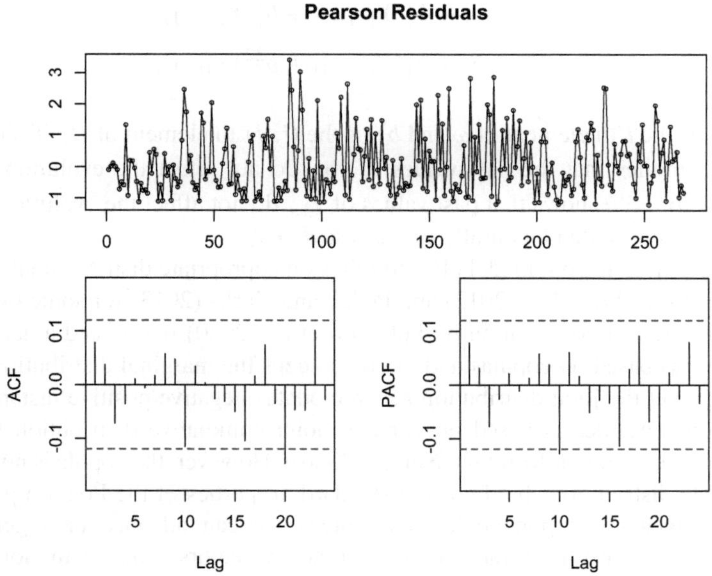

Fig. 10.4 Time series plot, acf (lower left plot) and partial acf (lower right plot) of Pearson residuals obtained by fitting model (10.10) with $p = q = 1$ to *Molothrus ater* abundance data

conform to real data properties; see Fokianos (2024) for a detailed discussion. However, we will see in this section that both linear and log-linear models can be extended to the case of multivariate count data after some appropriate assumptions.

8.1 Linear Models

Similarly to the case of univariate models, consider the vector autoregressive model of order one but under the framework of a multivariate Poisson autoregression. In this case, the observations are vectors of dimension d and they will be denoted by $\{\boldsymbol{Y}_t = (Y_{i,t})_{i=1}^d\}$. Let the d-dimensional mean process again defined by $\{\boldsymbol{\lambda}_t = (\lambda_{i,t})_{i=1}^d\}$. With this notation, and setting $\underline{\boldsymbol{Y}}_t = (\boldsymbol{Y}_t^{'}, \boldsymbol{Y}_{t-1}^{'}, \ldots, \boldsymbol{Y}_{t-(p-1)}^{'})$ the vector Poisson AR(1) model is defined by

$$Y_{i,t} \mid \underline{\boldsymbol{Y}}_{t-1} \sim \text{ marginally Poisson}(\lambda_{i,t}), \qquad \lambda_t = \omega + B_1 Y_{t-1}, \tag{10.18}$$

where ω is a d-dimensional vector and B_1 is a $d \times d$ unknown matrix. The elements of ω and B_1 are assumed to be positive such that $\lambda_{i,t} > 0$. for all i and t. It is instructive to consider (10.18) in more detail. For the simple case $d = 2$, it implies that

$$\lambda_{1,t} = \omega_1 + b_1^{11} Y_{1,(t-1)} + b_1^{12} Y_{2,(t-1)},$$
$$\lambda_{2,t} = \omega_2 + b_1^{21} Y_{1,(t-1)} + b_1^{22} Y_{2,(t-1)},$$

where ω_i is the i'th element of ω and b_1^{ij} is the (i, j)'th element of B_1. Then setting $b_1^{12} = 0$ implies that the past values of $Y_{2,t}$ do not affect the evolution of $Y_{1,t}$. Similarly, $b_1^{21} = 0$ shows that past values of $Y_{1,t}$ do not affect the evolution of $Y_{2,t}$. These arguments extend naturally to the case $d > 2$.

Data generation from model (10.18) relies on appropriate distributional assumptions; see Joe (1997) , Liu (2012) and Pedeli and Karlis (2013) for some examples. However, a recent contribution by Fokianos et al. (2020) introduced a data generating process based on copulas and in effect keeps the marginal distributions of $\boldsymbol{Y}_t$ as Poisson yet the joint distribution accommodates negative/positive instantaneous correlation. The idea is based on copulas (joint cumulative distribution function whose marginals are uniform; see Nelsen (2006)). However, the copula is not unique for discrete distributions but by using standard properties of the Poisson processes (waiting times are exponentially distributed) we can advance data generating processes for multivariate integer-valued random vectors. This methodology can be extended to include other marginal distributions like the negative binomial distribution. We refer to the articles by Fokianos et al. (2020) and Fokianos (2024) for more details.

The first order model (10.18) generalizes straightforwardly to vector Poisson autoregressive models of order p

$$\boldsymbol{\lambda}_t = \omega + \sum_{i=1}^{p} B_i \boldsymbol{Y}_{t-i},$$

or the vector version of (10.10) (see corresponding notation)

$$Y_{i,t} \mid \mathscr{F}_{t-1} \text{ is marginally Poisson}(\lambda_{i,t}), \quad \boldsymbol{\lambda}_t = \omega + \sum_{i=1}^{p} B_i \boldsymbol{Y}_{t-i} + \sum_{j=1}^{q} A_j \boldsymbol{\lambda}_{t-j}, \tag{10.19}$$

where $(A_j)_{j=1}^q$, $(B_i)_{i=1}^p$ are $d \times d$ unknown matrices and all the elements of ω, $(A_j)_{j=1}^q$, $(B_i)_{i=1}^p$ are positive such that $\lambda_{i,t} > 0$, for all i and t, are studied along the previous arguments. Interpretation of such models is straightforward.

8.2 Log-Linear Models

The log-linear model we consider is the direct analog of (10.5). In a more general form, assume that for each $i = 1, 2, \ldots, d$

$$Y_{i,t} \mid \underline{\boldsymbol{Y}}_{t-1} \text{ is marginally Poisson}(\lambda_{i,t}), \quad \boldsymbol{\nu}_t = \omega + \sum_{i=1}^{p} B_i \log(\boldsymbol{Y}_{t-i} + \mathbf{1}_d) \tag{10.20}$$

where $\boldsymbol{\nu}_t \equiv \log \boldsymbol{\lambda}_t$ is defined component wise (i.e. $\nu_{i,t} = \log \lambda_{i,t}$) and $\mathbf{1}_d$ denotes the d-dimensional vector which consists of ones. For this model, there is no need to impose any constraints on the matrix coefficients. Data generation from model (10.20) proceeds the same way as in Fokianos et al. (2020) and keeps marginally the Poisson distributional property. Additionally, the log-linear model accommodates covariates much easier than the linear model which requires that any such inclusion has to satisfy $\lambda_{i,t} > 0$. In the case of (10.20) though, if $\boldsymbol{Z}_t$ is a covariate vector of dimension d, then the second equation of (10.20) becomes $\nu_t = \omega + \sum_{i=1}^{p} B_i \log(Y_{t-i} + 1_d) + C\boldsymbol{Z}_t$ for a $d \times d$ matrix C. Interpretation of model parameters for the log-linear model (10.20) is identical to the case of the linear model but in terms of the vector process ν_t.

Remark 2 Models (10.19) and (10.20) are related to Hawkes processes (Hawkes 1971a,b) because they can be obtained after suitable discretization of the corresponding continuous time process. This connection has been explored in detail by Kirchner (2016) for univariate models in the context of INAR(∞) process which in turn is related to the linear model (see Ferland et al. 2006). Hawkes processes have been found useful in modeling and inference in several scientific areas—a review is out of the scope of this paper. In particular, multivariate Hawkes processes have been employed extensively in finance; see Embrechts et al. (2011) and Bacry et al. (2015) among others. For a recent application of multivariate Hawkes process in ecology, see Nicvert et al. (2024) who analyze interactions between several species using camera trap data.

8.3 Quasi-Likelihood Inference

Suppose that $\{\boldsymbol{Y}_t, t = 1, 2, \ldots, n\}$ denotes a sample from a count time series and for the sake of presentation assume model (10.18). Inference is analogously developed to the case of log-linear model and for $p, q > 1$. Following Fokianos et al. (2020), the estimation problem is approached by employing the QMLE methodology as in Sect. 6.1. Consider the following conditional quasi-likelihood function for the parameter vector θ,

$$L(\theta) = \prod_{t=1}^{n} \prod_{i=1}^{d} \left\{ \frac{\exp(-\lambda_{i,t}(\theta)) \lambda_{i,t}^{y_{i,t}}(\theta)}{y_{i,t}!} \right\},$$

which is identical to consider (10.18) assuming independence among time series. This is a strong assumption yet it simplifies computation of estimators and their respective standard errors. At the same time, it guarantees consistency and asymptotic normality of the maximizer. Furthermore, the dependence structure is taken into account because of the dependence of the likelihood function on lagged values. The quasi log-likelihood function is equal to

$$l(\theta) = \sum_{t=1}^{n} \sum_{i=1}^{d} \Big(y_{i,t} \log \lambda_{i,t}(\theta) - \lambda_{i,t}(\theta) \Big).$$

We denote by $\widehat{\theta} \equiv \arg\max_{\theta} l(\theta)$, the QMLE of θ. In the case that of B_1 being a diagonal matrix then $\widehat{\theta}$ is computed by equation by equation using existing software. Under suitable conditions, Fokianos et al. (2020) show that the QMLE $\widehat{\theta}$ is consistent and asymptotically normally distributed. We refer the interested reader to this reference for additional results and details.

9 Further Topics

High-dimensional data is frequently observed in applications. In the context of time series Basu and Matteson (2021) provide an overview of several methods, in the context of large autoregression and stochastic regression, and Hallin et al. (2020) provide a concise overview of factor models. A recent methodological contribution related to inference for high-dimensional count time series observation-driven models is that of Hall et al. (2019) who consider the d-dimensional autoregressive model

$$Y_{i,t} \mid \underline{Y}_{t-1} \ \text{Poisson}(\lambda_{i,t}), \quad \nu_t = \omega + B_1 \boldsymbol{Y}_{t-1}, \tag{10.21}$$

where ν_t is defined as in Eq. (10.20). The constant term ω is assumed to be known and the $d \times d$ matrix B_1 belongs to a compact subset, say $\mathscr{B}$ of the set of all $d \times d$ matrices with real elements that satisfy a sparsity constraint. Comparing (10.21) to (10.20) we note that the restriction of B_1 belonging to a compact set assures the stability of the joint process Y_t because the components of Y_{t-1} are unbounded, in general, recall the discussion in page 2.2 about the univariate analogue of (10.21). To estimate the matrix B_1 when the dimension d is much larger than n, the authors suggest the l_1 regularized QMLE defined by

$$\widehat{B}_1 = \arg\max_{B_1 \in \mathscr{B}} \frac{1}{n} \sum_{t=1}^{n} \sum_{i=1}^{d} \Big(y_{i,t} \nu_{i,t}(B_1) - \exp(\nu_{i,t}(B_1)) \Big) + \tilde{\lambda} \sum_{l=1}^{d} \sum_{m=1}^{d} |B_1^{(l,m)}|,$$

where $\tilde{\lambda}$ is a regularization parameter and $\nu_{i,t}(.)$ is defined by (10.21). The authors study mean square error bounds for the proposed estimators and show that they are closely connected with the bounds obtained in the Gaussian case. Further work along these lines was developed (for AR(p) type models) by Pandit et al. (2020) where the interested reader can obtain more references.

In a related work, Armillotta and Fokianos (2024) studied network autoregressive models for high-dimensional count time series with a fixed neighborhood structure. The main aim is to assess the effect of a network to multivariate time series processes. This problem has attracted much attention recently; see Zhu et al. (2017) who have proposed a Network Autoregressive model (NAR) and studied least squares inference. In the work by Armillotta and Fokianos (2024), the authors have proposed linear and log-linear Poisson network autoregression for count processes and studied quasi-likelihood inference for these models. For a given known neighborhood structure, such models can be fitted by using the `R` software `PNAR` (Armillotta et al. 2024). For an alternative approach to modeling multivariate counts in continuous time see Veraart (2019), Bennedsen et al. (2023), Eyjolfsson and Tjøstheim (2023), and Fang et al. (2024) for a network model employing Hawkes processes which are related to the linear and log-linear—see Remark 2 of Sect. 8.2. Finally, we mention the recent work by Maletz et al. (2024) who develop models for spatio-temporal count time series.

10 Outlook

As the last two sections indicate the analysis of multivariate ecological time series, in the sense of addressing spatio-temporal variability, is a substantial problem. Such process can be observed in different temporal/spatial scales; see Preston (1960) for an early work on classification of temporal scales. In the context of ecology, see Dutilleul (2011) who discusses extensively models, processes and inference for spatio-temporal data. The book by Cressie and Wikle (2011) contains additional statistical methods. Such studies are motivated by determining spatial and temporal heterogeneity in forests and agricultural habitats for managing wildlife; see Verdade et al. (2020) for a recent study. Furthermore, Broekhuis et al. (2021) study the effect of resource pulses in agricultural/silvicultural landscapes in understanding wildlife abundance patterns in different temporal scales. A large-scale and long-term, ecologically diverse analysis of wombats has been developed by Carver et al. (2021) who show that spotlight monitoring is a simple and important survey method. Finally Vallecillo et al. (2021) study the reliability of count time series data from animal population surveys. These examples provide diverse research areas worth exploring for managing wildlife properly and develop new statistical methodology.

Acknowledgments I would like to thank Dr. Moral for his invitation to contribute to this volume and assisting me with obtaining relevant time series data. Many thanks to B. Corrigan, F. Zafeiriou and a reviewer who have made several constructive remarks in an earlier version of this manuscript.

References

Agosto A, Cavaliere G, Kristensen D, Rahbek A (2016) Modeling corporate defaults: Poisson autoregressions with exogenous covariates (parx). J Empir Finance 38:640–663

Ahmad A, Franq C (2016) Poisson QMLE of count time series models. J Time Ser Anal 37:291–314

Ainley DG, Lindke K, Ballard G, Lyver PO, Jennings S, Toniolo V, Pennycook J, Massaro M, Santora JA (2017) Spatio-temporal occurrence patterns of cetaceans near Ross Island, Antarctica, 2002–2015: implications for food web dynamics. Polar Biol 40:1761–1775

Akaike H (1973) Information theory and an extension of the maximum likelihood principle. In: Petroc BN, Kaski F (eds) Second international symposium in information theory, Akademiai Kiado, Budapest, pp 267–281

Aknouche A, Bendjeddou S, Touche N (2018) Negative binomial quasi-likelihood inference for general integer-valued time series models. J Time Ser Anal 39:192–211

Alzaid A, Al-Osh M (1988) First-order integer-valued autoregressive (INAR (1)) process: distributional and regression properties. Stat Neerl 42:53–61

Armillotta M, Fokianos K (2024) Count network autoregression. J Time Ser Anal 45:584–612

Armillotta M, Tsagris M, Fokianos K (2024) Inference for network count time series with the r package pnar. R J 15:255–269

Auger-Méthé M, Newman K, Cole D, Empacher F, Gryba R, King AA, Leos-Barajas V, Mills F J, Nielsen A, Petris G, Thomas L (2021) A guide to state?space modeling of ecological time series. Ecol Monogr 91:e01470

Bacry E, Mastromatteo I, Muzy JF (2015) Hawkes processes in finance. Market Microstruct Liquidity 01:1550005

Barreto-Souza W, Ombao H (2022) The negative binomial process: a tractable model with composite likelihood-based inference. Scand J Stat Theory Appl 49:568–592

Barreto-Souza W, Piancastelli LSC, Fokianos K, Ombao H (2025) Time-varying dispersion integer-valued GARCH models. J Time Series Anal. https://onlinelibrary.wiley.com/toc/14679892/0/0

Basu S, Matteson DS (2021) A survey of estimation methods for sparse high-dimensional time series models. Available at https://arxiv.org/abs/2107.14754

Bennedsen M, Lunde A, Shephard N, Veraart AE (2023) Inference and forecasting for continuous-time integer-valued trawl processes. J Econom 236:105476

Blasco-Moreno A, Pérez-Casany M, Puig P, Morante M, Castells E (2019) What does a zero mean? understanding false, random and structural zeros in ecology. Methods Ecol Evol 10(7):949–959

Box GEP, Tiao GC (1975) Intervention analysis with applications to economics and environmental problems. J Am Stat Assoc 70:70–79

Bracher J, Held L (2022) Endemic-epidemic models with discrete-time serial interval distributions for infectious disease prediction. Int J Forecast 38:1221–1233

Breslow N (1990) Tests of hypotheses in overdispersed poisson regression and other quasi-likelihood models. J Am Stat Assoc 85:565–571

Brockwell PJ, Davis RA (1991) Time series: data analysis and theory, 2nd edn. Springer, New York

Broekhuis F, Elliot NB, Keiwua K, Koinet K, Macdonald DW, Mogensen N, Thuo D, Gopalaswamy AM (2021) Resource pulses influence the spatio-temporal dynamics of a large carnivore population. Ecography 44:358–369

Cameron AC, Trivedi PK (1998) Regression analysis of count data. Cambridge University Press, Cambridge

Cameron AC, Trivedi PK (2013) Regression analysis of count data, Econometric society monographs, vol 53, 2nd edn. Cambridge University Press, Cambridge

Carver S, Charleston M, Hocking G, Gales R, Driessen MM (2021) Long-term spatiotemporal dynamics and factors associated with trends in bare-nosed wombats. J Wildl Manag 85:449–461

Chen C, Liu LM (1993) Joint estimation of model parameters and outlier effects in time series. J Am Stat Assoc 88:284–297
Christou V, Fokianos K (2014) Quasi-likelihood inference for negative binomial time series models. J Time Ser Anal 35:55–78
Christou V, Fokianos K (2015) On count time series prediction. J Stat Comput Simul 2:357–373
Cox DR (1981) Statistical analysis of time series: some recent developments. Scand J Stat 8:93–115
Cressie N, Wikle CK (2011) Statistics for spatio-temporal data. Wiley series in probability and statistics. Wiley, Hoboken
Czado C, Gneiting T, Held L (2009) Predictive model assessment for count data. Biometrics 65:1254–1261
Davis RA, Holan SH, Lund R, Ravishanker N (eds) (2016) Handbook of discrete-valued time series. Handbooks of modern statistical methods. Chapman & Hall/CRC, London
Davis RA, Fokianos K, Holan SH, Joe H, Livsey J, Lund R, Pipiras V, Ravishanker N (2021) Count time series: a methodological review. J Am Stat Assoc 116:1533–1547
Dean CB, Eaves DM, Martinez CJ (1995) A comment on the use of empirical covariance matrices in the analysis of count data. J Stat Plan Inference 48:197–205
Dornelas M, ao LHA, Moyes F, Bates AE, Magurran AE, san A D, et al AAA (2018) Biotime: a database of biodiversity time series for the anthropocene. Glob Ecol Biogeogr 27:760–786
Douc R, Doukhan P, Moulines E (2013) Ergodicity of observation-driven time series models and consistency of the maximum likelihood estimator. Stoch Process Appl 123:2620–2647
Doukhan P, Fokianos K, Tjøstheim D (2012) On weak dependence conditions for Poisson autoregressions. Stat Probab Lett 82:942–948. With a correction in Vol. 83, pp. 1926–1927
Doukhan P, Fokianos K, Rynkiewicz J (2021) Mixtures of nonlinear poisson autoregressions. J Time Ser Anal 42:107–135
Dutilleul PRL (2011) Spatio-temporal heterogeneity: concepts and analyses. Cambridge University Press, Cambridge
Embrechts P, Liniger T, Lin L (2011) Multivariate Hawkes processes: an application to financial data. J Appl Probab 48A(New frontiers in applied probability: a Festschrift for Søren Asmussen):367–378
Eyjolfsson H, Tjøstheim D (2023) Multivariate self-exciting jump processes with applications to financial data. Bernoulli 29:2167–2191
Fang G, Xu G, Xu H, Zhu X, Guan Y (2024) Group network hawkes process. J Am Stat Assoc 119:2328–2344
Ferland R, Latour A, Oraichi D (2006) Integer-valued GARCH processes. J Time Ser Anal Anal 27:923–942
Fokianos K (2012) Count time series models. In: Rao TS, Rao SS, Rao CR (eds) Handbook of statistics: time series analysis–methods and applications, vol 30. Elsevier B. V., Amsterdam, pp 315–347
Fokianos K (2024) Multivariate count time series modelling. Econom Stat 31:100–116
Fokianos K, Fried R (2010) Interventions in INGARCH processes. J Time Ser Anal 31:210–225
Fokianos K, Fried R (2012) Interventions in log-linear Poisson autoregression. Stat Model 12:299–322
Fokianos K, Tjøstheim D (2011) Log-linear Poisson autoregression. J Multivar Anal 102:563–578
Fokianos K, Rahbek A, Tjøstheim D (2009) Poisson autoregression. J Am Stat Assoc 104:1430–1439
Fokianos K, Støve B, Tjøstheim D, Doukhan P (2020) Multivariate count autoregression. Bernoulli 26:471–499
Franchi G, Truquet L (2025) Time series models on compact spaces, with an application to dynamic modeling of relative abundance data in ecology. J Time Ser Anal. https://onlinelibrary.wiley.com/toc/14679892/0/0
Gneiting T, Balabdaoui F, Raftery AE (2007) Probabilistic forecasts, calibration and sharpness. J R Stat Soc: Ser B (Stat Methodol) 69:243–268

Hall EC, Raskutti G, Willett RM (2019) Learning high-dimensional generalized linear autoregressive models. IEEE Trans Inf Theory 65:2401–2422
Hallin M, Lippi M, Barigozzi M, Forni M, Zaffaroni P (2020) Time series in high dimensions: the general dynamic factor model. World Scientific, Singapore
Hawkes AG (1971a) Point spectra of some mutually exciting point processes. J R Stat Soc Ser B Methodol 33:438–443
Hawkes AG (1971b) Spectra of some self-exciting and mutually exciting point processes. Biometrika 58:83–90
Heinen A (2003) Modelling time series count data: an autoregressive conditional poisson model. Tech. Rep. MPRA Paper 8113, University Library of Munich. Available at http://mpra.ub.uni-muenchen.de/8113/
Hinde J, Demétrio CG (1998) Overdispersion: models and estimation. Comput Stat Data Anal 27:151–170
Inouye DI, Yang E, Allen GI, Ravikumar P (2017) A review of multivariate distributions for count data derived from the Poisson distribution. Wiley Interdiscip Rev Comput Stat 9(3):e1398, 25. https://doi.org/10.1002/wics.1398
Joe H (1997) Multivariate models and dependence concepts. Chapman & Hall, London
Joe H (2016) Markov models for count time series. In: Davis R, Holan S, Lund R, Ravishanker N (eds) Handbook of discrete-valued time series. CRC Press, Boca Raton, pp 29–49
Jung R, Tremayne A (2011) Useful models for time series of counts or simply wrong ones? Adv Stat Anal 95:59–91
Karlis D, Xekalaki E (2005) Mixed poisson distributions. Int Stat Rev 73:35–58
Kedem B, Fokianos K (2002) Regression models for time series analysis. Wiley, Hoboken
Kirchner M (2016) Hawkes and INAR(∞) processes. Stoch Process Appl 126:2494–2525
Lambert D (1992) Zero-inflated Poisson regression with an application to defects is manufacturing. Technometrics 34:1–14
Lawless JF (1987) Negative binomial and mixed Poisson regression. Can J Stat 15:209–225
Liboschik T, Kerschke P, Fokianos K, Fried R (2016) Modelling interventions in INGARCH processes. Int J Comput Math 93:640–657
Liboschik T, Fokianos K, Fried R (2017) tscount: an R package for analysis of count time series following generalized linear models. J Stat Softw 82:1–51. https://doi.org/10.18637/jss.v082.i05
Liu H (2012) Some models for time series of counts. PhD thesis, Columbia University
Maletz S, Fokianos K, Fried R (2024) Spatio-temporal count autoregression. Data Sci Sci 3:2425171. https://doi.org/10.1080/26941899.2024.2425171
McClintock BT, Langrock R, Gimenez O, Cam E, Borchers DL, Glennie R, Patterson TA (2020) Uncovering ecological state dynamics with hidden markov models. Ecol Lett 23:1878–1903
McCullagh P, Nelder JA (1989) Generalized linear models, 2nd edn. Chapman & Hall, London
McGrane-Corrigan B, Mason O, Moral RA (2024) Inferring stochastic group interactions within structured populations via coupled autoregression. J Theor Biol 584:111793
Nelsen RB (2006) An introduction to Copulas, 2nd edn. Springer, New York
Neumann M (2011) Absolute regularity and ergodicity of Poisson count processes. Bernoulli 17:1268–1284
Nicvert L, Donnet S, Keith M, Peel M, Somers MJ, Swanepoel LH, Venter J, Fritz H, Dray S (2024) Using the multivariate Hawkes process to study interactions between multiple species from camera trap data. Ecology 105:e4237
Pan W (2001) Akaike's information criterion in generalized estimating equations. Biometrics 57:120–125
Pandit P, Sahraee-Ardakan M, Amini AA, Rangan S, Fletcher AK (2020) Generalized autoregressive linear models for discrete high-dimensional data. IEEE J Sel Areas in Inf Theory 1:884–896
Pedeli X, Karlis D (2013) Some properties of multivariate INAR(1) processes. Comput Stat Data Anal 67:213–225
Preston FW (1960) Time and space and the variation of species. Ecology 41:612–627

Rissanen J (1978) Modelling by shortest data description. Automatica 14:465–471
Rydberg TH, Shephard N (2000) A modeling framework for the prices and times of trades on the New York stock exchange. In: Fitzgerlad WJ, Smith RL, Walden AT, Young PC (eds) Nonlinear and nonstationary signal processing. Isaac Newton Institute and Cambridge University Press, Cambridge, pp 217–246
Schwarz G (1978) Estimating the dimension of a model. Ann Stat 6:461–464
Shumway RH, Stoffer DS (2011) Time series analysis and its applications, 3rd edn. Springer, New York, with R examples
Sim T, Douc R, Roueff F (2021) General-order observation-driven models: ergodicity and consistency of the maximum likelihood estimator. Electron J Stat 15:3349–3393
Streett S (2000) Some observation driven models for time series of counts. PhD thesis, Colorado State University, Department of Statistics
Tjøstheim D (2012) Some recent theory for autoregressive count time series. TEST 21:413–438
Tong H (1990) Nonlinear time series: a dynamical system approach. Oxford University Press, New York
Vallecillo D, Gauthier-Clerc M, Guillemain M, Vittecoq M, Vandewalle P, Roche B, Champagnon J (2021) Reliability of animal counts and implications for the interpretation of trends. Ecol Evol 11:2249–2260
Veraart AE (2019) Modeling, simulation and inference for multivariate time series of counts using trawl processes. J Multivar Anal 169:110–129
Verdade LM, Moral RA, Calaboni A, do Amaral MV, Martin PS, Amorim LS, Gheler-Costa C, na CIP (2020) Temporal dynamics of small mammals in eucalyptus plantations in southeast Brazil. Glob Ecol Conserv 24:e01217
Wang C, Liu H, Yao JF, Davis RA, Li WK (2014) Self-excited threshold Poisson autoregression. J Am Stat Assoc 109:777–787
Wedderburn RWM (1974) Quasi-likelihood functions, generalized linear models and the Gaussian method. Biometrika 61:439–447
Weiß CH (2008) Thinning operations for modeling time series of counts–a survey. Adv Stat Anal 92:319–341
Weiß CH (2018) An introduction to discrete-valued time series. Jonh Wiley, Hoboken
Weiß CH, Jahn M (2024) Soft-clipping ingarch models for time series of bounded counts. Stat Modelling 24:295–319
West M, Harrison P (1997) Bayesian forecasting and dynamic models, 2nd edn. Springer, New York
White H (1982) Maximum likelihood estimation of misspecified models. Econometrica 50:1–25
Zeger SL, Liang KY (1986) Longitudinal data analysis for discrete and continuous outcomes. Biometrics 42:121–130
Zhu F (2011) A negative binomial integer-valued GARCH model. J Time Ser Anal 32:54–67
Zhu F (2012) Zero-inflated poisson and negative binomial integer-valued garch models. J Stat Plann Inference 142:826–839
Zhu F, Li Q, Wang D (2010) A mixture integer-valued ARCH model. J Stat Plann Inference 140:2025–2036
Zhu X, Pan R, Li G, Liu Y, Wang H (2017) Network vector autoregression. Ann Stat 45:1096–1123
Zucchini W, MacDonald IL (2009) Hidden Markov models for Time Series: an introduction using R. CRC Press, Boca Raton

Chapter 11
Urbanisation and Bird Assemblages: Insights from Temporal Beta Diversity

Hideyasu Shimadzu

Abstract Urbanisation is a major driver of unidirectional ecological change, yet its long-term impacts on biodiversity remain difficult to quantify. This study proposes and applies a model-based framework to assess temporal beta diversity over a 35-year period in a suburban bird assemblage in Tokyo, Japan. By estimating species abundances, the framework derives species-specific contribution ratios, enabling the identification of those species and groups most responsible for changes in assemblage over time. This contribution ratio approach facilitates the decomposition of temporal beta diversity, more specifically changes in assemblage composition, into interpretable components, thereby offering insight into ecological dynamics under anthropogenic pressure. This framework provides a practical tool for biodiversity monitoring and highlights the importance of long-term, high-resolution data for understanding assemblage responses to environmental change.

Keywords Avifauna · Assemblages · Biodiversity · Birds · Population dynamics · Temporal beta diversity · Urbanisation

1 Introduction

Understanding the extent to which biodiversity is undergoing unprecedented transformation in the Anthropocene is crucial for successful ecosystem restoration—a global objective articulated by the United Nations and targeted for achievement by 2030 (FAO, IUCN CEM and SER 2021). Biodiversity *change*, as distinct from *difference*, refers specifically to temporal shifts in the abundances of co-occurring species at a given location over time. This dynamism, encompassing turnover in both species identity and abundance, is conceptualised as *temporal beta diversity*, defined as the dissimilarity in community composition across two or more points in

H. Shimadzu (✉)
Department of Data Science, Kitasato University, Kanagawa, Japan
e-mail: shimadzu.hideyasu@kitasato-u.ac.jp

L. Martins Verdade, R. A. Moral (eds.), *Wildlife Monitoring*,
https://doi.org/10.1007/978-3-032-05821-8_11

time (Magurran et al. 2019; McGlinn et al. 2025). In contrast, the widely appreciated concept of alpha diversity characterises a community's state at a single point in space and time. For a comprehensive overview of modern biodiversity studies and challenges, see Magurran and McGill (2013).

Much of the recent emerging evidence suggests that ecological assemblages across the globe are undergoing marked compositional changes (Dornelas et al. 2014; Vellend et al. 2017; Blowes et al. 2019; Xu et al. 2023), often exhibiting rates of temporal beta diversity that exceed those predicted by ecological theory (Gotelli et al. 2017). Capturing and interpreting temporal beta diversity is therefore integral to robust biodiversity monitoring and conservation. However, conventional dissimilarity indices widely used in spatial contexts, while informative for descriptive analysis, typically lack the capacity to account for temporal dynamics or their associated uncertainty, and they cannot partition community change into ecologically interpretable components (Shimadzu et al. 2015). To address these limitations, recent developments in probabilistic and model-based approaches provide more flexible frameworks for understanding community dynamics through the lens of temporal beta diversity.

This chapter revisits the framework developed by Shimadzu et al. (2015), which investigates temporal beta diversity by explicitly incorporating changes in species abundances over time, thereby establishing a theoretical linkage to the field of population ecology. Their approach decomposes observed dissimilarity (i.e. shifts in species abundances) into stochastic and deterministic components, allowing genuine ecological signals to be distinguished from background noise in time series data—an essential step towards addressing the fundamental challenges of ecology (May 2020). A salient feature of temporal ecological data, as opposed to spatial ones, is their inherent unidirectionality; time proceeds only from past to future. Analytical approaches should thus respect this asymmetry, which naturally suggests formulations grounded in differential equations, amongst other methods.

To illustrate this framework, the current study examines long-term observations on a suburban bird assemblage. Suburban environments function as zones where natural and anthropogenic processes intersect, providing an exemplary context in which to investigate biodiversity dynamics. Birds, as widely recognised bioindicators, are especially informative in this regard. Understanding how avian assemblages respond over time to urban expansion via a temporal beta diversity concept is essential for effective wildlife assessment and management in human-modified landscapes.

Model-based frameworks offer distinct advantages in the context of long-term ecological monitoring, where issues such as variation in sampling effort and imperfect detection can obscure underlying ecological patterns (Shimadzu and Darnell 2015; Shimadzu et al. 2016). By applying this framework to a uniquely long-term dataset from a suburban bird assemblage, the present study seeks to elucidate patterns of assemblage change associated with urbanisation, as expressed through temporal beta diversity. The modelling approach not only quantifies the magnitude of compositional change, but also reveals mechanistic insights into which species or functional groups drive these temporal changes. In doing so, this study sheds light on the value of integrating temporal beta diversity with model-based inference to enhance our understanding of ecological responses to urbanisation, ultimately informing more effective conservation strategies in dynamic, anthropogenically influenced ecosystems.

2 Framework

2.1 Additive Structure in Ecological Community Abundance

Ecological communities, or assemblages, comprise species that co-occur at a given location and time. To characterise such communities and their temporal dynamics mathematically, let $N_i(t) \geq 0$ denote a discrete random variable representing the numerical abundance of species i at time t. The total abundance of the community is then given by the sum of the abundances of its constituent species as

$$N(t) = \sum_i N_i(t),$$

or, in terms of expected values,

$$\lambda(t) = \sum_i \lambda_i(t),$$

where $\lambda(t) = \mathbb{E}[N(t)]$ and $\lambda_i(t) = \mathbb{E}[N_i(t)] > 0$.

Temporal shifts in ecological communities arise cumulatively through changes in the abundances of individual species. In differential form, this can be expressed as

$$d\lambda(t) = \sum_i d\lambda_i(t) = \sum_i (\lambda_i(t+dt) - \lambda_i(t)). \tag{11.1}$$

It is often insightful to quantify temporal change in terms of community growth—i.e. the relative change over an infinitesimal time increment dt. This growth can be expressed as a weighted sum of the natural logarithmic differences in species abundances at two time points as

$$\begin{aligned} \alpha(t)dt = \frac{d\lambda(t)}{\lambda(t)} &= \sum_i \frac{d\lambda_i(t)}{\lambda_i(t)} \frac{\lambda_i(t)}{\lambda(t)} \\ &= \sum_i d\log(\lambda_i(t))\, p_i(t) = \sum_i \log\left(\frac{\lambda_i(t+dt)}{\lambda_i(t)}\right) p_i(t), \end{aligned} \tag{11.2}$$

where the weight $p_i(t) = \lambda_i(t)/\lambda(t)$ is known as the relative abundance of species i at time t, that is, species compositional representation as $\sum_i p_i(t) = 1$.

Both expressions, absolute change (Eq. 11.1) and relative change (Eq. 11.2), serve as the basis for quantifying temporal beta diversity while maintaining a direct connection to population size in terms of abundance. When a species shows no change over time, both absolute and relative changes are obviously zero as $d\lambda_i(t) = d\log(\lambda_i(t)) = 0$.

Temporal change is intrinsically unidirectional (Shimadzu et al. 2015), progressing from past to future, and its sign indicates whether a population is increasing or declining. This interpretability renders abundance-based approaches especially informative, in contrast to species richness, which merely provides a binary consequence of abundance (Shimadzu et al. 2016). It is also worth noting that abundances across subregions exhibit a similar additive structure to that of the broader geographical area under investigation. However, this property does not generally extend to other widely used biodiversity measures, such as species richness (Shimadzu 2018).

2.2 *Additive Decomposition of Species and Environmental Contributions to Biodiversity Change*

The estimation of species abundances $\lambda_i(t)$ has long been amongst the most active and advanced areas of research in quantitative ecology over recent decades; see Ovaskainen and Abrego (2020) for modern (joint) species distribution models (JSDMs). A commonly adopted approach employs regression-based models that describe functional relationships between the response variable (i.e. species abundance) and explanatory variables (e.g. environmental or auxiliary variables). These models aim to quantify how species abundances vary in response to environmental change.

Standard modelling frameworks, widely available for practitioners through software and packages, include generalised linear models (GLMs; McCullagh and Nelder 1989) and generalised additive models (GAMs; Hastie and Tibshirani 1990), in which the natural logarithm is typically a choice for the link function to model the abundance $N_i(t)$. For example, a GLM can be expressed as a linear combination of explanatory variables (often with an intercept term but omitted here and throughout for an ease of exposition),

$$\log(\lambda_i(t)) = \sum_j \beta_{ij} x_j(t),$$

where $\{x_j(t)\}$ denotes a set of environmental or auxiliary variables, which are usually assumed to be measured without error. Addressing measurement error in explanatory variables from a regression model perspective—commonly known as the errors-in-variables problem—poses additional challenges; see Carroll et al. (2006) for a comprehensive overview of the challenge and Foster et al. (2012), Stoklosa et al. (2014) for further discussion in the ecological modelling context. The coefficients $\{\beta_{ij}\}$ represent model parameters to be estimated from data.

Under the GLM framework, community growth (Eq. 11.2) can be re-expressed as a weighted sum involving both species i and explanatory variables j,

$$\begin{aligned}\alpha(t)dt &= \sum_i \sum_j \beta_{ij} \left(x_j(t+dt) - x_j(t)\right) p_i(t) \\ &= \sum_i d_i(t) = \sum_j d_j(t),\end{aligned}$$

where

$$d_i(t) = \sum_j \beta_{ij} \left(x_j(t+dt) - x_j(t)\right) p_i(t), \quad \text{and}$$

$$d_j(t) = \sum_i \beta_{ij} \left(x_j(t+dt) - x_j(t)\right) p_i(t).$$

This decomposition enables the contributions of individual species and environmental variables to community growth to be partitioned in a manner that is additive.

The relative contributions of each species i or environmental factor j to temporal beta diversity can then be quantified using the following expressions as

$$r_i(t) = \frac{|d_i(t)|}{\sum_i |d_i(t)|}, \quad \text{and}$$

$$r_j(t) = \frac{|d_j(t)|}{\sum_j |d_j(t)|},$$

such that $\sum_i r_i(t) = \sum_j r_j(t) = 1$. These ratios indicate which species or environmental factors exert the strongest influence on temporal beta diversity at a given point in time. Examples will be presented in a later section.

As a measure of community change between two time points, temporal beta diversity can be calculated in various pairwise ways. A common choice is to compare all time points with a fixed reference point (e.g. the initial or final time point), a practice likely inherited from spatial analysis. However, in a temporal context, it is often more intuitive and ecologically meaningful to compare adjacent time points, thereby aligning more naturally with population dynamics and the concept of community growth. The choice of comparison method should, however, depend upon the specific ecological question under scrutiny.

Where communities, or assemblages, of interest exhibit strong seasonal dynamics, it is often helpful to compute temporal beta diversity at an annual scale for each calendar month. This strategy controls for seasonal effects and allows longer-term biodiversity trends to be more clearly identified. This approach will also be illustrated in this study.

Although taxonomic species are commonly used as the unit of biodiversity analysis (indexed by i), broader groupings, such as spatial guilds (Shimadzu et al. 2013), ecological archetypes (Dunstan et al. 2011) or canonical classifications based on the ordered sequence of presences and absences (Gotelli et al. 2021), may also serve this purpose and yield ecologically relevant insights. The framework presented here is equally applicable to such aggregated categories and can reveal which groups of species contribute most significantly to observed temporal beta diversity. This application will be explored in detail in a later section focusing on temporal change amongst migratory bird groups.

3 Data and Methods

Suburban bird assemblage data were collected monthly over a 35-year period at a single site in the northwestern part of Tokyo, Japan. The survey recorded more than 60 species and approximately 65,000 individual birds. This exceptionally well-documented long-term census was conducted by the private school *Jiyu Gakuen* (Fig. 11.1), covering the entirety of its 100,000 m^2 campus from 1964 to 1998, a period during which progressive urbanisation transformed the surrounding landscape.

This urbanisation led to a marked reduction of agricultural land and the expansion of residential areas, particularly between 1965 and 1980 (Fig. 11.2). These changes resulted in substantial habitat loss, gain and fragmentation for some suburban bird species. For detailed descriptions of the census methodology and campus environment, see Kira (2000), Uchida et al. (2003), Shimadzu and Shibata (2005).

Each species' abundance time series is modelled individually using a GAM, incorporating year and month as explanatory variables. The model is specified as

$$\log(\lambda_i(t)) = s_i(\texttt{Year}) + \gamma_{i,\,\texttt{Month}}$$

where s_i is a smooth function, capturing long-term annual trends, and $\gamma_{i,\,\texttt{Month}}$ is a mean effect representing the average seasonal variation across months. Here, the smooth function chosen is a thin plate regression spline, with a maximum of 9 degrees of freedom by default, as specified for the `gam` function in the `mgcv` package (Wood 2017) in R. Although other environmental variables, such as the area of residential or agricultural land shown in Fig. 11.2, could be included as explanatory variables, they are omitted in this analysis to facilitate a more straightforward application of the methodological framework.

To evaluate temporal beta diversity, each species' contribution ratio $r_i(t)$ is calculated by comparing adjacent time points. This approach preserves the direct mathematical relationship with community growth (Eq. 11.2). The trajectory of each $r_i(t)$ for each month is then illustrated using a stacked area plot, where the shaded

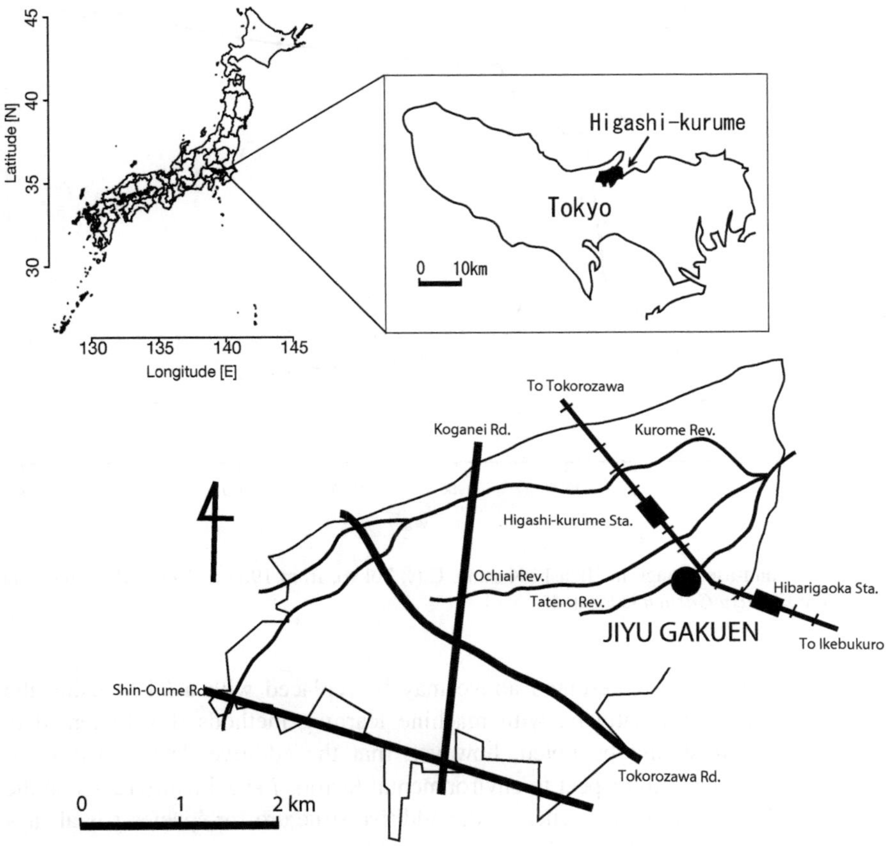

Fig. 11.1 The geographical location of the data collection site *Jiyu Gakuen* in Tokyo, Japan

regions represent the standardised contributions of each species or group to temporal beta diversity at time t. By plotting these trajectories separately for each month, seasonal fluctuations are controlled for, allowing clearer comparison of long-term trends at the annual scale.

To facilitate interpretation, the analysis focuses on the ten most abundant species, treated as individual units. In contrast, winter and summer migratory species (Uchida et al. 2003), as well as several less common species, are grouped accordingly. This grouping enables clearer comparison of the influence of individual species and broader species groups on temporal beta diversity.

The analytical procedure here comprises two primary steps: (1) estimating the expected species abundance $\lambda_i(t)$; and (2) calculating the contribution ratio of species $r_i(t)$ or, where applicable, the environmental contribution ratio $r_j(t)$. These steps are relatively straightforward and can be implemented using standard statistical software such as R. A collection of R functions for calculating contribution ratios is also available from Shimadzu et al. (2015). For estimating expected species

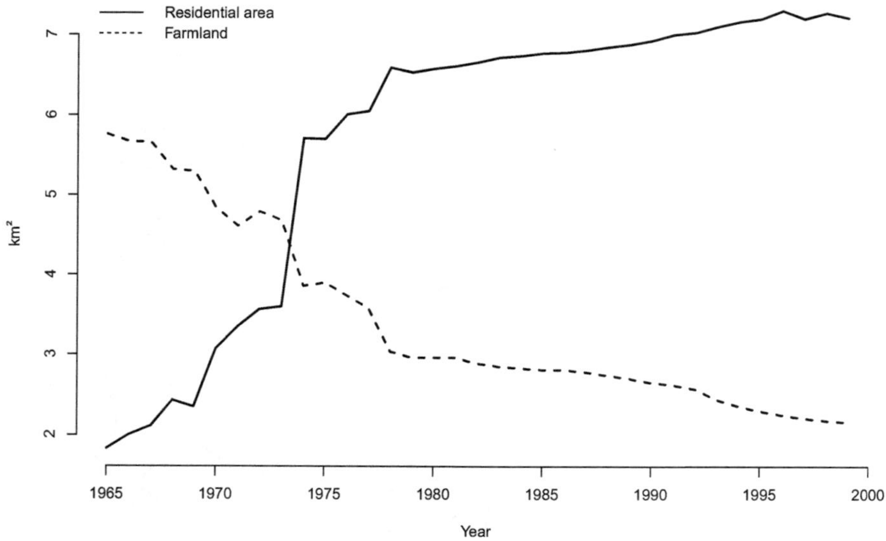

Fig. 11.2 Land-use change in Higashi-Kurume City, Tokyo, from 1965 to 1998, where the data collection site *Jiyu Gakuen* is located

abundances, the GAM presented above may be replaced with a GLM using the `glm` function or, alternatively, with machine learning methods if sufficient data are available. It should be noted, however, that the additive decomposition of contribution ratios with respect to environmental factors j specifically relies on the use of the log link function, whereas the additive structure for species i is always preserved irrespective of the link function (Eq. 11.2).

4 Results

Each panel in Fig. 11.3 presents the contribution ratios $r_i(t)$ for individual species or designated groups across each calendar month throughout the observation period. The stacked area plots provide a comprehensive illustration of temporal beta diversity, revealing dynamic shifts in the influence of different species over time. These plots reveal several key patterns, offering insight into the ecological processes underlying long-term assemblage change in this suburban landscape.

Larger colour-shaded regions indicate species exerting greater influence on temporal beta diversity, typically corresponding to either highly abundant species undergoing population shifts or seasonally dominant migrants. The plots also reveal pronounced seasonal turnover between winter and summer birds, reflecting established migratory patterns. In addition, a number of species gradually increased their contributions during the period of urban expansion, suggesting longer-term responses to changes in land use. Conversely, some species either disappeared

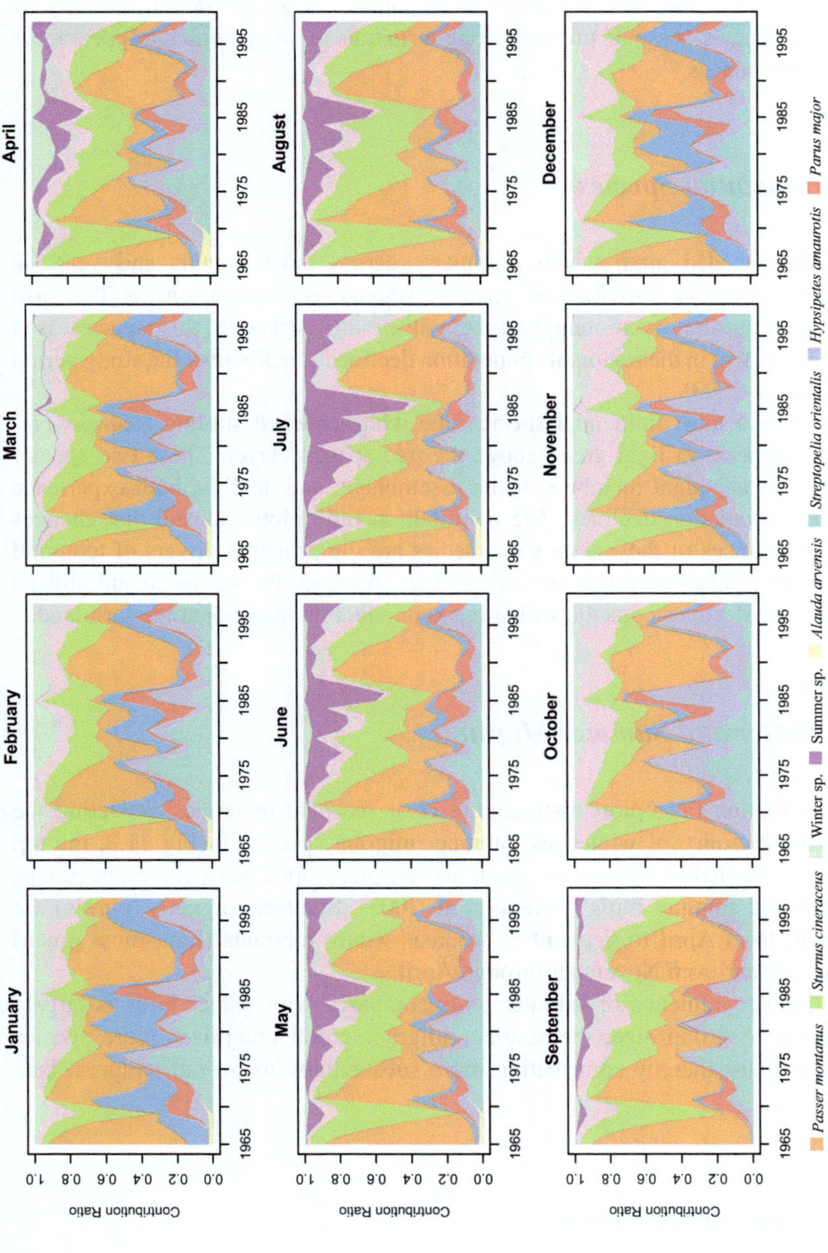

Fig. 11.3 Monthly contribution ratio plots over 35 years. Each colour-shaded area represents the contribution of a species or group to temporal beta diversity

or became newly established during the study period, independent of seasonal migration. These patterns underscore how changes in both resident and migratory bird populations contributed to the overall dynamics of assemblage composition.

The following subsections explore these patterns in greater detail, focusing on influential species, seasonal migrants, long-term transitions and notable appearances and disappearances.

4.1 Influential Species

The orange-shaded area, which dominates across most months and years in Fig. 11.3, corresponds to the tree sparrow (*Passer montanus*), one of the most influential contributors to temporal beta diversity. Although this species is a common resident in the region, its population declined steadily over the study period (Uchida et al. 2003).

The second most influential species, the white-cheeked starling (*Sturnus cineraceus*), appears in light green above the tree sparrow layer. These two species are the most abundant members of the assemblage, and notably, both experience long-term population declines. The contribution ratio plots indicate that changes in the abundances of these dominant species are the principal drivers of temporal beta diversity in this suburban bird assemblage. As noted by Shimadzu and Shibata (2005), these declines coincide with a marked reduction in surrounding farmland.

4.2 Winter and Summer Migrants

The contribution ratio plots also capture clear seasonal dynamics, reflecting the alternating presence of winter and summer migrant species. In Fig. 11.3, the top two layers, shaded in light moss green and purple, respectively, represent groups of winter and summer birds (Uchida et al. 2003). Summer migrants (purple) are prominent from April to September, whereas winter migrants (light moss green) typically appear from November through April.

The greater influence of summer birds compared with winter birds is largely attributable to group size; the summer migrant group comprises more species and individuals, thereby contributing more substantially to overall temporal beta diversity.

4.3 Disappearing Species

The Eurasian skylark (*Alauda arvensis*) exhibits a distinctive pattern in the contribution ratio plots. The lemon-yellow shaded area in the lower left portion of the plots corresponds to this species and is visible only between April and July in the early years, up until the late 1960s. Thereafter, the species ceases to appear in the plots, consistent with field observations indicating that skylarks have not been recorded at the site since that time. Kira (2000) suggests a possible connection to the decline of farmland area (Fig. 11.2).

4.4 Newly Established Resident Species

A major community shift occurred in the late 1960s, during which three lower colour layers in the plots began to emerge and grow more pronounced (Fig. 11.3). These correspond to the oriental turtle dove (*Streptopelia orientalis*), the brown-eared bulbul (*Hypsipetes amaurotis*) and the great tit (*Parus major*). This shift coincides with the disappearance of skylarks. Uchida et al. (2003) report that brown-eared bulbuls, previously non-resident in the area, became year-round residents during this period. Similarly, Shimadzu and Shibata (2005) note that these three species increased in abundance as residential development expanded around the observation site.

4.5 Late-Appearing Species

In the late 1980s, the carrion crow (*Corvus corone*) began to emerge as a notable contributor to temporal beta diversity, as evidenced by the gradual expansion of the light grey shaded area, underneath the winter and summer migratory species layers, in the contribution plots (Fig. 11.3).

5 Discussion and Conclusion

This study has applied a model-based framework to assess long-term patterns of temporal beta diversity in a suburban bird assemblage undergoing rapid urbanisation. By quantifying the species-specific contributions to the assemblage change over time, the analysis has revealed not only seasonal migratory dynamics but also the lasting impacts of land-use transformation. The results demonstrate that temporal beta diversity in this setting is driven largely by shifts in the abundance of a few dominant species, particularly *Passer montanus* and *Sturnus cineraceus*, both of which decline steadily as farmland is replaced by residential development. This finding reinforces the view that common and abundant species may have a disproportionate influence on biodiversity change and that their trajectories can serve as early indicators of broader shifts in ecological structure.

The contribution ratio plots also captured seasonal patterns in temporal beta diversity, reflecting the rhythmic alternation of winter and summer migrant species. These migratory pulses contributed significantly to community variation throughout the year, with summer migrants showing greater influence due to their higher abundances. Such seasonal dynamics are essential for understanding community turnover, especially in regions where migratory species constitute a substantial portion of avian biodiversity. Importantly, these patterns are not masked by long-term trends, owing to the use of month-specific visualisations that disentangle seasonal variation from directional change.

Beyond seasonal and dominant species, the analysis has identified notable transitions in species presence and residency status. The disappearance of *Alauda arvensis* and the concurrent emergence of species such as *Hypsipetes amaurotis*, *Streptopelia orientalis* and *Parus major* reflect a restructuring of the avifauna linked to urban development. Some of these species, previously migratory or occasional visitors, became established residents, highlighting the adaptive responses of certain species to anthropogenic environments. Similarly, the gradual rise in influence of *Corvus corone* in the late 1980s suggests delayed but notable changes in assemblage composition as urban pressures intensified (Uchida et al. 2003).

The current study has demonstrated the utility of a probabilistic and abundance-based framework for dissecting the dynamics of temporal beta diversity. By decomposing community change into species-level contributions and linking them to ecological or anthropogenic drivers, this approach offers both mechanistic insight and conceptual clarity. As urbanisation continues to reshape landscapes globally, such methods are essential for detecting biodiversity responses over time and for informing evidence-based conservation strategies. In this context, long-term monitoring, especially at fine temporal resolution, remains a critical foundation for understanding ecological resilience and change in human-dominated environments.

Acknowledgments This work was partially supported by JSPS KAKENHI 25K09165.

References

Blowes SA, Supp SR, Antão LH, Bates A, Bruelheide H, Chase JM, Moyes F, Magurran A, McGill B, Myers-Smith IH, Winter M, Bjorkman AD, Bowler DE, Byrnes JEK, Gonzalez A, Hines J, Isbell F, Jones HP, Navarro LM, Thompson PL, Vellend M, Waldock C, Dornelas M (2019) The geography of biodiversity change in marine and terrestrial assemblages. Science 366(6463):339–345

Carroll RJ, Ruppert D, Stefanski LA, Crainiceanu CM (2006) Measurement error in nonlinear models, 2nd edn. CRC Press LLC, London

Dornelas M, Gotelli NJ, McGill B, Shimadzu H, Moyes F, Sievers C, Magurran AE (2014) Assemblage time series reveal biodiversity change but not systematic loss. Science 344(6181):296–299

Dunstan PK, Foster SD, Darnell R (2011) Model based grouping of species across environmental gradients. Ecol Modell 222(4):955–963

FAO, IUCN CEM and SER (2021) Principles for ecosystem restoration to guide the United Nations Decade 2021–2030. Rome, FAO

Foster SD, Shimadzu H, Darnell R (2012) Uncertainty in spatially predicted covariates: Is it ignorable? J R Stat Soc Ser C: Appl Stat 61(4):637–652

Gotelli NJ, Shimadzu H, Dornelas M, McGill B, Moyes F, Magurran AE (2017) Community-level regulation of temporal trends in biodiversity. Sci Adv 3(7):e1700315

Gotelli NJ, Moyes F, Antão LH, Blowes SA, Dornelas M, McGill BJ, Penny A, Schipper AM, Shimadzu H, Supp SR, Waldock CA, Magurran AE (2021) Long-term changes in temperate marine fish assemblages are driven by a small subset of species. Glob Change Biol 28(1):46–53

Hastie TJ, Tibshirani RJ (1990) Generalized additive models. Chapman & Hall/CRC

Kira Y (2000) The natural history of Minamisawa. Jiyu Gakuen Shuppankyoku, Tokyo

Magurran AE, McGill BJ (eds) (2013) Biological diversity: frontiers in measurement and assessment. Oxford University Press, Oxford

Magurran AE, Dornelas M, Moyes F, Henderson PA (2019) Temporal βdiversity—a macroecological perspective. Glob Ecol Biogeogr 28(12):1949–1960

May MR (2020) What determines population density? In: Unsolved problems in ecology. Princeton University Press, Princeton, pp 67–75

McCullagh P, Nelder J (1989) Generalized linear models, 2nd edn. Chapman and Hall/CRC, Boca Raton

McGlinn DJ, Blowes SA, Dornelas M, Engel T, Martins IS, Shimadzu H, Gotelli NJ, Magurran A, McGill BJ, Chase JM (2025) Disentangling nonrandom structure from random placement when estimating β-diversity through space or time. Ecosphere 16(3):e70061

Ovaskainen O, Abrego N (2020) Joint species distribution modelling: with applications in R. Cambridge University Press, Cambridge

Shimadzu H (2018) On species richness and rarefaction: size- and coverage-based techniques quantify different characteristics of richness change in biodiversity. J Math Biol 77:1363–1381

Shimadzu H, Darnell R (2015) Attenuation of species abundance distributions by sampling. R Soc Open Sci 2:140219

Shimadzu H, Shibata R (2005) Analysis of bird count series by local regression to explore environmental changes. J Jpn Stat Soc 34(2):187–207

Shimadzu H, Dornelas M, Henderson PA, Magurran AE (2013) Diversity is maintained by seasonal variation in species abundance. BMC Biol 11(1):98

Shimadzu H, Dornelas M, Magurran AE (2015) Measuring temporal turnover in ecological communities. Methods Ecol Evol 6(12):1384–1394

Shimadzu H, Foster SD, Darnell R (2016) Imperfect observations in ecological studies. Environ Ecol Stat 23(3):337–358

Stoklosa J, Daly C, Foster SD, Ashcroft MB, Warton DI (2014) A climate of uncertainty: accounting for error in climate variables for species distribution models. Methods Ecol Evol 6(4):412–423

Uchida Y, Shimadzu H, Sekimoto T (2003) The relationship between the avifauna and environmental changes at Jiyu-Gakuen in Tokyo: a statistical analysis of the bird-census data for 35 years. Strix 21:53–70

Vellend M, Dornelas M, Baeten L, Beauséjour R, Brown CD, De Frenne P, Elmendorf SC, Gotelli NJ, Moyes F, Myers-Smith IH, Magurran AE, McGill BJ, Shimadzu H, Sievers C (2017) Estimates of local biodiversity change over time stand up to scrutiny. Ecology 98(2):583–590

Wood SN (2017) Generalized additive models, 2nd edn. CRC Press, Taylor & Francis Group, London

Xu WB, Blowes SA, Brambilla V, Chow CFY, Fontrodona-Eslava A, Martins IS, McGlinn D, Moyes F, Sagouis A, Shimadzu H, van Klink R, Magurran AE, Gotelli NJ, McGill BJ, Dornelas M, Chase JM (2023) Regional occupancy increases for widespread species but decreases for narrowly distributed species in metacommunity time series. Nat Commun 14(1)

Chapter 12
Counting Animals We Can't See: A Triple Count Model for Scarce Vestige Data

Niamh Mimnagh, Iuri Ferreira, Luciano Martins Verdade, and Rafael de Andrade Moral

Abstract Estimating animal abundance is fundamental to ecological research and wildlife management but is often hindered by the high cost, logistical difficulty, and invasiveness of direct observation methods. This Chapter presents the triple Poisson model, a novel hierarchical model for estimating animal abundance from vestige count data, such as scats, under conditions of data scarcity. The model accounts for group size, number of groups, vestige production and decay, and transect coverage, and is flexible enough to incorporate overdispersion through negative binomial extensions. Through extensive simulation studies and comparison with traditional distance sampling methods, we demonstrate that the triple Poisson model yields robust abundance estimates even with limited data, particularly when some prior information is available. We further illustrate its real-world applicability through case studies on collared peccaries, sika deer, and red foxes. This framework provides a cost-effective, non-invasive alternative for estimating wildlife abundance and has potential in ecological monitoring where data are sparse or difficult to collect.

Keywords Wildlife abundance estimation · Vestige data · Triple Poisson model · Bayesian hierarchical model · Scat surveys · Animal population monitoring · Overdispersion · Sparse data

N. Mimnagh (✉)
Maynooth University, Maynooth, Ireland

I. Ferreira
Federal University of São Carlos, São Carlos, Brazil

L. Martins Verdade
Wildlife Monitoring Consultancy, Campinas, São Paulo, Brazil

R. A. Moral
Department of Mathematics and Statistics, Maynooth University, Maynooth, Co. Kildare, Ireland

L. Martins Verdade, R. A. Moral (eds.), *Wildlife Monitoring*,
https://doi.org/10.1007/978-3-032-05821-8_12

1 Introduction

Estimating wildlife abundance may be relatively expensive and time-consuming. Therefore, whenever possible monitoring population fluctuation by an abundance index tends to be more cost-effective (Nichols 2014). However, in order to understand evolutionary-ecological processes and make decisions concerning wildlife management (i.e., conservation, use, coexistence, and monitoring, according to Caughley (1994)), one might need to estimate actual species abundance (Verdade et al. 2014). In addition, methods that involve capturing or even direct sightings of animals can be invasive and pose risks to both wildlife and humans (Verdade et al. 2013).

In this Chapter, we propose a modelling framework based on a triple Poisson hierarchy, which allows for the estimation of animal abundance from animal vestige count data, where a vestige may include any trace that an animal leaves behind as it moves through a study area. In the simulation studies and case studies examined in this paper, the vestiges examined are scats, that we assume are produced at a constant rate. Advantages associated with the use of vestige data to estimate animal abundance include the reduced cost and labour required to carry out the survey, as well as the reduced risk of disturbance caused to the animal, when compared with direct methods of estimating animal abundance (Verdade et al. 2014). In addition, we show that the modelling framework proposed here can be useful for estimating animal abundance from small-scale monitoring programmes, which may have very few transects and so produce data that is scarce.

The remainder of this Chapter is organised as follows. In Sect. 3 we introduce the modelling framework to estimate animal abundance from vestige count data. In Sect. 4 we present simulation studies, which were carried out to assess the estimates of abundance under a range of scenarios and with varying levels of prior information. Finally, in Sect. 5 we present a number of case studies which we use to illustrate how this modelling approach may have real-world applications.

2 Related Works

Several methods have been previously developed for estimating animal relative abundance based on vestige count data.

Distance sampling (Buckland et al. 1993) is a methodology that, while originally proposed to estimate animal density using animal count data, may also be used to estimate animal abundance using vestige data by making some small modifications (Marques et al. 2001). Distance sampling using vestige data involves modelling, with a detection function, the assumption that the detectability of vestiges decreases with increasing distance from a transect. This is done to obtain an estimate of vestige density, which may then be used to obtain an estimate of animal density and finally animal abundance. Marques et al. (2001) estimate the abundance of sika deer using this distance sampling methodology. In Sect. 4 we provide details of a simulation study which involves the comparison of abundance estimates obtained using a distance sampling model with those obtained using the novel triple Poisson model that we propose in this paper.

Becker (1991), Becker et al. (1998), and Patterson et al. (2004) propose a model for estimating abundance based on the observation of animal tracks in snow. This model assumes that the number of animal groups in the area may be obtained by following tracks and locating each group. For this reason, this methodology could prove time- and resource-intensive. The triple Poisson model that we propose in this paper allows us to assign a prior distribution to group size, which means that we may estimate an unknown number of groups, and true values do not need to be obtained.

The Formozov–Malyshev–Pereleshin formula for estimating animal abundance using animal track data is described by Stephens et al. (2006). This formula, originally proposed and published through Russian in Chelintsev (1995), involves estimating the probability that a transect will intersect an animal's track. This probability is then used to estimate the total number of track crossings, which can be used to estimate animal density. The successful implementation of this formula requires both estimates of animal daily travel distances and counts of animal tracks whose age is known.

3 Methods

3.1 Model Formulation

Our proposed model assumes that we are examining closed populations of mammals which move around randomly in G groups of size N_i, $i = 1, \ldots, G$, within a study area which is homogeneous in terms of habitat use. We assume that

$$G \sim \text{Poisson}(\lambda_G),$$

and therefore the total number of animals in the area is

$$T = \sum_{i=1}^{G} N_i.$$

By assuming the group sizes N_i are independent and distributed as

$$N_i \sim \text{Poisson}(\lambda_N),$$

we can derive the conditional distribution of $T|G$

$$T|G \sim \text{Poisson}(G\lambda_N).$$

However, the data that we examine is not composed of realisations of the abundance T. Instead, we observe a fraction of the number of vestiges left in the study area by a certain species. Let V_t represent the total vestige count in the area at time t. We assume that the number of vestiges left at time $t = 1$ has mean βT, i.e. depends on the individual vestige production rate β. By letting $V_1 \sim \text{Poisson}(\beta T)$ and assuming vestiges left in the environment disappear exponentially over time with some constant rate, we may write

$$V_t|V_1, \ldots, V_{t-1} \sim \text{Poisson}\left(\beta T + \sum_{j=1}^{t-1} \beta T e^{-\delta(t-j)}\right).$$

where δ is a vestige decay parameter.

We are interested, however, in the limiting distribution of V_t, and for that we assume that after a short period of time, the total number of vestiges produced plus old vestiges that remain in the environment from previous time points will become constant. We next examine the rate term in the Poisson distribution above, which can be rewritten as follows:

$$\beta T + \beta T \sum_{j=1}^{t-1} e^{-j\delta} = \beta T e^{-(0)\delta} + \beta T \sum_{j=1}^{t-1} e^{-j\delta} = \beta T \sum_{j=0}^{t-1} e^{-j\delta}.$$

This allows us to calculate

$$\lim_{t\to\infty} \beta T \sum_{j=0}^{t-1} e^{-j\delta} = \frac{e^{\delta}\beta T}{e^{\delta} - 1}.$$

We may then define a new parameter $\alpha = \left(\frac{e^{\delta}\beta}{e^{\delta}-1}\right)$ which we call the individual vestige surplus, given that it accounts for the new vestiges produced per individual, plus vestiges that were produced at previous time points, minus those that have decayed. We can then write the marginal distribution of $V = V_t$ as $V \sim \text{Poisson}(\alpha T)$.

By assuming random deposition of vestiges, the distribution of the observed number of vestiges will depend on the coverage level of the sampling method (e.g. transects). We refer to this coverage as $\nu \in (0, 1)$. Given that study area and transect length are generally known, we assume that ν can be calculated and is thus a known parameter, and that every vestige within the covered area is detected. Therefore, the distribution of the observed number of vestiges Y, alongside the full modelling hierarchy, can be written using the following triple Poisson hierarchical model:

$$Y|T, G, \nu \sim \text{Poisson}(\alpha T \nu) \quad (12.1)$$

$$T|G \sim \text{Poisson}(G\lambda_N) \quad (12.2)$$

$$G \sim \text{Poisson}(\lambda_G) \quad (12.3)$$

The above expression for observed vestiges Y may involve only a single visit per transect (i.e., Y as a vector of length R), but may also involve multiple visits per transect (i.e., Y as an $R \times S$ matrix, where S is the number of visits to each transect). In each case, the true abundance T is assumed not to change between transect visits. Abundance estimates obtained from single-visit data are compared to those obtained from data involving temporal replication in the Appendix.

The performance of the model is highly dependent on how well α, the individual vestige surplus, is estimated. Therefore we may either opt to set up an informative prior for α, or simply fix it as a "known" value. To estimate λ_G and λ_N we may use informative or non-informative priors, depending on the level of prior information that we possess.

It would also be reasonable to assume an aggregated process of resource allocation and/or aggregated animal behaviour, which may result in vestiges which are produced at a non-constant rate. This would, in turn, affect the number of vestiges present at each transect. A simple extension that would accommodate this assumption would be to treat the top tier of the hierarchy as an over-dispersed process, e.g.

$$Y|T, G, \nu \sim \text{Negative Binomial}(\alpha T \nu, \phi) \quad (12.4)$$

and use a non-informative gamma prior to estimate ϕ, the over-dispersion parameter. Analogously, if the group sizes are believed to be highly variable, a reasonable modification to the model above would use $N_i \sim \text{Negative Binomial}(\lambda_N, \phi_N)$, which results in $T \sim \text{Negative Binomial}(G\lambda_N, G\phi_N)$. Finally, if numbers of groups are highly variable when comparing across different regions, we may also assume $G \sim \text{Negative Binomial}(\lambda_G, \phi_G)$. This introduces a family of models that can accommodate different scenarios, depending on the species being monitored.

4 Simulation Studies

In this section we describe simulation studies which were conducted with the aim of determining the effect on abundance estimates of varying the choice of prior distribution used for individual vestige surplus α. To this end, vestige count data was simulated from a triple Poisson model, and models were fitted in JAGS (Plummer et al. 2003) (using the R statistical software version 4.0.2 (R Core Team 2022)) with different options for α. We wished to compare abundance estimate from scenarios when α is given a weakly-informative Uniform prior to scenarios when α is known and supplied correctly to the model as data, and when α is "known" but is incorrect, and is supplied as data.

The results are presented in Fig. 12.1, in which the true value for α is 35 vestiges and the true abundance is 86. The small crossbar in Fig. 12.1 represents the 95% credible interval for the estimate of abundance when α is known correctly. From this we can see that when the true value for α is known, estimates of abundance are highly accurate and precise. The large crossbar represents the 95% credible interval for the estimate of abundance when a Uniform(10, 50) prior is supplied for α. When α is unknown and is given a Uniform prior, the estimate for abundance has still got a high degree of accuracy, but is now imprecise. The ribbon represents the 95% credible interval when incorrect values in the range (20, 50) are supplied for α. In this case, the estimate for abundance is still precise (i.e., the credible interval is still quite narrow), but is now highly inaccurate.

Another possible scenario that we have assessed is that in which no prior information is available for the value of α, and a non-informative prior must be used. Results of these simulations are available in the Appendix, and from these results we conclude that if no information is available for the value of α, then informative prior distributions are required for λ_N and λ_G.

Further to this, simulation studies wherein data is simulated from a triple Poison model were run to assess model estimation for different prior distributions on λ_G and λ_N, the effects of using a Poisson distribution on vestige count versus

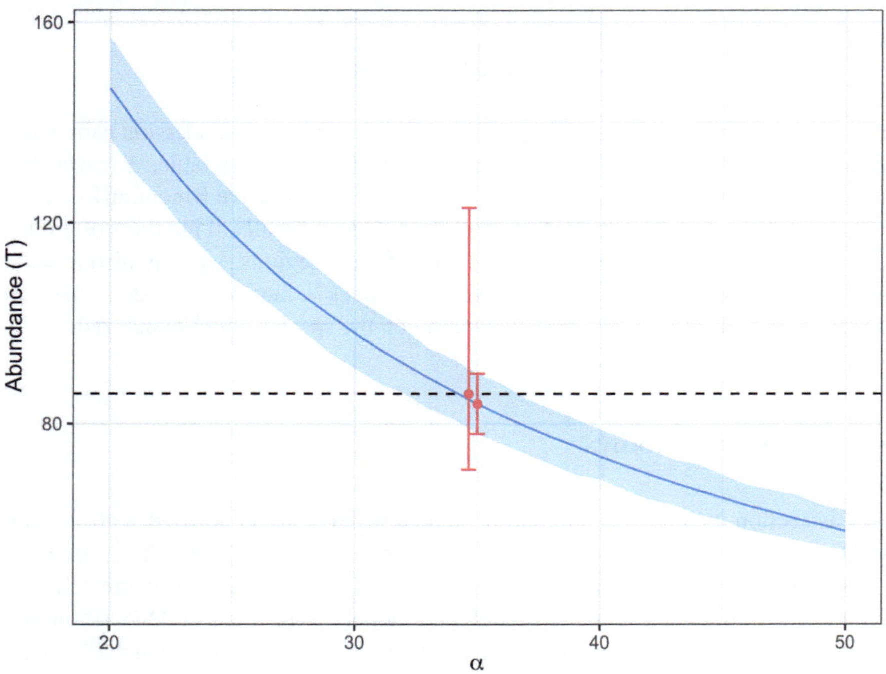

Fig. 12.1 Abundance mean estimates and 95% credible intervals when the true value of α is known correctly (short crossbar), when α is known incorrectly (ribbon) and when α is supplied a Uniform prior (long crossbar), with the true abundance denoted by the dashed horizontal line

using a negative binomial distribution, and the effect on abundance estimation of using temporally replicated vestige counts (i.e. transects that are visited on multiple occasions). The full details and results of all simulation studies can be found in the Appendix.

In addition to the simulation studies described above, studies were run to compare the accuracy of abundance estimates produced by the triple Poisson model to those produced by a distance sampling model (both in a frequentist and Bayesian framework), when data is scarce.

The Bayesian distance sampling model was implemented following the example of Kéry and Royle (2015). The distance sampling model involves employing counts of observed vestiges Y to estimate the true number of vestiges per transect z, and subsequently estimating vestige density, animal density, and finally animal abundance. Data used for the distance sampling model is composed of J total vestiges, each collected from one of R transects, along with the distance between each vestige and the transect. The observed vestige count at transect i ($i = 1, ..., R$) is given by

$$Y_i \sim \text{Binomial}(z_i, p_i)$$
$$z_i \sim \text{Poisson}(\beta_i)$$

where z_i us the actual number of vestiges present, and p_i is the vestige detection probability at transect i.

The maximum distance from each transect at which vestiges might be observed is denoted as B, and this distance B is divided into $k = 1, ..., K$ intervals. $\pi_{i,k}$ is the probability of detecting a vestige within each of these K intervals around transect i, and is as follows.

$$\pi_{i,k} = \left(-\frac{m_k^2}{2\sigma_i^2}\right)\left(\frac{d}{B}\right)$$

where m is the midpoint of each interval, σ is the half-normal scale parameter, $-\frac{m_k^2}{2\sigma_i^2}$ is the half-normal detection function evaluated at m, and d is the length of each interval. The overall probability of detection for transect i is then:

$$p_i = \sum_{k=1}^{K} \pi_{i,k}$$

Because we are working with K distance intervals (labelled $1, \ldots, K$) the original continuous distance values must be converted into the corresponding distance interval. For observation j, we model this distance interval C_j using a multinomial distribution as follows:

$$C_j \sim \text{multinomial}\left(\hat{\pi}*_j\right)$$

where $\hat{\pi}*_j$ is a vector of length K that contains the probabilities associated with vestige j being observed within each interval. Given that we know which transect each vestige is observed at, this parameter is calculated for all R transects as the $R \times K$ matrix whose (i,k) entries are $\frac{\pi_{i,k}}{p_i}$, and the probabilities associated with the appropriate transect (the k-th row of the matrix) are then utilised in the multinomial distribution above.

The vestige density can be calculated as $D_{\text{vestiges}} = \frac{\sum_i z_i}{A}$, where A is the area of the sample space. This vestige density can now be used to determine animal abundance (Marques et al. 2001). The total number of vestiges produced per day is calculated as

$$T_{\text{daily vestiges}} = \frac{D_{\text{vestiges}}}{\delta},$$

where δ is the time until vestige decay. The animal density D can then be given as:

$$D = \frac{T_{\text{daily vestiges}}}{\lambda},$$

where λ is the vestige production rate. This density can then be used to calculate abundance as

$$T = D \times A,$$

Data was simulated from this distance sampling model, which required the specification of the size of the study area, the number of vestiges within the study area, the distance between transects, the truncation distance (the distance from a transect within which vestiges may be observed), the detection function (which is used to model the distribution of vestiges given their distance from a transect), the vestige production rate (which was specified as 15 vestiges per day), and the time to vestige decay (which was specified as 10 days). Data was simulated using the `DSsim` package (Marshall 2020). Each simulation had a total of 5000 vestiges contained within a 2×5 km area, with vestige density constant across the study area. A truncation distance of 10 m was chosen, and transects were specified at a distance of 1 km from each other, which ensured that each simulation contained only two transects, each 5 km long. As a result, the data used in the triple Poisson model is comprised of just two values, i.e., the total number of vestiges observed at each transect. A half-normal detection function was used to model the probability of observing vestiges, given their distance from the transect.

Triple Poisson models, with combinations of informative and non-informative λ_G and λ_N were fitted to these simulated datasets. Simulation studies typically involve the true abundance given as a fixed value across all simulated datasets. However, because we are simulating from a distance sampling model, we first simulate 5000 total vestiges, and then use a probability detection function to determine which of these are visible near the transect, and finally calculate the true abundance. The result is that the true abundance T was slightly different for each simulated dataset ($T \in (25, 55)$). For this reason, in order to provide informative priors for λ_N and λ_G, it was assumed that the mean number of animals per group might be between three and seven, and the mean number of groups in the area might be between one and ten. This allowed us to use the informative priors $\lambda_N \sim \text{gamma}(5, 1)$ and $\lambda_G \sim \text{gamma}(10, 1)$.

The non-informative priors for λ_N or λ_G were specified using a gamma(0.01, 0.01) distribution. Data was simulated with 5000 vestiges present in the study area, so the individual vestige surplus of the triple Poisson model α was given a weakly-informative Uniform(10, 10000) prior, where α represents the number of new vestiges produced per individual per day plus the number of vestiges still present in the area from previous days. These models were implemented through JAGS (Plummer et al. 2003), using the `R2jags` package (Su and Yajima 2020).

Subsequently, distance sampling models were fitted to these simulations, using both frequentist and Bayesian frameworks. The frequentist model was fitted using the `Distance` (Miller et al. 2019) package, with combinations of δ and λ specified correctly and incorrectly. The aim of the frequentist application of the distance sampling model was to determine the effect on abundance estimates of slight inaccuracies in the values supplied for these parameters. The Bayesian implementation of the distance sampling model was then performed so that the distance sampling model might be fairly compared with the Bayesian implementation of the triple Poisson model, and this involved the use of informative and non-informative gamma prior distributions for δ and λ.

The full list of models fitted are given in Table 12.1. The accuracy of abundance estimates was assessed using relative bias for the true abundance T, averaged over all simulations, calculated as

$$\text{Relative mean bias} = \frac{\hat{T} - T}{T}.$$

The smaller the value for relative bias, the closer to the true value our estimated abundance values were.

In Figs. 12.2, 12.3 and 12.4, we present the results of the comparison of true abundances with estimates obtained using triple Poisson models and distance sampling models. In the case of the Bayesian implementation of the distance sampling model (Fig. 12.4), the model in which vestige production rate (λ) and time to vestige decay (δ) are assigned non-informative priors is not displayed. This is because—with estimates of abundance of approximately 12,000 individuals—it produces estimates of abundance that are on a much larger scale than the other Bayesian distance sampling models, which makes results from the other models difficult to read.

Table 12.1 (a) Triple Poisson models fit with informative and non-informative gamma priors on mean group size λ_N and mean number of groups λ_G. (b) Distance Sampling models fit using a frequentist framework, with individual vestige production (λ) and time to vestige decay (δ) supplied correctly ($\lambda = 15, \delta = 10$) and incorrectly, with incorrect values for $\lambda \in (16, 19)$ and incorrect values for $\delta \in (11, 14)$. (c) Distance Sampling models fit using a Bayesian framework fit with informative and non-informative gamma priors on individual vestige production λ and time to vestige decay δ

(a) Triple Poisson models

Model	λ_N	λ_G	Abundance relative bias
TP1	Informative	Informative	0.109
TP2	Informative	Non-informative	0.142
TP3	Non-informative	Informative	0.186
TP4	Non-informative	Non-informative	0.104

(b) Frequentist distance sampling models

Model	λ	δ	Abundance relative bias
DS1	15	10	0.000
DS2	16	10	0.091
DS3	15	11	0.062
DS4	16	11	0.147
DS5	17	12	0.265
DS6	18	13	0.359
DS7	19	14	0.436

(c) Bayesian distance sampling models

Model	λ	δ	Abundance relative bias
DS8	Known	Known	0.003
DS9	Informative	Informative	0.022
DS10	Informative	Non-informative	1.592
DS11	Non-informative	Informative	2.523
DS12	Non-informative	Non-informative	36.146

5 Case Studies

In this section, we will describe three different case studies with which we demonstrate various aspects of the triple Poisson model. Initially this modelling framework was applied to each case study, assuming that the observed vestige count Y was appropriately described using a Poisson distribution. Subsequently the models were re-run, assuming a negative binomial distribution for Y. For each case study the results of each of these models was compared using BIC and DIC values.

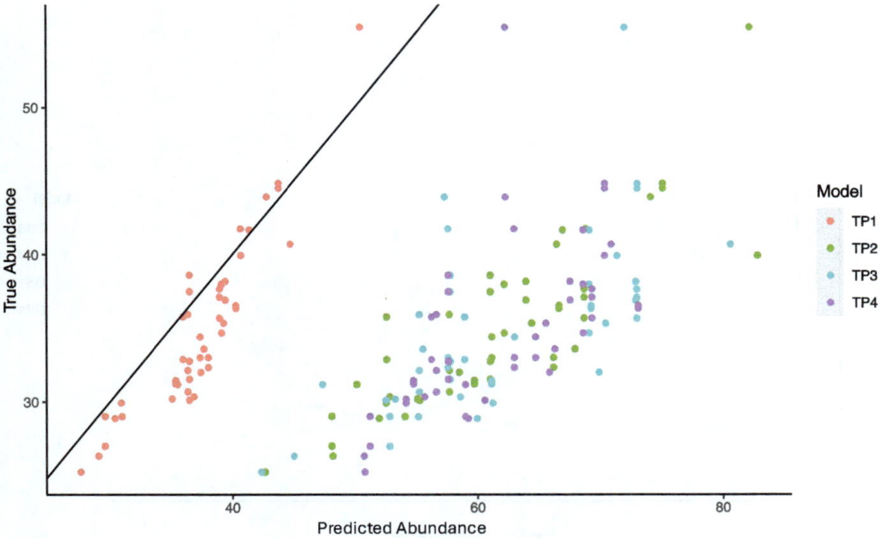

Fig. 12.2 A scatter plot comparing the abundance estimates from triple Poisson models (TP1–TP4) with varying priors on λ_N and λ_G to the true abundances

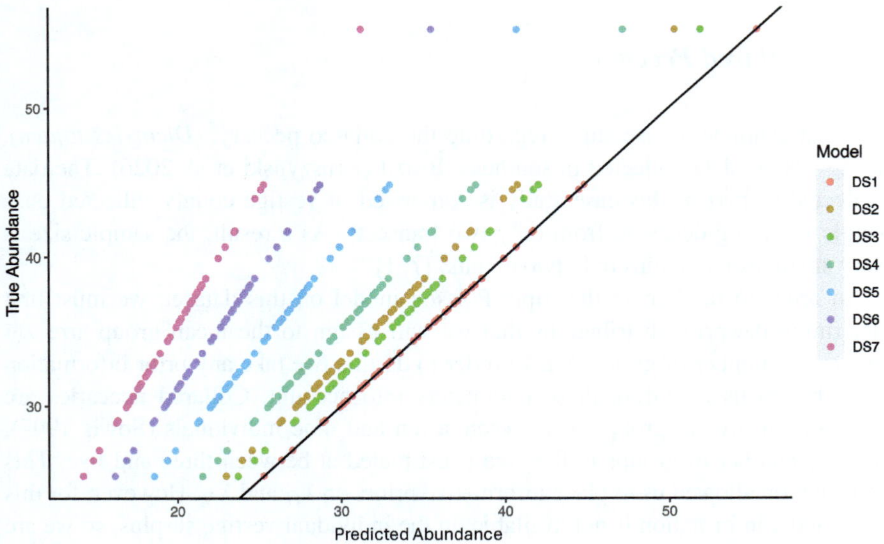

Fig. 12.3 A scatter plot comparing the abundance estimates from frequentist implementations of the distance sampling model (DS1–DS7, as described in Table 12.1) to the true abundances

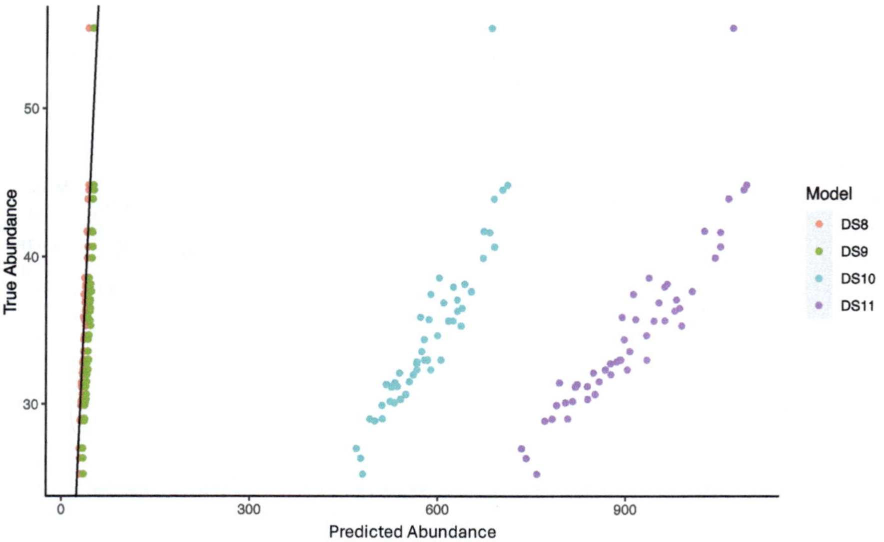

Fig. 12.4 A scatter plot comparing the abundance estimates from Bayesian implementations of the distance sampling model (DS8–DS11, as described in Table 12.1) to the true abundances

5.1 Collared Peccary

We first examine a case study regarding the collared peccary, (*Dicotyles tajacu*), using vestige data collected in southeast Brazil (Kruszynski et al. 2026). The data collected as part of this case study is comprised of vestige counts collected on a single sampling occasion from only two transects. As a result, the sample size is very small, as it contains only two counts: {7, 1}.

In order to implement the triple Poisson model on this dataset, we must first determine the prior distributions that we will assign to the mean group size λ_N and mean number of groups λ_G. In order to do this, we take any prior information available to us regarding these parameters into account. Collared peccaries are estimated to live in groups of between seven and nine individuals (Sowls 1997), and the number of groups in the area is estimated at between three and five. This information allowed us to place informative priors on λ_G and λ_N. However, for this case study, information is not available on the individual vestige surplus, so we are unable to place an informative prior on α in this case. For this reason we place a non-informative prior on vestige surplus: $\alpha \sim \text{gamma}(0.01, 0.01)$. As is demonstrated in Appendix, if a non-informative prior distribution must be used for vestige surplus α, and particularly if sample size is small, it is required that we specify informative prior distributions for λ_N and λ_G.

The transect along which seven vestiges were found was 8 km long, while the other transect was 12 km long. In order to estimate υ, the transect coverage rate, we require the distance around the transect within which vestiges might be observed.

This distance was not measured, so we make a conservative estimate that vestiges within 2 m on either side of the transect are visible. This results in transects whose area are $32{,}000\,\text{m}^2$ and $48{,}000\,\text{m}^2$ respectively. The study area is $43.65\,\text{km}^2$, and we can therefore estimate the transect coverage rate as $\upsilon \in \{0.00073, 0.0011\}$.

Our model was fitted with first a Poisson distribution on the vestige count Y, and then a negative binomial distribution, and the results were compared using DIC values. The model with the lower DIC value was the one with a negative binomial distribution on the observed number of vestiges Y. This model was chosen as the best fit, of the models available. The results obtained from this model were a mean estimate of a population of 44 collared peccaries within the study area, with a 95% credible interval of (16, 87).

5.2 Sika Deer

The next case study we examine involves vestige data collected from the sika deer (*Cervus nippon*) by Marques et al. (2001) in eight different regions in Scotland, as well as the distance from each vestige to the transect, measured in centimetres. Because this distance information is available, this case study provides an ideal opportunity to compare estimates produced using a triple Poisson model to those produced using a modification of a distance sampling model, as described by Marques et al. (2001).

As detailed briefly in Sect. 2, the implementation of this modified distance sampling model to estimate abundance using vestige data requires estimates of the rate at which vestiges are produced, and time to vestige decay. Our implementation of this modelling framework follows one described by Marques et al. (2001) and so we utilise their provided estimate that sika deer produce approximately 25 vestiges per day. The paper by Marques et al. (2001) provides estimates and standard errors for the time to vestige decay for this data. However, they provide these estimates based on the habitat groups for the data, and the month during which data is collected, neither of which we have access to within the dataset provided as part of the `Distance` R package (Miller et al. 2019). For this reason, we base this value upon analysis presented in Rexstad (2022) and obtain an estimate of 163.4 days (with a standard error of 14.2) for vestiges to decay.

We then fit a triple Poisson model to this data, and can compare estimates provided by these two methodologies. In order to implement our triple Poisson model, we must determine the prior distributions that should be assigned to estimate the mean group size λ_N and the mean number of groups λ_G. The prior information that we possess is that sika deer may live in groups of up to ten animals (Ratcliffe 1987), and that territory size of sika deer is in the range of $0.02 - 0.12\,\text{km}^2$. For this reason, we place an informative prior on $\lambda_N \sim \text{gamma}(5, 1)$ which can allow for group sizes of up to approximately 15 individuals. Our knowledge of the typical territory size allows us to also place an informative prior on λ_G. Each of the eight regions has a different area, ranging from 8 to $15.2\,\text{km}^2$. For this reason, we obtain

Table 12.2 This table presents the area of each of the eight regions in the sika deer dataset, as well as the theoretical maximum number of groups of sika deer that each area could contain, and the prior assigned for λG during our analyses

Region	Area	Max. groups	Prior
A	$13.9\,\text{km}^2$	700	gamma(1, 72)
B	$10.3\,\text{km}^2$	500	gamma(1, 45)
C	$8.6\,\text{km}^2$	430	gamma(1, 40)
E	$8\,\text{km}^2$	400	gamma(1, 40)
F	$14\,\text{km}^2$	700	gamma(1, 72)
G	$15.2\,\text{km}^2$	760	gamma(1, 75)
H	11.3km^2	565	gamma(1, 52)
J	$9.6\,\text{km}^2$	480	gamma(1, 50)

a separate theoretical group maximums per area. In order to determine the prior to use for λ_G, we use the size of the region, and prior knowledge that the territory of sika deer lies between two and twelve hectares. Table 12.2 contains the area of each region, the theoretical maximum number of groups it might contain, and the gamma prior chosen for our analysis.

From the estimates provided for the distance sampling implementation, we know that sika deer produce approximately 25 vestiges per day. We also assume that vestiges may take several months to decompose, and so we place a weakly informative prior distribution on α, i.e. $\alpha \sim \text{Uniform}(0, 4000)$. The use of this prior distribution implies that we assume that there are between 0 and 4000 vestiges present in the study area due to each individual, which, at a rate of 25 vestiges produced per day, allows vestiges to remain for up to approximately 5 months before becoming close to fully decomposed.

We know that the largest distance at which vestiges were observed from the transect was two metres. For this reason, the triple Poisson models are run assuming that vestiges are visible within two metres on either side of the transect, which allows us to estimate the transect coverage rate ν for each area.

Initially, each model was run assuming that the vestige count is adequately described using a Poisson distribution. This was then compared to a model in which the vestige count is assigned a negative binomial distribution. The model of best fit for each area was chosen using DIC values, and a comparison of abundance estimates produced using these triple Poisson models of best fit and the distance sampling model can be found in Table 12.3. In this table we see that point estimates obtained using a triple Poisson model for areas A and B are very close to those obtained using a distance sampling model, while there is a greater difference in point estimates obtained for other areas. In particular, we see a large difference in abundance estimates for areas H and J. For these areas, the distance sampling model produces quite small estimates, while the triple Poisson model produces larger point estimates. This can be explained by a combination of factors. The first is the amount of data available for these areas. In each of areas H and J, only one transect is surveyed. This means that the triple Poisson models fitted to these regions are fitted to data that contains only one observation. The simulation studies that we have performed for scarce data contain data collected from two transects, and so the

Table 12.3 Mean estimates and their 95% confidence (distance sampling) or credible (triple Poisson) intervals for sika deer abundance per area from a distance sampling model and the triple Poisson model

Area	Distance sampling		Triple Poisson	
	Estimate	95% CI	Estimate	95% CI
A	1027	(690, 1528)	1028	(584, 1952)
B	382	(219, 667)	409	(167,982)
C	33	(15, 74)	106	(20, 451)
E	29	(8, 99)	71	(19,312)
F	209	(173, 252)	340	(128, 1036)
G	125	(18, 856)	281	(78, 834)
H	17	(14, 21)	95	(11, 431)
J	69	(57, 83)	165	(35, 622)

models fitted to areas H and J are run using data that is more scarce than even the most scarce simulations performed. A further reason for the inflated point estimates obtained for these areas is the very small transect coverage rate associated with them. Areas A and B, for which triple Poisson point estimates are closest to distance sampling point estimates, have a coverage rate of 0.00048 and 0.00042 respectively. The transect coverage rates associates with areas A and B are twice as large as the next largest transect coverage rate (associated with area C), and over 10 times larger than those associated with areas H and J, which have coverage rates of 0.000071 and 0.000041 respectively. Due to the formulation of this model, the abundance T is confounded with transect coverage ν, and very small values for ν may lead to inflated estimates for abundance T.

5.3 *Red Foxes*

The final case study examined involves vestige data collected from the red fox (*Vulpes vulpes*) by Cavallini (1994). This data was collected from nine transects in central Italy. In the paper by Cavallini (1994), an index of abundance is obtained, estimated as the number of vestiges per kilometre. This data was collected once every month over the course of a year, and so it is a good candidate for our triple Poisson modelling framework that takes into account temporal replication. As detailed in Sect. 3, here the observed vestige count Y is allowed to vary by transect and month. However, in our model implementation we assume that the true abundance T varies only by transect.

Before we can implement the triple Poisson model on this case study, we must first determine the prior distributions that to be assigned to the mean group size λ_N and mean number of groups λ_G. We possess prior knowledge on common group sizes for red foxes, as well as their territory size, which will allow us to place informative priors on λ_G and λ_N.

Red foxes may have as many as nine cubs in a litter, which means that a group of red foxes may contain as many as 11 individuals (Baker and Harris 2004). Their territory ranges from 5 to 12 km^2. This case study is contained within a study area of size 2448 km^2, and so in this area there could be as many as 490 groups of red foxes. We can now specify that $\lambda_G \sim \text{gamma}(1, 50)$, which allows a maximum of between 400 and 500 groups.

Transect coverage in this case was again calculated using transect length and study area, assuming that vestiges within 2 m of the transect are visible. To inform our prior distribution of vestige surplus α, we know that red foxes may produce eight vestiges per day (Webbon et al. 2004). This data was collected monthly over a 12 month period, so we remove the initial observations (April 1992) for each region, as we do not know how long these vestiges may have been present in the environment. The original paper (Cavallini 1994) mentions that at each sampling occasion the vestiges observed were collected and taken for laboratory analysis. For this reason, counts collected disregard counts initially collected in previous months, and so we assume that the longest time that the vestiges observed may have been present in the environment was approximately 30 days. The prior distribution used for vestige surplus α reflects this belief that each individual might produce eight vestiges per day, and these vestiges are only present for approximately one month until the following sampling occasion: $\alpha \sim \text{Uniform}(0, 240)$.

The model was fitted initially with a Poisson distribution on the vestige count, and then a negative binomial distribution. The results were compared using DIC values, and the model with the lower DIC value was the one with a negative binomial distribution on the vestige count. This model was chosen as the best fit for this data. The result obtained from this model was a mean estimate of 3304 red foxes in the study area, with a 95% credible interval of (2603, 4442).

Table 12.4 provides the DIC and BIC values associated with a model with a Poisson vestige count and a negative binomial vestige count for each of the case studies examined in this chapter. In each comparison, the DIC values and BIC values were in agreement as to the model of best fit. For each case study, the model with the lower DIC/BIC value was chosen as the optimal fit.

Table 12.4 DIC and BIC values for models fitted to case studies in which we compare a Poisson vestige count to a Negative binomial vestige count

	DIC		BIC	
Model	Poisson	Negative binomial	Poisson	Negative binomial
Collared Peccary	16.12	15.14	15.54	14.44
Red Fox	1195.94	629.33	1427.36	759.74
Sika Deer A	555.63	145.10	561.37	151.39
Sika Deer B	283.9	94.40	288.82	99.69
Sika Deer C	14.92	15.52	16.37	17.62
Sika Deer E	46.25	36.34	48.41	34.94
Sika Deer F	7.29	8.41	5.32	6.26
Sika Deer G	34.839	27.52	36.22	25.72
Sika Deer H	5.07	5.50	3.34	3.52
Sika Deer J	5.82	6.34	3.96	4.35

6 Conclusion

In this Chapter, we have presented a novel modelling framework that allows for the estimation of animal abundance using vestige data. We have presented the results of an extensive simulation study, which demonstrates the performance of this model under a range of conditions. Full details of these simulations are available in the Appendix.

As part of the simulation study, we simulated scarce data from a distance sampling model, and fitted various distance sampling and triple Poisson models to this data to compare abundance estimates. While distance sampling models consistently produce accurate estimates of abundance when vestige production rate λ and time to vestige decay δ are known precisely (Table 12.1(b)), small inaccuracies in λ and δ produce estimates of abundance with relative bias that increases quite rapidly. Triple Poisson models were fitted with priors on λ_N and λ_G both informative and non-informative. These models produced estimates of abundance with relative bias $\in (0.1, 0.2)$ regardless of prior choice for λ_N or λ_G. In Fig. 12.2, the triple Poisson model with informative priors on λ_G and λ_N better reflects the true abundances, while models for which information is unavailable to inform λ_N and λ_G do not perform quite as well.

While these scarce data simulation studies (detailed in Sect. 4) reveal that the triple Poisson model can produce unbiased abundance estimates when even very few transects are available for surveying, having more data is always preferable. Simulation studies designed to examine the effect of sample size on abundance estimates reveal that scenarios in which a greater number of sites are available for survey produce less biased abundance estimates, as might be expected. Model formulation suggests that abundance estimates T may be confounded with transect coverage rate ν. The results obtained using the sika deer case study in Sect. 5 demonstrate this. When the amount of the study area covered by transects is very small, estimates of abundance will be inflated, as can be seen in Table 12.3. This highlights the need at the experimental design phase to ensure the use of appropriately sized study areas for which adequate transect coverage can be obtained.

Simulations reveal that vestige surplus α is confounded with abundance T, which is also to be expected given model formulation. If α is known, estimates for abundance are not dependent on prior distributions used for λ_G and λ_N. When α is not known precisely, but enough information is available to provide a prior distribution that is weakly informative, dependence on λ_G and λ_N increases. Finally, if no information is available and a non-informative prior must be supplied for α, to avoid producing abundance estimates that are highly biased, informative priors are required for both λ_G and λ_N.

The assumption that vestige count Y can be adequately described using a Poisson distribution relies on vestige deposition constant across the study area. In many cases, this can prove to be an unrealistic assumption. In such cases, a simple alternative is the use of a negative binomial distribution to describe the vestige count. This allows for overdispersion in vestige counts, which could be attributed to the occurrence of vestiges at higher rates in some areas, due to factors such as resource allocation, animal behaviour, or non-constant rates of vestige production. The use of a negative binomial distribution for vestige counts was examined by simulation study. A notable finding of this study was that even a small level of overdispersion in vestige counts can cause quite a large increase in relative bias of abundance estimates, when compared to the Poisson model for vestige count. This impact is seen most strongly when both sample size is small and non-informative prior distributions are used for both λ_G and λ_N. This occurs even when α is known. Consequently, if overdispersion is expected in vestige count, and prior knowledge that may inform λ_G and λ_N is not available, it is advisable that larger sample sizes are collected, either by increasing numbers of transects established, or by visiting transects on multiple occasions.

In this study we assumed vestiges experience an exponential rate of decay. However, in the future we plan to examine alternative scenarios, in which vestiges may decay according to different distributions, to allow us to model the transient phase of the population, as mentioned in Sect. 3. In the future, we intend to perform agent-based modelling simulations of animal movement, while incorporating vestige deposition and decay probabilities. By assuming different sampling designs, we will then be able to estimate the abundance T and compare with the true number of animals in the environment. This might facilitate field work on data collection and improve cost-efficiency of animal counting.

The decision-making process concerning wildlife management is usually based on the cost-efficiency relations of survey/monitoring methods available (Nichols 2014). Traditional methods involving capturing and/or direct sight seen of animals tend to be invasive, time-consuming, and relatively expensive (Verdade et al. 2013, 2014). In addition, traditional methods of animal counting tend to have low precision and unknown accuracy (Verdade et al. 2014). The use of vestiges on the estimation of actual abundance or population density simplifies the process of animal counting; therefore, improving the decision-making process concerning wildlife management.

Gallant et al. (2007), Murray et al. (2005), and Barnes (2001) examine the limitations associated with the use of scat surveys to estimate abundance. Among these limitations is the possibility for vestiges to be concentrated at certain sites. As described in Eq. 12.4, the use of a negative binomial distribution for vestige counts provides a possible solution to this issue. Another of these limitations is the relatively small amount of a sample area that tends to be covered by transects. The triple Poisson model as detailed in Eq. 12.2 incorporates the transect coverage rate in the vestige count Y, which allows us to take into account the fact that very small sections of the study area are visible during transect surveys, though as we discovered in Sect. 5.2, very small transect coverage may lead to inflated estimates of abundance, and this should be taken into account at the experimental design phase. A final limitation is the possibility for vestiges produced at certain times of the year to decompose more quickly than those produced at other times. It is true that the triple Poisson model as described in this paper assumes that vestiges decay at a constant rate. A subject of planned future work will allow vestiges to decay at different rates, possibly incorporating the effect of covariates into the rate of decay, which may allow us to model the transient phase of the population.

Acknowledgments Niamh Mimnagh's work was supported by a Science Foundation Ireland grant number 18/CRT/6049. The opinions, findings and conclusions or recommendations expressed in this material are those of the author(s) and do not necessarily reflect the views of the Science Foundation Ireland.

Appendix: Simulation Studies

In this appendix, we present the results of simulations that were run to examine the effect on abundance estimates of varying parameter values and prior distributions. To assess abundance estimates for different population sizes, we assigned true values for group size $\lambda_N \in \{5, 10\}$, and number of groups $\lambda_G \in \{5, 10\}$. We also examine the effect of our level of prior knowledge of vestige surplus α, using scenarios in which α is presumed somehow "known", and can be supplied to the triple Poisson model as data, and comparing this to scenarios in which a level of prior information allows us to place weakly informative uniform priors on α, and scenarios in which we possess no prior information for α and must assign it a non-informative gamma prior. To examine the effect of sample size, we present abundance estimates when data is collected on a single visit to 10 transects, and compare these to estimates obtained using data collected on a single visit to 100 transects. We then briefly examine the effect of temporal replication by estimating abundance using data collected on 10 visits to 10 transects.

The effect on abundance estimates of prior information on mean group size λ_N and mean number of groups λ_G is assessed using combinations of informative gamma(7,1) and non-informative gamma(0.01, 0.01) prior distributions, where the gamma distributions are parameterised in terms of their shape and rate.

To assess abundance estimates when overdispersion is present in our data, we first perform simulations which allow for overdispersion in the number of vestiges observed at each transect Y as follows.

$$Y|T, G, \nu \sim \text{Negative Binomial}(\alpha T \nu, \phi)$$

where $\phi \in \{0.2, 2\}$ is the overdispersion parameter. The smaller the value of ϕ, the greater the degree of overdispersion present in the vestige counts. This allows us to account for situations in which an aggregated process of resource allocation or animal behaviour cause vestiges to be produced at a non-constant rate.

Finally, we present simulation results for a "triple negative binomial" model, formulated as follows.

$$\begin{aligned} Y|T, G, \nu &\sim \text{Negative Binomial}(\alpha T \nu, \phi) \\ T|G &\sim \text{Negative Binomial}(G\lambda_N, G\phi_N) \\ G &\sim \text{Negative Binomial}(\lambda_G, \phi_G) \end{aligned}$$

This formulation allows for overdispersion in all layers of the model hierarchy, which allows for a non-constant vestige production, and allows for variability in group size and numbers of groups.

Overall, the simulation study examined for the triple Poisson model contained a total of 288 variable combinations, each of which was simulated 100 times. In each case we estimated mean group size λ_N, mean number of groups λ_G, abundance T and the vestige surplus α.

The accuracy of abundance estimates was assessed using relative bias for the true abundance T, averaged over all simulations, calculated as follows.

$$\text{Relative bias} = \frac{\hat{T} - T}{T}.$$

The smaller the value for relative bias, the closer to the true value our estimated abundance values were. Figures 12.5, 12.6, 12.7, 12.8, 12.9, 12.10, and 12.11 present results for each of the previously described simulations. Each is divided into four sub-figures, (a) to (d) corresponding to varying prior distributions assigned to λ_N and λ_G.

(a) This sub-figure corresponds to a scenario in which prior information is available for group size ($\lambda_N \sim \text{Gamma}(7, 1)$) but not number of groups ($\lambda_G \sim \text{Gamma}(0.01, 0.01)$)
(b) This sub-figure corresponds to a scenario in which prior information is available for number of groups ($\lambda_G \sim \text{Gamma}(7, 1)$) but not group size ($\lambda_N \sim \text{Gamma}(0.01, 0.01)$)
(c) This sub-figure corresponds to a scenario in which prior information is available for both group size ($\lambda_N \sim \text{Gamma}(7, 1)$) and number of groups ($\lambda_G \sim \text{Gamma}(7, 1)$)
(d) This sub-figure corresponds to a scenario in which prior information is not available for group size ($\lambda_N \sim \text{Gamma}(0.01, 0.01)$) or number of groups ($\lambda_G \sim \text{Gamma}(0.01, 0.01)$)

Within each sub-figure we present relative biases for the mean estimate and the 95% credible interval for the abundance, using data collected on a single visit to 10 transects, data collected on a single visit to 100 transects, and data collected on 10 visits to 10 transects. The true value for vestige surplus α in all cases is 20 vestiges.

In Fig. 12.5, we present results when observed vestige count $Y \sim \text{Poisson}(\alpha T \nu)$, and vestige surplus α is known and correctly supplied to the model as data. Relative bias in abundance estimates in each case are very small, and there are no obvious differences in relative bias across sub-figures (a) to (d). This suggests that when α is known, prior distributions used for λ_N and λ_G have little effect on abundance estimation. In each case, relative bias is largest when using data collected on a single visit to 10 transects, and results for data collected on a single visit to 100 transects are comparable to those collected on 10 visits to 10 transects.

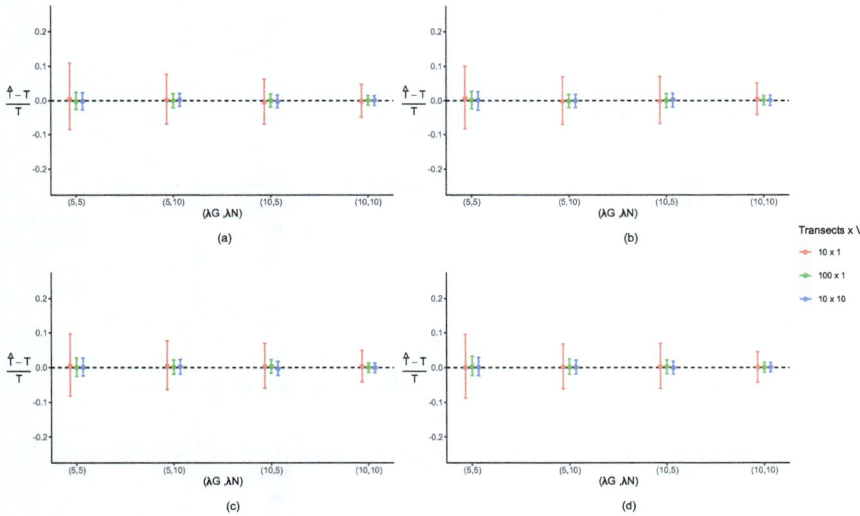

Fig. 12.5 Relative bias in mean estimate and 95% credible intervals for abundance T, for Poisson Y and α known, with true values for λ_N and $\lambda_G \in \{5, 10\}$, at varying sample sizes, and for different prior distributions on λ_N and λ_G, from 100 simulations. (**a**) $\lambda_N \sim \text{Gamma}(7, 1)$, $\lambda_G \sim \text{Gamma}(0.01, 0.01)$. (**b**) $\lambda_N \sim \text{Gamma}(0.01, 0.01)$, $\lambda_G \sim \text{Gamma}(7, 1)$. (**c**) $\lambda_N \sim \text{Gamma}(7, 1)$, $\lambda_G \sim \text{Gamma}(7, 1)$. (**d**) $\lambda_N \sim \text{Gamma}(0.01, 0.01)$, $\lambda_G \sim \text{Gamma}(0.01, 0.01)$

In Fig. 12.6, we present results when observed vestige count $Y \sim \text{Poisson}(\alpha T \nu)$, and vague prior knowledge of vestige surplus allows us to specify $\alpha \sim \text{Uniform}(1, 100)$. Relative biases are larger when α must be estimated, when compared with those in Fig. 12.5. Relative biases for Fig. 12.6a–c are very similar to each other, and display very little difference due to sample size. However, a difference in relative bias is obvious in Fig. 12.6d, when non-informative gamma prior distributions are used for both λ_N and λ_G. In this case, relative biases are larger, and appear dependent on sample size, with the largest relative bias associated with the smallest sample size, which contains data collected on a single visit to 10 transects, and the smallest relative bias associated with data collected over 10 visits to 10 transects. This suggests that when α must be estimated with a weakly-informative uniform prior, best results are obtained when information is available for at least one of λ_G and λ_N.

In Fig. 12.7 we present results when observed vestige count $Y \sim \text{Poisson}(\alpha T \nu)$, and no prior information is available for vestige surplus, and so it must be specified as $\alpha \sim \text{Gamma}(0.01, 0.01)$. In this case we can see that there is large variability present in relative bias for abundance estimates,and that relative biases are much larger than the scenarios presented in Figs. 12.5 and 12.6. This is particularly evident in Fig. 12.7d, which corresponds to a situation in which no information is available for α, λ_N or λ_G, and all are estimated using non-informative gamma priors. In this case, relative biases of up to approximately 110 tell us that abundance estimates are unreliable when no information is available for α, λ_N or λ_G, and the triple Poisson

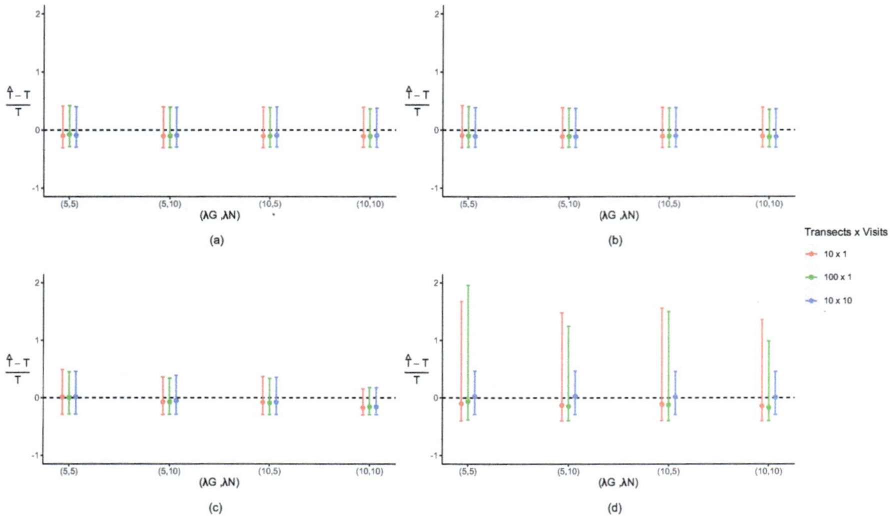

Fig. 12.6 Relative bias in mean estimate and 95% credible intervals for abundance T, for Poisson Y and uniform α, with true values for λ_N and $\lambda_G \in \{5, 10\}$, at varying sample sizes, and for different prior distributions on λ_N and λ_G. (**a**) $\lambda_N \sim$ Gamma(7, 1), $\lambda_G \sim$ Gamma(0.01, 0.01). (**b**) $\lambda_N \sim$ Gamma(0.01, 0.01), $\lambda_G \sim$ Gamma(7, 1). (**c**) $\lambda_N \sim$ Gamma(7, 1), $\lambda_G \sim$ Gamma(7, 1). (**d**) $\lambda_N \sim$ Gamma(0.01, 0.01), $\lambda_G \sim$ Gamma(0.01, 0.01)

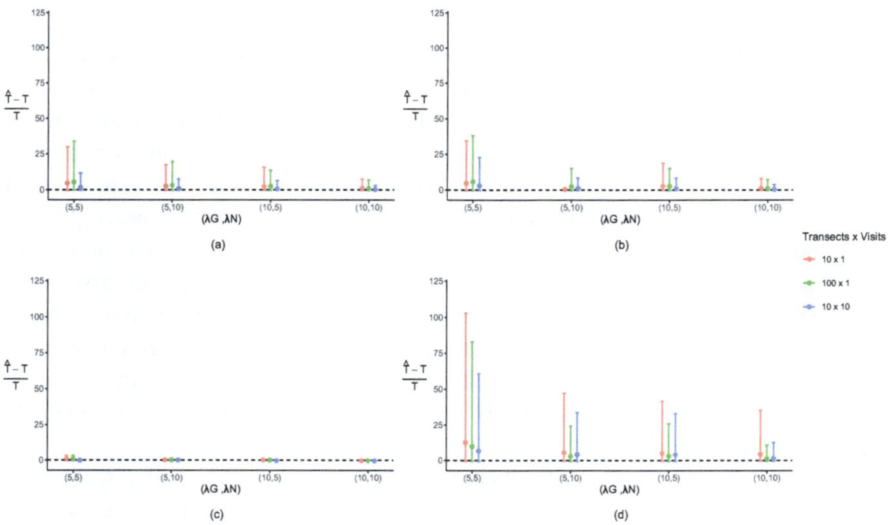

Fig. 12.7 Relative bias in mean estimate and 95% credible intervals for abundance T, for Poisson Y and α estimated using a non-informative gamma distribution, with true values for λ_N and $\lambda_G \in \{5, 10\}$, at varying sample sizes, and for different prior distributions on λ_N and λ_G. (**a**) $\lambda_N \sim$ Gamma(7, 1), $\lambda_G \sim$ Gamma(0.01, 0.01). (**b**) $\lambda_N \sim$ Gamma(0.01, 0.01), $\lambda_G \sim$ Gamma(7, 1). (**c**) $\lambda_N \sim$ Gamma(7, 1), $\lambda_G \sim$ Gamma(7, 1). (**d**) $\lambda_N \sim$ Gamma(0.01, 0.01), $\lambda_G \sim$ Gamma(0.01, 0.01)

model should not be used to produce abundance estimates in this case. Relative biases are also very high in Fig. 12.7a and b, when information is only available for either λ_N or λ_G, but not both. Figure 12.7c, which corresponds to a situation in which prior information is available for both λ_N and λ_G produces much smaller relative biases, and is the only scenario in which abundance estimates may be relied upon. From this we may conclude that if no information is available for α, the triple Poisson model may be used to estimate abundance, provided that information is available for λ_N and λ_G, and both can be specified by informative prior distributions.

In Fig. 12.8, we present results when $Y \sim$ Negative Binomial$(\alpha T \nu, \phi)$, overdispersion levels in vestige counts are small ($\phi = 2$), and vestige surplus α is known and correctly supplied to the model as data. Results in Fig. 12.8a–c are very similar to those observed in Fig. 12.5, when $Y \sim$ Poisson$(\alpha T \nu)$ and α is known. However, relative bias presented in Fig. 12.8d is much larger than relative bias presented in the corresponding Fig. 12.5d. This is most evident when data collected on a single visit to 10 transects means that sample size is small. This suggests that when overdispersion is present in vestige counts, even if levels of overdispersion are small, that larger sample sizes are required for reliable abundance estimation.

In Fig. 12.9, we present results obtained when $Y \sim$ Negative Binomial$(\alpha T \nu, \phi)$, overdispersion levels in vestige counts are large ($\phi = 0.2$), and vestige surplus α is known and correctly supplied to the model as data. Results are similar to those presented in Fig. 12.8, though the issues with relative bias encountered in

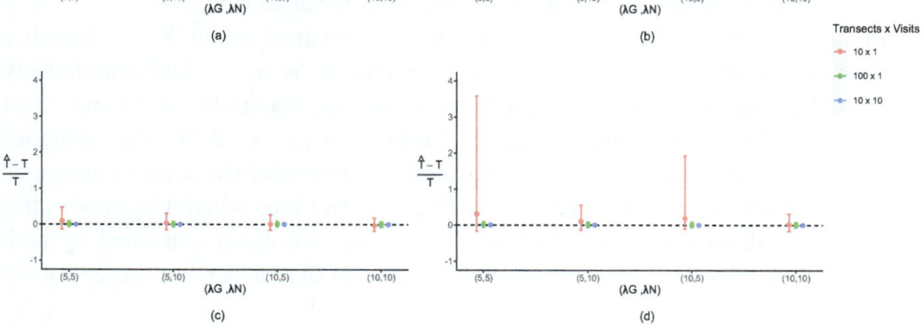

Fig. 12.8 Relative bias in mean estimate and 95% credible intervals for abundance T, for negative binomial Y with small levels of overdispersion and α known, with true values for λ_N and $\lambda_G \in \{5, 10\}$, at varying sample sizes, and for different prior distributions on λ_N and λ_G. (**a**) $\lambda_N \sim$ Gamma(7, 1), $\lambda_G \sim$ Gamma(0.01, 0.01). (**b**) $\lambda_N \sim$ Gamma(0.01, 0.01), $\lambda_G \sim$ Gamma(7, 1). (**c**) $\lambda_N \sim$ Gamma(7, 1), $\lambda_G \sim$ Gamma(7, 1). (**d**) $\lambda_N \sim$ Gamma(0.01, 0.01), $\lambda_G \sim$ Gamma(0.01, 0.01)

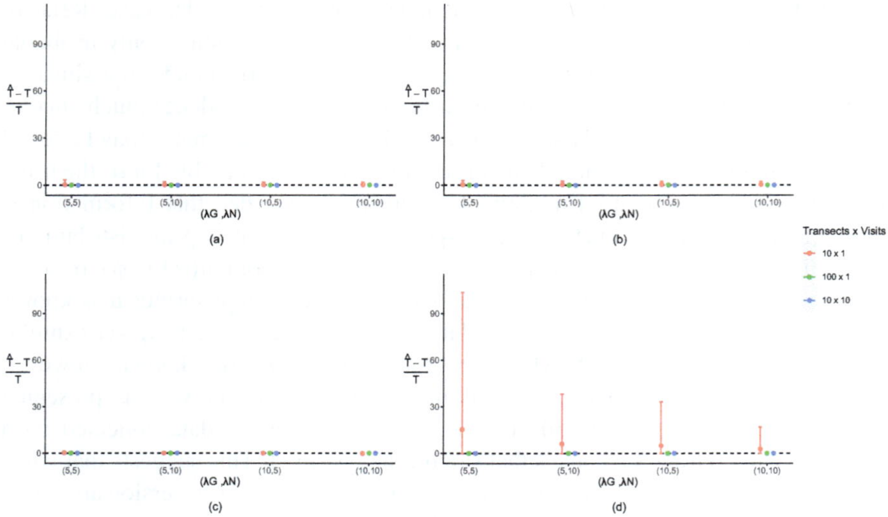

Fig. 12.9 Relative bias in mean estimate and 95% credible intervals for abundance T, for negative binomial Y with large levels of overdispersion and α known, with true values for λ_N and $\lambda_G \in \{5, 10\}$, at varying sample sizes, and for different prior distributions on λ_N and λ_G. (**a**) $\lambda_N \sim \text{Gamma}(7, 1)$, $\lambda_G \sim \text{Gamma}(0.01, 0.01)$. (**b**) $\lambda_N \sim \text{Gamma}(0.01, 0.01)$, $\lambda_G \sim \text{Gamma}(7, 1)$. (**c**) $\lambda_N \sim \text{Gamma}(7, 1)$, $\lambda_G \sim \text{Gamma}(7, 1)$. (**d**) $\lambda_N \sim \text{Gamma}(0.01, 0.01)$, $\lambda_G \sim \text{Gamma}(0.01, 0.01)$

Fig. 12.8d are worsened in Fig. 12.9 by the greater levels of overdispersion present in this data. The conclusions that we can draw from this simulation are similar to those from Fig. 12.8. When overdispersion is present in vestige counts Y, due to animal behaviour or non-constant vestige production, the ideal situation involves prior information available to inform prior distributions for λ_N and/or λ_G, and larger samples are preferable to obtain accurate estimates for abundance.

In Figs. 12.10 and 12.11, we present results obtained when $Y \sim$ Negative Binomial$(\alpha T \nu, \phi)$, and vestige surplus is estimated as $\alpha \sim \text{Uniform}(1, 100)$. Figure 12.10 contains small overdispersion in vestige counts ($\phi = 2$) and 12.11 contains vestige counts with greater overdispersion ($\phi = 0.2$). The scenarios presented here are similar to the previous negative binomial simulations presented in Figs. 12.8 and 12.9. The issue of very large relative bias when non-informative gamma prior distributions are used for λ_N and λ_G are again worsened by both the estimation of α and the increase in the degree of overdispersion present in the counts.

Finally, simulation studies were performed to assess estimates obtained using a triple negative binomial model, in which the observed vestige count Y, the number of groups G and the abundance T are all assumed to contain overdispersion. Data was simulated from three different models: a triple Poisson model (TP), a triple negative binomial model with small levels of overdispersion in Y, T and G (TNB-1) and finally a triple negative binomial model with large levels of overdispersion

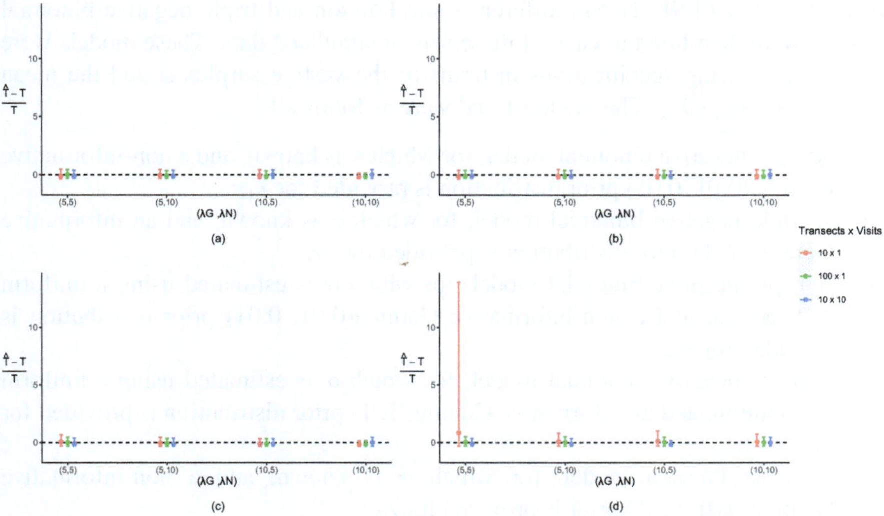

Fig. 12.10 Relative bias in mean estimate and 95% credible intervals for abundance T, for negative binomial Y with small levels of overdispersion and α estimated with a uniform distribution, with true values for λ_N and $\lambda_G \in \{5, 10\}$, at varying sample sizes, and for different prior distributions on λ_N and λ_G. (**a**) $\lambda_N \sim \text{Gamma}(7, 1)$, $\lambda_G \sim \text{Gamma}(0.01, 0.01)$. (**b**) $\lambda_N \sim \text{Gamma}(0.01, 0.01)$, $\lambda_G \sim \text{Gamma}(7, 1)$. (**c**) $\lambda_N \sim \text{Gamma}(7, 1)$, $\lambda_G \sim \text{Gamma}(7, 1)$. (**d**) $\lambda_N \sim \text{Gamma}(0.01, 0.01)$, $\lambda_G \sim \text{Gamma}(0.01, 0.01)$

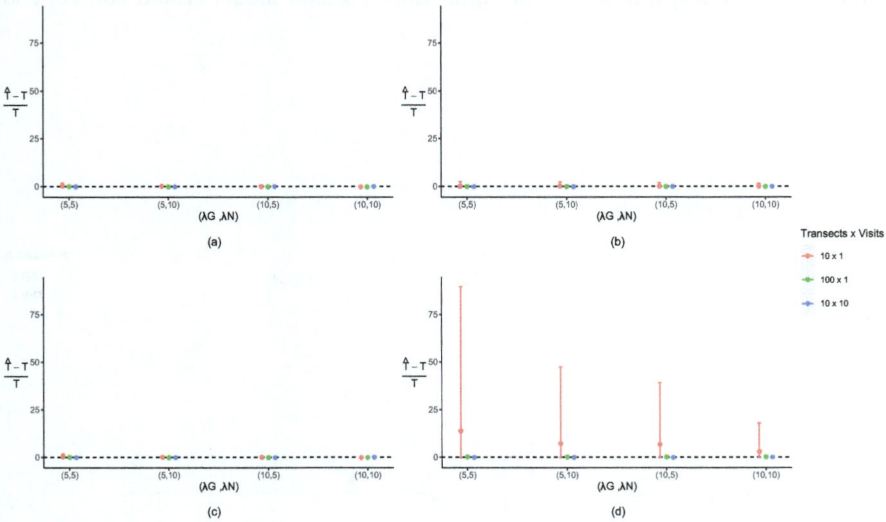

Fig. 12.11 Relative bias in mean estimate and 95% credible intervals for abundance T, for negative binomial Y with large levels of overdispersion and information available to inform a uniform α, with true values for λ_N and $\lambda_G \in \{5, 10\}$, at varying sample sizes, and for different prior distributions on λ_N and λ_G. (**a**) $\lambda_N \sim \text{Gamma}(7, 1)$, $\lambda_G \sim \text{Gamma}(0.01, 0.01)$. (**b**) $\lambda_N \sim \text{Gamma}(0.01, 0.01)$, $\lambda_G \sim \text{Gamma}(7, 1)$. (**c**) $\lambda_N \sim \text{Gamma}(7, 1)$, $\lambda_G \sim \text{Gamma}(7, 1)$. (**d**) $\lambda_N \sim \text{Gamma}(0.01, 0.01)$, $\lambda_G \sim \text{Gamma}(0.01, 0.01)$

in Y, T and G (TNB-2). Six different triple Poisson and triple negative binomial models were then fitted to each of these sets of simulated data. These models were fitted with varying specifications in terms of the vestige surplus α and the mean number of groups λ_G. The models fitted were as follows:

(A) A triple negative binomial model, for which α is known, and a non-informative Gamma(0.01, 0.01) prior distribution is provided for λ_G;
(B) A triple negative binomial model, for which α is known, and an informative Gamma(7, 1) prior distribution is provided for λ_G;
(C) A triple negative binomial model, for which α is estimated using a uniform distribution, and a non-informative Gamma(0.01, 0.01) prior distribution is provided for λ_G;
(D) A triple negative binomial model, for which α is estimated using a uniform distribution, and an informative Gamma(7, 1) prior distribution is provided for λ_G;
(E) A triple Poisson model, for which α is known, and a non-informative Gamma(0.01, 0.01) prior is provided for λ_G;
(F) A triple Poisson model, for which α is estimated using a uniform distribution, and a non-informative Gamma(0.01, 0.01) prior is provided for λ_G.

In Fig. 12.12, results for data simulated from a triple Poisson model (TP) are very similar to results for data simulated from a triple negative binomial model with small overdispersion (TNB-1). This indicates that if the true data had small amounts of overdispersion, that fitting a triple Poisson model would suffice, and

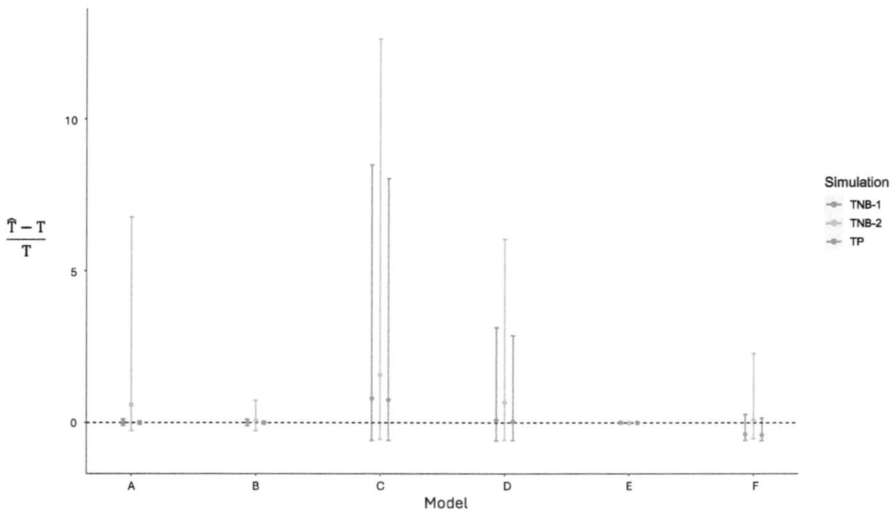

Fig. 12.12 Relative mean bias in abundance estimates for triple negative binomial models (A–D) and triple Poisson models (E–F), fitted to data simulated from triple negative binomial models (denoted by the green and red lines) and triple Poisson models (denoted by the blue line)

that if the true data were triple Poisson, that fitting a triple negative binomial model would provide similar results. As we have seen in simulation studies described above, relative bias in abundance estimates is larger when α is estimated (models C and D) than when it is known and provided as data. When α is estimated using a Uniform distribution, an informative gamma prior on λ_G (model D) produces less uncertainty in abundance estimates than a non-informative gamma prior on λ_G (model C). For both of these simulations, when α is known and provided as data, relative bias in abundance estimates is very small regardless of the model fitted or the prior distribution provided to λ_G.

When data is simulated from a triple negative binomial model with higher levels of overdispersion (TNB-2), relative bias in abundance estimates are larger in almost all scenarios. When α must be estimated (models C, D and F), relative bias in abundance estimates are considerably higher for this data. When α is known and an informative gamma prior is supplied for λ_G (model B), relative bias in abundance estimates are similar to those associated with the triple Poisson and triple negative binomial with small overdispersion simulations. However, when α is known and a non-informative prior is supplied for λ_G (model A), relative bias in abundance estimates increases again. From this we can conclude that when overdispersion levels in our data are high, that an informative prior for λ_G and prior information on α are required to reduce uncertainty in abundance estimates.

References

Baker PJ, Harris S (2004) Red Foxes. The behavioral ecology of red foxes in urban Bristol. In: Macdonald DW, Sillero-Zubiri C (eds) Biology and conservation of wild canids. Oxford University Press, Oxford

Barnes RFW (2001) How reliable are dung counts for estimating elephant numbers? Afr J Ecol 39(1):1–9

Becker EF (1991) A terrestrial furbearer estimator based on probability sampling. J Wildl Manag 55:730–737

Becker EF, Spindler MA, Osborne TO (1998) A population estimator based on network sampling of tracks in the snow. J Wildl Manag 62:968–977

Buckland ST, Anderson DR, Burnham KP, Laake JL (1993) Distance sampling: estimating abundance of biological populations. Springer Netherlands

Caughley G (1994) Directions in conservation biology. J Anim Ecol 63:215–244

Cavallini P (1994) Faeces count as an index of fox abundance. Acta Theriol 39(4):417–424

Chelintsev NG (1995) Mathematical principles of winter censuses of mammals. Biull Mosk obshch ispyt prir 100:3–19

Gallant D, Vasseur L, Bérubé CH (2007) Unveiling the limitations of scat surveys to monitor social species: a case study on river otters. J Wildl Manag 71(1):258–265

Kéry M, Royle JA (2015) Applied hierarchical modeling in ecology: analysis of distribution, abundance and species richness in R and BUGS. Elsevier, Amsterdam

Kruszynski C, Moral RA, Diniz-Reis TR, Piña CI, Verdade LM (2026) Estimating social species number of groups and population density based on vestiges in human-altered environments: the case study of collared-peccaries (pecari tajacu). In: Verdade LM, Moral RA (eds) Wildlife management: integrating conservation and innovation in human-altered landscapes. Springer, Cham

Marques FFC, Buckland ST, Goffin D, Dixon CE, Borchers DL, Mayle BA, Peace AJ (2001) Estimating deer abundance from line transect surveys of dung: sika deer in southern Scotland. J Appl Ecol 38:349–363
Marshall L (2020) DSsim: distance sampling simulations. R package version 1.1.5
Miller DL, Rexstad E, Thomas L, Marshall L, Laake JL (2019) Distance sampling in r. J Stat Softw 89(1):1–28
Murray D, Ellsworth E, Zack A (2005) Assessment of potential bias with snowshoe hare fecal pellet-plot counts. J Wildl Manag 69(1):385–395
Nichols JD (2014) The role of abundance estimates in conservation decision-making. In: Applied ecology and human dimensions in biological conservation. Springer, Berlin, pp 117–131
Patterson BR, Quinn NWS, Becker EF, Meier DB (2004) Estimating wolf densities in forested areas using network sampling of tracks in snow. Wildl Soc Bull 32(3):938–947
Plummer M et al (2003) Jags: a program for analysis of Bayesian graphical models using Gibbs sampling. In: Proceedings of the 3rd international workshop on distributed statistical computing, Vienna, vol 124, pp 1–10
R Core Team (2022) R: a language and environment for statistical computing. R Foundation for Statistical Computing, Vienna
Ratcliffe P (1987) Distribution and current status of Sika deer, Cervus nippon, in Great Britain. Mamm Rev 17(1):39–58
Rexstad E (2022) Multipliers and indirect surveys: dung surveys including estimates of production and decay rates. https://examples.distancesampling.org/Distance-mult/multipliers-distill.html. Accessed 05-02-2023
Sowls LK (1997) Javelinas and other peccaries: their biology, management and use, 2nd edn. Texas A&M University Press, College Station
Stephens PA, Zaumyslova OY, Miquelle DG, Myslenkov AI, Hayward GD (2006) Estimating population density from indirect sign: track counts and the Formozov–Malyshev–Pereleshin formula. Anim Conserv 9(3):339–348
Su Y-S, Yajima M (2020) R2jags: using R to run 'JAGS'. R package version 0.6-1
Verdade LM, Moreira JR, Ferraz KMPMB (2013) Counting Capybaras. In: Capybara. Springer, Berlin, pp 357–370
Verdade LM, Lyra-Jorge MC, Pina CI (2014) Applied ecology and human dimensions in biological conservation. Springer, Berlin
Webbon CC, Baker PJ, Harris S (2004) Faecal density counts for monitoring changes in red fox numbers in rural Britain. J Appl Ecol 41(4):768–779

Chapter 13
The Influence of Fragmented Landscapes on Population Density Studies Using Camera Trapping

Vittoria Façanha Bassani, Giulia Canali Forace, David Fernando Luz Rodrigues, Maria Fernanda Rolim Lourenço, Thiago Philipe de Camargo e Timo, Alexandra Sanches, Niamh Mimnagh, Rafael de Andrade Moral, and Iuri Emmanuel de Paula Ferreira

Abstract An intricate interplay between fauna and flora is required to maintain and enhance the essential ecological attributes such as structure, biodiversity, and functionality. However, nowadays, the availability of natural resources to sustain the fauna has dropped considerably due to the extent of land use and occupation by human activities, which are often associated with habitat degradation. Furthermore, many invasive species have become major threats to native fauna, mainly due to competition for resources and disease transmission. Understanding trends and fluctuations in populations of wild and invasive species is essential for developing biodiversity conservation policies. Most monitoring programs rely on counts obtained by camera traps, but estimating density when individuals can't be individually identified is challenging. Methods are often based on fragile assumptions such as spatial homogeneity, representative sampling across habitats, and independent movement patterns for the animals—assumptions that don't hold

V. F. Bassani · G. C. Forace · D. F. L. Rodrigues · M. F. R. Lourenço · T. P. de Camargo e Timo
Research Group on Statistical Modeling in Environmental Sciences (CeMECA), Center for Natural Sciences (CCN), Lagoa do Sino Campus, Federal University of São Carlos (UFSCar), Buri, SP, Brazil

A. Sanches · I. E. de Paula Ferreira
Research Group on Statistical Modeling in Environmental Sciences (CeMECA), Center for Natural Sciences (CCN), Lagoa do Sino Campus, Federal University of São Carlos (UFSCar), Buri, SP, Brazil

Graduate Program in Wildlife Conservation (PPGCFau), Federal University of São Carlos (UFSCar), Buri, SP, Brazil
e-mail: iuri@ufscar.br

N. Mimnagh
Department of Mathematics and Statistics, Maynooth University, Maynooth, Ireland

R. A. Moral (✉)
Department of Mathematics and Statistics, Maynooth University, Maynooth, Co. Kildare, Ireland
e-mail: Rafael.DeAndradeMoral@mu.ie

L. Martins Verdade, R. A. Moral (eds.), *Wildlife Monitoring*,
https://doi.org/10.1007/978-3-032-05821-8_13

for fragmented landscapes. In this study, an agent-based simulation platform was developed in R, where users can import realistic landscapes, set up habitat weights for landscape classes, simulate trajectories across the landscape, and collect data from virtual camera traps. The program exports detection tables, density estimates from the Random Encounter Model (REM), and descriptive statistics for the simulated paths. As a case study, we simulated the movement dynamics of the invasive species *Sus scrofa* (wild boar) in the Southwest of São Paulo, Brazil, one of the worst invasive species in the world, associated with habitat destruction, strong competitive pressure on native pigs, crop destruction, and the transmission of several diseases. The study aimed to evaluate the performance of REM estimates on a heterogeneous fragmented landscape.

Keywords Wildlife monitoring · Agent-based simulation · Wild boar · Random encounter models · Invasive species

1 Introduction

Fauna and flora are part of the world's ecological balance and one depends on the other. Today, the natural resources available to fauna are increasingly scarce and restricted, for reasons such as habitat fragmentation (Løvschal et al. 2017) and resource scarcity resulting from human actions (Ogutu et al. 2016). Conservation of these habitats and animals is essential for the balance of the biosphere, which is maintained through feedback mechanisms (Lovelock and Margulis 1974).

Wildlife monitoring programs are essential for environmental and biodiversity conservation, since the abundance of the target species indicates trophic imbalances resulting from environmental changes. Monitoring and obtaining population estimates of invasive species are key actions for biodiversity conservation plans, as they are among the main causes of species extinction, directly affecting native species and causing changes in ecosystems through disturbances (Clavero Pineda and García-Berthou 2005; Doherty et al. 2016). Wildlife monitoring usually focuses on small populations, with evasive animals that are difficult to identify visually, especially when the landscape presents obstacles and refuges. Data can come from captures, sightings, photographs, or traces of animal presence (Jourdain et al. 2020).

Camera traps are photographic devices placed in concealed, sealed, and waterproof boxes. They capture images and videos when activated by a detection system (Srbek-Araujo and Chiarello 2007). This equipment is used primarily for the observation of medium and large mammals, including nocturnal species, providing satisfactory results in the monitoring of fauna (Tobler and Powell 2013; Potter et al. 2019).

Quantifying animal abundance or density depends on counts of occurrences, and records may be of marked or unmarked animals. Classical methods for estimating abundance are based, for example, on counting marked animals, where the identification of individuals provides a high degree of reliability to the estimates; this is the case of capture-recapture methods, for example Tourani (2022). However, there are target species that do not have characteristics that allow individual identification, such as the rosette patterns of jaguars and leopards, nor is it possible to collect the animals for marking, due to their size, evasiveness, or aggressive behavior.

In studies with camera trap monitoring, one of the simplest ways to estimate the density of unmarked populations is to apply the Random Encounter Model (REM) (Rowcliffe et al. 2008). In this approach, population density is estimated based on the expected encounter rate between animals and traps, assuming random and independent movement of the animals, as if they were particles of an ideal gas (Hutchinson and Waser 2007). REM estimation considers the collective viewshed approach, where the result is obtained by aggregating different detection zones (camera traps, for instance). Despite its simplicity, REM presents many strong assumptions, such as the random positioning of cameras with respect to habitat use, closed population, independently taken photographs, and the random movement of animals. For gregarious species, the number of groups is counted, and the abundance is corrected by the average group size (Gilbert et al. 2021). In addition, information is required regarding the average movement speed of the animals and the geometry of the detection zones (Rowcliffe et al. 2008).

Gilbert et al. (2021) alerted to the possibility of bias in REM estimates arising from the non-random allocation of cameras or even from the animal movement pattern being influenced by landscape characteristics. To better understand this possibility, they suggested simulation studies in a virtual environment considering heterogeneous landscapes, movement patterns, and computational representations of camera traps and data collection processes.

Following the suggestion of Gilbert et al. (2021), this study introduces an agent-based platform to simulate camera trap studies. Agent-based models are characterized by hypothetical entities (the agents) that are free to move in a simulated space and interact with each other and with the environment based on well-established rules (DeAngelis and Diaz 2019). However, despite the simplicity of these models, chaotic behavior and complex configurations can arise as emergent properties of these systems (Wolfram 2002). The agent-based platform was written in R (R Core Team 2021) and was inspired by the monitoring of the European wild boar (*S. scrofa*) in Southeast Brazil.

The European wild boar (*S. scrofa*) is listed among the 100 worst invasive species in the world. Its presence causes significant damage, such as crop and habitat destruction, disturbance of nesting sites and plant assemblages, competition with native fauna, disease dissemination, and even a risk to human life (Sordi 2015; Risch et al. 2021; McDonough et al. 2022). Depending on its behavior, *S. scrofa* can act as predators, agricultural pests, and even seed dispersers (Genov 1981; Geisser and Reyer 2004; Bueno et al. 2011; O'Connor and Kelly 2012). In Brazil, this animal has a long and ancient history of invasion, mostly in southern regions (Hegel et al. 2022; La Sala et al. 2023). They are opportunistic omnivores with a great ability to adapt their feeding (Genov 1981; Rosell et al. 2001; Baubet et al. 2004; Irizar et al. 2004; Schley and Roper 2003; Massei and Genov 2004; Herrero et al. 2005; Hafeez et al. 2011). When it comes to habitat preference, the species chooses open environments that are predominantly herbaceous habitats interspersed with forest patches (Lewis et al. 2017; Cordeiro et al. 2018). These animals need surroundings near crop areas where they can find food as well as places to shelter against intense heatwaves and find protection from predators (Choquenot and Ruscoe 2003).

Our simulation studies were carried out considering a real landscape as a reference, centered in the Brazilian Southeast and with an area of 2661 km^2, in an ecotone of the Atlantic Forest and Cerrado biomes with a high level of fragmentation resulting from human actions, such as livestock farming and grain production. The landscape used in the simulations was imported from the *Bioma Stats* program (de Paula Ferreira et al. 2023), a computational program that processes historical land use and occupation data, providing customized visualizations and statistics on landscape classes.

The simulations aimed to answer three main questions: (i) Are REM estimates reliable in heterogeneous landscapes? (ii) What is the appropriate exclusion window to avoid problems with dependent samples (i.e. photographs)? (iii) What is the influence of camera positioning on REM estimates?

2 Study-Area Characterization

The study area was a square-shaped region in southwest São Paulo, Brazil, covering 2661 km^2, centered at the farm 'Lagoa do Sino" (23°35'57.3"S, 48°31'46.6"W), where a campus of the Federal University of São Carlos (UFSCar) is located (Fig. 13.1). The study area is an ecotone between two of the most human-impacted Brazilian biomes, the Atlantic Forest and Cerrado, and the region is characterized by several human activities, especially grain farming, forest plantation and extensive livestock farming. Similar to many regions in Brazil, the area is negatively affected by uncontrolled wild boar populations.

In our approach, land use and land cover maps were used as the background for the simulations. Landscape data were obtained from MapBiomas Project (MapBiomas Project 2023) and analyzed using the *Bioma Stats* R package (de Paula Ferreira et al. 2023). The *Bioma Stats* program offers an end-to-end solution for studying Brazilian landscapes, providing functions to define study area boundaries, download and store geospatial data, summarize and visualize landscape data and their historical series, reclassify maps and calculate landscape metrics, and export results in ready-to-analyze formats.

In 2020, the study area was predominantly occupied by soybean crops (22.1% of the total area), followed by Atlantic forest (20.8% forest formation; 17.1% forest plantation), mosaic of uses (14.9%) and pastures (14.3%). The forest formation class, the main habitat for *S. scrofa*, contains 4468 fragments in the study area, with an average area of 12.9 ha (0.13 km^2), an edge density of 43.9 m/ha (edge length per hectare), and an aggregation index of 85.0%. If we also consider forest

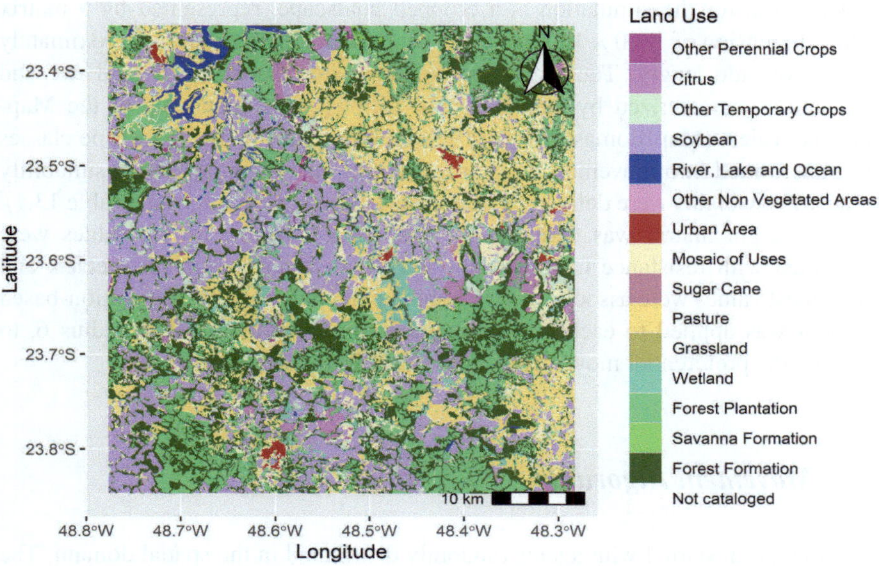

Fig. 13.1 Study area situated in Southwest São Paulo, Brazil, in the year 2020. Land use and land cover map generated by the *Bioma Stats* program (de Paula Ferreira et al. 2023). Data: MapBiomas Project (MapBiomas Project 2023)

planting as additional habitats for pigs, then these values rise to 35.0 ha for the average area and 93.0% for the aggregation index, while the number of fragments and the edge density drop to 2993 and 36.9 m/ha, respectively. In both cases, the average fragment area is inferior to the species' home range needs (Martins et al. 2019). There are many suitable habitats depicted in the southern or northern regions of the study area, but the central portion lacks forest fragments and, potentially, causes habitat disconnection for populations in the edges of the landscape induced by soybean plantations.

3 Agent-Based Platform

3.1 Landscape and Habitat Weights

Our approach consists of simulating wild boar trajectories in a realistic landscape and then collecting detection data from virtual camera traps that can be assigned at any point in space. Simulations are agent-based, and agents represent groups of wild boars determined by their centroids.

The basis for the simulation is a cropped landscape, represented by a matrix with a dimension of 1800×1800 pixels, where each pixel represents approximately 28.7 m of side length. The landscape contains several classes of land use and occupation characterized by annual cycles, following the standards of the MapBiomas Project (MapBiomas Project 2023). In our approach, the landscape classes are transformed into movement resistance weights and potential habitat suitability values. The weights were defined by biologists from our research team (Table 13.1).

The weight matrix was obtained in such a way that the lowest values were associated with resistance to movement and the permanence of the species, and the highest values were associated with habitat potential. Then, a convolution-based gradient was applied to each pixel, using a Moore neighborhood of radius 6, to estimate the preferential movement vector (direction vector, $\mathbf{v}$).

3.2 Movement Algorithm

The simulation started with agents randomly distributed in the spatial domain. The time horizon was discrete as $\Delta t = 1\,\text{min}$, totaling $4 \times 7 \times 12 \times 60 = 20,160$ iterations (four weeks, considering 12 h of movement per day). The first week constituted a burn-in period, discarded from the analysis to eliminate the effects of initial conditions outside the habitat area. In total, 40,000 independent wild boar groups were simulated from the same landscape.

The empirical mean distance of $15\,\text{km}\,\text{day}^{-1}$ was scaled to the raster coordinate system, resulting in a velocity of $0.73\,\text{px}\,\text{min}^{-1}$. The trajectory of each agent was modeled as a biased random walk, whose directionality is given by the vector field derived from the matrix of habitat weights (Algorithm 1). At each step, the movement angle is determined by

$$\theta_t = \theta_{det} + \epsilon_t, \tag{13.1}$$

where θ_{det} is the deterministic component aligned to the field vector and ϵ_t is the stochastic disturbance $\epsilon_t \sim N(0, \sigma_\theta^2)$, with

$$\sigma_\theta = \pi\big(1 - \|\mathbf{v}\|\big), \qquad 0 < \|\mathbf{v}\| < 1 \tag{13.2}$$

where $||\mathbf{v}||$ is the Euclidean norm of the direction vector. Thus, when $||\mathbf{v}|| \approx 0$, the direction is purely random; as $||\mathbf{v}||$ approaches 1, the motion becomes entirely guided by the vector field. Thus, a spatial stochastic process is obtained that combines exploratory behavior with attraction to areas of greater suitability.

Table 13.1 Habitat classes, suitability weights and brief ecological rationale used in the simulation study

Main class	Subclass	Weight	Brief rationale
Not Catalogued	Not Catalogued	−8	Area lacking clear information; assumed low suitability due to uncertainty and possible scarcity of resources.
Forest	Forest Formation	10	Ideal habitat: shelter, food variety, and water availability.
Forest	Savanna Formation	8	Good resource availability, but vegetation cover is less dense than in closed forests.
Forest	Forest Plantation	7	Monospecific plantations provide cover and some food despite structural homogeneity.
Non-Forest Natural Formation	Wetland	5	Wet areas supply water and mud, but generally offer less vegetative cover.
Non-Forest Natural Formation	Grassland	4	Very open areas; used occasionally for movement or opportunistic feeding.
Farming	Pasture	−3	Little cover; exploited sporadically, not preferred habitat.
Farming	Sugarcane	4	Provides food (stems, residues) and some cover, yet is an intensive monoculture.
Farming	Agriculture–Pasture Mosaic	2	Mixed crops and open patches; attracts by partial food availability.
Non-Vegetated Area	Urban Area	−10	High disturbance and human presence; virtually unsuitable as habitat.
Non-Vegetated Area	Other Non-Vegetated Areas	−7	Typically bare soil, mining sites, etc.; few resources or shelter.
Water Body	River, Lake, Ocean	3	Important for drinking and wallowing; used more sporadically.
Farming	Soybean	2	Seasonal food source; limited cover outside growing season.
Farming	Other Temporary Crops	4	Includes maize or sorghum; high food supply and some cover while standing.
Farming	Citrus	2	Fruit availability, but orchard structure offers limited cover.
Farming	Other Perennial Crops	4	May offer fruit and some shelter depending on crop type and management.

Algorithm 1 Stochastic biased random walk of wild boar groups

Require: $R = 40{,}000$ ▷ Number of independent simulations
Require: $T = 4 \times 7 \times 12 \times 60$ ▷ Total time steps (minutes)
Require: $T_{\text{burn}} = 1 \times 7 \times 12 \times 60$ ▷ Burn-in (adaptation) period
Require: $v = 0.73$ ▷ Mean velocity (px/min)
Require: $\alpha = 5$ ▷ Pareto shape parameter
Require: $raster_\theta, raster_{\text{norm}}$ ▷ Movement vector field (direction and norm)
Require: L ▷ Landscape side length (pixels)
Ensure: Collection of trajectories (post burn-in)
for $r = 1$ **to** R **do**
  Initialize agent at a random location $(x, y) \in [0, L] \times [0, L]$
  traj $\leftarrow \emptyset$
  for $t = 1$ **to** T **do**
    $(i, j) \leftarrow t(\lfloor x, y \rfloor)$ ▷ Transform coordinates to indexes
    $\mu \leftarrow raster_\theta[i, j]$ ▷ Get mov. direction
    $\nu \leftarrow raster_{\text{norm}}[i, j]$ ▷ Get mov. intensity
    $\sigma_\theta \leftarrow \pi\,(1 - \nu)$ ▷ Angular noise
    $\theta \leftarrow \theta_{det} + \epsilon_t$
    $\Delta\ell \leftarrow \text{Pareto}(\alpha,\ x_m = (\alpha - 1)\,v/\alpha)$
    $x' \leftarrow x + \Delta\ell\,\cos\theta, \quad y' \leftarrow y + \Delta\ell\,\sin\theta$
    if (x', y') outside $[0, L] \times [0, L]$ **then**
      Reflect (x', y') back into $[0, L]$ boundaries
    $(x, y) \leftarrow (x', y')$
    if $t > T_{\text{burn}}$ **then**
      Append $\left(t - T_{\text{burn}}, x, y\right)$ to *traj*
  Store *traj* in *results*$[r]$
return *results*

In this model, the step length $\Delta\ell$ follows a Pareto distribution with shape parameter $\alpha = 5$ and scale parameter

$$x_m = \frac{(\alpha - 1)\,\bar{\ell}}{\alpha} = \frac{4\,\bar{\ell}}{5}, \tag{13.3}$$

where $\bar{\ell}$ is the average step length

$$\bar{\ell} \approx 0.726 \text{ pixels/min}. \tag{13.4}$$

The probability density function is then

$$f_{\Delta\ell}(x) = \frac{\alpha\, x_m^\alpha}{x^{\alpha+1}}, \quad x \geq x_m. \tag{13.5}$$

Such a heavy-tailed distribution allows for occasional long steps while maintaining an average step length of $\bar{\ell}$.

The agents are free to move within the landscape, a bi-dimensional square-shaped region with reflexive boundaries; as an agent approaches the boundary, a repulsive force is applied to maintain it within the simulation domain. The side of the square landscape is approximately 52 km long and its size was chosen based on the probability that the simulated trajectories will be affected by the boundary effect (BE). This probability is given by:

$$P(\text{BE}) = \frac{L^2 - (L - hr)^2}{L^2}, \quad L > 0 \text{ and } hr < L, \tag{1}$$

Where L is the side length of the landscape and hr is the estimated home range of the species, both in the same unit of length. Consider the agents moving in coordinates $(x, y) \in [0, L] \times [0, L]$ and that $hr/2$ is the typical limit for the radius of movement. So, we can define that the agents moving within the coordinates $hr/2$ and $L - hr/2$ are free from the boundary effect. Thus, Eq. (1) is a simple area relation where the numerator is the total area minus the edge-free area (i.e., the core square-sized $L - hr$), and the denominator is the total area. Many studies indicate a home range (diameter) that varies between 2.5 and 3.0 km for wild boars, resulting in $P(BE)$ approximately 0.1 in our simulations (or 10% of the simulated trajectories).

3.3 Camera Traps and the Sampling Process

The system allows the user to allocate virtual camera traps for simulated studies of animal density estimation. The traps can be centered at any position in the landscape, and the user is free to define their attributes, such as positioning angle, maximum detection distance, field-of-view angle, detection probability curve, and duration of the exclusion window (to avoid repeated/depended records). In this study, all camera traps were settled considering a random position angle, with 30 m of maximum detection distance and 90° of field of view.

The platform also has a function to integrate data from selected trajectories and virtual camera traps, which accounts for intersections and provides simulated results representing detections as output. The image sampling frequency was set to one frame per time step, that is, one frame per minute.

In our platform, the user can choose the output format, with identified and unidentified photographs, perfect and imperfect detection, recorded or unrecorded distances. In this study, we considered unidentified perfect detections, without distance records.

To evaluate the influence of landscape characteristics on REM estimation, we deployed virtual cameras at each node of a 17 × 17 square grid of monitoring stations. Each station was positioned at least 2.8 km from its nearest neighbor and from the boundary of the grid. The simulations considered 40,000 independent groups (i.e. the agents) and their individual trajectories, as the simulated landscape density was 15.03 agents km^{-2}.

Note that the trajectories can be resampled to simulate scenarios with a lower population density. In fact, stations can also be rearranged to represent different sample designs, varying the representativeness of the landscape classes (forest, pasture, etc.) and the total sample effort.

3.4 Density Estimation via Random Encounter Models

In this study, we used random encounter models (REM) to estimate animal densities. This approach provides a simple yet effective solution for calculating densities when individual recognition is not possible and has become widely used with camera-trap data. REM was originally proposed to model the rate of encounter between molecules of an ideal gas, based on the density and velocity of the particles and the collision cross-section among them. Hutchinson and Waser (2007) adapted the REM to model the encounter between animals and passive monitoring devices. Rowcliffe et al. (2008) then established the theoretical basis for its application in camera trap studies. Rowcliffe et al. showed how to estimate the animal density from the expected rate of encounters between animals and traps, and corrected the result by considering the conical geometry of the detection zone. According to Rowcliffe et al. (2008), the density is given by:

$$\hat{D}_{REM} = \frac{y}{t} \cdot \frac{\pi}{v \cdot r \cdot (2 + \theta)} \tag{13.6}$$

where y is the number of independent records; t is the sampling effort (total monitoring time); v is the average velocity; r is the camera detection radius; and θ is the camera angle of view ($rad.$). For valid results, the units for length and time must be compatible.

Despite its simplicity, there are many assumptions for the REM application: (i) the encounters must be independent and, thus, multiple occurrences of the same event must be filtered by an exclusion window; (ii) the animal's movement must be independent to the cameras; (iii) the geometric parameters of the field of view must be known (or measured) previously; (iv) the average velocity of the animals must be known a priori or properly estimated; (v) density must be constant during the monitoring time, which implies population closure.

3.5 Analysis of Simulated Trajectories

Our platform provides statistical tools for analyzing the simulated trajectories. These data can be filtered by time periods (e.g., weeks), and the program then outputs key movement features such as the centroid of movement, the radius of activity, and the home range based on the life area.

Our platform uses the *sf* package for spatial data manipulation (Pebesma 2018). Based on the recorded locations, Minimum Convex Polygons (MCPs) are constructed to represent the area occupied by each individual. In our approach, the diameters of the MCPs are considered measurements of home range (in km), while the MCP areas reflect the life-area (in km^2), i.e. the spatial extent used by an agent during the period of camera-trap sampling. In addition, the number of MCPs overlapping the monitoring stations is considered the ground truth for the local abundance.

We also derived a ground truth for the local population density from the trajectories data. This quantity is obtained from the perspective of point processes theory:

(i) A representative sample of animal positions is retrieved from the simulated trajectories;
(ii) For each station, we draw a circular area covering $1\ \text{km}^2$;
(iii) Then, we compute the points within this area and their relative frequency;
(iv) Lastly, we multiply the population size by the relative frequency of each station, deriving its local density.

3.6 *Evaluating the REM Estimators*

In different scenarios, the REM estimates were compared with the ground truth density. For example, for studies considering only one camera trap, the ground truth was the local density, as detailed in the previous section. In studies considering multi-camera arrangements, the ground truth was the averaged local density at the selected monitoring stations. And in global landscape studies, considering the aggregated results from the 289 stations, the ground truth density was the ratio between the number of agents and the landscape area. To assess accuracy and precision, we use the following statistical indicators.

Bias

$$\text{Bias} = \hat{D}_{\text{REM}} - D_{\text{true}} \tag{13.7}$$

where $\hat{D}_{\text{REM}}$ is the density estimated by REM and D_{true} is the known reference density.

Root Mean Squared Error (RMSE)

$$\text{RMSE} = \sqrt{\frac{1}{n}\sum_{i=1}^{n}\left(\hat{D}_i - D_{\text{true}}\right)^2} \tag{13.8}$$

where n is the number of simulations, $\hat{D}_i$ is the i-th REM estimate, and D_{true} is the true density.

Mean Absolute Percentage Error (MAPE)

$$\text{MAPE} = \frac{100}{n} \sum_{i=1}^{n} \left| \frac{\hat{D}_i - D_{\text{true}}}{D_{\text{true}}} \right| \tag{13.9}$$

Lin's Concordance Correlation Coefficient (CCC)
Lin's CCC evaluates both precision and accuracy by combining Pearson's correlation with a bias correction term:

$$\rho_c = \frac{2\,\sigma_{xy}}{\sigma_x^2 + \sigma_y^2 + (\mu_x - \mu_y)^2} \tag{13.10}$$

where μ_x and μ_y are the means of $\{\hat{D}_i\}$ and $\{D_{\text{true}}\}$; σ_x^2 and σ_y^2 are their variances; σ_{xy} is the covariance between the REM estimates and the true values.

3.7 Bouncing-Particle Experiment: Evaluation of the Exclusion Window Duration

This study was designed to test the performance of REM considering an ideal situation, where the assumptions of random and independent movement, homogeneous space, and closed population are satisfied. The study considered a closed box with a side equal to 400 m, with particles inside, moving randomly, and colliding with the box's walls elastically. The movement of the particles was modeled with a Random Walk, with discrete steps drawn from the Pareto distribution (Algorithm 1). The particles were simulated asynchronously, one at a time, during a total of 201,160 iterations (min), corresponding to 30 consecutive weeks of monitoring. The particles' speed was 0.347 m s^{-1}, equivalent to 15 km traveled per day during 12 h of movement.

A virtual camera with a maximum detection distance of 30 m and a field of view of 90° was installed in the center of the box. In each simulation round, the number of times the particle was found in the detection area was recorded. The simulations considered five different values for the exclusion windows (EW: 0, 5, 15, 30, and 60 min) associated with three different values of population density (3.75, 11.25, and 20.00 agents km^{-2}). Then, the values of the REM estimates obtained from the counts were compared with the real density, calculated from the number of simulated particles and the total area occupied by the box.

4 Results and Discussion

4.1 On the Simulation of Bouncing Particles in a Box and the Exclusion Window Effect

In the REM application, some authors suggest the use of an exclusion window to avoid duplicate counts, which results from animals staying in the detection area. The particle study in a closed box was conducted to evaluate the duration effect of the exclusion window in the density estimation using REM, with an initial objective to find appropriate values from the exclusion window to deal with counting individuals, taking into account the movement speed and three hypothetic levels of population density (low, medium and high). The simulations of the particle box considered six situations, REM estimation with and without the exclusion window, at the three distinct densities, with exclusion window durations varying up to 1 h.

Without the exclusion window, it is possible that the same animal crosses the detection zone and is captured many times during a short period of time. With the exclusion window, the animal is allowed to move away and return from the camera during the latency period, avoiding duplicate counts, although it does not stop the same animal from being detected more than once in distinct moments, such as on different days or weeks. Regardless of whether or not the exclusion window is applied, repeated counts can intensify when the animal's range of action is close to the camera's positioning.

Rowcliffe et al. (2008) emphasize the independence of records as a fundamental assumption of the method. That does not mean that the same animal cannot be recorded in distinct moments, but the problem lies in taking many pictures of one encounter and considering them as independent detections. For the purpose of avoiding pseudo-replications, Rowcliffe et al. recommended the use of exclusion window information that reflect the behavior of the target species and the remaining time in front of the cameras. For example, Cusack et al. (2015) applied the exclusion window in density estimations of lions in the Serengeti National Park after observing some groups' resting habits in detection cameras' areas. Cusack et al. (2015) debate the importance of considering social and territorial behaviors in determining the minimal time of detection, observing that shorter times can overestimate the encounter rates, while long periods of time can dispose of potentially valid information and underestimate the encounter rates. Rovero and Zimmermann (2016) also warn about the need to filter photographs of the same animal in a sequence of movements within the detection area, which would bias the estimates of the encounter rate. However, Sollmann (2018) reports that the use (or lack) of the exclusion window depends on the target species, with a recommendation of windows of 30 min to 1 h for animals that remain in front of the cameras for long periods, while for more mobile species, shorter intervals (5 or 1 min) may be sufficient.

The movement algorithm used in this work does not incorporate important behaviors such as resting or patrolling, social interactions, and other patterns that could alter the time spent in front of cameras. In the simulations, the agents are constantly moving, randomly or following the movement vectors, with the possibility of returning to the detection zone after the first encounter. Furthermore, agents move in the simulations using discrete steps drawn from the Pareto distribution, in which small steps are more frequent than larger ones. The steps are simulated with an average size of 20.82 m and a α parameter equal to 5.0, so that 94.77% of the steps are shorter than the sector radius of 30 m, producing many dependent photographs in sequence. Sequential photographs could be avoided with smaller detection areas, but this would lead to fewer counts available to estimate animal abundance.

Analyzing the simulated data, we observed that sequential photographs are informative and help prevent underestimation of animal density, which is the opposite of the recommendations of Rowcliffe et al., Cusack et al., and Rovero and Zimmermann. In the simulations, the REM estimates were compatible with the actual density values in scenarios without the exclusion window. However, when the exclusion window was applied, population density estimates were underestimated, with high discrepancies occurring for longer exclusion windows (Fig. 13.2a). Lin's CCC was used to compare results between methods and against a gold standard for analyzing precision and accuracy, combining Pearson's linear correlation with a bias quantification of the estimates. REM estimates showed a high level of agreement when no exclusion window was applied (87–94%, n = 50 simulations). However, with the exclusion window, the agreement decreased significantly, ranging from 54% to 66% for a 5-min window to just 16–24% for a 1-h window (Fig. 13.2b).

The disagreement between the REM measurements when considering the exclusion window was due to bias, which increased as the window length increased. Without an exclusion window, the bias corresponded to a maximum of 6.7% of the true population density value, while with the exclusion window, the bias varied between −38% and −72% of the true population density. RMSE and MAPE also presented higher values for the condition in which the exclusion window was applied: for high density, of 20 ind km^{-2}, the RMSE varied between 3.28 and 3.91 ind km^{-2} without the application of the exclusion window; however, when the window was applied, the RMSE reached values between 8.66 and 14.47 ind km^{-2}, that is, it was up to 3.7 times higher. Without the exclusion window, MAPE varied between 30% and 40% for low densities, 16–23% for medium densities, and between 12% and 16% for high densities. However, with the application of the exclusion window, MAPE increased to values between 38% and 72% and the pattern of association with population density was reversed: the higher the population density, the higher the MAPE obtained (Table 13.2).

The percentage of excluded photographs was directly related to the duration of the exclusion window (GLM, quasibinomial, $F(3, 596) = 1349.2$, $P < 0.0001$), with exclusion values of 39% for 5-min windows, 55% for 10-min windows, 61% for 30-min windows, and 67% for 1-h windows. The exclusions were mostly made up of data from the same individual—a sequence of photographs of the same particle at a crossing of the detection area or the return of the particle to the detection area in a short time interval. With increasing density, it is expected that more photographs will be excluded due to the simultaneous encounter of two or more animals;

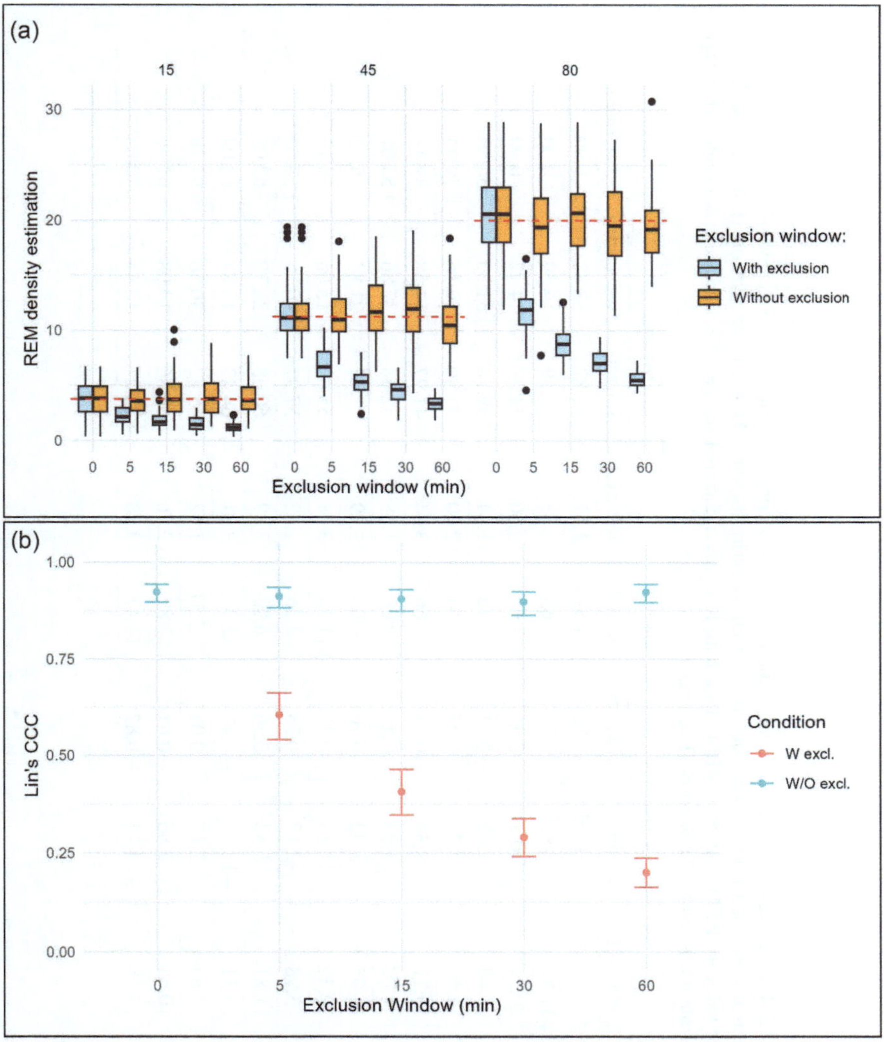

Fig. 13.2 (**a**, top) Estimates of population density (REM) in the simulation study with and without the application of the exclusion window. The study was conducted with N particles moving randomly (Random Walk) in a square box of 250 m, where a camera trap with a detection radius of 30 m and an opening angle of 90° was placed (N = 15, 45, and 80 particles). The particles moved for 151,200 time steps (minutes), with an average speed of 1.25 km/h, and the estimation process was repeated 50 times for each experimental condition. Red dashed lines indicate the true simulated population density. (**b**, bottom) Lin's Concordance Correlation Coefficient between the true values of population density and the REM estimates, with and without the exclusion window (w excl. and w/o excl.)

Table 13.2 Random Encounter Models (REM) estimates for the particle simulation study considering scenarios with exclusion window (w excl.) and without exclusion window (w/o excl.). Statistics obtained from 50 simulations for each scenario, showing the exclusion window (EW), true simulated density of particles (True dens.), the average value for REM estimates (REM), their bias, Root Mean Square Error (RMSE) and Mean Absolute Percentage Error (MAPE), besides the mean exclusion rate produced by the window (Excl. rate)

		REM		BIAS		RMSE		MAPE (%)		
EW (min)	Dens. (km^{-2})	w excl.	w/o excl.	w excl.	w/o excl.	w excl.	w/o excl.	w excl.	w/o excl.	Excl. rate (%)
0	3.75	3.77	3.77	0.02	0.02	1.51	1.51	32.82	32.82	0.00
	11.25	11.51	11.51	0.26	0.26	2.63	2.63	16.59	16.59	0.00
	20.00	20.31	20.31	0.31	0.31	3.60	3.60	14.72	14.72	0.00
5	3.75	2.27	3.64	−1.48	−0.11	1.72	1.44	39.78	31.80	36.86
	11.25	6.88	11.50	−4.37	0.25	4.64	2.70	38.80	19.15	39.94
	20.00	11.61	19.21	−8.39	−0.79	8.66	3.84	41.97	14.58	39.42
15	3.75	1.81	4.00	−1.94	0.25	2.08	1.88	52.48	37.86	53.01
	11.25	5.33	11.95	−5.92	0.70	6.05	3.19	52.62	21.83	55.02
	20.00	8.87	20.32	−11.13	0.32	11.23	3.74	55.67	15.08	56.10
30	3.75	1.58	3.98	−2.17	0.23	2.26	1.82	57.97	40.25	58.60
	11.25	4.42	11.81	−6.83	0.56	6.90	3.19	60.68	22.70	62.05
	20.00	7.11	19.51	−12.89	−0.49	12.93	3.91	64.43	15.84	63.04
60	3.75	1.28	3.83	−2.47	0.08	2.51	1.52	65.82	30.56	65.38
	11.25	3.39	10.64	−7.86	−0.61	7.89	2.76	69.86	19.08	67.46
	20.00	5.55	19.18	−14.45	−0.82	14.47	3.28	72.27	12.44	70.83

however, the influence of density on the percentage of excluded photographs was small within each level of the exclusion window (e.g., for $EW = 5$ min, the mean exclusion rate ranged from 36.9% to 39.9%; for $EW = 15$ min, from 53.0% to 56.1%; for $EW = 30$ min, from 58.6% to 63.0%; and for $EW = 60$ min, from 65.4% to 70.8%) (n = 50).

Here, we consider the hypothesis that the excluded photographs are informative. Thus, it is expected that we could correct the bias in the REM estimates simply by reconstructing the sample. This can be done from the percentage of photographs excluded in each scenario, ER(%), from the expression:

$$\mathrm{REM_{adjusted}} \approx \frac{\mathrm{REM_{with\ exclusion}}}{1 - \mathrm{ER}/100} \tag{13.11}$$

In this way, an animal could be counted several times at different occasions, even if close together, occasionally returning to the detection site due to its erratic movement, and from our perspective it would not be considered pseudo-replication. In the following analysis, we just recovered the information excluded by the window applying the correction (13.11), and the matching was close enough to provide evidence for this point (Fig. 13.3).

The particle box study supports the hypothesis that repeated photographs are informative for estimating population density and that the usual recommendation to apply the exclusion window biases density estimates, producing underestimated values. Here, only photographs of animals in motion, in a situation analogous

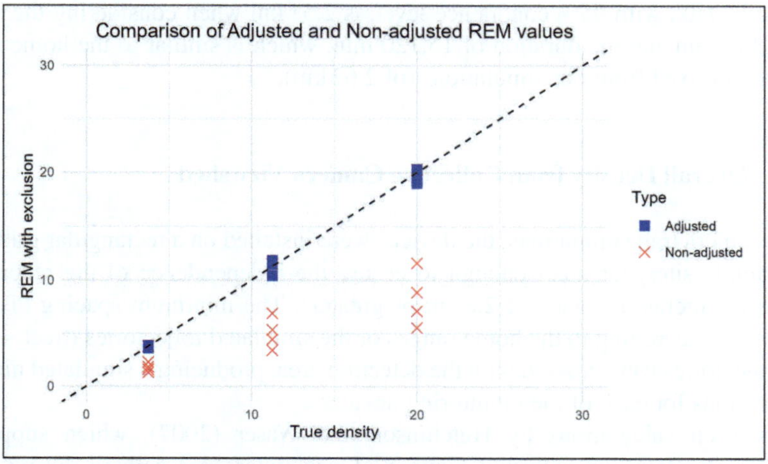

Fig. 13.3 Density estimates provided by REM calculated with the exclusion window, adjusted (blue) and unadjusted (red) for the recovery of information from the excluded photographs (based on Eq. 13.11). The dotted line indicates the equivalence between the REM estimates and the true population density

to the simulated one, are considered. More studies are needed to understand the implications of excluding photographs of animals resting in the detection area.

4.2 *Simulations on an Heterogeneous Landscape*

4.2.1 Description of the Simulated Trajectories

This study was based on 40,000 independent trajectories in the study area, representing a population density of 15 groups per km^2. Initially, a random sample of 1000 independent trajectories was selected to characterize the pattern of movement in the landscape. The home range was estimated at 2.62 ± 0.3 km, while the life area (MCP) was about $2.89 \pm 0.06\,km^2$ (mean $\pm$ SE), but few agents exhibited an action radius of up to 8 km and $11.8\,km^2$ of MCP. Our study considered only 21 consecutive days and, therefore, a smaller life area was expected compared to studies that included longer monitoring periods or different seasons.

In comparison, Silva et al. (2023) measured the home ranges (MCP) from 11 monitored individuals (*S. scrofa*, male and female) tracked by radio collar devices and observed over varying time periods (from 7 up to 154 days) in the states of Santa Catarina and Rio Grande do Sul of Brazil. The authors found a MCP between 2.51 and $46.2\,km^2$, with approximately 55% of values falling below $7\,km^2$.

Although home range is considered a key point in understanding the movement patterns and behavior of the species, this attribute did not need to be included in the movement algorithm's rules. The maximum dispersal radius expected for a random isotropic walk, with 95% confidence level, is 2.51 km when considering the speed of 20.8 m/min and the duration of 15,120 min, which is similar to the home range statistics derived from our simulations (of 2.62 km).

4.2.2 Overall Density from Collective Camera Viewshed

In the camera trap simulations, the devices were installed on a rectangular grid with 289 sample sites, far apart enough to ensure the independence of the records of adjacent cameras (distance of 2.8 km or greater). The minimum spacing distance was chosen according to the home ranges of the simulated trajectories (Sect. 4.2.1). The trajectories were crossed with the detection area, producing a simulated number of detections for each of the monitoring devices.

Based on calculations by Hutchinson and Waser (2007), which supported the REM models, the expected number of encounters per camera during three consecutive weeks was 161.4, or approximately 7.7 counts per day. However, these calculations consider spatial homogeneity as if the agents were ideal gas molecules. In *S. scrofa* simulations, the average number of counts was close to expected, adding up to 166.8 records per camera, a value only 3.4% above the predicted by the REM assumptions. However, applying the 1-h exclusion window, as recommended by the field authors (Sollmann 2018), changes the average number of counts per camera, decreasing it to 59.0, a 35.4% decrease.

The simulated trajectories were analyzed using the point processes theory, which allowed us to calculate the level of occupation in the surroundings of each camera (local density). The local density is used in this study as the gold standard to evaluate the performance of REM estimators. Taking into account all sample sites, the average (overall) density was 15.2 agents km^{-2}. In comparison, the REM estimation by the combined viewshed of all cameras, without the exclusion window, was 13.9 agents km^{-2}—thus, this presents a bias of 8.9% below the reference value. However, the REM estimate with an exclusion window of 1 h presented a significant discrepancy, totaling 4.9 agents km^{-2}. Note that this value, when adjusted by the proportion of excluded detections (Eq. 13.11), results in 14.0 agents km^{-2}, which is considerably close to the result obtained without the exclusion window. This fact reinforces the importance of repeated records in REM estimation, as previously discussed.

4.2.3 Standalone Cameras and Landscape Influence

In simulations, the number of monitoring sites is unlimited, and a significantly large number of camera traps can be deployed in the landscape at once. However, in practice, researchers work with constraints in terms of funding, time, and personnel, and most monitoring studies are conducted with a small number of devices. Therefore, evaluating the performance of standalone cameras and the influence of the landscape is crucial to determine the sample design and effort required.

In our simulations, the REM estimates for standalone cameras had a large dispersion, with an asymmetric distribution subject to outliers (Min: 0.00, Q1: 0.93, Q2: 5.22, Q3:16.68, Max: 131.17 agents km^{-2}; mean = 13.87 agents km^{-2}; skewness = 2.70; skewness test Z_{cal} = 18.89, $P < 0.001$). Compared to the overall density (averaged local densities), a reasonable bias was obtained, from -1.16 agents km^{-2} (7.7% below the average), but the results were highly inaccurate, with RMSE of 21.2 agents km^{-2} and MAPE of 98.3%. In that way, it is observed that standalone cameras are not able to accurately reflect the average density of the landscape.

In the study of the bouncing particle in a box (Sect. 4.1), which was conducted without the influence of the landscape, the RMSE and MAPE of the standalone cameras had not exceeded 3.91 agents km^{-2} and 22.7% at medium and high densities (Table 13.2). The higher error (RMSE) found in the heterogeneous landscape can be associated with the patterns of animal movements, since cameras are assigned to different areas (resistances, habitats, and corridors). However, it should also be considered that the study of particles inside the box used an effort sample 10 times larger (30 consecutive weeks), and, for this reason, 3.16 times lower error margins would be expected.

We also study the distribution of the local density values in the neighborhoods of each camera. Likewise for the standalone REM estimates, the local density was

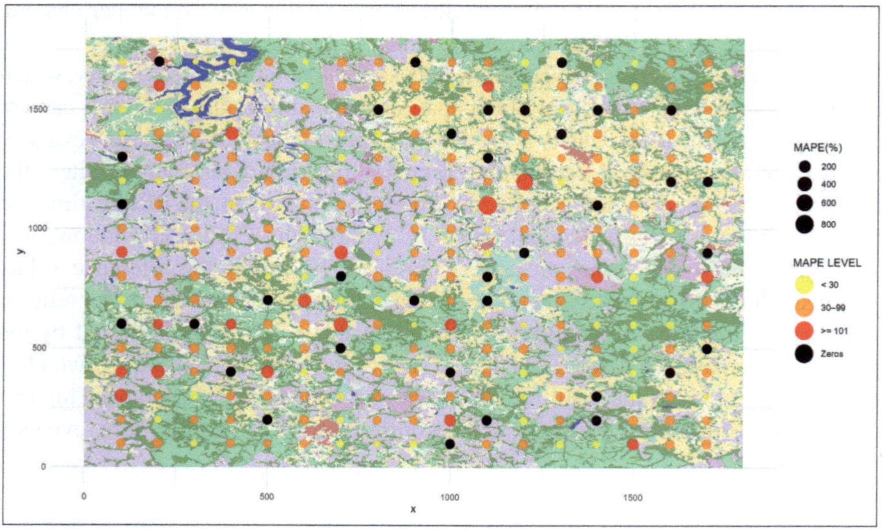

Fig. 13.4 Density estimates performance over a 289 camera-trap grid (Mean Absolute Percentage Error: MAPE). Random Encounter Model estimates are compared to the local densities derived from simulated trajectories. The overall density in the landscape was 15.03 agents km^{-2}. Land cover map colored according to the legend of Fig. 13.1

asymmetrically distributed and showed outliers (Min: 0.20, Q1: 6.20, Q2: 12.80, Q3: 20.80, Max: 88.00 agents km^{-2}; mean = 15.22 agents km^{-2}; skewness = 1.62; skewness test $Z_{cal} = 11.36$, $P < 0.001$). It was observed that, for 25% of the sample sites, the local density was 20.8 agents km^{-2} or higher. Furthermore, although the net average density is only 15.0 agents km^{-2}, a camera, which was installed in a position of high animal aggregation, resulted in a local density of 88.0 agents km^{-2}.

We divided the standalone camera-trap performance into four categories according to the MAPE: moderate accuracy (<30%), low accuracy (30–99%), extremely low accuracy (>100%) and noninformative (zeros) (Fig. 13.4). In simulations, about 25.3% of the traps showed moderate levels of accuracy, and 55.0% showed low accuracy, in such a way that 75.3% had MAPE inferior to 100%. However, 12% the traps were noninformative and about 7.6% showed a extremely low level of accuracy (MAPE $> 100\%$).

In order to identify where the devices perform better, we spliced the habitat into three main categories: natural habitat areas (weights equal to 7 or greater); matrix area, the dominant land cover type that surrounds and connects habitat patches (weights between 3 and 6); and resistance area, the portions of the landscape that impede movement of organisms due to low habitat suitability or high energetic/mortality cost of traversing them (weights below 3). Taking into account this arbitrary division, the landscape is composed of 24.2% habitats, 23.9% matrix areas, and 51.9% resistance terrains, and the grid of 289 camera traps was distributed in accordance with these proportions.

Table 13.3 Distribution of Mean Absolute Percentage Error (MAPE) categories across landscape components (n = 289)

Category	Habitat	Matrix	Resistance	Total
Zeros	1 (0.3%)	1 (0.3%)	33 (11.4%)	35 (12.1%)
<30%	18 (6.2%)	38 (13.2%)	17 (5.9%)	73 (25.3%)
30–99%	36 (12.5%)	28 (9.7%)	95 (32.9%)	159 (55.0%)
≥101%	15 (5.2%)	2 (0.7%)	5 (1.7%)	22 (7.6%)
Total	**70 (24.2%)**	**69 (23.9%)**	**150 (51.9%)**	**289 (100.0%)**

As expected, most of the noninformative results fell into the resistance area (94.3%; 33 of 35) (Table 13.3). However, few cameras in the resistance area were associated with excessive errors—only 4.4% of the informative cameras (nonzero results) showed MAPE exceeding 100%. In resistance areas, low-accuracy REM estimates were the predominant result, totaling 63.3% of the cameras installed in these areas, and only 11.3% (17 of 150) of the devices showed reasonable results with moderate accuracy.

REM estimates with moderate accuracy (MAPE $<$ 30%) were practically twice as frequent in the matrix areas compared to the natural habitat areas. These categories have almost the same representativeness; however, the matrix areas had 13.2% of the best results while the natural habitat areas had only 6.2% (Table 13.3). This is a counterintuitive result since we expect more counts in natural habitat areas, which are preferred by biologists to concentrate the sample efforts (Mentges et al. 2021).

For the 70 devices settled in the natural habitat areas, 21.4% showed an extremely low level of accuracy, perhaps associated with an excess of counts. However, in the matrix area the percentage of extremely low accuracy estimates was about 2.9%. In summary, the odds of extremely low accuracy REM estimates were approximately 7.39 times higher in natural habitat areas compared to matrix areas.

According to the simulations, the better places to allocate cameras are the matrix areas. This choice avoids elevated extremely low accuracy levels for MAPE associated with natural habitat areas and uninformative samplings associated with resistance areas. However, the results are obtained from a reductionist model, with landscape weights empirically chosen and a simplistic algorithm to simulate the animal's movement and behavior, and, thus, these results would need to be validated by a real life observational study.

The simulations revealed important insights that can contribute to efficient sample designs in fauna monitoring and conservation studies. However, future studies are required to verify the consistency of these simulations and assess the utility of our approach. Such works should consider improvements to our movement algorithm, inclusion of additional landscape features (barriers, elevation, buildings, water bodies) and their influences, interspecific interactions, calibration of detection areas for cover and visibility, and comparisons with other models (REST (Nakashima et al. 2018), N-mixture (Royle 2004), Distance Sampling (Buckland et al. 2015)). However, this study provides an important theoretical frame of reference to inspire future validation studies relying on real life sampling. The results of these validation studies, combined with the framework developed here, have the potential to change how camera-trap studies and REM analyses are performed.

5 Conclusion

In this study, we explored the agent-based model as a tool to evaluate the performance of camera-trapping sample designs and REM estimates in a fragmented landscape. This simulation approach can easily represent complex monitoring sets that are unfeasible and too expensive to be implemented in real situations. The results indicated a great variation for the REM estimates of the standalone camera, and their accuracy was related to location and landscape composition (habitat weights). Interestingly, the devices located in the matrix area performed better than those allocated in the natural habitat areas, where some observations with excessive counts were associated with density overestimation. Furthermore, the use of the exclusion window to retrieve dependent detections was associated with an underestimation of population density. These results challenge common practices in wildlife monitoring, and more studies are required to validate these findings under real-life conditions.

Acknowledgments IEPF acknowledges the financial support provided by the National Council for Scientific and Technological Development (CNPq), Brazil, under Grant No. 406540/2023-3. VFB and GCF acknowledge the financial support provided by the CNPq, under grants No. 149025/2024-6 and No. 181497/2025-5, respectively.

References

Baubet E, Bonenfant C, Brandt S (2004) Diet of the wild boar in the French Alps. Galemys 16(1): 101–113

Buckland ST, Rexstad EA, Marques TA, Oedekoven CS, et al (2015) Distance sampling: methods and applications, vol 431. Springer, Berlin

Bueno CG, Reiné R, Alados CL, Gómez-García D (2011) Effects of large wild boar disturbances on alpine soil seed banks. Basic Appl Ecol 12(2):125–133

Choquenot D, Ruscoe WA (2003) Landscape complementation and food limitation of large herbivores: habitat-related constraints on the foraging efficiency of wild pigs. J Anim Ecol 72(1):14–26

Clavero Pineda M, García-Berthou E (2005) Invasive species are a leading cause of animal extinctions. © Trends Ecol Evol 20(3):110

Cordeiro JLP, Hofmann GS, Fonseca C, Oliveira LFB (2018) Achilles heel of a powerful invader: restrictions on distribution and disappearance of feral pigs from a protected area in northern Pantanal, western Brazil. PeerJ 6:e4200

Cusack JJ, Swanson A, Coulson T, Packer C, Carbone C, Dickman AJ, Kosmala M, Lintott C, Rowcliffe JM (2015) Applying a random encounter model to estimate lion density from camera traps in Serengeti National Park, Tanzania. J Wildl Manag 79(6):1014–1021

de Paula Ferreira IE, Octaviano Fontana K, de Souza CA, Santa MF, Souza RS, Molin PG, de Camargo Gomes V, Beatriz Oliveira Prado A, de Paula Ferreira RE, Marques GB, Stefanini JVV (2023) biomastats: a package for the analysis of land use and land cover data in Brazilian territory. https://github.com/iep-ferreira/biomastats. R package, available under the Affero GPL 3.0 license. Data sourced from MapBiomas Collection 7 v.1

DeAngelis DL, Diaz SG (2019) Decision-making in agent-based modeling: a current review and future prospectus. Front Ecol Evol 6:237

Doherty TS, Glen AS, Nimmo DG, Ritchie EG, Dickman CR (2016) Invasive predators and global biodiversity loss. Proc Natl Acad Sci 113(40):11261–11265

Geisser H, Reyer H-U (2004) Efficacy of hunting, feeding, and fencing to reduce crop damage by wild boars. J Wildl Manag 68(4):939–946

Genov P (1981) Food composition of wild boar in North-Eastern and Western Poland. Acta Theriol 26(10):185–205

Gilbert NA, Clare JDJ, Stenglein JL, Zuckerberg B (2021) Abundance estimation of unmarked animals based on camera-trap data. Conserv Biol 35(1):88–100

Hafeez S, Abbas M, Khan ZH, Rehman E (2011) Preliminary analysis of the diet of wild boar (*Sus scrofa l., 1758*) in Islamabad, Pakistan. Turk J Zool 35:115–118

Hegel CGZ, Faria GMM, Ribeiro B, Salvador CH, Rosa C, Pedrosa F, Batista G, Sales LP, Wallau M, Fornel R et al (2022) Invasion and spatial distribution of wild pigs (*Sus scrofa* l.) in Brazil. Biol Invasions 24(12):3681–3692

Herrero J, Irizar I, Laskurain NA, García-Serrano A, García-González R (2005) Fruits and roots: wild boar foods during the cold season in the southwestern pyrenees. Ital J Zool 72(1):49–52

Hutchinson JMC, Waser PM (2007) Use, misuse and extensions of "ideal gas" models of animal encounter. Biol Rev 82(3):335–359

Irizar I, Amaia Laskurain N, Herrero J (2004) Wild boar Frugivory in the Atlantic basque country. Galemys 16(Special Issue):125–134

Jourdain NOAS, Cole DJ, Ridout MS, Rowcliffe JM (2020) Statistical development of animal density estimation using random encounter modelling. J Agric Biol Environ Stat 25:148–167

La Sala LF, Burgos JM, Caruso NC, Bagnato CE, Ballari SA, Guadagnin DL, Kindel A, Etges M, Merino ML, Marcos A, et al (2023) Wild pigs and their widespread threat to biodiversity conservation in South America. J Nat Conserv 73:126393

Lewis JS, Farnsworth ML, Burdett CL, Theobald DM, Gray M, Miller RS (2017) Biotic and abiotic factors predicting the global distribution and population density of an invasive large mammal. Sci Rep 7(1):44152

Lovelock JE, Margulis L (1974) Atmospheric homeostasis by and for the biosphere: the Gaia hypothesis. Tellus 26(1–2):2–10

Løvschal M, Bøcher PK, Pilgaard J, Amoke I, Odingo A, Thuo A, Svenning J-C (2017) Fencing bodes a rapid collapse of the unique greater mara ecosystem. Sci Rep 7(1):41450

MapBiomas Project (2023) Collection 7 of Brazilian annual land use and land cover mapping series. https://mapbiomas.org/en. Accessed 2025-06-22

Martins F, Mourão GM, Campos Z, Pellegrin A, Silva VS, et al (2019) Activity pattern and habitat selection by invasive wild boars (*Sus scrofa*) in Brazilian agroecosystems. Mastozool Neotrop 26(1):129–141

Massei G, Genov PV (2004) The environmental impact of wild boar. Galemys 16(1):135–145

McDonough MT, Ditchkoff SS, Smith MD, Vercauteren KC (2022) A review of the impacts of invasive wild pigs on native vertebrates. Mamm Biol 102(2):279–290

Mentges A, Blowes SA, Hodapp D, Hillebrand H, Chase JM (2021) Effects of site-selection bias on estimates of biodiversity change. Conserv Biol 35(2):688–698

Nakashima Y, Fukasawa K, Samejima H (2018) Estimating animal density without individual recognition using information derivable exclusively from camera traps. J Appl Ecol 55(2): 735–744

O'Connor S-J, Kelly D (2012) Seed dispersal of Matai (Prumnopitys taxifolia) by Feral pigs (*Sus scrofa*). N Z J Ecol 36:228–231
Ogutu JO, Piepho H-P, Said MY, Ojwang GO, Njino LW, Kifugo SC, Wargute PW (2016) Extreme wildlife declines and concurrent increase in livestock numbers in kenya: What are the causes? PloS One 11(9):e0163249
Pebesma E (2018) Simple features for R: standardized support for spatial vector data. The R J 10(1):439–446. https://doi.org/10.32614/RJ-2018-009
Potter LC, Brady CJ, Murphy BP (2019) Accuracy of identifications of mammal species from camera trap images: a Northern Australian case study. Austral Ecol 44(3):473–483
R Core Team (2021) R: a language and environment for statistical computing. R Foundation for Statistical Computing, Vienna. https://www.R-project.org/
Risch DR, Ringma J, Price MR (2021) The global impact of wild pigs (*Sus scrofa*) on terrestrial biodiversity. Sci Rep 11(1):13256
Rosell C, Fernández-Llario P, Herrero J (2001) El jabalí (*Sus scrofa* linnaeus, 1758). Galemys 13 (2):1–25
Rovero F, Zimmermann F (eds) (2016) Camera trapping for wildlife research. Pelagic Publishing, Exeter
Rowcliffe JM, Field J, Turvey ST, Carbone C (2008) Estimating animal density using camera traps without the need for individual recognition. J Appl Ecol 45:1228–1236
Royle JA (2004) N-mixture models for estimating population size from spatially replicated counts. Biometrics 60(1):108–115
Schley L, Roper TJ (2003) Diet of wild boar *Sus scrofa* in Western Europe, with particular reference to consumption of agricultural crops. Mamm Rev 33(1):43–56
Silva VS, Tortato MA, Salvador CH (2023) Densidade populacional e área de vida de javalis em áreas estratégicas para vigilância e defesa agropecuária. Technical Report (CNPSA) 613, Embrapa Suínos e Aves, Concórdia. https://www.embrapa.br/busca-de-publicacoes/-/publicacao/1150234/
Sollmann R (2018) A gentle introduction to camera-trap data analysis. Afr J Ecol 56(4):740–749
Sordi C (2015) Guerra ao javali: invasão biológica, feralização e domesticação nos campos sulinos. Rev Antropol UFSCar 5(1):59–77
Srbek-Araujo AC, Chiarello AG (2007) Armadilhas fotográficas na amostragem de mamíferos: considerações metodológias e comparação de equipamentos. Rev Bras Zool 24:647–656
Tobler MW, Powell GVN (2013) Estimating jaguar densities with camera traps: problems with current designs and recommendations for future studies. Biol Conserv 159:109–118
Tourani M (2022) A review of spatial capture–recapture: ecological insights, limitations, and prospects. Ecol Evol 12(1):e8468
Wolfram S (2002) A new kind of science. Wolfram Media, Inc., Champaign

Chapter 14
Statistical Model Adequacy, Goodness-of-Fit, and Model Selection: A Theoretical Review

Rachel McInerney, Sergio F. D. Reis, and Rafael de Andrade Moral

Abstract The concepts of model adequacy, model goodness-of-fit, and model selection are often mischaracterised, likely due to being ineluctably intertwined. As a result, some descriptions of these ideas are imprecise and/or incomplete in ecological literature. The aim of this chapter is to foster a nexus between the theoretical concepts of model adequacy, goodness-of-fit, and selection, and their application to complex problems in statistical ecology. We will facilitate a progression from novice to expert knowledge by (i) unambiguously defining model adequacy, model goodness-of-fit, and model selection from a statistical standpoint, (ii) emphasising the importance of understanding the integration of these concepts and their differences, with the main associated benefit being obtaining more reliable biological inferences in ecological research, and (iii) succinctly outlining what each analytical step entails to help eradicate future misuse.

Keywords Model adequacy · Model goodness-of-fit · Model selection · Statistical ecology · Ecological statistics · Half normal plots

1 Introduction

> Inferences made from... models are only as good as the models themselves (Warren and Seifert 2011).

Making informed decisions to address ecological challenges is essential to solve practical problems. The consequences of incorrect decisions can be detrimental and costly in terms of finances, time, natural resources, equipment, or data. The implementation of statistical models can help wildlife and ecosystem management

R. McInerney (✉) · R. A. Moral
Department of Mathematics and Statistics, Maynooth University, Maynooth, Co. Kildare, Ireland
e-mail: rachel.mcinerney.2021@mumail.ie

S. F. D. Reis
Departamento de Biologia Animal, State University of Campinas (UNICAMP), Campinas, Brazil

L. Martins Verdade, R. A. Moral (eds.), *Wildlife Monitoring*,
https://doi.org/10.1007/978-3-032-05821-8_14

and conservation biology needs, but with the suite of statistical tools, techniques, and models available, choosing an appropriate model can be a daunting task. We wish to pave a less overwhelming route so that applied ecologists can make appropriate scientific judgements.

While there is no tailor-made, step-by-step procedure for statistical analysis, we provide an introductory guide based on solid statistical principles. We promote an accessible tool, i.e., the hnp R package (Moral et al. 2017; R Core Team 2025), that can assess the adequacy of any statistical model from which we can calculate a diagnostic metric and simulate new samples. We discuss other statistical techniques to determine adequacy and compare models but refer the curious reader to more in-depth reviews that are beyond the scope of this chapter.

Some consulted literature in environmental and other applied areas mention the same statistical concepts with different terminology. For instance, some articles mention 'adequacy' in terms of "capturing... data" (Cabras et al. 2021), "a necessary condition for... a reasonable fit to the data" (Refenes and Zapranis 1999), or "the absolute fit of a model to the data" (Rice and Mayrose 2021); and these are mere fragments among a wealth of others. As such, terminology is not perfectly aligned across different disciplines. At times, it is unclear whether "fit" refers to how data points compare to fitted means, forecasts or predictions, or to how well the fitted distribution represents the empirical data.

Model adequacy, goodness-of-fit, and model selection are closely related but distinct and depend on each other but serve different purposes. They do not always follow a strict order (such as adequacy first, then fit, then selection) in the literature. Just because a model is adequate does not mean it is the best choice, and selecting a model does not guarantee that it is adequate.

In statistics, the data-generating process is the true, often unknown, probabilistic mechanism that produces observed data. It describes the stochastic, structural, and, when applicable, causal relationships among variables, including their probability distributions, dependencies, random components, and governing parameters (e.g., means, variances). The data-generating process is a theoretical construct, different from the observed data points themselves, and represents how observations arise from a system, whether through randomized experiments, observational studies, or natural processes (Casella and Berger 2002).

Model adequacy, goodness-of-fit, and model selection should be applied sequentially and recursively to enhance statistical models. Each concept can be described using a core idea: *representation* for model adequacy, *closeness* for goodness-of-fit, and *balance* for model selection, which interface these three concepts hierarchically. In Box 14.1, we provide precise and complete definitions for the terms "model adequacy", "goodness-of-fit", and "model selection", inspired partly by definitions in Hastie et al. (2009), Kutner et al. (2005) and Burnham and Anderson (2002).

Box 14.1 Definitions of Model Adequacy, Goodness-of-Fit, and Model Selection

Model adequacy: the extent to which a statistical model approximates the true data-generating process, as assessed primarily through adherence to assumptions.

Goodness-of-fit: a statistical measure of how closely predictions from a theoretically adequate model match observed data.

Model selection: the process of selecting the most appropriate statistical model from a set of candidates by balancing goodness-of-fit and model complexity, relying on prior assessments of adequacy and fit to support consistency with the data-generating process.

The consecutive procedure from model adequacy to goodness-of-fit to model selection should occur cumulatively in terms of the core ideas defined for each concept: representation establishes a sound model structure, then prediction accuracy can be quantified across competing models, which supports balance to identify the most appropriate model while highlighting its limitations. Limitations in model selection, tied to representation, closeness, and balance, prompt recursive re-evaluation of earlier steps. This ensures that model inadequacies, inaccuracies, and poor structure specifications are addressed systematically to improve statistical modelling.

Next, we briefly discuss the preliminary stages of model adequacy assessment, goodness-of-fit quantification, and selection. In Box 14.2, we provide a short glossary of commonly-used terminology in statistical modelling, which we mention throughout the chapter.

Box 14.2 A Glossary of Relevant Statistical Terminology

Bootstrapping: a way to estimate the distribution of a statistic using data resampling or simulations from a specified model.

Cross validation: a generalised method to assess predictive performance using unseen data.

Deviance: a measure of goodness-of-fit; calculated as minus two times the difference between the log likelihood of a fitted model and the relevant saturated model.

Diagnostic metric: a quantitative measure or statistic derived from a statistical model that helps to evaluate its performance.

Empirical data: observations.

Forecast: a future observation that is predicted by a model (extrapolation).

(continued)

Fitted value: an estimate predicted by a model.

Frequentist statistics: a statistical paradigm that utilises repeated sampling and long-term event frequencies to draw conclusions from data.

Information criterion: a statistic based on an information theoretic approach that can be used to compare models.

Lack-of-fit error: arises when the model structure fails to capture the systematic patterns in the data.

Likelihood: a quantification of how plausible the observed data is, given a fitted model.

Loss function: measures lack-of-fit; the discrepancy between observed and fitted values.

Model evaluation: an umbrella term for assessing model adequacy, goodness-of-fit, and predictive performance.

Model performance: a combination of model goodness-of-fit and predictive performance quantifications.

Objective function: a mathematical expression that evaluates the quantity to be maximised or minimised in an optimisation problem, subject to specified constraints.

Overfitting: occurs when a statistical model excessively captures noise or random variability in the data, rather than the true underlying relationship between the response variable and the predictor variables, resulting in an overly complex model that fits the observed data well but fails to generalise accurately.

Out-of-sample data: any observation that was not used to fit the model.

Overdispersion: a phenomenon that occurs when the variability of the data is greater than assumed by the fitted model.

Penalty function: a constraint that penalises an objective function for undesirable properties.

Predictive performance: a quantification of how well a model can estimate the response variable for out-of-sample (new) data. Predictive performance using in-sample training data is an assessment of goodness-of-fit.

Predictor: explanatory variable, also referred to as independent variable.

Principle of Parsimony: simpler models with fewer parameters are favoured to their more complex counterparts when they perform similarly (Occam's razor).

Pure error: unpredictable, non-systematic variability in data due to, for instance, measurement errors or random fluctuations.

Residual: a function of the difference between a fitted value and its associated observed data point.

Response variable: the dependent variable of interest.

(continued)

Saturated model: an overfitted model that reproduces all the data points exactly (interpolation).

Spurious predictor: a variable that appears to be important but has no true causal or meaningful relationship with the response.

Systematic error: a consistent bias in measurements or estimates due to flaws in the measurement process, experimental design, or data collection, causing predictable deviations from the true values.

2 Data Collection and Visualisation

Data that are accurate, complete, relevant, consistent, reliable, and representative, can support robust statistical analyses and valid answers to research questions. Therefore, before collection, the analyst must first decide what data is required, how much, and how it will be obtained to safeguard against futile fieldwork efforts. Such experiments and observational studies should be well-designed to obtain the appropriate granularity of data.

Large datasets are desired in wildlife monitoring, when possible, as they can alleviate uncertainty arising from incomplete knowledge about the system being studied, such as environmental factors (Gupta et al. 2012), pure error (Kelchner 2009), and imprecisions in statistical analyses such as biased parameter estimates or unreliable conclusions due to small sample sizes (Janková et al. 2020). Another advantage of plentiful data relates to cross validation, in which data is randomly split into at least two types of sub-datasets: within-sample data for fitting the model and out-of-sample data to assess predictive performance. However, independence issues have been raised regarding this (Warren and Seifert 2011); can we say that the unseen data is truly 'new'?

Sparse data poses a challenge due to the numerous assumptions required to model it effectively. In contrast, rich data (with minimal missing values and accurate extreme observations, if present) reduces these assumptions, allows to fit more complex models, and optimises the bias-variance trade-off in parameter estimation. It should be noted that precision is reduced when more parameters are estimated from the same data source (Kelchner 2009).

Initial plots containing the response variable can be very beneficial (Stasinopoulos 2025, personal communication) for revealing insights about the relationships between variables. For instance, they can suggest the need for variable transformations or polynomials beyond the first degree for continuous predictors. The types of visualisations in exploratory data analysis depend on the research aim: comparison, decomposition, distributional assessments, etc., and range from scatter plots with relevant facets, line plots for longitudinal data, bar plots for categories, and maps or bubble plots for spatial components, among many others. See Wilke (2019) for a comprehensive review of visualisation techniques.

3 Model Construction

All models are wrong, but some are useful (Box 1976).

The essence of model construction in Ecology is to approximate a biological process. The true model is the exact mathematical representation of the data-generating process, encompassing all systematic patterns, relationships, and stochastic components that produce the observed data. The true model is unknowable for two reasons: firstly, we lack complete information about the data-generating process; secondly, observed data represent only a finite sample of possible realisations, subject to noise and sampling variability. Thus, a proposed model will be inexact (Kelchner 2009) but not necessarily useless (Yahyah et al. 2010). A statistical model whose predictions (interpreted with their associated standard errors) imply relationships that are biologically unrealistic or inconsistent with established ecological knowledge should not be relied on for inferential purposes (Johnson and Omland 2004).

When formulating models, one must balance many things: complexity, interpretation, numerical stability, computational tractability, procedural efficiency, and available techniques. Models should be simple and comprehensible (Getz et al. 2018) but not omit key variables or critical patterns (i.e., underfit the data). They should also address the research question in a pragmatic and realistic way.

A statistical model is based upon ideas and questions and both can be updated. Gupta et al. (2012) describe the modelling process as starting with a conceptual model, then configuring it using mathematical equations, and finally fitting it using computational software. *A priori* consideration of which predictors should be included in a model is a critical step to ensure scientific validity. Cautious variable choice helps to avoid spurious covariates that inflate the variances of parameter estimates (Ripplinger and Sullivan 2010).

If p is the number of observed variables, then 2^p additive linear models can be fitted (Garcia-Angulo and Claeskens 2024). If we consider two-way or higher order interactions, the number of options grows even faster. Fitting all possibilities, therefore, is computationally expensive, particularly in high dimensional settings, and model selection becomes impractical. In such cases, we recommend that highly correlated variables are removed from the model specification and/or regularization is imposed to shrink unimportant variables, which alleviates multicollinearity (Hooten and Hobbs 2015) and overfitting (Yates et al. 2022). Different regularization strategies can be used, such as ridge (Hoerl and Kennard 1970), LASSO (Least Absolute Shrinkage and Selection Operator; Tibshirani 1996), elastic-net (Zou and Hastie 2005), horseshoe-type penalties (Carvalho et al. 2009), among other frameworks.

We now expand on the concepts of model adequacy, goodness-of-fit, and model selection in the following sections.

4 Model Adequacy

We begin with the origin of the term "adequacy", which comes from the Latin word *adaequatus* meaning "equalized", which was thought to mean "equal to what is needed or desired" in the seventeenth century (etymonline.com, accessed on 17/04/2025). One interesting lexical definition of 'adequacy' is "the state of being sufficient for the purpose concerned" (Vocabulary.com, accessed on 17/04/2025).

Baumberger et al. (2017) described three types of adequacy regarding climate models: adequacy for purpose, meaning suitability to the specific modelling task; structural adequacy, meaning satisfying the model's distributional assumptions, and adequacy for prediction, meaning accurately forecasting outcomes. These three meanings of adequacy (purpose, structure, and prediction) demonstrate that the term is not singular in its application but has varied interpretations depending on the context.

Some researchers believe that "a model should be tested for the purpose for which it was designed" (Sony and Marriapan 2021) as "a model that is . . . adequate for one purpose may not be adequate for another" (Wang et al. 2020). However, the phrase 'structural adequacy' is the closest match for what statisticians term 'adequacy' because a statistical model has underlying assumptions that must be verified to ensure its correct use. Assumptions can be implicit in the modelling process (Kelchner 2009) or explicitly outlined by model properties.

If the model assumptions are inconsistent with the statistical properties of the data it is fitted to, such as its mean, variance, or distribution, the inferences drawn from it cannot be trusted (Wang et al. 2020). For instance, overdispersion is a widespread phenomenon in wildlife monitoring count data. This is a direct violation of the assumed equidispersion property of the Poisson distribution. If a fitted model does not properly accommodate such extra-variability, then standard errors will be underestimated and model selection can be problematic (Demétrio et al. 2014). Thus, checking model adequacy merits sufficient attention; it should not be viewed as a burden or impediment to grapple with during satisfactory model identification. We advocate against the unprincipled use of inadequate models as they often yield inflated inferential errors (Rice and Mayrose 2021).

The tools, i.e., software, specific statistical techniques, or else, to determine model adequacy depend on the type of model fitted. Graphical analyses can disclose patterns in an interpretable (Yahyah et al. 2010) but also subjective way. Residual diagnostics can reveal non-random departure from the assumed distribution (do Nascimento et al. 2023), if any, that could be caused by a missing important effect (Ponciano et al. 2009). Other possible model inadequacies such as violations of assumptions of independence, linearity, normality, or homoscedasticity can manifest in visualisations or more objective quantifications from hypothesis tests. There are no fixed assumptions for all models, for instance, when checking the adequacy of a fitted classical linear model, one would expect to see homoscedasticity, however, other types of models may assume heteroscedasticity of specific forms. For example, the Negative Binomial Type 1 model assumes that the variance is a linear function of the mean.

4.1 Half Normal Plots

Sony and Marriapan (2021) claim that "model adequacy is not a single step process... rather it is a complex amalgamation of several statistical tests which should be applied methodologically." Although there are many different types of complementary adequacy checks for specific model assumptions, it is possible to empirically assess the distributional adequacy of a model, from which new data samples can be simulated, by comparing the expected and realised behaviour of model diagnostics.

When released, the hnp R package (Moral et al. 2017) was marketed as being an indicator of overdispersed count and discrete proportion data, even though examples showcased in the original paper also included continuous and censored data. However, its functionalities go far beyond detecting overdispersion, as it can collectively assess all the assumptions of a model at once.

The methodology for half normal plots with a simulated envelope was originally proposed by Atkinson (1985). In summary, a discrepancy metric i.e., a type of residual, is calculated for each observation that the model is fitted to and sorted in ascending order. Parametric bootstrapping is performed many times (99 is the default in hnp) to simulate new response variables from the fitted model distribution. The model is then refitted using each new response and the discrepancy metrics are calculated again. The diagnostics for each refitted model are also sorted in ascending order and combined column-wise in a matrix. Then, for each row of this matrix, the 2.5th, 50th (i.e., median) and 97.5th percentiles are calculated to constitute the simulated envelope for each order statistic. The expected (i.e., theoretical) quantiles of the normal distribution (calculated using the inverse of the normal cumulative distribution function) comprise the x-axis. For a half-normal plot, the absolute values of the diagnostics are sorted and the x-axis labels are obtained via the inverse of the *half*-normal cumulative distribution function. It is expected that, for an adequate model, most observed diagnostics will fall within the simulated envelope.

The contents of these plots can be quantified as an adequacy metric by calculating the percentage of residuals that fall outside the simulated envelope. If all residuals are enclosed by the simulated envelope, then the data is a plausible realisation of the fitted model. Half normal plots are easy to apply and available in open-source software, making them attractive to practitioners. Additional information can also be revealed: (i) the residual normality assumption for Gaussian generalised linear (mixed) models. holds if the fitted model's residuals lie on the identity line, and (ii) overdispersion often displays as the residuals increasingly diverging from (and above) the median line as the x-axis progresses from left to right.

As with most statistical techniques, there are limitations to this method. Firstly, it can be computationally intensive for models that are slow to fit. Also, hnp is not deterministic; it is a simulation-based approach and so, a different plot or percentage of residuals outside the simulation envelope is produced each time. Therefore, we recommend the mean percentage approach of several hnp runs and the decision rules proposed by Mainguy and Moral (2021). Furthermore, hnp does not tell the user which assumption of the posited model is failing, if any. However, we see this as a heuristic attribute as the onus is on the user to inspect this. Other visualisations can help here, such as worm plots (van Buuren and Fredriks 2001) or bucket plots (see Chap. 9, this book, for more information). Lastly, as residuals are always negligible for close-to saturated models and always exactly zero for saturated models, hnp will fail to detect distributional departures here and so, these models may be guised as adequate. Concomitant methods can help to determine adequacy in such cases and selection based on information-theoretic approaches, as well as quantification of out-of-sample predictive performance, can detect the presence of unnecessary excess parameters.

5 Goodness-of-Fit

It is not guaranteed that a fitted model will describe the observed data well. In fact, some mismatches are anticipated as a reasonable model will not trace the trajectory of the data points exactly, like a 'join the dots' exercise. There is pure error and lack-of-fit error associated with any statistical model (Sony and Marriapan 2021). The literature defines goodness-of-fit in a multitude of ways; some examples include the "degree of accuracy of a model" (Teferra et al. 2014) or "model compatibility" (Cabras et al. 2021). We define goodness-of-fit as a quantification of how well a model fits the in-sample data it was trained on. A first paramount point is that a model is fit to data and not vice versa. This is analogous to fitting a shoe to one's foot; we aim to find a snug resemblance with ample wiggle room.

There are more enlightening ways to discover if a model captures the core features of observations than superimposing its fitted values to the data and eyeballing under- or over-estimation. Measures of goodness-of-fit include a suite of distance metrics such as the residual sum of squares or mean squared error, the Chi-squared goodness-of-fit test, bootstrapping approaches, or visualisations such as rootograms (Kleiber and Zeileis 2016). However, simple measures may not suffice when model complexity grows (Tedeschi 2019). Akin to model adequacy, these tools depend on the type of model fitted.

The coefficient of determination, R^2, computes the proportion of variance in the response variable that is sufficiently explained by the model. This, and its adjusted version which penalises model complexity, are popular methods in Statistical Ecology (Sony and Marriapan 2021). Ideally, we want to maximise the variance explained (Laubach et al. 2021). Alas, for linear models, R^2 (i.e., the correlation between the empirical data and their fitted values squared) becomes artificially larger as more regression coefficients are added to the model (Tedeschi 2019). This increase is often due to fitting inherent noise, that is, random variability or pure error, in the data as R^2 does not respect the Principle of Parsimony. We render it a misleading statistic as a "good fit" could be due to spurious predictors (Hamer et al. 2012), but if used regardless, we would recommend the implementation of its *adjusted* analogue in combination with other metrics.

Lack-of-fit warrants model improvement, however (i) small differences between empirical data and model predictions can stem from redundant variables and (ii) a model that closely explains actual data will likely not predict out-of-sample data well (Mac Nally et al. 2018; Shmueli 2010). It is worth noting that some models are highly data driven and prone to overfit (Falk and Muthukrishna 2023).

6 Model Selection

When an adequate inferential or forecasting model has been found, it is not guaranteed to be the most optimal one. Additionally, models should fit the data well so qualifying the goodness-of-fit is paramount (Symonds and Moussalli 2011), but the phenomenon of overfitting proves that goodness-of-fit considered in isolation is not sufficient to ascertain which model to employ. Adequate models can have reasonable goodness-of-fit, but this is not always definite. Model selection is a companion task to both adequacy and goodness-of-fit assessments and should be used in conjunction to seek the "best approximating" model (Bozdogan 1987). Bozdogan's conception of 'best' refers to a model that minimizes the expected discrepancy between itself and the true underlying process, balancing fit with simplicity (fewer parameters). Bozdogan's conception of 'approximation' refers to a model that serves as a simplified representation of the true, unknown process, capturing essential relationships between variables (i.e., how predictors influence the response) but not fully replicating the data-generating process, as all models are inherently imperfect.

If more than one predictor is tenable, numerous models should be considered (Symonds and Moussalli 2011). Model selection should involve identifying the top-ranking model among a set of *adequate* competing candidates. Mac Nally et al. (2018) recommend that the null model (if adequate) should be included in the candidate set as a control. There are different ways to compare model suitability and this becomes more challenging as models become more complex (e.g., when interactions or non-linear terms are involved). One method is an information-theoretic approach, that is ubiquitous and understandable but difficult to calculate in some settings (e.g, for generalised additive models involving smooth functions).

Information criteria (IC) adopt a "discrimination approach" (Refenes and Zapranis 1999); there is a trade-off regarding goodness-of-fit and model complexity. Models that fit the data well are rewarded but unnecessary parameters are penalised to curb overfitting (Spanos 2010). The balance of a penalty function (Mi et al. 2015) with a loss function (Yates et al. 2022) is a more conservative way to measure model performance than goodness-of-fit alone. Relative scores are used to rank models from most to least "statistically supported" (Mainguy and Moral 2021) with lower values suggesting a preferred model. Analogues that correct for small sample sizes also exist. Uncertainty can be addressed by computing, for instance, standard errors and confidence intervals for IC scores. This is often done in the Bayesian paradigm via the Widely Applicable Information Criterion and its standard error (Watanabe 2013), however it can also be done in the frequentist paradigm via non-parametric bootstrapping.

Akaike's information criterion (AIC; Akaike 1973) is an estimate of the Kullback–Leibler divergence, which is a measure of the distance between the tentative model and a statistical abstraction of the data's probabilistic structure (Giacopetti et al. 2016). This underscores the fact that AIC evaluates models based on statistical fit rather than their ability to fully capture complex biological mechanisms. It is calculated as the model deviance plus a penalty term that is proportional to the number of model parameters used. Therefore, the smaller the AIC value, the more favourable the model is. How close candidate models are (Schennach and Wilhelm 2016) to the smallest AIC value should also be addressed; see Burnham et al. (2011) or Anderson and Burnham (2002) for recommendations.

Kelchner (2009) comments on how AIC differences complicate choosing one model only. In such a case, prediction weights can be calculated for models with similar AIC scores (the model with the smallest AIC will be given the largest weight, and so on). These Akaike weights can be interpreted as "the probability that a given model is the best approximating model" (Symonds and Moussalli 2011) so higher weights imply better models. When models have similar Akaike weights, we are still unsure which model is best (Schorning et al. 2016). One option here is to perform model averaging[1] using the weighted support of the individual fits to obtain more robust inferences (Johnson and Omland 2004). Variable importance can also be disclosed by summing the Akaike weights of models that contain the variable of interest.

While AIC can be inclined to select an overfitted model, especially for few competitors (Sony and Marriapan 2021; Link and Barker 2006), the Bayesian information criterion (BIC; Schwarz 1978) often selects a more parsimonious model due to its stricter penalty term for model complexity. The practitioner must decide which information criterion to apply. A limitation of IC is that they are only approximations in some cases, especially when random effects or semi-parametric terms are included in the candidate models.

The chosen model should transfer well to unseen data to avoid under- or over-fitting (Piironen and Vehtari 2017). We can estimate a model's predictive performance using cross validation (CV); the model with the most accurate predictive score while incorporating uncertainty errors is preferred (Hastie et al. 2009). Stone (1977) showed an asymptotic equivalence between leave-one out cross validation[2] and AIC. Mac Nally et al. (2018) showed that when log-likelihoods are used, CV

---

[1] In model averaging, the predictions are averages weighted by the Akaike weights; the estimated model coefficients are not averaged.

[2] Leave-one out cross validation (LOO-CV) is a process in which a model is fitted to the data (n-1) times, where n is the sample size. One observation is left out each time, a discrepancy metric is calculated for the predicted missing observation, and all discrepancy metrics are pooled into one LOO-CV score.

and IC results often concur. CV is versatile and it can be used in the absence of a derived likelihood. See Yates et al. (2022) for more information.

Likelihood-ratio testing (Neyman and Pearson 1928) uses a hypothesis testing framework to decide if specific predictors are worth keeping in the model (Kelchner 2009). However, a key caveat is that the compared models must be nested i.e., there exists a simpler model in the candidate set that is a reduced version of a more complex model also in the candidate set. Mixed models must have the same random structure (van Rij et al. 2022). Thought-provoking limitations of hypothesis testing include the dichotomous significance level of p-values and the following misbelief (Spanos 2010). The "fallacy of acceptance" concerns failing to reject the null hypothesis; this should not be misconstrued as evidence supporting the null hypothesis.

Evidence ratios are another option for model comparison (Symonds and Moussalli 2011), as is step-wise variable selection (Efroymson 1960) but the latter has its limitations. It cannot span the entire set of covariates, see Whittingham et al. (2006) for more details.

Model selection methods themselves do not validate model assumptions (Sony and Marriapan 2021). Hence, model adequacy should be assessed before selection and not the converse; this view aligns with Mac Nally et al. (2018) and other authors who have shown that inadequate models can be selected (Duchêne et al. 2016; Kelchner 2009; Ripplinger and Sullivan 2010; Refenes and Zapranis 1999). We advise that models found to be inadequate (via hnp or else) should subsequently be discarded as potential candidates within selection procedures.

7 Discussion and Conclusion

First and foremost, we would like to mention that Statistics is a fluid field. Researchers are currently creating novel techniques and relentlessly advancing existing methods to relax the limitations associated with them. Other model adequacy, goodness-of-fit, and selection methods will be introduced in the future.

Some areas of model evaluation are currently lacking and pose platforms for future investigation. It should be addressed here that there are no methods for assessing the adequacy of a model fitted to a binary or multinomial response variable with one trial. Even with contrasting predictors, each of these models will invariably reproduce the marginal probability of success, meaning assessment of their adequacy (and goodness-of-fit) is not informative.

This chapter was an overview (albeit a non-exhaustive one) of model adequacy, goodness-of-fit, and model selection. We have shown that these cohesive concepts are interlinked but emphasised their differences. Model adequacy and selection should be implemented to increase confidence in model inferences and projections and to strengthen decision making. Utilising different methodologies provides multiple lenses to base reasoning upon.

It is tempting for analysts to be misled by lax model acceptance (Kelchner and Thomas 2007) but this can vitiate model estimates. The conundrum of choosing an adequate statistical model can be streamlined with tools such as hnp and other software. Our takeaway message is for readers to adopt the overall philosophy behind model adequacy assessment, goodness-of-fit testing, and selection, and also understand that there is a myriad of other useful tools available which we have not fully discussed.

Due to the intricacies of real-world data, even when a statistically-minded modeller is equipped with a diverse toolbox and the accumulated dataset is of a high standard, sometimes the available predictors cannot capture sufficient observed variation in the response within an adequate model, meaning that reliable predictions cannot be obtained. Nevertheless, we remain optimistic and believe that harnessing synergy in Statistical Ecology will lessen disciplinary silos and thus improve the models used in this area.

Acknowledgements R. McInerney received financial support from Research Ireland and the Ministère de l'Environnement, de la Lutte contre les changements climatiques, de la Faune et des Parcs, in Québec, Canada under grant number EPSPG/2024/48. We would like to thank Dr. Julien Mainguy in particular for his critical eye and attention to detail, for challenging existing methods, and for his diligent efforts that help to further bridge the gap between Statistics and Ecology.

Author Contributions This chapter was conceptualised by R. Moral. R. McInerney performed the literature review and wrote the first draft of this chapter. S. Reis and R. Moral reviewed and edited the text and provided their expertise.

References

Akaike H (1973) Information theory and an extension of the maximum likelihood principle. In: Petrov BN, Csaki F (eds) Proceedings of the 2nd international symposium on information theory, pp 267–281. Akadémiai Kiadó

Anderson DR, Burnham KP (2002) Avoiding pitfalls when using information-theoretic methods. J Wildl Manag 66(3):912–918

Atkinson AC (1985) Plots, transformations, and regression: an introduction to graphical methods of diagnostic regression analysis. Clarendon Press

Baumberger C, Knutti R, Hirsch Hadorn G (2017) Building confidence in climate model projections: an analysis of inferences from fit. Wiley Interdiscip Rev Clim Chang 8(3):e454

Box GEP (1976) Science and statistics. J Am Stat Assoc 71(356):791–799

Bozdogan H (1987) Model selection and Akaike's information criterion (AIC): the general theory and its analytical extensions. Psychometrika 52(3):345–370

Burnham KP, Anderson DR (2002) Model selection and multimodel inference: A practical information-theoretic approach, 2nd edn. Springer

Burnham KP, Anderson DR, Huyvaert KP (2011) AIC model selection and multimodel inference in behavioral ecology: some background, observations, and comparisons. Behav Ecol Sociobiol 65(1):23–35

Cabras S, Castellanos ME, Ratmann O (2021) Goodness of fit for models with intractable likelihood. TEST 30(3):713–736

Carvalho CM, Polson NG, Scott JG (2009, April) Handling sparsity via the horseshoe. In: Artificial intelligence and statistics. PMLR, pp 73–80

Casella G, Berger RL (2002) Statistical inference, 2nd edn. Duxbury Press

Demétrio CGB, Hinde J, Moral RA (2014) Models for overdispersed data in entomology. In: Ecological modelling applied to entomology. Springer, pp 219–259

do Nascimento RLS, Fagundes RA d A, de Souza RMCR, Cysneiros FJA (2023) Interval regression model adequacy checking and its application to estimate school dropout in Brazilian municipality educational scenario. Pattern Anal Applic 26(1):39–59

Duchêne S, Di Giallonardo F, Holmes EC (2016) Substitution model adequacy and assessing the reliability of estimates of virus evolutionary rates and time scales. Mol Biol Evol 33(1):255–267

Efroymson MA (1960) Multiple regression analysis. In: Ralston A, Wilf HS (eds) Mathematical methods for digital computers. John Wiley

Falk CF, Muthukrishna M (2023) Parsimony in model selection: tools for assessing fit propensity. Psychol Methods 28(1):123–136

Garcia-Angulo AC, Claeskens G (2024) Bootstrap for inference after model selection and model averaging for likelihood models. Metrika 88:311–340

Getz WM, Marshall CR, Carlson CJ, Giuggioli L, Ryan SJ, Romañach SS, Boettiger C, Chamberlain SD, Larsen L, D'Odorico P, O'Sullivan D, Coulson T (2018) Making ecological models adequate. Ecol Lett 21(2):153–166

Giacopetti M, Crestaz E, Materazzi M, Pambianchi G, Posavec K (2016) A multi-model approach using statistical index and information criteria to evaluate the adequacy of the model geometry in a fissured carbonate aquifer (Italy). Water 8(7):271

Gupta HV, Clark MP, Vrugt JA, Abramowitz G, Ye M (2012) Towards a comprehensive assessment of model structural adequacy. Water Resour Res *48*(8):Article W08503

Hamer CE, Halbert BE, Webster M, Scharer JM (2012) Assessment of model adequacy and parameter identifiability for predicting contaminant transport in the Beaverlodge Lake area, Canada. WIT Trans Ecol Environ 164:159

Harper D (n.d.) Etymology of adequacy. Online Etymology Dictionary. Retrieved April 17, 2025, from https://www.etymonline.com/word/adequacy

Hastie T, Tibshirani R, Friedman J (2009) The elements of statistical learning, 2nd edn. Springer

Hoerl AE, Kennard RW (1970) Ridge regression: biased estimation for nonorthogonal problems. Technometrics 12(1):55–67

Hooten MB, Hobbs NT (2015) A guide to Bayesian model selection for ecologists. Ecol Monogr 85(1):3–28

Janková J, Shah RD, Bühlmann P, Samworth RJ (2020) Goodness-of-fit testing in high dimensional generalized linear models. J Royal Stat Soc Ser B (Stat Methodol) 82(3):773–795

Johnson JB, Omland KS (2004) Model selection in ecology and evolution. Trends Ecol Evol 19(2):101–108

Kelchner SA (2009) Phylogenetic models and model selection for noncoding DNA. Plant Syst Evol 282(3–4):109–126

Kelchner SA, Thomas MA (2007) Model use in phylogenetics: nine key questions. Trends Ecol Evol 22(2):87–94

Kleiber C, Zeileis A (2016) Visualizing count data regressions using rootograms. Am Stat 70:296–303

Kutner MH, Nachtsheim CJ, Neter J, Li W (2005) Applied linear statistical models, 5th edn. McGraw-Hill

Laubach ZM, Murray EJ, Hoke KL, Safran RJ, Perng W (2021) A biologist's guide to model selection and causal inference. Proc R Soc B Biol Sci 288(1943):20202815

Link WA, Barker RJ (2006) Model weights and the foundations of multimodel inference. Ecology 87(10):2626–2635

Mac Nally R, Duncan RP, Thomson JR, Yen JDL (2018) Model selection using information criteria, but is the "best" model any good? J Appl Ecol 55(3):1441–1444

Mainguy J, Moral RA (2021) An improved method for the estimation and comparison of mortality rates in fish from catch-curve data. N Am J Fish Manag 41(5):1436–1453

Mi G, Di Y, Schafer DW (2015) Goodness-of-fit tests and model diagnostics for negative binomial regression of RNA sequencing data. PLoS One 10(3):e0119254
Moral RA, Hinde J, Demétrio CGB (2017) Half-normal plots and overdispersed models in R: the hnp package. J Stat Softw 81(10)
Neyman J, Pearson ES (1928) On the use and interpretation of certain test criteria for purposes of statistical inference part I. Biometrika 20(1–2):175–240
Piironen J, Vehtari A (2017) Comparison of Bayesian predictive methods for model selection. Stat Comput 27(3):711–735
Ponciano JM, Taper ML, Dennis B, Lele SR (2009) Hierarchical models in ecology: confidence intervals, hypothesis testing, and model selection using data cloning. Ecology 90(2):356–362
R Core Team (2025) R: A language and environment for statistical computing. R Foundation for Statistical Computing
Refenes APN, Zapranis AD (1999) Neural model identification, variable selection and model adequacy. J Forecast 18(5):299–332
Rice A, Mayrose I (2021) Model adequacy tests for probabilistic models of chromosome-number evolution. New Phytol 229(6):3602–3613
Ripplinger J, Sullivan J (2010) Assessment of substitution model adequacy using frequentist and Bayesian methods. Mol Biol Evol 27(12):2790–2803
Schennach SM, Wilhelm D (2016) A simple parametric model selection test. J Am Stat Assoc 112(520):1663–1674
Schorning K, Bornkamp B, Bretz F, Dette H (2016) Model selection versus model averaging in dose finding studies. Stat Med 35(22):4021–4040
Schwarz G (1978) Estimating the dimension of a model. Ann Stat 6(2):461–464
Shmueli G (2010) To explain or to predict? Stat Sci 25(3):289–310
Sony M, Marriapan V (2021) A methodological approach to assessment and reporting of the model adequacy in simulation studies. Int J Operat Res Informat Syst 12(4):1–17
Spanos A (2010) Akaike-type criteria and the reliability of inference: model selection versus statistical model specification. J Econ 158(2):204–220
Stone M (1977) An asymptotic equivalence of choice of model by cross-validation and Akaike's criterion. J Royal Stat Soc Ser B (Methodol) 39(1):44–47
Symonds MRE, Moussalli A (2011) A brief guide to model selection, multimodel inference and model averaging in behavioural ecology using Akaike's information criterion. Behav Ecol Sociobiol 65(1):13–21
Tedeschi LO (2019) Assessing the predictive adequacy of simple and complex mathematical models. J Anim Sci 97(Suppl 3):24
Teferra K, Shields MD, Hapij A, Daddazio RP (2014) Mapping model validation metrics to subject matter expert scores for model adequacy assessment. Reliab Eng Syst Safety 132:9–19
Tibshirani R (1996) Regression shrinkage and selection via the lasso. J Royal Stat Soc Ser B Stat Methodol 58(1):267–288
van Buuren S, Fredriks M (2001) Worm plot: A simple diagnostic device for modelling growth reference curves. Stat Med 20(8):1259–1277
van Rij J, Wieling M, Baayen R, van Rijn H (2022) Itsadug: interpreting time series and autocorrelated data using GAMMs (R package version 2.4.1)
Vocabulary.com (n.d.) Adequacy. In Vocabulary.com dictionary. Retrieved April 17, 2025, from https://www.vocabulary.com/dictionary/adequacy
Wang C, Butts CT, Lakon CM (2020) Model adequacy checking/goodness-of-fit testing for behavior in joint dynamic network/behavior models, with an extension to two-mode networks. Sociol Methods Res 51(4):1886–1919
Warren DL, Seifert SN (2011) Ecological niche modeling in Maxent: the importance of model complexity and the performance of model selection criteria. Ecol Appl 21(2):335–342
Watanabe S (2013) A widely applicable Bayesian information criterion. J Mach Learn Res 14(1):867–897
Whittingham MJ, Stephens PA, Bradbury RB, Freckleton RP (2006) Why do we still use stepwise modelling in ecology and behaviour? J Anim Ecol 75(5):1182–1189

Wilke C (2019) Fundamentals of data visualization: A primer on making informative and compelling figures, 1st edn. O'Reilly Media

Yahyah M, Baines A, Joanes DN (2010) Graphical approach to model adequacy based on exact and near replicates. J Appl Stat 37:121–129

Yates LA, Aandahl Z, Richards SA, Brook BW (2022) Cross validation for model selection: a review with examples from ecology. Ecol Monogr 93

Zou H, Hastie T (2005) Regularization and variable selection via the elastic net. J Royal Stat Soc Ser B Stat Methodol 67(2):301–320

Chapter 15
Mean-Variance Modelling of Neotropical Passerine Bird Songs

Rafael de Andrade Moral, Mateus S. Mendes, and Luciano Martins Verdade

Abstract In this study, we evaluated the possible influence of evolutionary and ecological processes on the fundamental frequency of Neotropical passerine bird songs. We selected 27 species from three families, distributed across three distinct guilds. Their song files in WAV format downloaded from WikiAves were compared with available audio from the Macaulay Library, from the Ornithology laboratory at Cornell University, to double-check species. We then obtained sonograms for each song file and computed different variables related to the duration and the musical pitch of the bird songs (fundamental frequency, minimum and maximum frequencies at the start and end of each song snippet). We fitted gamma generalized linear models to the fundamental frequency data, incorporating covariates in both the mean and dispersion parameters (i.e., guild, family, and the interaction between them) to perform tests to assess the significance of the model effects. We also fitted multivariate covariance generalized linear models to the duration and frequency-related variables. We found no significant effect of guild and family or their interaction in the mean fundamental frequency. However, there was a significant effect of the interaction between guild and family in the dispersion. For the multivariate analyses, there were significant global effects of guild, family and their interaction for both the mean and dispersion. Although limited in number of species and families, the present study corroborates the hypothesis that phylogeny limits the scope of bird songs in terms of musical pitch. The present results are potentially relevant, since they stress the importance of behavioral-ecological processes that occur at the individual level on the adaptive pressures concerning song skills in passerine birds.

R. A. Moral (✉)
Department of Mathematics and Statistics, Maynooth University, Maynooth, Co. Kildare, Ireland
e-mail: rafael.deandrademoral@mu.ie

M. S. Mendes
University of São Paulo/CENA/Isotopic Ecology Lab, Piracicaba, SP, Brazil

L. Martins Verdade
Wildlife Monitoring Consultancy, Campinas, São Paulo, Brazil

L. Martins Verdade, R. A. Moral (eds.), *Wildlife Monitoring*,
https://doi.org/10.1007/978-3-032-05821-8_15

Keywords Biomusicology · Double generalized linear models · Fundamental frequency · Musical pitch · Guild

1 Introduction

Musical activity is inherent to the human species, but recently new studies have shown the evolution of this trait in animal species (Brown et al. 2001). The production and emission of sound signals appears to bring great advantages to animals, such as territorial defense, group cohesion, mating, and many other functions (Griffin 1992). Through the emission of sound signals, a receiver may be able to take advantage of them to extract useful information. Species coevolve in environments with particular soundscapes and thus develop and learn to use survival mechanisms, which may be phylogenetically acquired or learned culturally (Bro-Jorgensen 2009). This occurs with some animal groups, for example birds, whose songs may be learned after birth, and vary according to population (Henry and Lucas 2008).

The study of animal singing in biological populations encompassing music comes under the name of biomusicology, coined by the Swedish researcher Nils Wallin, who introduced this definition and popularized the idea (Wallin et al. 2001). As living beings capable of vocalizing, one can study from the anatomical and physiological structure to the cognitive mechanisms that allowed for the development of the receiving apparatus, the emitter and the interpreter of sound signals, enabling the individual to recognise information useful for their survival, distinguishing them from noises. Going beyond, we can also analyze the ecology of these behaviors and the intra- and interspecific interactions, thus promoting the understanding of why other species do something that we call music.

Finally, the means to evaluate these behaviors can be diverse, including aspects of population density, diversity, ecological relations, and include sound aspects such as their physical characteristics (e.g., height, amplitude, frequency), as well as musical aspects (e.g., timbre, texture, tempo, mode, rhythm, form, density and projection). As a relatively new branch of ecological studies, biomusicology represents an interdisciplinary area of science that presents great potential for growth, providing new tools with potentially increased efficacy against already existing sampling methods. The study of sound landscapes can provide complementary information to other ecological studies that results in increasingly accurate diagnostics and thus contribute to conservation and management policies, for example. As an exploratory initial study, we evaluated the possible influence of evolutionary and ecological processes on the fundamental frequency of Neotropical passerine bird songs. We then modelled the multivariate relationships between duration, and minimum and maximum frequencies at the start and end of a bird song to uncover potential relationships between these variables and bird functional (guild) and taxonomic (family) characteristics.

2 Material and Methods

We selected 27 species from three bird families (Emberizidae, Thraupidae, and Tyrannidae), with nine species per family, distributed across three distinct guilds (insectivores, frugivores, and omnivores) (Table 15.1). These families were chosen because they present species that belong to the three guilds, and the species were chosen based on literature that presented a more robust classification within the guilds. We downloaded the song files in WAV format from the sound library WikiAves (2008), one per each species, with no sex distinction. We compared these files with available audio from the Macaulay Library, from the Ornithology laboratory at Cornell University (www.macaulaylibrary.org), to double-check species identification. We then obtained sonograms for each song file, using the bioacoustics software Avisoft SASLab (Specht 2002), as well as data on the fundamental frequencies of each bird song, which is related to the musical pitch. The Avisoft software splits each song file into elements based on an automatic threshold set for the amplitude. To standardize the data collection process, we used the fundamental frequencies calculated from the first ten fragments for each species.

Since the fundamental frequency data consist of positive values and were highly skewed, we fitted gamma generalized linear models (McCullagh and Nelder 1989) with a log link, including in the linear predictor for the mean the effects of guild, family, the interaction between guild and family, and a random effect per species, since observations within the same song file are correlated. Moreover, since the dispersion varied across guild:family combinations, we also modelled the dispersion parameter with covariates, i.e. included the effects of guild, family, and the interaction between them (this type of model can be referred to as a double GLM). In statistical notation, let Y_{ijk} be the fundamental frequency observed for the k-th bird species, which belongs to the i-th guild and j-th family. We then have

$$Y_{ijk} \mid S_k \sim \text{Gamma}\left(\mu_{ijk}, \phi_{ij}\right)$$

Table 15.1 Bird species from which song files were obtained, organised by family (Emberizidae, Thraupidae, and Tyrannidae) and guild (insectivores, frugivores, and omnivores)

Guild	Emberizidae	Thraupidae	Tyrannidae
Insectivore	*Saltator atricolis*	*Conirostrum speciosum*	*Colonia colonus*
	Sporophila caerulescens	*Dacnis cayana*	*Xolmis cinereus*
	Volatinia jacarina	*Tangara seledon*	*Xolmis velatus*
Frugivore	*Arremon flavirostris*	*Hemithraupis ruficapilla*	*Elaenia flavogaster*
	Emberizoides herbicola	*Neothraupis fasciata*	*Tyrannus melancholicus*
	Schistochlamys ruficapillus	*Piranga flava*	*Tyrannus savana*
Omnivore	*Cyanoloxia brissonii*	*Euphonia chlorotica*	*Empidomus varius*
	Saltator fuliginosus	*Euphonia violacea*	*Megarhynchus pitangua*
	Saltator similis	*Pipraeidea melanonota*	*Myiozetes similis*

$$S_k \sim \mathrm{N}\left(0, \sigma_s^2\right),$$

with

$$\log\left(\mu_{ijk}\right) = \beta_0 + g_i + f_j + (gf)_{ij} + s_k$$

$$\log\left(\phi_{ij}\right) = \gamma_0 + g_i^* + f_j^* + \left(g^* f^*\right)_{ij},$$

where β_0 and γ_0 are the intercepts for the mean and dispersion parameters, respectively, g_i and g_i^* are the effects of the i-th guild, f_j and f_j^* are the effects of the j-th family, $(gf)_{ij}$ and $(g^*f^*)_{ij}$ are the effects of the interaction between the i-th guild and the j-th family, and s_k is the k-th species random intercept. This is an example of distributional modelling, reviewed in detail in Stasinopoulos et al. (2026, this volume). Under our parameterization of the gamma models, we have,

$$\mathrm{Var}\left(Y_{ijk} | S_k = s_k\right) = \mu_{ijk}^2 \phi_{ij}^{-1},$$

and hence a larger value for the dispersion parameter ϕ_{ij} corresponds to a smaller variance. We then performed likelihood-ratio (LR) tests to assess the significance of the model effects. Finally, we fitted nested models to group the parameter estimates of categories with similar behavior. We used package glmmTMB (Brooks et al. 2017) to fit the models and a function based on package coefplot2 (Bolker and Su 2011) to produce the coefficient plots, for R software (R Core Team 2019).

To model the multivariate relationship between the variables duration, and minimum and maximum frequencies at the start and end of the bird songs, we used the framework of multivariate covariance generalized linear models (McGLM; Bonat and Jørgensen 2016). This is a marginal modelling framework, and therefore there is no assumption of a probability distribution, only first and second moments (i.e. mean and variance). Now let $Y_{ijk,r}$ denote the observations made for the r-th response variable, where $r = 1, 2, 3, 4, 5$. Our assumption for the mean $\mu_{ijk,r}$ is

$$\log\left(\mu_{ijk,r}\right) = \beta_{0,r} + g_{i,r} + f_{j,r} + (gf)_{ij,r},$$

where the effects of guild, family and their interaction are now allowed to differ between response variables. Our assumption for the variance-covariance matrix for the full matrix of responses $\boldsymbol{Y}$ is

$$\mathrm{Var}(\boldsymbol{Y}) = \mathrm{Bdiag}\left(\tilde{\boldsymbol{\Sigma}}_1, \tilde{\boldsymbol{\Sigma}}_2, \tilde{\boldsymbol{\Sigma}}_3, \tilde{\boldsymbol{\Sigma}}_4, \tilde{\boldsymbol{\Sigma}}_5\right)\left(\boldsymbol{\Sigma}_b \otimes \mathbf{I}_5\right)\mathrm{Bdiag}\left(\tilde{\boldsymbol{\Sigma}}_1^T, \tilde{\boldsymbol{\Sigma}}_2^T, \tilde{\boldsymbol{\Sigma}}_3^T, \tilde{\boldsymbol{\Sigma}}_4^T, \tilde{\boldsymbol{\Sigma}}_5^T\right),$$

where Bdiag$(\cdot)$ is an operator that creates a block-diagonal matrix, $\boldsymbol{\Sigma}_b$ is the variance-covariance matrix between response variables. $\tilde{\boldsymbol{\Sigma}}_r$ is the lower triangular part of the Cholesky decomposition of $\mathbf{\Sigma}_r$, which is the variance-covariance matrix of the r-th response and takes the form

$$\mathbf{\Sigma}_r = \mathbf{V}_r^{1/2} \mathbf{\Omega}_r (\boldsymbol{\tau}) \mathbf{V}_r^{1/2},$$

where $\mathbf{V}_r$ is a diagonal matrix incorporating the variance function for the r-th response variable and $\mathbf{\Omega}_r(\boldsymbol{\tau})$ is the matrix that incorporates the effects of potential covariates. Since all variables are continuous and right-skewed, we assumed a Tweedie variance function, which translates into specifying the diagonal elements of $\mathbf{V}_r$ as

$$\vartheta_r = \mu_{ijk,r}^{p_r},$$

where p_r is the power parameter to be estimated for each response variable. This power parameter makes this type of regression model very flexible, and allows it to incorporate different density shapes in terms of skewness and kurtosis (i.e., heavy-tailedness). Finally, the $\mathbf{\Omega}_r(\boldsymbol{\tau})$ matrix is modelled with an exponential-matrix link function to include the additive effects of guild and family in the dispersion. We assessed the significance of the effects on the mean and variance for each response via Wald tests based on the asymptotic normality of the estimators. The model was fitted using package mcglm (Bonat 2018), for R software.

All data and R scripts are made available at https://www.github.com/rafamoral/meanvar_birdsongs.

3 Results

3.1 Fundamental Frequencies

The fundamental frequency data was highly skewed to the right, and observations from omnivores appeared to present less variability overall (Fig. 15.1). We found no significant effect of the interaction between guild and family in the mean fundamental frequency (LR = 6.01, d.f. = 4, p = 0.1983), and also no significant main effects of guild (LR = 0.42, d.f. = 2, p = 0.8105) or family (LR = 0.89, d.f. = 2, p = 0.6415), (Fig. 15.2a). However, there was a significant effect of the interaction between guild and family in the dispersion (LR = 49.58, d.f. = 4, $p < 0.0001$). The combinations of guild and family that presented a statistically different dispersion were omnivores with Thraupidae and Emberizidae (grouping all other seven combinations yielded LR = 7.46 associated with 6 d.f., p = 0.2806), (Fig. 15.2b).

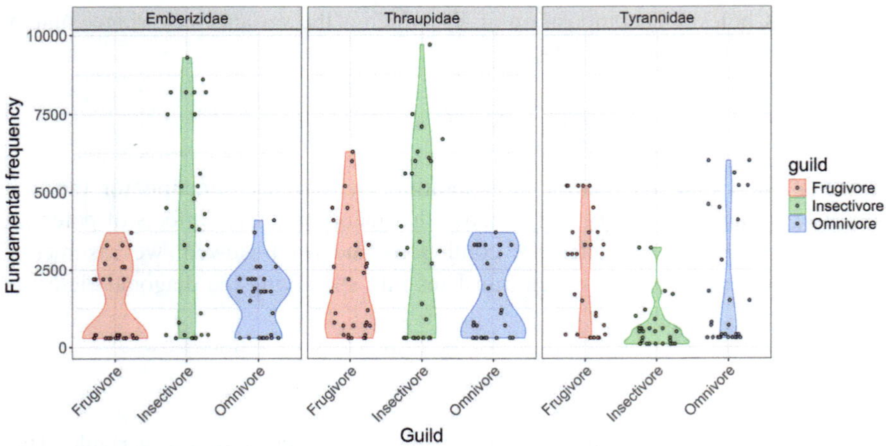

Fig. 15.1 Violin plots of the observed fundamental frequencies calculated for each bird song, split by family and guild. Points are jittered to aid visualization

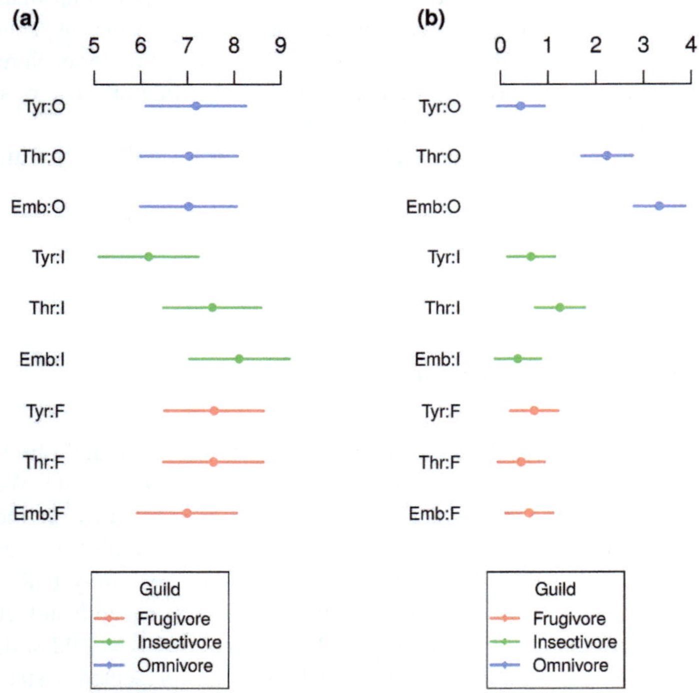

Fig. 15.2 Estimated coefficients and respective 95% Wald confidence intervals for the (**a**) parameters in the linear predictor for the mean (**b**) parameters in the linear predictor for the dispersion. Tyr Tyrannidae, Thr Thraupidae, Emb Emberezidae, O omnivores, I insectivores, F frugivores

3.2 Multivariate Modelling

The five variables analysed all presented a highly right-skewed distribution, with the presence of outlying observations (Fig. 15.3). It was found that guild membership played a significant role in the variance of song duration, but neither guild nor family significantly affected the mean duration (Fig. 15.4). The means of the maximum frequencies, both at the start and end of the song snippets, were significantly affected by the interaction between guild and family, whereas the variance was also significantly affected by guild and family (Fig. 15.4). For the minimum frequencies, however, this was the case only for the mean, whereas family and guild had little effect on the variance (Fig. 15.4). After taking into account the effects of guild and family in the mean and variance, there was little leftover correlation between duration and the frequency-related variables, but the minimum and maximum frequencies at the start and end of the song snippets were positively correlated (Fig. 15.5). Finally, the power parameter estimates for all variables but maximum frequency at the start were positive and significant, and allowed for the accommodation of the skewness and kurtosis in the data.

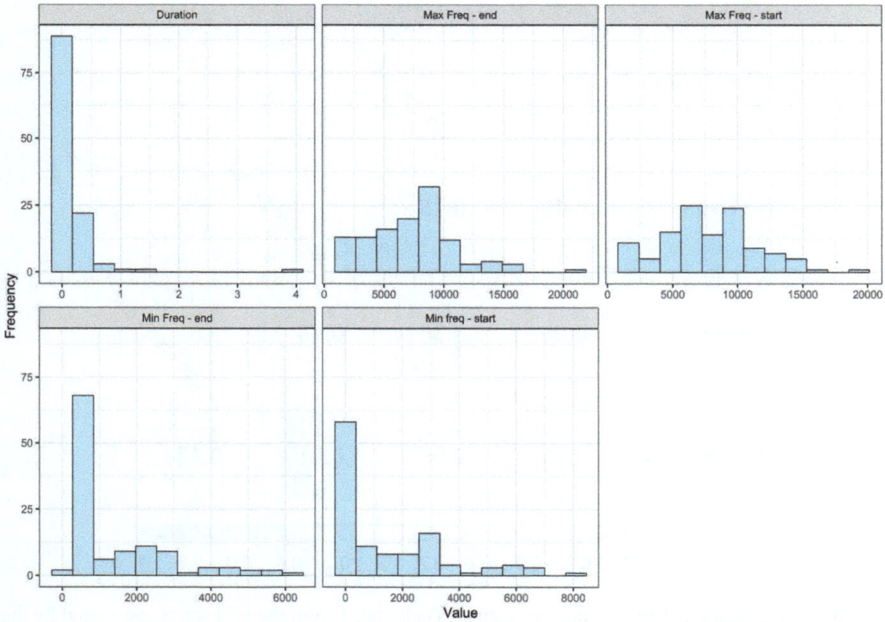

Fig. 15.3 Histograms of the five variables analysed via the multivariate covariance generalized linear modelling framework. Duration is computed in seconds, whereas the frequency-related variables are computed in Hz

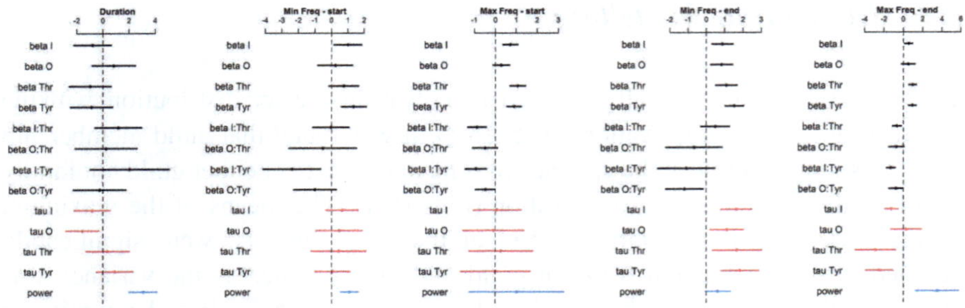

Fig. 15.4 Coefficient estimates (points) and 95% confidence intervals (line segments) obtained from the multivariate covariance generalized linear model. The "beta" coefficients (black) relate to the mean (I insectivores, O omnivores, Thr Thraupidae, Tyr Tyrannidae, I:Thr interaction between insectivores and Thraupidae, O:Thr interaction between omnivores and Thraupidae, I:Tyr interaction between insectivores and Tyrannidae, O:Thr interaction between omnivores and Tyrannidae), whereas the "tau" coefficients (red) relate to the dispersion. Frugivores and Emberezidae were the baseline levels. The "power" coefficient (blue) is the power parameter of the Tweedie-family variance function, which accommodates flexibility in terms of skewness and kurtosis

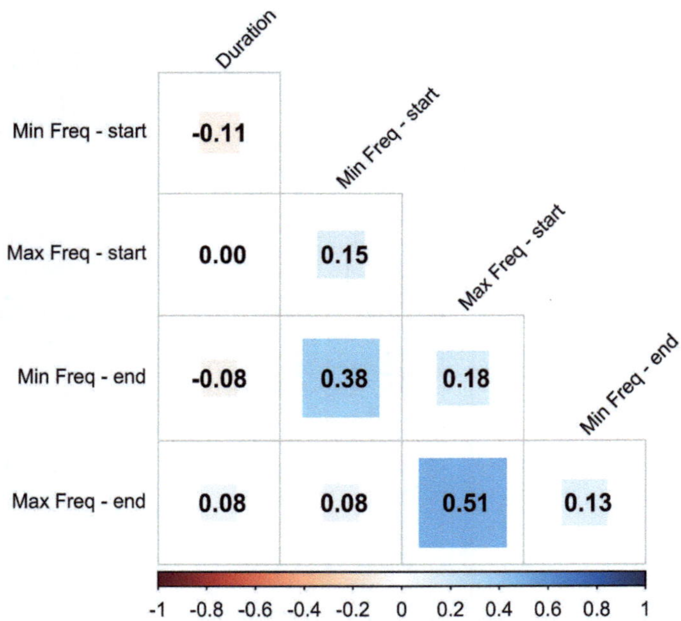

Fig. 15.5 Correlations between the five variables calculated from the bird songs, estimated by the multivariate covariance generalized linear model, after taking into account the effects of guild and family in both the mean and dispersion

4 Discussion and Conclusion

Although limited in number of species and families, the present study corroborates the hypothesis that phylogeny limits the scope of bird songs in terms of musical pitch. However, ecological processes related, for instance, with functional groups might have a relevant role in microevolutionary processes concerning pitch variability. In this study, this variability appears to be smaller for omnivores. However, the categorization within guilds might be eventually dubious in the sense that it can be based on an incomplete knowledge of the species' diet and foraging behavior (Simberloff and Dayan 1991). Also, foraging behavior may be associated with habitat, which is usually related with the nature of physical barriers in the soundscape. For example, ground and canopy birds can be omnivorous, however their soundscape would likely differ, and so would the evolutionary pressures on their song skills.

The double GLM framework played an important role in uncovering the relationship between family and guild and the fundamental frequency data. Frequently, statistical analyses focus solely on making inference about the mean, while the dispersion (or other) parameters are treated as "nuisance", only serving the purpose of fitting a particular distribution and accounting, for instance, for extra-variability (or over-dispersion). In this work, modelling the dispersion with covariates was crucial to understand how family and guild were influencing the distribution of the fundamental frequencies, which was not at the mean level (or location of the distribution), but at the variance level (or scale/spread of the distribution). The McGLM framework was also useful in jointly modelling all five variables whilst taking into account the correlation between them, as well as the effects of guild and family in both the mean and dispersion. Even though there are limitations in the dataset, these modelling frameworks proved promising tools for future studies in this area.

The present results are potentially relevant, since they stress the importance of ecological processes that occur at the individual level on the adaptive pressures concerning song skills in passerine birds. In addition, the present methodology, based on the joint modelling of mean and dispersion, can be crucial to understand how novel anthropic pressures can affect song development in passerine birds (e.g., Mendes et al. 2011). For this reason, future studies should consider habitat as a possible source of variation in bird song features, which can be particularly relevant for monitoring programs based on songbirds. Future studies should not only include a larger number of species and families at crossing spatial scales (from local to regional) and different structuring of functional groups, but also measurements of other musical aspects, such as density, whereby the interaction between individuals' songs and the environment can also be studied. Moreover, studies involving a single species and how its vocalization changes in terms of the harmonic spectrum and projection would be useful to understand intraspecific and individual behavioral variation.

Acknowledgments This study was financially supported by Fundação de Amparo à Pesquisa do Estado de São Paulo (FAPESP)/Biota Program (Proc. No. 2017/01304-4). LMV held a research productive fellowship from CNPq (Proc. No. 312049/2015-3).

References

Bolker B, Su Y (2011) coefplot2: coefficient plots. R package version 0.1.3.2

Bonat WH (2018) Multiple response variables regression models in R: the mcglm package. J Stat Softw 84(1):1–30

Bonat WH, Jørgensen B (2016) Multivariate covariance generalized linear models. J R Stat Soc Ser C Appl Stat 65(5):649–675

Bro-Jorgensen J (2009) Dynamics of multiple signalling systems: animal communication in a world in flux. Trends Ecol Evol 25:292–300

Brooks ME, Kristensen K, van Benthem KJ, Magnusson A, Berg CW, Nielsen A, Skaug HJ, Maechler M, Bolker BM (2017) glmmTMB balances speed and flexibility among packages for zero-inflated generalized linear mixed modeling. R J 9:378–400

Brown S, Merker B, Wallin NL (2001) An introduction to evolutionary musicology. In: Wallin NL, Merker B, Brown S (eds) The origins of music. The MIT Press

Griffin DR (1992) Animal minds. Chicago University Press, 376 pages

Henry KS, Lucas JR (2008) Coevolution of auditory sensitivity and temporal resolution with acoustic signal space in three songbirds. Anim Behav 76:1659–1671

McCullagh P, Nelder JA (1989) Generalized linear models. CRC Press, 532 pages

Mendes S, Colino-Rabanal VJ, Peris SJ (2011) Bird song variations along an urban gradient: the case of the European blackbird (Turdus merula). Landsc Urban Plan 99:51–57

R Core Team (2019) R: a language and environment for statistical computing. R Foundation for Statistical Computing, Vienna. https://www.R-project.org/

Simberloff D, Dayan T (1991) The guild concept and the structure of ecological communities. Annu Rev Ecol Syst 22:115–143

Specht R (2002) Avisoft-saslab pro: sound analysis and synthesis laboratory. Avisoft Bioacoustics, Berlin

Stasinopoulos DM, De Bastiani F, Burke K, Merder J (2026) Distributional regression techniques for wildlife management. In: Martins Verdade L, Moral RA (eds) Wildlife monitoring: integrating conservation and innovation in human-altered landscapes. Springer, Cham

Wallin NL, Merker B, Brown S (2001) The origins of music. MIT Press, 512 p

WikiAves (2008) WikiAves, a enciclopédia das aves do Brasil. Available at: http://www.wikiaves.com.br/. Access in: 15/11/2018

Chapter 16
Monitoring with Machines: A Review of Computational Bioacoustics

Anthony Gibbons, Ian Donohue, and Andrew Parnell

Abstract Ecological monitoring is essential for understanding biodiversity amid rapid global change. Within wildlife monitoring, Passive Acoustic Monitoring (PAM) has become a powerful and commonplace tool. Advances in computational methods—fuelled by affordable data collection and machine learning—have accelerated its growth, yet challenges persist, particularly in data integration, species coverage, and classification model evaluation. In this paper, we systematically review the state of PAM, analysing approximately 80 studies (2005–2024) spanning diverse taxa and ecosystems. We evaluate its strengths, limitations, and applications. Our findings highlight significant increases in dataset sizes over time, with spectrograms dominating preprocessing methods. However, there remains a lack of consensus on evaluation methods for machine learning models, and code sharing has outpaced data availability. We propose a roadmap for future research for closing identified gaps, which includes advocating for open-sourced data and code, improving data preprocessing, and providing accessible model hosting. Broad uptake of these recommendations will enhance impact of computational bioacoustics on ecological science.

Keywords Bioacoustics · Passive Acoustic Monitoring · Species recognition · Machine learning

A. Gibbons (✉)
Department of Mathematics and Statistics, Hamilton Institute, Maynooth University, Kildare, Ireland
e-mail: anthony.gibbons.2022@mumail.ie

I. Donohue
Zoology, School of Natural Sciences, Trinity College Dublin, Dublin, Ireland

A. Parnell
School of Mathematics and Statistics, University College Dublin, Dublin, Ireland

L. Martins Verdade, R. A. Moral (eds.), *Wildlife Monitoring*,
https://doi.org/10.1007/978-3-032-05821-8_16

1 Introduction

Wildlife monitoring is critical for understanding ecological dynamics, both on land and in marine environments. Tracking bird species diversity can act as a proxy for habitat health and is a commonly monitored metric for signaling broader environmental stress (Shirley and Janke 2023). Bat surveys safeguard vital roosting habitats, such as caves and forests, protecting vulnerable species from threats (Jones et al. 2009). Insect monitoring can function both to support pollinator restoration efforts or with pest control (Forup and Memmott 2005). Similarly, marine surveys can serve either tracking invasive species or detect overfishing or impacts of pollution in the water (Harris and Tyrrell 2001).

Traditionally, species monitoring surveys have relied on human-intensive methods such as camera trapping (Cao et al. 2022), DNA barcoding (Lahaye et al. 2008), capture-recapture (Converse et al. 2022), and radio tracking (Gottwald et al. 2019). However, the planning and environmental impact assessment of infrastructural developments such as wind farms often necessitate even more detailed monitoring, especially in Special Areas of Conservation where species like birds and bats are of particular concern due to potential mortality risks from wind turbine collisions (Wood et al. 2017).

In recent years, acoustic monitoring has become a vital tool in wildlife research, driven by advancements in hardware that have enhanced the efficiency and scale of data collection. *Bioacoustics*, the scientific study of animal sound, involves analysing animal vocalizations and soundscapes and investigating their ecological behaviour. Meanwhile, *Passive Acoustic Monitoring (PAM)* refers to the use of acoustic sensors to non-invasively capture environmental sounds. Although the terms are occasionally used interchangeably in this review, we distinguish Bioacoustics as pertaining to the signal processing and analysis of audio datasets, while PAM specifically denotes the surveillance technique itself.

The adoption of PAM has accelerated the shift toward big data in bioacoustics, enabling researchers to handle increasingly large and complex datasets. Moreover, citizen science initiatives such as Xeno-canto[1] and Cornell Lab's eBird[2] have contributed to this trend by expanding the scale and scope of data collection (Fierro et al. 2016).

The integration of Artificial Intelligence (AI) and Deep Learning (DL) has shifted bioacoustic research away from traditional methods, improving large-scale data processing and species identification accuracy with minimal human effort (Andaryani et al. 2021). This review assesses these advancements, their impact on wildlife monitoring, and future directions, focusing on the migration towards more fully autonomous methods. Four specific research questions are addressed:

1. What are the strengths and weaknesses of PAM and bioacoustics compared with traditional (manual) species monitoring methodologies?
2. What signal processing tools and methods are used to analyse bioacoustic data?

[1] https://xeno-canto.org/.

[2] https://ebird.org.

3. How do the diversity, use and performance of automatic monitoring methods compare?
4. What is the level of open-source code and data practices in bioacoustics research?

The structure of this paper is as follows. In Sect. 2, we describe the systematic process used to select and analyse the literature. In Sect. 3, we present the primary findings of the review based on the above research questions. We conclude with some recommendations of best practice moving forward in using autonomous (in particular machine learning) methods in bioacoustics.

2 Survey Methodology

Our review employed a systematic approach to identify and evaluate relevant studies, guided by established protocols for comprehensive literature searches. The methodology encompassed a broad search strategy with a preference towards articles in peer-reviewed academic journals. The process involved defining clear inclusion and exclusion criteria, followed by a structured assessment to investigate variables related to the research questions across the resulting body of literature.

PAM is now a staple of ecological monitoring by utilizing low cost recording and data storage equipment. While audio processing and machine learning methods did not gain much prominence in the bioacoustics field until the late 2000s (Jerome Sueur and Simonis 2008; Ren et al. 2009), and deep learning not until 2016 (Goëau et al. 2016), manual surveying by expert ecologists and ornithologists has remained common in fieldwork since the early 2000s (Camphuysen et al. 2004; Lattemann and Höpner 2008; Drewitt and Langston 2006). We have therefore constrained our search to papers published since 2000.

We used Google Scholar, Scopus and Web of Science for the initial literature search, which was carried out on 19th November 2024, with the following search string:

```
("passive acoustic monitoring" OR "species monitoring"
OR bioacoust* OR ecoacoust* OR vocali* OR biodiversity
OR "soundscape" OR "biodiversity assessment" OR "species
detection") AND (animal OR bat* OR bird* OR cetacean* OR
insect* OR mammal*) AND (acoustic OR ultrasonic OR harmonic
OR "signal processing" OR software OR "data analysis"
OR "classification accuracy" OR "machine learning" OR
"deep learning" OR "statistics" OR "acoustic indices")
AND ("automatic monitoring" OR "automatic detection" OR
"automated detection" OR "manual methods" OR "traditional
methods" OR "traditional monitoring")
```

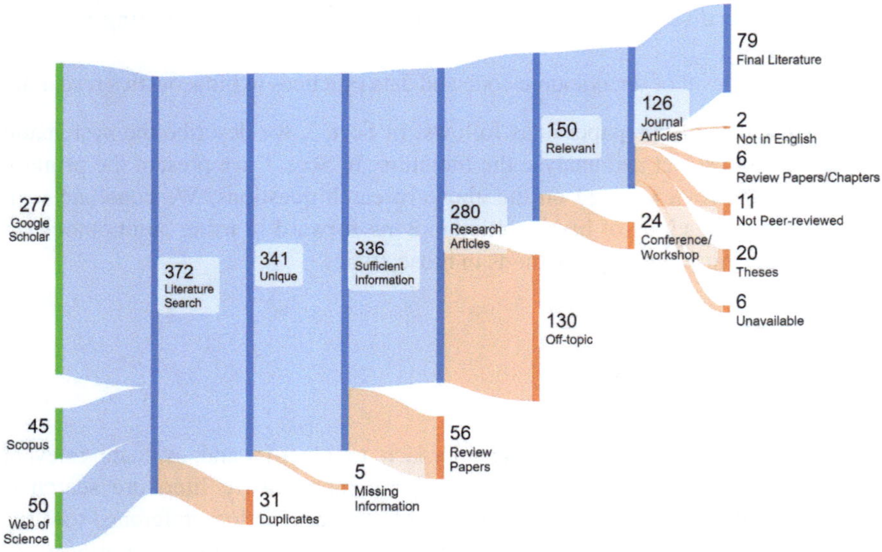

Fig. 16.1 Systematic review collection and filtering stages (created with https://sankeymatic.com/)

This search returned 277, 45 and 50 entries on Google Scholar, Scopus and Web of Science respectively, or 372 entries in total. The literature was then filtered by removing duplicates, entries with missing abstracts, those not in English, and papers unavailable or tied to workshops and conferences. Further screening excluded literature reviews, off-topic papers, and items like theses or non-peer-reviewed preprints. In the final manual filtering stage, some exclusion criteria, such as not being in English or being review papers, were applied again since some results were not identified initially through metadata. Figure 16.1 gives a visual breakdown of inclusion and exclusion criteria.

This finally yielded 79 entries, ranging from 1 entry in 2005 to 16 entries in 2024. Five years (2006, 2008, 2010, 2011, and 2012) had no entries. The `.bib` file published along with this paper lists all articles, books, proceedings, and reports resulting from this screening process, along with other articles added for context while writing the review. See https://github.com/gibbona1/pam_lit_review. Figure 16.2 provides a detailed depiction of the geographic distribution of study sites referenced in the literature. The points are coloured by their primary ecosystem types studied, with a notable concentration of sites along coastal regions. All the continents, excluding Antarctica, have some level of representation among the included literature.

From each article in the resulting literature list, the following key information was extracted: publication year, monitoring year, location of the study site, monitored taxa, dataset size and duration, preprocessing and analysis methods employed, programming languages used, data and code availability, reported metrics and their

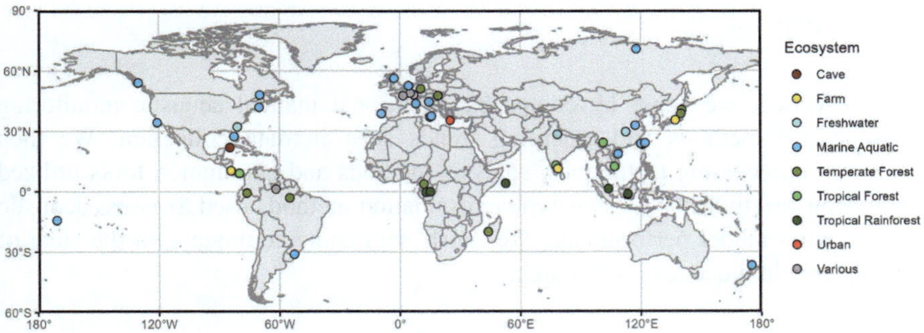

Fig. 16.2 Geographic distribution of the study sites in the literature. The map has been cropped to remove everything below 60°S, which included Antarctica, as there were no studies present

Table 16.1 Description of variables extracted in this review

Variable	Description
`Source Type`	Document type, e.g., book chapter, journal article.
`Year`	The year the study was published.
`Monitoring Year`	The year or range of years when data was collected.
`Study Sites`	Description of where the research took place.
`Latlong`	Coordinates (latitude,longitude) of the study sites.
`Dataset Hours`	Total hours of data collection or monitoring time.
`Dataset GB`	Size of the dataset in gigabytes.
`Taxa`	The taxonomic group(s) monitored.
`Analysis`	Automated, Semi-automated or Manual.
`Preprocessing`	Data cleaning or preparation steps before analysis.
`Analysis Methods`	Methodologies and techniques used.
`Automation Tools`	Automation tools or software used in analysis.
`Programming Language`	Programming language(s) used.
`Code Availability`	Whether the code is publicly available (Yes, No).
`Data Availability`	Whether the data can be accessed (Yes, No).
`Evaluation Method`	Method used to assess results or model performance.

corresponding values, evaluation methods used, and research gaps/limits identified. Some of this information was not available in the papers themselves or their supplementary information, so estimates were made when feasible, otherwise fields were left blank. In particular, dataset size often had to be estimated using the length (in hours) of the data collected and the recorder's chosen sampling rate and bit depth. Similarly, the latitude and longitude often had to be estimated based on Google Map searches of the reported sites. Furthermore, a number of columns in the extracted data available in the Supplementary Material are not included here as they are free text fields that we used as reference for calculation or categorizing discrete values. Table 16.1 gives a full breakdown of the chosen variables.

3 Results

In this section, we begin by comparing traditional manual acoustic monitoring with passive methods and emerging trends seen in data collection. We then analyse preprocessing techniques, analysis methods and automation tools utilized in bioacoustic studies. Next, we explore evaluation methods used to assess acoustic classification model performance. We finish with some comments on the state of open-source in bioacoustics research.

3.1 Traditional Versus Passive Acoustic Monitoring

Traditional acoustic monitoring surveys were heavily reliant on human expertise and simple tools (e.g., listening or non-programmable recorders), with a history of use akin to traditional trapping. While fully manual data collection and annotation is occasionally still done, it is often in a lab environment (Görres and Chesmore 2019 for example).

By manual monitoring, we refer to studies where humans were responsible for collecting and processing all the data. Autonomous methods rely entirely on computational techniques, such as machine learning or other software, to process nearly all the data. Semi-autonomous methods combine elements of both approaches. The vast majority of the literature required at least some level of human annotation and inference for at least part of the data collected to act as a ground-truth evaluation method, so we did not categorize papers using this approach as wholly manual. Thus, papers identified as using automated or semi-automated acoustic analysis can be classified as utilizing PAM.

Traditional manual monitoring exhibits several strengths, arising from a longer history of use than PAM. It is effective in targeting individual or a small list of taxa (Prendergast et al. 2020). There is a low technological barrier, usually requiring minimal equipment such as handheld recorders or simply documenting by ear. The data collection process is thus less likely to fail when compared with passive methods. Direct human observation allows for real-time interpretation of sounds, which supports immediate management decisions like crop pest control (Lagos-Kutz et al. 2020). Lastly, maintaining the ability to distinguish species by ear or through basic analysis with trained listeners is needed to validate data (Drew 2011).

Disadvantages of traditional acoustic monitoring methods stem from the limitations of human surveyors. These methods are labour-intensive, requiring active human presence and listening effort, similar to the demands of sweep netting (Prendergast et al. 2020). They rely heavily on taxonomic expertise, to identify dozens of species by sound, a scarce resource (Williams et al. 2018). Even so, human observation introduces subjectivity, leading to observer bias and inconsistent, often unverifiable, results, much like transect walks, which hampers reproducibility (Prendergast et al. 2020). Manual monitoring offers limited scalability due to the

Table 16.2 Strengths and weaknesses of manual acoustic monitoring

Aspect	Strengths	Weaknesses
Effectiveness	Works well in targeted settings	Limited scale
Accessibility	Low tech barrier	Labor intensive
Real-time use	Direct observation	Temporal constraints
Precision	Species specificity	Expertise dependent
Consistency	–	Subjectivity

increased labour cost, rendering it ineffective for large areas or extended time periods. Finally, temporal constraints mean traditional methods miss ecological activity outside observation windows, such as at night, due to safety and effectiveness risks for surveyors (Prendergast et al. 2020). These strengths and limitations are summarized in Table 16.2.

The four papers tagged as performing manual monitoring addressed some drawbacks or limitations to their methods, namely:

- The study by Picciulin et al. (2022) analysed the impact of boat noise on fish and dolphin biophony. Future research aims to mitigate anthropogenic noise and expand to species-specific contributions to the soundscape.
- Papin et al. (2018) developed an acoustic localization method for monitoring grey wolves at large scales. It mentions integration of automatic detection and localization, as well as expanding to more species.
- Sanchez et al. (2017) recorded and analysed the echolocation calls of *Natalus primus* bats to support conservation. The research was constrained by a limited study area and the authors seek to next explore diverse habitats.
- Milanelli et al. (2024) examined temporal patterns in the soundscape of a port area within an urban estuary. Plans for further research includes understanding the influence of soundscape on species and assessing long-term anthropogenic noise impacts.

PAM has become the prevailing analysis setup for acoustic-based wildlife monitoring. Figure 16.3 shows the dominance of automated and semi-automated methods as opposed to purely manual monitoring for the included literature, growing stronger across time. Manual monitoring features sparingly throughout the study period, appearing only four times. These studies focused on bats, carnivores, fish and marine mammals. Semi-automated analysis has been used relatively consistently over time, though at relatively low levels. Both fish and marine mammals had roughly equal amounts of papers using automated and semi-automated analysis, owing to the more difficult data collection process associated with ocean ecosystems. Automated methods are by far the most popular analysis method since 2020. The trend appears to show that the field will continue to grow in line with those using computational methods.

Anthropogenic noise can be reduced with passive acoustic recorders due to the lack of a requirement for a human or machine/vehicle present. Expanding the space-

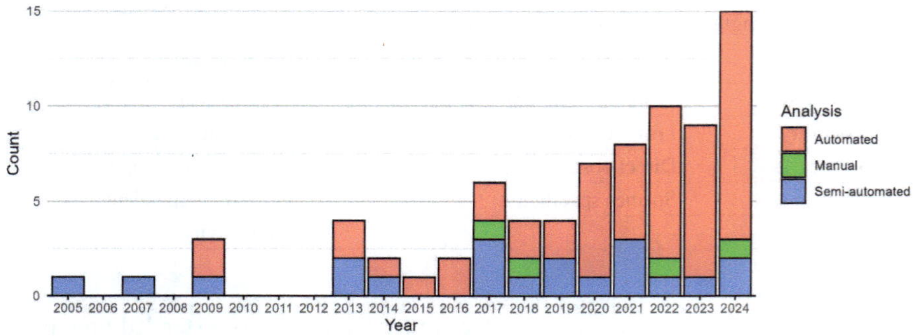

Fig. 16.3 Analysis type by year

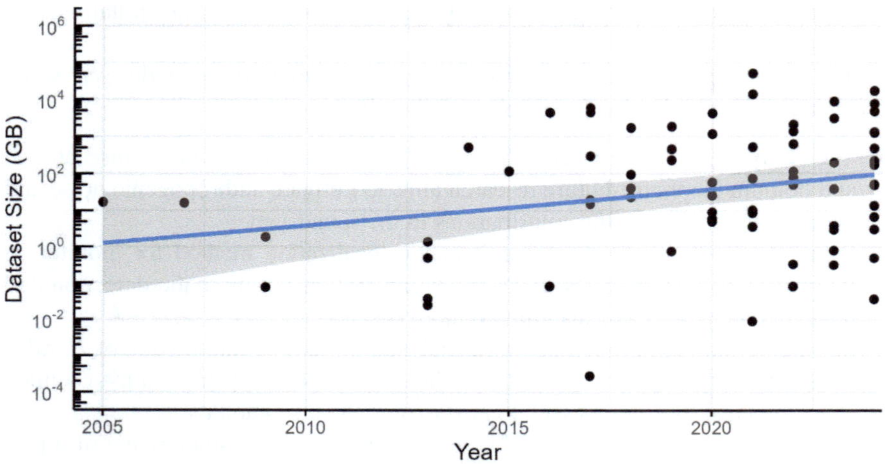

Fig. 16.4 Dataset size of studies, in log scale. The dataset sizes in gigabytes, when available, is shown by year. The data size ranges from under 1 MB to 50 TB. The slope of the fitted line is 0.226 (95% CI: [0.016, 0.436]) shows a significant positive trend. Given the log scale, this suggests exponential growth in dataset size over time

and time-scale of studies quickly outpaces the labour budget and so autonomous data collection is a worthwhile step (Oswald et al. 2022; Sebastián-González et al. 2015). The ease of expanding PAM projects compared to the costs of purely manual monitoring efforts allows for greater efficiency in gathering data, calculating species diversity metrics by leveraging economies of scale. The growth in data collection over time exemplified this migration to acoustic monitoring as a 'big data' problem. Exponential dataset growth, facilitated by cheaper audio data collection and storage hardware, quickly surpasses what can feasibly be processed by humans alone, and so automatic methods are employed to perform inference to match the scope and granularity of data collection. Figure 16.4 shows the size and length of the datasets used in the studies, when available (note the log scale on the vertical axis).

Several of PAM's weaknesses, however, can be regarded as strengths of traditional methods, several of which were mentioned above. The migration towards deep learning (DL) has raised the technological barrier for non-technical ecologists, particularly in terms of data quality, microphone range (Gillespie et al. 2020), reliance on a singular mode (sound) compared to humans (sight and sound), and high capital expense of equipment. Studies often employ classifiers with a static taxa list and frequently mention future plans to expand it (Sun et al. 2022; Risch et al. 2013), whereas expert ecologists demonstrate greater flexibility. The question of recorder placement and management often arises (Papin et al. 2018; Putland et al. 2018). Data management poses additional challenges (Trapanotto et al. 2022; Arnaud et al. 2022).

3.2 Signal Processing Tools and Software

The standard acoustic monitoring pipeline involves some level of preprocessing before the bulk of the computation (e.g. DL Models, species classification) is carried out. The spectrogram is by far the most common of these techniques used in bioacoustics, as this visualization of sound can be treated like an image which can then draw from decades of computer vision research. Spectrograms can be further modified with the mel scale (Jeantet and Dufourq 2023), log magnitude (Trapanotto et al. 2022), per channel energy normalization (PCEN), and extracting Mel Frequency Cepstral Coefficients (MFCC) (Varun Prakash et al. 2024). Data augmentation is still somewhat rare in bioacoustic studies (Jeantet and Dufourq 2023; Owens et al. 2024) but is widely used in Computer Vision.

A comparison of preprocessing methods used in each decade in the study period is given in Fig. 16.5. 'Other Methods' are those that only appeared once across the surveyed literature. Frequency Filters, including High-pass (Fundel et al. 2023) and Band-pass (Rivas et al. 2024) filters, feature in each decade. Low-pass filters are much less common in acoustic monitoring studies since the vast majority of biophony occurs at higher frequencies and so that signal should be maintained. Segmentation, in which audio clips are split into (possibly overlapping) segments isolate specific sounds or events. It is a standard preprocessing technique in bioacoustics and may not always be explicitly mentioned in the data or methods sections of the studies; its relative prevalence is, therefore, likely to be underestimated here.

Bioacoustic researchers utilise a variety of programming languages for their data analysis, with Python, MATLAB and R being the most commonly used. They feature in 35, 24 and 16 of the featured literature items, respectively, although MATLAB has become less prominent since the 2020s. MATLAB offers excellent audio functions, signal processing capabilities, and integrated development environment, making it intuitive for beginners. However, it is proprietary and it is unclear if the scripts used could be run in Octave, its open-source equivalent. Python and R are both open-source. Python excels in deep learning applications and provides extensive libraries for audio analysis and signal processing. It has

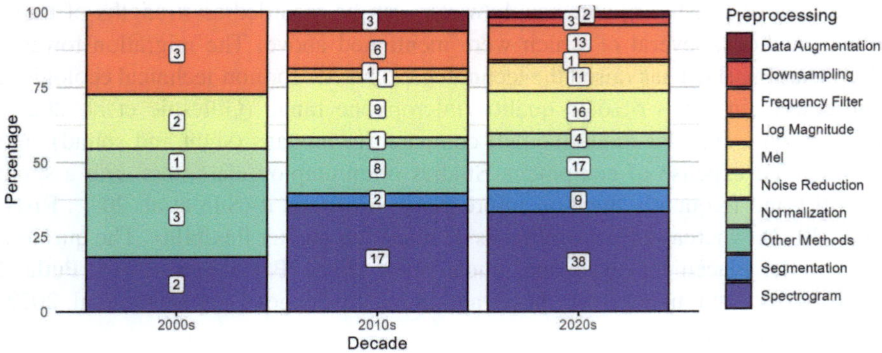

Fig. 16.5 Preprocessing methods per decade. The y-axis shows the percentage breakdown of total preprocessing methods used in each decade. Each stacked bar shows the number of studies using each method. Since some papers use multiple methods, the bar totals do not match the decade's paper count. Spectrograms are the most common method overall, though Frequency Filters (which includes manual, band-pass and high-pass filters) were slightly more common in the 2000s. Normalization, common across computer vision generally, was common in each decade. The category 'Other Methods' increases in each decade showing the occasional need for niche techniques. We see the number of unique methods (including 'Other Methods') per decade increasing from 7 in the 2000s to 26 in the 2020s. Papers published in 2024 alone used 14 distinct methods. Mel-based methods (both mel spectrogram and MFCC) showed up once in the 2010s then increased significantly to 11 times in the 2020s

become the default language for deep learning in bioacoustics. The most commonly used Convolutional Neural Network (CNN) architectures such as ResNet, VGG, MobileNet and Inception all have implementations, some even have possibility of loading weights from instances trained on the ImageNET dataset (Deng et al. 2009). R is particularly strong in statistical analysis and offers several widely used audio packages. Figure 16.6 gives the distribution of programming languages used over the survey period.

The counts of analysis methods and automation tools used is shown in Fig. 16.7. Many papers feature more than one analysis method or automation tool and so the sum of counts may not equal the number of papers analysed. 'Analysis methods' describes the procedures or techniques employed in the studies, not the software itself. This includes classification methods, Machine Learning models (CNN, Random Forest, Hidden Markov Model), sub-fields of deep learning (Transfer learning, Data Augmentation, Ensemble learning), and clustering noise reduction. CNNs dominate here as classification tasks have leapfrogged off computer vision enhancements signalled by the AlexNet (Krizhevsky et al. 2012) paper. Species accumulation curves plot the cumulative number of species observed with increasing sampling effort and tend to asymptote as most species in an area are documented. They are used to indicate the rate at which a particular site or temporal resolution gains species richness (Metcalf et al. 2021). 'Other Methods', a collection of approaches that each only appeared once in the literature (once or twice for automation tools), shows a large amount of rare applications. Papers tagged with

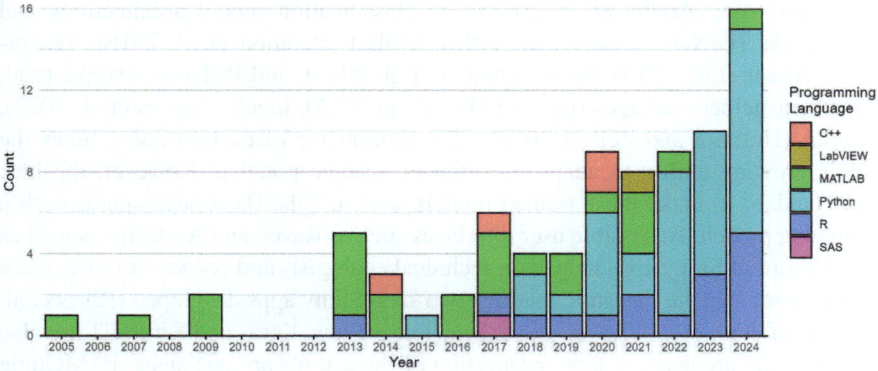

Fig. 16.6 Programming language used per year. There is a clear movement from MATLAB and C++ to more data science oriented languages, such as Python and R. The existence of many machine learning frameworks such as PyTorch and TensorFlow dramatically reduce the time taken to train ML models. R also has a range of packages for bioacoustics research in particular such as `seewave` (Sueur et al. 2008). SAS and LabVIEW were each featured once and so it is difficult to say whether they are likely to gain popularity

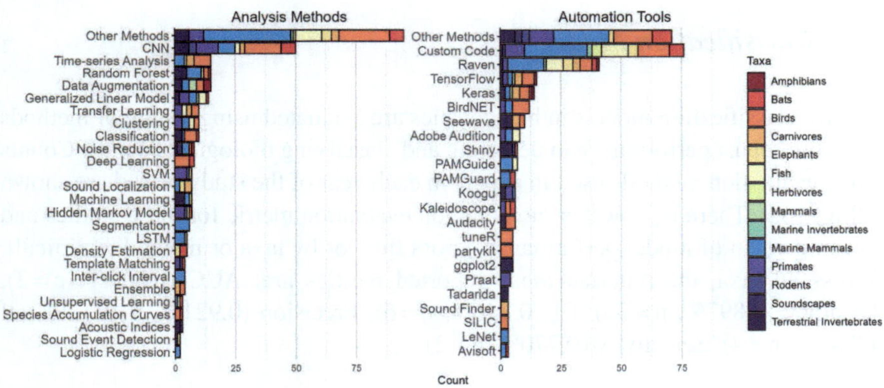

Fig. 16.7 Popular analysis methods and automation tools by taxa. Most of the less popular analysis methods were only used on birds and marine animals. 'Other Methods' was by far the largest category, with a large proportion going towards fish, marine mammals and birds. For Automation Tools, 'Custom code' features the most in the studies, showing the lack of out of the box solutions available for the range of analyses carried out. Marine Mammals require the largest amount of custom code

'Custom Code' refer to studies where the authors developed their own code for specific parts of their work, as opposed to using only readily available software.

Looking at automation tools, these are software used to process the data in some way. The tools were used to facilitate data annotation (Raven (K. Lisa Yang Center for Conservation Bioacoustics at the Cornell Lab of Ornithology 2024), BirdNET (Kahl et al. 2021), Koogu (Madhusudhana 2023), Tadarida (Bas et al. 2017), Audacity (Audacity Team 2025), Kaleidoscope (Wildlife Acoustics 2025),

BatDetect (Mac Aodha et al. 2018)); as classification model architectures and frameworks (LeNet (Lecun et al. 1998), YOLO (Redmon et al. 2016), TensorFlow (Abadi et al. 2015), Keras (Chollet et al. 2015), and PyTorch (Paszke et al. 2019)); or helper packages (ggplot2 (Wickham 2016), tuneR (Ligges et al. 2023), partykit (Hothorn and Zeileis 2015)). The dominance of custom code shows the field's growing ability to adapt to computer science practice. However, the lack of a standard to make these trained models work off the shelf necessitates such a tailored approach. Accessible user interfaces such as Raven and Audacity as well as the plethora of Shiny apps helps non-technical ecologists and conservationists work on such projects; the literature features two such Shiny apps developed (Rivas et al. 2024; Ruff et al. 2023). Raven/RavenPro, TensorFlow, Keras and BirdNET are also popular software used. A large proportion of these tools are packages: PAMGuide and PAMGuard are MATLAB; tuneR, Shiny and seewave are R packages; Keras, TensorFlow, partykit and koogu are python packages. Almost all of the automation tools were used on birds or Marine Mammals, as these are the most common taxa in bioacoustics studies. 'Other Methods' here combine those tools used two times or less across the literature in order to simplify the figure.

3.3 Classification Models

Species classification models in bioacoustics are evaluated using a suite of methods to ensure robust performance in detecting and classifying biological sounds. Counts of the evaluation methods used in papers in each year of the study period are shown in Fig. 16.8. There is, however, no common evaluation metric for these studies and so comparison of model performance across time or by taxa or method is difficult. Across all taxa, the reported mean reported metrics are: AUC (0.853%, n = 2), Accuracy (0.897%, n = 34), F1 (0.846%, n = 6), Precision (0.921%, n = 6), Recall (0.943%, n = 4), Sensitivity (0.770%, n = 2).

3.4 Open Science Practices

The landscape of processing PAM datasets is evolving, with increasing amounts of data and code being openly available, but still falling short of being fully accessible or compatible. Initiatives like DCASE (Stowell et al. 2015) and Kaggle competitions such as BirdCLEF (Klinck et al. 2024) have led the way in promoting open-source practices, requiring participants to share code in their submissions. However, full open-source—requiring model architectures, weights, training code and data being open—remains rare outside of these competitions. Model hosting, on platforms such as Hugging Face,[3] has seen only limited uptake in bioacoustics (Smith et al. 2024)

[3] https://huggingface.co/.

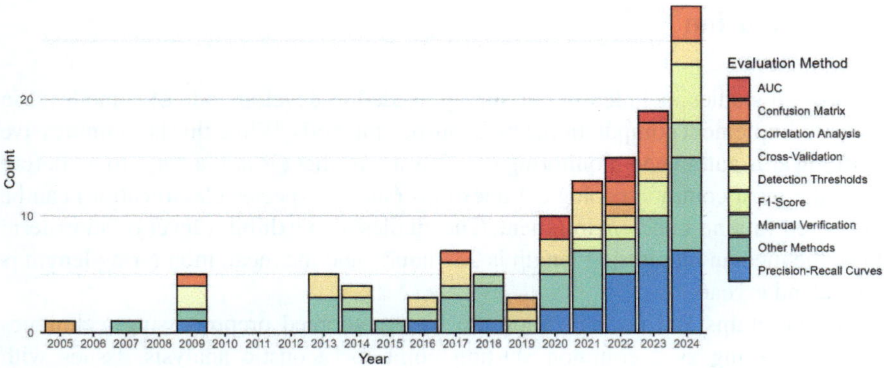

Fig. 16.8 Evaluation methods. There is no clear consensus in evaluation methods over time. AUC and cross-validation are good measures of overall performance, while confusion matrices detail prediction outcomes across categories. Detection thresholds optimise sensitivity, and manual verification confirms results against expert checks. Again the 'Other Methods' category is a collection of methods that only appeared once across the study period. Precision-Recall curves are the most common since 2020

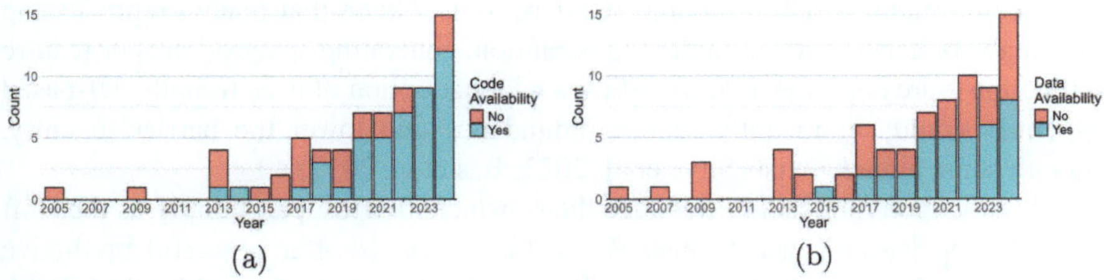

Fig. 16.9 Code and data availability trends by year. (**a**) Code Availability by Year shows a steady increase in the proportion of studies sharing code over time. (**b**) Data Availability by Year indicates a slower but positive trend in data sharing practices. With increasing dataset size, it may be less feasible to host large datasets. By 'available code and data' here we mean open-source materials. If papers mention availability upon request, access is limited or the repository was not found during the review, then this was marked as a No

despite its potential to streamline access and use of such state of the art models. Open-source repositories such as GitHub (GitHub Inc. 2025), Dryad (Dryad 2025) and Zenodo (European Organization for Nuclear Research and OpenAIRE 2013) provide platforms for sharing, preserving, and accessing scientific code, data, and research outputs, fostering collaboration and reproducibility across the bioacoustics field. Figure 16.9 gives some insight to the adoption of open code and data over our study period.

4 Conclusion

Almost all studies included in our survey needed to develop code of some level in order to implement computational bioacoustics methods. While this is an impressive shift towards automated monitoring of ecosystems, there is still a ways to go before even the most common ecological questions (such as species classification) can be answered in a no-code environment. The studies also exhibit a level of short-term bias; the median monitoring length is 237 hours, and the mean monitoring length is 2.6 calendar years.

Spectrograms and normalization are widely adopted preprocessing techniques in PAM, acting as a common starting point for acoustic analysis. Issues with the quantity and quality of labelled data persist, and techniques such as data augmentation have not been widely adopted to address this. The use of synthetic data as an augmentation technique, which is emerging as a promising approach in computer vision and large language models, is even less popular in bioacoustics, but has been shown to be fruitful (Fu et al. 2023). Furthermore, the field often relies on custom code for preprocessing and classifier training or inference, despite there being many widely-available standard tools. Given that many preprocessing and analysis steps—such as audio segmentation, generating spectrograms or feature extraction—are common across studies, a wider adoption of user-friendly, UI-based pipelines could significantly reduce redundancy and lower the barrier to entry, though some already exist (Silva et al. 2022; Bas et al. 2017).

Classification models in PAM face their own challenges, particularly as the shift toward deep learning accelerates. While these models offer powerful predictive capabilities, there is a lack of standardised evaluation methods, making it difficult to compare performance across studies or ensure reproducibility. This move toward deep learning also sacrifices explainability, a trade-off that complicates ecological interpretation and undermines its ability to be used in tackling broader ecological research questions. To address these issues, the field would benefit from greater consensus on metrics and benchmarks, which would be helped with the open data and hosted models as mentioned above. As PAM continues to mature, bridging these gaps will be key to unlocking its full potential for wildlife monitoring and beyond.

References

Abadi M, et al (2015) TensorFlow: large-scale machine learning on heterogeneous distributed systems. arXiv preprint arXiv:1603.04467. Software available from tensorflow.org. https://arxiv.org/abs/1603.04467

Andaryani S, et al (2021) The utility of a hybrid GEOMOD-Markov Chain model of land-use change in the context of highly water-demanding agriculture in a semi-arid region. Ecol Inform 64:101332. ISSN: 1574-9541. https://doi.org/10.1016/j.ecoinf.2021.101332. https://www.sciencedirect.com/science/article/pii/S1574954121001230

Arnaud V, et al (2022) Improving the workflow to crack small, unbalanced, noisy, but genuine (SUNG) datasets in bioacoustics: the case of Bonobo calls. bioRxiv. https://doi.org/10.

1101/2022.06.26.497684. eprint: https://www.biorxiv.org/content/early/2022/06/29/2022.06.26.497684.full.pdf. https://www.biorxiv.org/content/early/2022/06/29/2022.06.26.497684

Audacity Team (2025) Audacity®. Copyright © 1999–2025. Accessed 25 March 2025. https://audacityteam.org/

Bas Y, Bas D, Julien J-F (2017) Tadarida: a toolbox for animal detection on acoustic recordings. J Open Res Softw 5(1):6–6

Camphuysen K, et al (2004) Towards standardised seabirds at sea census techniques in connection with environmental impact assessments for off-shore wind farms in the UK. Tech. rep. COWRIE-BAM-02-2002. Royal Netherlands Institute for Sea Research (NIOZ), Apr. 2004. https://doi.org/10.13140/RG.2.1.2230.0244. https://www.researchgate.net/publication/287490365_Towards_standardised_seabirds_at_sea_census_techniques_in_connection_with_environmental_impact_assessments_for_offshore_wind_farms_in_the_UK

Cao H, et al (2022) Wavelet analysis reveals phenology mismatch between leaf phenology of temperate forest plants and the Siberian Roe deer molting under global warming. Remote Sens 14(16). ISSN: 2072-4292. https://doi.org/10.3390/rs14163901. https://www.mdpi.com/2072-4292/14/16/3901

Chollet F, et al (2015) Keras. https://keras.io. Software available from keras.io

Converse, SJ, McClintock, BT, Conn PB (2022) Special feature: linking capture–recapture and movement. Ecology 103(10):e3770. https://doi.org/10.1002/ecy.3770. eprint: https://esajournals.onlinelibrary.wiley.com/doi/pdf/10.1002/ecy.3770. https://esajournals.onlinelibrary.wiley.com/doi/abs/10.1002/ecy.3770

Deng J, et al (2009) Imagenet: a large-scale hierarchical image database. In: 2009 IEEE conference on computer vision and pattern recognition. IEEE, pp 248–255

Drew LW (2011) Are we losing the science of taxonomy?: as need grows, numbers and training are failing to keep up. BioScience 61(12):942–946. ISSN: 0006-3568. https://doi.org/10.1525/bio.2011.61.12.4. eprint: https://academic.oup.com/bioscience/article-pdf/61/12/942/19404798/61-12-942.pdf. https://doi.org/10.1525/bio.2011.61.12.4

Drewitt AL, Langston RHW (2006) Assessing the impacts of wind farms on birds. Ibis 148(s1):29–42. https://doi.org/10.1111/j.1474-919X.2006.00516.x. eprint: https://onlinelibrary.wiley.com/doi/pdf/10.1111/j.1474-919X.2006.00516.x. https://onlinelibrary.wiley.com/doi/abs/10.1111/j.1474-919X.2006.00516.x

Dryad (2025) Dryad digital repository. https://datadryad.org. Accessed 26 March 2025

European Organization for Nuclear Research and OpenAIRE (2013) Zenodo. en. 2013. https://doi.org/10.25495/7GXK-RD71. https://www.zenodo.org/

Fierro P, et al (2016) Rainbow Trout diets and macroinvertebrates assemblages responses from watersheds dominated by native and exotic plantations. Ecol Indic 60:655–667. ISSN: 1470-160X. https://doi.org/10.1016/j.ecolind.2015.08.018. https://www.sciencedirect.com/science/article/pii/S1470160X15004392

Forup ML, Memmott J (2005) The restoration of plant–pollinator interactions in hay meadows. Restor Ecol 13(2):265–274

Fu Y, et al (2023) Classification of birdsong spectrograms based on DR-ACGAN and dynamic convolution. Ecol Inform 77:102250. ISSN: 1574-9541. https://doi.org/10.1016/j.ecoinf.2023.102250. https://www.sciencedirect.com/science/article/pii/S1574954123002790

Fundel F, Braun, DA, Gottwald S (2023) Automatic bat call classification using transformer networks. Ecol Inform. ISSN: 1574-9541. https://doi.org/10.1016/j.ecoinf.2023.102288. https://www.sciencedirect.com/science/article/pii/S1574954123003175

Gillespie D, et al (2020) Passive acoustic methods for tracking the 3D movements of small cetaceans around marine structures. PLoS One 1–16. https://doi.org/10.1371/journal.pone.0229058

GitHub Inc. (2025) GitHub. https://github.com. Accessed 26 March 2025

Goëau H, et al 2016 LifeCLEF bird identification task 2016: the arrival of deep learning. In: Working notes of CLEF 2016 - conference and labs of the evaluation forum. CEUR Workshop Proceedings. 1609. Évora, Sept. 2016, pp 440–449. https://hal.science/hal-01373779

Görres C-M, Chesmore D (2019) Active sound production of scarab beetle larvae opens up new possibilities for species-specific pest monitoring in soils. Sci Rep 9(1):10115

Gottwald J, et al (2019) Introduction of an automatic and open-source radio-tracking system for small animals. Methods Ecol Evol 10(12):2163–2172. https://doi.org/10.1111/2041-210X.13294. eprint: https://besjournals.onlinelibrary.wiley.com/doi/pdf/10.1111/2041-210X.13294. https://besjournals.onlinelibrary.wiley.com/doi/abs/10.1111/2041-210X.13294

Harris LG, Tyrrell MC (2001) Changing community states in the Gulf of Maine: synergism between invaders, overfishing and climate change. Biol Invasions 3:9–21

Hothorn T, Zeileis A (2015) partykit: a modular toolkit for recursive partytioning in R. J Mach Learn Res 16:3905–3909. https://jmlr.org/papers/v16/hothorn15a.html

Jeantet L, Dufourq E (2023) Improving deep learning acoustic classifiers with contextual information for wildlife monitoring. Ecol Inform. ISSN: 1574-9541. https://doi.org/10.1016/j.ecoinf.2023.102256. https://www.sciencedirect.com/science/article/pii/S1574954123002856

Jerome Sueur TA, Simonis C (2008) See-Wave, a free modular tool for sound analysis and synthesis. Bioacoustics 18(2):213–226. eprint: https://doi.org/10.1080/09524622.2008.9753600

Jones G, et al (2009) Carpe noctem: the importance of bats as bioindicators. Endangered Species Res 8(1–2):93–115

K. Lisa Yang Center for Conservation Bioacoustics at the Cornell Lab of Ornithology (2024) Raven pro: interactive sound analysis software (version 1.6.5). Computer Software, Ithaca. https://www.ravensoundsoftware.com/

Kahl S, et al (2021) BirdNET: a deep learning solution for avian diversity monitoring. Ecol Inform 61:101236

Klinck H, et al (2024) BirdCLEF 2024. https://kaggle.com/competitions/birdclef-2024. Kaggle

Krizhevsky A, Sutskever I, Hinton GE (2012) Imagenet classification with deep convolutional neural networks. In: Advances in neural information processing systems, 25

Lagos-Kutz D, et al (2020) The soybean aphid suction trap network: sampling the aerobiological "soup". Am Entomol 66(1):48–55. ISSN: 1046-2821. https://doi.org/10.1093/ae/tmaa009. eprint: https://academic.oup.com/ae/article-pdf/66/1/48/32895020/tmaa009.pdf. https://doi.org/10.1093/ae/tmaa009

Lahaye R, et al (2008) DNA barcoding the floras of biodiversity hotspots. Proc Natl Acad Sci 105(8):2923–2928. https://doi.org/10.1073/pnas.0709936105. eprint: https://www.pnas.org/doi/pdf/10.1073/pnas.0709936105. https://www.pnas.org/doi/abs/10.1073/pnas.0709936105

Lattemann S, Höpner T (2008) Environmental impact and impact assessment of seawater desalination. Desalination 220(1). European Desalination Society and Center for Research and Technology Hellas (CERTH), Sani Resort 22–25 April 2007, Halkidiki, pp 1–15. ISSN: 0011-9164. https://doi.org/10.1016/j.desal.2007.03.009. https://www.sciencedirect.com/science/article/pii/S0011916407006005

Lecun Y, et al (1998) Gradient-based learning applied to document recognition. Proc IEEE 86(11):2278–2324. https://doi.org/10.1109/5.726791

Ligges U, et al (2023) tuneR: analysis of music and speech. https://CRAN.R-project.org/package=tuneR

Mac Aodha O, et al (2018) Bat detective—deep learning tools for bat acoustic signal detection. PLoS Comput Biol. Fenton B (ed): e1005995. ISSN: 1553-7358. https://doi.org/10.1371/journal.pcbi.1005995

Madhusudhana S (2023) shyamblast/Koogu: v0.7.2. https://doi.org/10.5281/ZENODO.8254287. https://zenodo.org/record/8254287

Metcalf OC et al (2021) Optimizing tropical forest bird surveys using passive acoustic monitoring and high temporal resolution sampling. In: Pettorelli N, Astaras C (eds) Remote sensing in ecology and conservation, pp 45–56. ISSN: 2056-3485. https://doi.org/10.1002/rse2.227

Milanelli AM, et al (2024) Temporal patterns in the soundscape of the port area in an urban estuary. Estuar Coast Shelf Sci 297. ISSN: 0272-7714. https://doi.org/10.1016/j.ecss.2023.108596

Oswald JN, et al (2022) A collection of best practices for the collection and analysis of bioacoustic data. Appl Sci 12(23). ISSN: 2076-3417. https://doi.org/10.3390/app122312046. https://www.mdpi.com/2076-3417/12/23/12046

Owens AF, et al (2024) Automated detection of Bornean white-bearded gibbon (Hylobates albibarbis) vocalisations using an open-source framework for deep learning. bioRxiv. https://doi.org/10.1101/2024.04.15.589517. eprint: https://www.biorxiv.org/content/early/2024/07/21/2024.04.15.589517.full.pdf. https://www.biorxiv.org/content/early/2024/07/21/2024.04.15.589517

Papin M, et al (2018) Acoustic localization at large scales: a promising method for grey wolf monitoring. Front Zool. ISSN: 1742-9994. https://doi.org/10.1186/s12983-018-0260-2

Paszke A, et al (2019) PyTorch: an imperative style, high-performance deep learning library. In: Wallach H, et al. (eds) Advances in neural information processing systems 32 (NeurIPS 2019). Software available from pytorch.org. 2019, pp 8024–8035. https://papers.nips.cc/paper/9015-pytorch-an-imperative-style-high-performance-deep-learning-library.pdf

Picciulin M, et al (2022) A fish and dolphin biophony in the boat noise-dominated soundscape of the Cres-Lošinj Archipelago (Croatia). J Mar Sci Eng 2. ISSN: 2077-1312. https://doi.org/10.3390/jmse10020300

Prendergast KS, et al (2020) The relative performance of sampling methods for native bees: an empirical test and review of the literature. Ecosphere 11(5):e03076. https://doi.org/10.1002/ecs2.3076. eprint: https://esajournals.onlinelibrary.wiley.com/doi/pdf/10.1002/ecs2.3076. https://esajournals.onlinelibrary.wiley.com/doi/abs/10.1002/ecs2.3076

Putland RL, et al (2018) A hidden Markov model approach to indicate Bryde's whale acoustics. Ecol Indic 479–487. ISSN: 1470-160X. https://doi.org/10.1016/j.ecolind.2017.09.025. https://www.sciencedirect.com/science/article/pii/S1470160X17305939

Redmon J, et al (2016) You only look once: unified, real-time object detection. In: Proceedings of the IEEE conference on computer vision and pattern recognition, pp 779–788

Ren Y, et al (2009) A framework for bioacoustic vocalization analysis using hidden Markov models. Algorithms 2(4):1410–1428. ISSN: 1999–4893. https://doi.org/10.3390/a2041410. https://www.mdpi.com/1999-4893/2/4/1410

Risch D, et al (2013) Minke whale acoustic behavior and multi-year seasonal and diel vocalization patterns in Massachusetts Bay, USA. In: Marine ecology progress series, pp 279–295. ISSN: 1616-1599. https://doi.org/10.3354/meps10426

Rivas F, et al (2024) Rthoptera: standardised insect bioacoustics in R. bioRxiv. https://doi.org/10.1101/2024.11.04.621915. eprint: https://www.biorxiv.org/content/early/2024/11/08/2024.11.04.621915.full.pdf. https://www.biorxiv.org/content/early/2024/11/08/2024.11.04.621915

Ruff ZJ, et al (2023) PNW-Cnet v4: automated species identification for passive acoustic monitoring. SoftwareX 23. ISSN: 2352-7110. https://doi.org/10.1016/j.softx.2023.101473

Sanchez L, Moreno CR, Mora EC (2017) Echolocation calls of Natalus primus (Chiroptera: Natalidae): implications for conservation monitoring of this species. Cogent Biol Burda H (ed) ISSN: 2331-2025. https://doi.org/10.1080/23312025.2017.1355027

Sebastián-González E, et al (2015) Bioacoustics for species management: two case studies with a Hawaiian forest bird. Ecol Evol 5(20):4696–4705. https://doi.org/10.1002/ece3.1743. eprint: https://onlinelibrary.wiley.com/doi/pdf/10.1002/ece3.1743. https://onlinelibrary.wiley.com/doi/abs/10.1002/ece3.1743

Shirley TR, Janke AK (2023) Ring-necked pheasant nest site selection in a landscape with high adoption of fall-seeded cover crops. Wildl Soc Bull 47(1):e1394. https://doi.org/10.1002/wsb.1394. eprint: https://wildlife.onlinelibrary.wiley.com/doi/pdf/10.1002/wsb.1394. https://wildlife.onlinelibrary.wiley.com/doi/abs/10.1002/wsb.1394

Silva B, et al (2022) soundClass: an automatic sound classification tool for biodiversity monitoring using machine learning. Methods Ecol Evol 2356–2362. ISSN: 2041-210X. https://doi.org/10.1111/2041-210x.13964

Smith RD, et al (2024) teticio/audio-diffusion. Sept. 25, 2024. https://github.com/teticio/audio-diffusion

Stowell D, et al (2015) Detection and classification of acoustic scenes and events. IEEE Trans Multimedia 17(10):1733–1746

Sun Y, et al (2022) Classification of animal sounds in a hyperdiverse rain- forest using convolutional neural networks with data augmentation. Ecol Indic. ISSN: 1470-160X. https://doi.org/10.1016/j.ecolind.2022.109621

Sueur J, Aubin T, Simonis C (2008) Seewave: a free modular tool for sound analysis and synthesis. Bioacoustics 18:213–226

Trapanotto M, et al (2022) Convolutional neural networks for the identification of African lions from individual vocalizations. J Imag ISSN: 2313-433X. https://doi.org/10.3390/jimaging8040096. https://www.mdpi.com/2313-433X/8/4/96

Varun Prakash R, et al (2024) Multi-level LSTM framework with hybrid sonic features for human-animal conflict evasion. Visual Comput. ISSN: 1432-2315. https://doi.org/10.1007/s00371-024-03588-9

Wickham H (2016) ggplot2: elegant graphics for data analysis. Springer, New York. ISBN: 978-3-319-24277-4. https://ggplot2.tidyverse.org

Wildlife Acoustics (2025) Kaleidoscope pro: bioacoustic analysis software. Computer Software. Version 5.4.8. Accessed 25 March 2025. https://www.wildlifeacoustics.com/products/kaleidoscope-pro

Williams EM, O'Donnell CFJ, Armstrong DP (2018) Cost-benefit analysis of acoustic recorders as a solution to sampling challenges experienced monitoring cryptic species. Ecol Evol 8(13):6839–6848

Wood H, Lindborg R, Jakobsson S (2017) European Union tree density limits do not reflect bat diversity in wood-pastures. Biol Conserv 210:60–71. ISSN: 0006-3207. https://doi.org/10.1016/j.biocon.2017.04.001. https://www.sciencedirect.com/science/article/pii/S0006320717305827

Chapter 17
Control Theoretic Perspectives on Wildlife Monitoring: A Tutorial Introduction to Observers for Population Models

Oliver Mason

Abstract In this chapter, we give a tutorial introduction to some key control theoretic concepts relevant to wildlife monitoring. In particular, we review the definitions of observability and detectability for linear and nonlinear systems. We then describe several approaches to the design of asymptotic observers for state estimation, paying particular attention to observers for nonlinear systems arising in models of population dynamics. Observers for discrete-time and continuous-time models are considered, and we illustrate the theoretical results on several well-known models from population biology.

Keywords Observability · Dynamic observers · Nonlinear observers · Nonlinear matrix models · Lotka-Volterra systems

1 Introduction

The use of mathematical models in the study of population biology has a long and interesting history. The book by Bacaër (2010) gives a very enjoyable and readable account of some of the major milestones in the field. Much of the research on applications of dynamical systems theory in this setting has been concerned with describing the long-term behaviour of populations: determining whether a population tends to extinction or persists, for instance see Smith and Thieme (2011). For a structured population, the principal object of study is the *state-vector* $x(t)$ which evolves in time. Here, *matrix models* play a key role (Cushing (2024), Logofet and Ulanova (2022)) in describing the dynamics. If the population structure has a finite number n of distinct classes, the vector $x(t)$ has n components with the ith component giving the size or density of the ith class at time t. Once a model to

O. Mason (✉)
Dept. of Mathematics & Statistics, Maynooth University, Co. Kildare, Ireland
e-mail: oliver.mason@mu.ie

L. Martins Verdade, R. A. Moral (eds.), *Wildlife Monitoring*,
https://doi.org/10.1007/978-3-032-05821-8_17

characterise how $x(t)$ evolves in time has been constructed, a key question is to understand the behaviour of solutions of this model for large values of t or as t tends to infinity. Many classical models are described by difference equations of the form $x(t+1) = f(x(t))$ or by ordinary differential equations $\dot{x}(t) = f(x(t))$. More complicated models can take aspects such as continuous spatial structure or stochasticity into account, and require more sophisticated mathematical techniques from partial or stochastic differential equations. In all the cases mentioned above, the object of study is the state vector $x(t)$, which describes how the population structure changes with time.

Over the past few decades, several authors have considered the use of control theoretic concepts in population biology: see Varga (2008) for an introduction to some of the main ideas. The notion of state is also one of the central concepts in control theory. However, in contrast to the discussion in the previous paragraph, control theory is concerned with *open* systems that interact with an external environment or actors. Control theoretic systems typically involve output variables which are measured to monitor the system's behaviour, and input variables which are used to manipulate how the system behaves. The standard formulation of a control system thus involves, at each time t, an input vector $u(t)$ and an output vector $y(t)$, in addition to the state $x(t)$. The output $y(t)$ is typically some function of the state vector $x(t)$ (it can also depend on the value of the input $u(t)$). For a wildlife population, $x(t)$ may be a vector containing the population densities for all groups in a population, while $y(t)$ may represent a selection of these, or the total population across some of the groups, for instance. It should be noted that the 'behavioral' approach due to Jan Willems, see Polderman and Willems (1997), takes a different view, and does not regard the separation of inputs and outputs as fundamental or intrinsic. However, we will work with the standard, input-state-output paradigm throughout the chapter. Ideally, the current value of the state $x(t)$ is used to determine a value of the input $u(t)$ which is then 'fed back' to achieve some objective. However, it is not always possible to measure the entire state, and instead some function of the state or only some of the state variables are measured. These measurements define the output vector $y(t)$. A central issue in control theory concerns determining whether or not the output of a system can be used to estimate its state and, when this is possible, how best to do this. These questions are at the heart of observability theory and the theory of observer design.

Alongside stability, which is arguably the most-studied system theoretic property in ecological applications, the notions of controllability and observability are fundamental in control theory. Informally, controllability is concerned with the ability to move the system from a given initial state to a desired state by appropriately selecting inputs or control actions. On the other hand, the notion of observability concerns the ability to reconstruct the state vector from an appropriate sequence of measured output measurements. The property of observability depends on both the system dynamics and the mapping defining the output. When a system is observable, the true value of the initial state can be determined from an appropriate sequence of measured outputs. In line with the theme of wildlife population monitoring, we shall focus on observability and the related concept of a dynamic observer for deterministic systems. An observer is another system that we construct based on the nominal one, where the state of the observer asymptotically converges to the state of the nominal system. While in theory the initial state can be reconstructed from output measurements for an observable system, observers are a more effective and reliable way of state estimation.

So, what does all this have to do with population monitoring? The remarkable fact is that when a dynamic observer exists, it can allow us to accurately estimate all components of the state vector based on the output $y(t)$. For many scenarios, this means that the entire state can be estimated using measurements of only some of its components. For instance, suppose that we have a system that models the dynamics of an age-structured population, but we can only measure certain age groups, such as the oldest one. If the overall system consisting of the state and output vectors is observable, then it is possible to estimate the sizes of *all age groups* while only measuring some of them over an appropriate time window. Similarly, if our system models interacting species, such as in a Lotka-Volterra model, then an observer would allow us to determine the densities of all species based on measurements of some species. In this context, the species whose populations are measured/monitored are referred to as *indicator species*.

Since the fundamental work of Luenberger (1966) on observer design for linear time-invariant (LTI) systems, there has been a significant amount of work on extending these ideas to classes of nonlinear systems. This line of work is particularly relevant to our discussion, as most of the models used in mathematical population biology are nonlinear. Examples include Lotka-Volterra models in continuous-time and nonlinear matrix models in discrete-time. In addition to Luenberger-type asymptotic observers, other classes such as interval observers or dead-beat/finite-time observers have also been investigated. While there are some general results available on observer design for nonlinear systems, the particular structure of a system plays a vital role. We will focus for the most part on results relevant for systems arising in ecological applications.

The aim of this chapter is to introduce, in a relatively informal way, some of the key ideas and results of observability and dynamic observers to researchers in ecology who have no prior familiarity with control theory and its underlying mathematics. As such, there is no intention to present an exhaustive survey of the state of the art in nonlinear observer design or to present novel results for this problem. Other authors have already written extensive and technically detailed surveys on the topic, and the interested reader could consult any of the following references (Bernard et al. 2022; Bernard 2019; Boutat and Zheng 2021). Further, as the chapter is meant to be tutorial in nature, we will focus for the most part on basic results and illustrate these on simple ecological models in order to get the core ideas and concepts across. We will also refer to some more technically sophisticated work where appropriate, but we will not make this our main focus.

The basic outline of the chapter is as follows. In Sect. 2, we introduce our notation, as well as some mathematical background, and key control theoretic concepts for linear systems. We will illustrate the main ideas on standard, well-known systems arising in ecology. We then move on to results on observers for discrete-time nonlinear systems in Sect. 3 before considering the corresponding problem for continuous-time systems in Sect. 4. We focus in particular on results for Lotka-Volterra and predator-prey systems. In the final section, we review and summarise the discussion and highlight some questions related to observer design for ecological systems.

2 Observability and Observers for Linear Systems

Before reviewing the basic results on observability and dynamic observers for linear systems, we briefly introduce the main mathematical notation, terminology, and background results needed for our later discussion.

We use $\mathbb{R}^n_+$ to denote the cone of *nonnegative* elements of Euclidean n-space $\mathbb{R}^n$. An element of $\mathbb{R}^n$ is typically written as $x = (x_1, x_2, \ldots, x_n)$. Then $\mathbb{R}^n_+ = \{x \in \mathbb{R}^n : x_i \geq 0, 1 \leq i \leq n\}$. $\mathbb{R}^{m\times n}$ denotes the space of $m \times n$ matrices with real entries, and $\mathbb{R}^{m\times n}_+$ the cone of nonnegative matrices. We use $x \geq 0$ to indicate that $x \in \mathbb{R}^n_+$, while $x \gg 0$ means that $x_i > 0$ for all $1 \leq i \leq n$.

For a square matrix $A \in \mathbb{R}^{n\times n}$, the eigenvalues of A are the solutions λ of the characteristic equation $\det(\lambda I - A) = 0$ where I is the $n \times n$ identity. The set of all eigenvalues of A is known as the spectrum of A and denoted $\sigma(A)$. The spectral radius $\rho(A)$ of A is the maximum modulus of all eigenvalues of A, $\rho(A) = \max_{\lambda\in\sigma(A)} |\lambda|$. If $\rho(A) < 1$, A is said to be Schur-stable. It is a classical result that $\rho(A) < 1$ is equivalent to the matrix powers of A tending to zero asymptotically, $A^t \to 0$ as $t \to \infty$ (every entry of A^t tends to zero asymptotically). This property is relevant to the stability of discrete-time systems as it ensures local asymptotic stability for equilibria of nonlinear systems. If A^t is actually equal to the zero matrix for some finite value of t, the matrix A is said to be *nilpotent*. The property for continuous-time systems corresponding to Schur-stability is that the real parts of all eigenvalues of A are (strictly) negative: such matrices are said to be *Hurwitz*.

A matrix A is positive (semi-)definite if it is symmetric, meaning that A is equal to its transpose A^T, and all the eigenvalues of A are strictly positive (nonnegative). This is denoted by $A \succ 0$ $A \succeq 0$. The definitions and notation for negative (semi-)definite matrices are analogous.

Background on Linear Systems

To begin, let's consider a simple linear time-invariant (LTI) system in discrete-time with outputs but no inputs. The standard form of such a system is shown below.

$$
\begin{aligned}
x(t+1) &= Ax(t) \\
y(t) &= Cx(t).
\end{aligned}
\tag{17.1}
$$

The *state* $x(t)$ is a vector in $\mathbb{R}^n$ at each time t starting from some initial value at time 0, $x(0)$. In an ecological context, the state often represents the structure of a population, with each component giving the size or density of a particular group or cohort. The vector $y(t)$ contains the *measured outputs*. Typically, this is of lower dimension, say $y(t) \in \mathbb{R}^m$ with $m < n$. For example, if each component of the state vector represented the size of some age cohort within a population, $y(t)$ may contain the size of a selection of these. If it is only feasible to measure the oldest group, for example, then $y(t)$ would simply be given by the nth component $x_n(t)$, so the output is 1-dimensional. To keep things simple, we have omitted inputs from our system model for the moment.

In models of population dynamics, the variables take nonnegative values. Moreover, the system must have the property that nonnegative initial conditions lead to nonnegative solutions. Mathematically, this means that $x(0) \in \mathbb{R}^n_+$ must imply that $x(t) \in \mathbb{R}^n_+$ for all $t \geq 0$. To ensure this, the matrices A and C are nonnegative, meaning that all of their entries are nonnegative real numbers. The system is then said to be a positive system.

The corresponding system class for continuous-time models takes the form

$$\begin{aligned} \dot{x}(t) &= Ax(t) \\ y(t) &= Cx(t), \end{aligned} \tag{17.2}$$

with $\dot{x}(t)$ denoting the derivative of x with respect to time. For the continuous-time system to be positive, it is necessary and sufficient that the matrix A is Metzler, meaning that its off-diagonal entries are nonnegative: more formally, $a_{ij} \geq 0$ for all $i \neq j$.

2.1 Observability for LTI Systems

Intuitively, observability is concerned with recovering the value of the initial state, $x(0) = x_0$, from a sequence of output measurements. More formally, the LTI system (17.1) is observable in time t_f if the outputs $y(0), y(1), \ldots, y(t_f - 1)$ uniquely determine the initial state x_0. Mathematically, this means that the mapping from x_0 to the vector

$$\begin{pmatrix} y(0) \\ y(1) \\ \vdots \\ y(t_f - 1) \end{pmatrix}$$

is 1-1 or injective. This mapping is given by the matrix

$$\begin{pmatrix} C \\ CA \\ CA^2 \\ \vdots \\ CA^{t_f-1} \end{pmatrix}. \tag{17.3}$$

which is in $\mathbb{R}^{mt_f \times n}$. The classical *rank-nullity* theorem from linear algebra can be used to show that observability is equivalent to the matrix in (17.3) having rank n. While we have introduced observability for a general time t_f, it is common to focus on the case $t_f = n$. It is a consequence of the Cayley-Hamilton theorem that for a discrete-time LTI system, observability in time n implies observability in time t_f for all $t_f \geq n$. This is equivalent to the matrix in (17.3) having rank n for $t_f = n$.[1] This property is often simply referred to as observability, and we say that the matrix pair (A, C) is observable when the matrix (17.3) has rank n for $t_f = n$. Interestingly, the same condition in terms of the observability matrix also applies in the continuous-time case.

2.2 *Dynamic Observers for LTI Systems*

In theory, if a system is observable, it is possible to recover the value of the initial state from the output measurements. Why is this? Consider the discrete-time case and suppose that the system is observable. Then the mapping from the initial state to the sequence of output values $y(0), \ldots, y(n-1)$ is injective. This means there is only one possible initial state that can produce the given sequence of outputs, so the outputs uniquely determine the initial state. Once you know the initial state, this can then be used to predict the values of $x(t)$ at all subsequent times. However, this approach is not really practical due to issues such as measurement error, and it is more common to use a *dynamic observer* for state estimation. For linear systems, the classical approach to observer design is due to Luenberger (1966). The key idea is to use the measured outputs, coupled with knowledge of the dynamics of the system, to construct an auxiliary system, known as the observer, whose state will asymptotically converge to the state of the original system. We then use the state of the observer to estimate the true state.

For the simple LTI case in discrete-time, the observer system takes the form

$$\begin{aligned} \hat{x}(t+1) &= A\hat{x}(t) - L(\hat{y}(t) - y(t)) \\ \hat{y}(t) &= C\hat{x}(t). \end{aligned} \tag{17.4}$$

[1] When we fix $t_f = n$, this matrix is referred to as the *observability matrix* and will be denoted by $\mathcal{O}$.

Here, $\hat{x}(t)$ denotes the state of the observer. This is used as an estimate of the true state $x(t)$. The essential idea is to construct a matrix L in the feedback loop and use the error $\hat{y}(t) - y(t)$ between the observer output and true output as an input to correct the estimate $\hat{x}(t)$ at each step. The matrix L is chosen to ensure that $\hat{x}(t)$ converges to $x(t)$ asymptotically *irrespective* of the initial values $x(0)$ and $\hat{x}(0)$ of the system and observer states.

Mathematically, the aim is to find a matrix L that guarantees that the error $e(t) = \hat{x}(t) - x(t)$ converges to 0 for any initial conditions x_0, $\hat{x}_0$. We can show that the error $e(t)$ satisfies $e(t+1) = (A - LC)e(t)$. Furthermore, substituting the equation for $\hat{y}(t)$ into the first equation in (17.4), we see that

$$\hat{x}(t+1) = (A - LC)\hat{x}(t) + Ly(t).$$

Two things are worth noting here.

1. First to ensure that $e(t)$ converges to 0 as $t \to \infty$, we need to choose L so that the matrix $A - LC$ is Schur stable, which means that all of its eigenvalues lie inside the open unit disk in $\mathbb{C}$. So we require $|\lambda| < 1$ for every eigenvalue of $A - LC$.
2. In applications where state variables are always nonnegative, it makes sense to require that the observer state $\hat{x}$ is also nonnegative. If we look at the right hand side of the equation for $\hat{x}(t+1)$, we can see that to ensure this holds for any nonnegative output signal $y(\cdot)$ and initial estimate $\hat{x}(0)$, we need both L and $A - LC$ to be nonnegative matrices.

With regard to the first of these points, the following result is well known in control theory and applies to both continuous-time and discrete-time systems.

Theorem 1 *Let $A \in \mathbb{R}^{n\times n}$ and $C \in \mathbb{R}^{m\times n}$ be given real matrices. Further, suppose that the pair (A, C) is observable. Then, for any set S of admissible eigenvalues, there exists a matrix L in $\mathbb{R}^{n\times m}$ such that the eigenvalues of $A - LC$ are given by S*

In the above result, an 'admissible' set of eigenvalues simply means that it is possible for S to be the set of eigenvalues of a real $n \times n$ matrix. So, complex eigenvalues must occur as conjugate pairs, and the number of elements in S cannot exceed n.

The important fact is that once the pair (A, C) is observable, we can find a matrix L such that the eigenvalues of $A - LC$ can be freely assigned. In particular, this means that we can choose L so that they all lie in the open unit disk, or the open left half-plane. These facts guarantee the existence of asymptotic observers in discrete-time and continuous-time respectively. Note that when (A, C) is observable, there are infinitely many matrices L that can be used to define an observer. In choosing an observer, it is natural to look for L for which convergence is as fast as possible. For discrete-time systems, this would mean choosing L so that the eigenvalues of $A - LC$ are as small in magnitude as possible.

However, for applications in population dynamics, there is another important consideration. Ideally, we would like to maintain positivity in the observer system.

The above theorem does not say that the matrix L can be chosen so that both L and $A - LC$ are nonnegative, which is necessary if we want to ensure nonnegativity of the estimate $\hat{x}(t)$. Adding the additional requirements for nonnegativity leads to the so-called *positive observer problem.*

2.3 Some Discrete-Time Ecological Examples

Now that we have reviewed some of the main ideas around observability and observer design for LTI systems, let us look at some examples from ecological models. For the most part, we have used the excellent book Cushing (2024) as a source for these models. However, we also give references to the original papers where these may be helpful.

Example 1 The simplest age-structured linear model consists of just a juvenile and adult class. Let x_1 denote the population density of the juvenile class and x_2 the density of the adult class. The state dynamics are given by a Leslie matrix of the form

$$A = \begin{pmatrix} 0 & a_{12} \\ a_{21} & a_{22} \end{pmatrix}.$$

Here $a_{12} > 0$ is the per capita reproduction rate for the adult class (informally, the mean number of new offspring generated per adult), while a_{21} represents the fraction of juveniles surviving to adulthood, and a_{22} the fraction of adults surviving to the next measurement time or census. The parameters satisfy $0 < a_{21} \leq 1$, $0 \leq a_{22} \leq 1$.

If we monitor such a population and can only measure the adult class, this corresponds to $y(t) = x_2(t)$ and an output matrix $C = (0 \quad 1)$. The observability matrix then takes the form

$$O = \begin{pmatrix} 0 & 1 \\ a_{21} & a_{22} \end{pmatrix}.$$

As $a_{21} < 0$, this always has rank equal to 2 so the system is observable.

To construct a positive observer for this system, we need to find a matrix

$$L = \begin{pmatrix} l_1 \\ l_2 \end{pmatrix}$$

such that $l_1 \geq 0$, $l_2 \geq 0$ and

$$A - LC = \begin{pmatrix} 0 & a_{12} - l_1 \\ a_{21} & a_{22} - l_2 \end{pmatrix} \geq 0$$

with $A - LC$ Schur-stable. For this simple system, we can choose $l_1 = a_{12}$ and $l_2 = a_{22}$. This leads to a matrix $A - LC$ that is not only Schur-stable, but *nilpotent*, as $(A - LC)^2$ will be the zero matrix. The practical meaning of this is that the approximation error will be zero in two steps, and the observer state $\hat{x}(t)$ will equal the true state for $t \geq 2$ irrespective of the initial conditions. An observer with this property, where its state converges to the true state in a finite number of steps is known as a *dead-beat* or *finite-time* observer. A key point is that the observer yields the densities/sizes of both the juvenile and adult populations based on measurements of the adult population only.

In the case where we can only measure the size of the juvenile population, the output matrix is $C = (1 \quad 0)$ and the observability matrix is now

$$O = \begin{pmatrix} 1 & 0 \\ 0 & a_{12} \end{pmatrix}$$

so O again has rank 2 and the system is observable in this case also as $a_{12} > 0$. For L to define a positive observer here, we need $A - LC$ to be Schur-stable and

$$A - LC = \begin{pmatrix} -l_1 & a_{12} \\ a_{21} - l_2 & a_{22} \end{pmatrix} \geq 0.$$

It follows that l_1 must be equal to 0. In this case, we cannot make $A - LC$ nilpotent unless $a_{22} = 0$ so it is not possible to construct a finite-time positive observer based on measurements of the juvenile population when $a_{22} > 0$.

Leslie and Usher Matrices

Leslie and Usher matrices arise very frequently in discrete-time models of population dynamics. It is quite common practice in simple models, see Cushing (2024), to break the dynamics into two processes: one related to fertility, the other to survival and progression. This naturally leads to a canonical representation of *population projection matrices* given by $A = F + T$. Here, F denotes the fertility matrix and T denotes the progression or transition matrix. Leslie matrices arise more commonly when we consider a population that is structured by age, while Usher matrices arise when the structure is related to size or mass. The key difference from a mathematical point of view concerns the diagonal entries of A. In a Leslie matrix, at most one of these is non-zero, while in an Usher matrix, all diagonal entries can be non-zero.

A generalised Leslie matrix takes the form

$$A = \begin{pmatrix} a_{11} & a_{12} & \dots & a_{1,n-1} & a_{1n} \\ a_{21} & 0 & \dots & 0 & 0 \\ 0 & a_{32} & \dots & 0 & 0 \\ \vdots & \vdots & \ddots & \vdots & \vdots \\ 0 & 0 & \dots & a_{n,n-1} & a_{nn} \end{pmatrix} \tag{17.5}$$

where the entries in the first row satisfy $a_{1i} \geq 0$, while all the other entries indicated satisfy $0 \leq a_{ii-1} \leq 1$ for $2 \leq i \leq n$, $0 \leq a_{nn} \leq 1$. The corresponding state vector $(x_1(t), x_2(t), \ldots, x_n(t))$ describes the age structure of the population, with $x_1(t)$ the density of the youngest age-group, $x_2(t)$ the density of the next youngest, and so on up to $x_n(t)$ giving the density of the oldest cohort. The entries in the first row are used to define the fertility matrix F while the other entries define the transition between age classes and are used to specify the matrix T.

As we mentioned above, Usher matrices arise in models of a population structured by size rather than age. Also, it is assumed that the time-step in the model corresponds exactly to the time needed to progress from one size class to another. In contrast to Leslie matrices, it is possible to remain in the same class at each time, leading to non-zero diagonal entries. The general form for an Usher matrix is shown below.

$$A = \begin{pmatrix} a_{11} & a_{12} & a_{13} & \ldots & a_{1,n-1} & a_{1n} \\ a_{21} & a_{22} & 0 & \ldots & 0 & 0 \\ 0 & a_{32} & a_{33} & \ldots & 0 & 0 \\ \vdots & \vdots & \vdots & \ddots & \vdots & \vdots \\ 0 & 0 & 0 & \ldots & a_{n-1,n-1} & 0 \\ 0 & 0 & 0 & \ldots & a_{n,n-1} & a_{nn} \end{pmatrix}. \tag{17.6}$$

As with Leslie matrices, the entries in the first row define birth rates for the fertility matrix, while the remaining entries define the transition rates for the matrix T.

If the growth of a population is described by an Usher or Leslie matrix, this specifies the system matrix A in the first of our state equations in (17.1). The matrix C is determined by which groups, defined by age or size, we can measure at each time t. If we only measure one group, then C will be a $1 \times n$ matrix, which is often written as a row vector $c^T = (c_1\ c_2\ \ldots\ c_n)$. For example, if we measure the oldest or largest group, then $C = \begin{pmatrix} 0\ 0 \ldots 1 \end{pmatrix}$. Now we look at some specific examples of linear models involving matrices of Usher or Leslie type.

Example 2 The following numerical examples of generalised Leslie and Usher matrices are taken from Example 2.1 in Cushing (2024).

The Leslie matrix

$$\begin{pmatrix} 0 & 0 & 1.893 \\ 0.631 & 0 & 0 \\ 0 & 0.0631 & 0.6272 \end{pmatrix}$$

was used in a model of stingray population dynamics constructed in Mollet and Cailliet (2002), while the following Usher matrix appears in a model for the population of the desert tortoise in Doak et al. (1994).

$$A = \begin{pmatrix} 0 & 0 & 0 & 0 & 0 & 0.042 & 0.069 & 0.069 \\ 0.716 & 0.567 & 0 & 0 & 0 & 0 & 0 & 0 \\ 0 & 0.149 & 0.567 & 0 & 0 & 0 & 0 & 0 \\ 0 & 0 & 0.149 & 0.605 & 0 & 0 & 0 & 0 \\ 0 & 0 & 0 & 0.235 & 0.560 & 0 & 0 & 0 \\ 0 & 0 & 0 & 0 & 0.225 & 0.678 & 0 & 0 \\ 0 & 0 & 0 & 0 & 0 & 0.249 & 0.851 & 0 \\ 0 & 0 & 0 & 0 & 0 & 0 & 0.016 & 0.860 \end{pmatrix}.$$

If we consider the Leslie matrix in the above example, it is a specific example of the general form shown below.

$$A = \begin{pmatrix} 0 & 0 & a_{13} \\ a_{21} & 0 & 0 \\ 0 & a_{32} & a_{33} \end{pmatrix}.$$

Matrices of this form arise in connection with the celebrated LPA (larva-pupa-adult) models in insect population dynamics. If we measure the size of the adult population, corresponding to x_3, the output matrix is $C = (0\ 0\ 1)$. Now choosing the observer gain to be

$$L = \begin{pmatrix} a_{13} \\ 0 \\ a_{33} \end{pmatrix}$$

we see that

$$A - LC = \begin{pmatrix} 0 & 0 & 0 \\ a_{21} & 0 & 0 \\ 0 & a_{32} & 0 \end{pmatrix}$$

which is nilpotent and satisfies $(A - LC)^3 = 0$. So, with this choice of L, we obtain another example of a finite-time observer which will exactly match all components of the population state vector after 3 steps.

3 Nonlinear Observers for Discrete-Time Systems

Having introduced some of the main concepts and ideas behind observability and observer design for LTI systems, we now discuss various nonlinear extensions of these. Of course, the general problem for nonlinear systems is considerably more complex, and most results either apply to specific system classes, or apply locally, rather than globally. For the most part, we will focus on results that are relevant to systems arising in ecological applications. In this section, we will consider discrete-time systems, while we discuss the continuous-time case in the next section. In keeping with the approach throughout, the discussion will be somewhat informal. Our aim is to explain the main ideas and results, and to illustrate these on some concrete, ecological examples.

We will consider nonlinear systems with output in the general form:

$$\begin{aligned} x(t+1) &= f(x(t)) \\ y(t) &= h(x(t)). \end{aligned} \tag{17.7}$$

For nonlinear systems, we will typically deal with *local* observers where convergence of the estimate to the true state is guaranteed in some neighbourhood of an equilibrium. With this in mind, let x^* denote an equilibrium of (17.7), so that $f(x^*) = x^*$.

Many of the results concerning observer existence for nonlinear systems, in continuous-time and discrete-time, rely on the *stability* properties of an equilibrium x^*. In keeping with the nature of this chapter, we'll avoid introducing a lot of technical definitions and instead explain the principal ideas needed to understand the results below. An equilibrium is *Lyapunov stable* if we can guarantee that any solution $x(t)$ remains as close as we wish to x^* by selecting an initial state $x(0)$ sufficiently close to x^*. For local asymptotic stability, we require in addition that the solution converges to x^* asymptotically. The explanations above apply for both continuous-time and discrete-time systems.

Throughout this section, we will use A to denote the Jacobian matrix of f evaluated at the positive equilibrium x^*, and C to denote the Jacobian of h at x^*.

The general problem of observer design for nonlinear systems was posed by Thau in Thau (1973). Since then, it has attracted the attention of several researchers and fundamental results have been derived in papers such as Krener and Respondek (1985), Kazantzis and Kravaris (2001), Sundarapandian (2004a). Recent comprehensive surveys of the problem, albeit with a focus on continuous-time systems, can be found in the references Boutat and Zheng (2021), Bernard (2019), Bernard et al. (2022). To keep the technical details to a minimum, we will follow the formulation of the problem for discrete-time systems given in Sundarapandian (2004a,b). Moreover, we will focus on the design of what that authors calls *identity* observers: this simply means that we are interested in estimating the state $x(t)$ rather than some function $\phi(x(t))$ of the state.

The goal is to construct an observer system of the form

$$\hat{x}(t+1) = g(\hat{x}(t), y(t)) \tag{17.8}$$

on $\mathbb{R}^n$ such that $\hat{x}(t)$ converges exponentially to the true state $x(t)$ as t tends to ∞. Exponential convergence means that there are positive constants $M > 0, r > 0$ such that $\|\hat{x}(t) - x(t)\| \leq Me^{-rt}$.

More formally, the system (17.8) is an (identity) exponential observer for (17.7) if the function g is continuously differentiable, $g(x^*, 0) = x^*$, and there is some $\epsilon > 0$ such that the following two conditions are satisfied for any $x(0)$, $\hat{x}(0)$ with $\|x(0) - x^*\| < \epsilon$ and $\|x^* - \hat{x}(0)\| < \epsilon$.

1. If $\hat{x}(0) = x(0)$, then $\hat{x}(t) = x(t)$ for all $t \geq 0$.
2. $x(t) - \hat{x}(t)$ tends to zero exponentially.

This means that if both the true state and the estimate start from values close to the equilibrium, then the estimate given by the observer state $\hat{x}(t)$ will converge at an exponential rate to the value of $x(t)$. It is important to emphasise that the observer introduced above is local in nature. It only works in some neighbourhood of the equilibrium x^*. This contrasts with the LTI system case, where convergence is global.

General conditions for the existence of nonlinear observers were given in the paper Sundarapandian (2004a). The most relevant result for us is Theorem 2 of the paper on the design of identity observers, which has been used or adapted to design observers for models in population dynamics in subsequent work. We will describe one such application for discrete-time predator-prey models from Gámez et al. (2017) below. We now recall the statement of Theorem 2 from Sundarapandian (2004a).

Theorem 2 *Assume that the system* (17.7) *has a Lyapunov stable equilibrium at* x^*. *Then* (17.7) *has an exponential identity observer if and only if there exists a matrix* L *such that* $A - LC$ *is Schur stable. Moreover, when such a matrix exists, the system*

$$\hat{x}(t+1) = f(\hat{x}(t)) + L(y(t) - h(\hat{x}(t))) \qquad (17.9)$$

is an exponential observer for (17.7)

3.1 Application to Population Dynamics Models

In this subsection, we will illustrate the relevance of Theorem 2 to wildlife population monitoring by outlining a couple of applications of the result to construct local observers for models in population dynamics.

A Nonlinear Juvenile Adult Model

Nonlinear matrix models are widely used in the study of structured populations when transition or fertility rates are density dependent. The general form of such models is

$$x(t+1) = A(x(t))x(t), \qquad (17.10)$$

for some matrix-valued function $A : \mathbb{R}^n_+ \to \mathbb{R}^{n\times n}_+$. We now illustrate the use of Theorem 2 on a simple nonlinear matrix model of a juvenile-adult system from Cushing (2024). As noted in that reference, this is a simplified version of the Ebenman model from Ebenman (1988) of competition for resources between adults and juveniles.

The model is given by

$$\begin{pmatrix} x_1(t+1) \\ x_2(t+1) \end{pmatrix} = \begin{pmatrix} 0 & \frac{b_2}{1+cx_2} \\ s_1 & 0 \end{pmatrix} \begin{pmatrix} x_1(t) \\ x_2(t) \end{pmatrix} \tag{17.11}$$

where the model parameters satisfy $b_2 > 0$, $c > 0$, $0 < s_1 \leq 1$.

To apply Theorem 2, we first need to identify a positive equilibrium. For this, we solve the equations

$$x_1 = \frac{b_2 x_2}{1 + cx_2}, \quad x_2 = s_1 x_1.$$

The only non-zero solution is given by

$$x_1 = \frac{b_2 s_1 - 1}{cs_1}, \quad x_2 = \frac{b_2 s_1 - 1}{c}.$$

So the system admits a positive equilibrium if and only if $b_2 s_1 > 1$. In keeping with our general notation, let's assume that $b_2 s_1 > 1$ and write x^* for the associated positive equilibrium. The Jacobian of this system is given by

$$J(x) = \begin{pmatrix} 0 & \frac{b_2}{(1+cx_2)^2} \\ s_1 & 0 \end{pmatrix}.$$

If we substitute in the equilibrium value $x_2 = \frac{b_2 s_1 - 1}{c}$, we find that the Jacobian at x^* is

$$J(x^*) = \begin{pmatrix} 0 & \frac{1}{b_2 s_1^2} \\ s_1 & 0 \end{pmatrix}.$$

This matrix has eigenvalues at $\pm \frac{1}{\sqrt{b_2 s_1}}$, so both of these have modulus less than 1 under the assumption $b_2 s_1 > 1$, and hence the positive, survival, equilibrium is Lyapunov stable, and we can apply the theorem to construct a local observer.

Suppose that we measure the adult population, $x_2(t)$. Then $h(x_1, x_2) = x_2$ and the matrix $C = (0 \;\; 1)$. The matrix A in the theorem is simply $J(x^*)$, so to design the observer we need a matrix

$$L = \begin{pmatrix} l_1 \\ l_2 \end{pmatrix}$$

such that the matrix

$$A - LC = \begin{pmatrix} 0 & \frac{1}{b_2 s_1^2} - l_1 \\ s_1 & -l_2 \end{pmatrix}$$

is Schur-stable. Remember that the Jacobian $A = J(x^*)$ is Schur-stable. This means that if we choose $l_2 = 0$ and any $0 \leq l_1 \leq \frac{1}{b_2 s_1^2}$ then $A - LC$ is nonnegative and Schur-stable. It makes intuitive sense to choose $l_1 = \frac{1}{b_2 s_1^2}$ as this would make $A-LC$ nilpotent and may lead to faster, if not finite-time, convergence.

Discrete-Time Predator-Prey Model

In Gámez et al. (2017), the authors consider a discrete-time system obtained by sampling a continuous-time Lotka-Volterra model. In particular, they consider an Euler discretisation of the standard system of differential equations for a Lotka-Volterra predator-prey model. The form of the discrete-time system studied is as follows.

$$\begin{aligned} x_1(t+1) &= r x_1(t) - s x_1^2(t) - b x_1(t) x_2(t) \\ x_2(t+1) &= d x_2(t) - e x_2^2(t) + c x_1(t) x_2(t), \end{aligned} \tag{17.12}$$

where all parameters are positive and $r > 1$, $d < 1$. We have altered the notation here from the original paper in the interests of simplicity. Also, we have not included the discretisation parameter as it is not relevant for our discussion.

An equilibrium of this system satisfies $f(x^*) = x^*$ where $f : \mathbb{R}^2 \to \mathbb{R}^2$ is the function defining the right-hand side. Positive equilibria are of particular interest as these correspond to co-existence, and the authors of Gámez et al. (2017) give conditions for such an equilibrium to exist and be Lyapunov stable.

For the observer problem, we need to specify a measurement function h. There are several natural possibilities here. In pest management scenarios, the predator is often introduced as a means of controlling the pest population. This means that it is reasonable to assume that the predator population $x_2(t)$ is measured. As the authors explain, it is more convenient to work with the deviation from the equilibrium population for purely technical reasons, so that $y(t) = x_2(t) - x_2^*$ is the output of the system. This corresponds to an observation function $h(x) = x_2 - x_2^*$. The Jacobian of h is then simply the matrix $C = (0 \;\; 1)$. On the other hand, if the deviation of the total population density from the equilibrium value is measured, then

$$y(t) = h(x(t)) = (x_1(t) + x_2(t)) - (x_1^* + x_2^*),$$

and the relevant Jacobian matrix is $C = (1 \;\; 1)$.

The authors of Gámez et al. (2017) first derive conditions for the existence of a positive, asymptotically stable equilibrium of the system. They then show that for the case where the predator population is measured, these same conditions imply that there exists an exponential observer. They also obtain a corresponding result for the case where the prey population is monitored.

4 Nonlinear Observers for Continuous-Time Lotka-Volterra Systems

There has been a significant parallel activity related to the design of observers for continuous-time nonlinear systems described by ordinary differential equations (ODEs). Here, the general model class is of the form below.

$$\begin{aligned} \dot{x}(t) &= f(x(t)) \\ y(t) &= h(x(t)). \end{aligned} \tag{17.13}$$

It is assumed that the functions $f : \mathbb{R}^n \to \mathbb{R}^n$ and $h : \mathbb{R}^n \to \mathbb{R}^m$ are continuously differentiable and that x^* is an equilibrium of the system. In this context, this means that $f(x^*) = 0$, $h(x^*) = 0$. We shall again use A and C to denote the Jacobian matrices of f and h at x^*. In Sundarapandian (2002), the author considered local exponential observers for the continuous-time system (17.13). The ideas and results of Sundarapandian (2002) mirror those for the discrete-time case discussed above.

A local exponential observer is a system

$$\dot{\hat{x}}(t) = g(\hat{x}(t), y(t))$$

with the two properties that $x(0) = \hat{x}(0)$ implies that $x(t) = \hat{x}(t)$ for all $t \geq 0$, and that $x(t) - \hat{x}(t)$ tends to zero exponentially for initial conditions $x(0)$, $\hat{x}(0)$ in some neighbourhood of x^*. The key point of the second condition is that the estimate $\hat{x}(t)$ converges exponentially to the true state $x(t)$ provided both trajectories start 'sufficiently close' to the equilibrium.

The following result from Sundarapandian (2002) gives local conditions for the existence of an exponential observer in a neighbourhood of a Lyapunov-stable equilibrium x^*.

Theorem 3 *Assume that the system* (17.13) *has a Lyapunov stable equilibrium at* x^*. *If there exists a matrix* L *such that* $A - LC$ *is Hurwitz, then the system*

$$\dot{\hat{x}}(t) = f(\hat{x}(t)) + L(y(t) - h(\hat{x}(t))) \tag{17.14}$$

is a local exponential observer for (17.13).

As in the discrete-time case, the interesting implication for wildlife population monitoring is that it is possible to estimate all states x_i by measuring only some of them in the output vector y. In addition, it is possible to improve the rate of convergence by choosing L so that the eigenvalues of $A - LC$ are further into the left half plane (essentially making the real parts of the eigenvalues of $A - LC$ 'more negative').

Example 3 As a simple illustration of how Theorem 3 can be applied, let's consider the basic predator-prey system of Lotka-Volterra type.

$$\begin{aligned}\dot{x}_1 &= ax_1 - bx_1x_2 \\ \dot{x}_2 &= -cx_2 + dx_1x_2\end{aligned} \tag{17.15}$$

where a, b, c, d are all positive. It is well known, and easy to show, that this system has a positive equilibrium at $x^* = (\frac{c}{d}, \frac{a}{b})$. Moreover, the Jacobian at x^* is given by

$$A = J(x^*) = \begin{pmatrix} 0 & -\frac{bc}{d} \\ \frac{da}{b} & 0 \end{pmatrix}$$

and x^* is Lyapunov stable, but not asymptotically stable. The stability can be shown using the form of the periodic solutions of the system in the interior of the positive quadrant, $\{x \in \mathbb{R}^2 : x \gg 0\}$. If we measure the size of the predator population, given by $y(t)$, this corresponds to an output matrix, $C = (0 \;\; 1)$. If we choose

$$L = \begin{pmatrix} l_1 \\ l_2 \end{pmatrix}$$

where both $l_1 > 0, l_2 > 0$, we see that

$$A - LC = \begin{pmatrix} 0 & -\frac{bc}{d} - l_1 \\ \frac{da}{b} & -l_2 \end{pmatrix}.$$

This matrix has negative trace and positive determinant, and hence is Hurwitz, for any choice of positive l_1, l_2. This means that Theorem 3 applies and the system given by (17.14) defines an exponential observer for (17.15).

In a later paper Sundarapandian (2010), the same author applied the above result to construct local observers for 2 and 3 dimensional Lotka-Volterra predator-prey systems. It has also been used in López et al. (2007) to design observers for a 5-species food web involving carnivores, herbivores, and plants.

More recently, the authors of López et al. (2019) have outlined how the basic approach of Sundarapandian (2002) can be adapted to design observers that are robust with respect to certain types of parameter uncertainty. The central idea is to assume that the Jacobian matrix $A = J(x^*)$ is not only Hurwitz, but is a *sign-stable* matrix. This means that any matrix with the same sign pattern as A is Hurwitz.

Two matrices M, N have the same sign pattern if the locations of negative, zero, and positive entries in M and N are identical. The idea behind designing a robust observer is then to require that the observer matrix L in Theorem 3 is chosen so that A and $A - LC$ are guaranteed to have the same sign pattern. As A is sign-stable, it is immediate that $A - LC$ is Hurwitz. The robustness comes from the observation that $M - LC$ will also be Hurwitz for any matrix M with the same sign pattern as A. Essentially, this means that if the numerical values in A are inaccurate, the matrix L will still define a local observer for the system provided their sign is correct. As noted in López et al. (2019), this is not an unreasonable assumption, as the signs of the entries in A are typically determined by the nature of interactions between species or cohorts in a system. For instance, a change in the sign of an entry in A may indicate that a cooperative interaction had been incorrectly identified as being competitive, or that a predator had been incorrectly identified as prey. While not impossible, such errors are far less likely than numerical inaccuracies in parameter values.

An Alternative Approach to Observer Design: State-Ratio Dynamics
A novel and distinct approach to the design of observers for nonlinear systems, specifically of Lotka-Volterra type, was recently developed in Badri (2022). The methodology makes use of the strict positivity of solutions and considers the ratios $\frac{x_i(t)}{\hat{x}_i(t)}$ of the true and estimated states rather than the error $x_i(t) - \hat{x}_i(t)$. Of course, it is crucial to design the observer in such a way that $\hat{x}_i(t)$ is always positive for the ratios to be well-defined. After describing the structure of the observer, the main result of Badri (2022) gives conditions under which the parameters of the observer will ensure the *global* convergence of all the ratios $\frac{x_i(t)}{\hat{x}_i(t)}$ to 1 as $t \to \infty$.

To give a little more detail, the class of system studied in Badri (2022) takes the form

$$\dot{x}_i(t) = x_i(t)\left(r_i + \sum_{j=1}^{n} a_{ij}x_j(t)\right), \quad 1 \le i \le n. \tag{17.16}$$

Here r_i denotes the intrinsic growth rate for the ith population while the matrix $A = (a_{ij})$ describes the interaction terms between the species in the system. The author assumes the existence of a unique positive equilibrium x^* so that the system can be rewritten in the form

$$\dot{x}_i(t) = x_i(t)\sum_{j=1}^{n} a_{ij}(x_j(t) - x_j^*). \tag{17.17}$$

Further, the measured output is assumed to be given by a function of the form

$$y(t) = k \prod_{i=1}^{n} x_i^{c_i}(t)$$

for some real numbers $k, c_1, c_2, \ldots c_n$. Write c for the column vector in $\mathbb{R}^n$ whose components are c_i. Note that this form for $y(t)$ includes the simple case where only 1 species is observed: for instance, if only the ith species is measured, this is modeled by setting $k = 1$, $c_i = 1$ and $c_j = 0$ for $j \neq i$.

The aim is to design an observer system that uses the ratios between the measured outputs of the original system and the outputs of the observer system to ensure that the state ratios $\frac{x_i(t)}{\hat{x}_i(t)}$ converge to 1. The observer system used is of the form:

$$\dot{\hat{x}}_i(t) = \hat{x}_i(t) \sum_{j=1}^{n} a_{ij}(\gamma_j(t)\hat{x}_j(t) - x_j^*) \tag{17.18}$$

with the observer output given by $\hat{y}(t) = k \prod_{i=1}^{n} \hat{x}_i^{c_i}(t)$. The terms c_i appearing here are the same as in the original system.

The term $\gamma_j(t)$ is used to feed back the ratio between the measured output of the true system and that of the observer system (17.18). To define $\gamma_j(t)$, you first compute

$$r_0 = \frac{y(t)}{\hat{y}(t)}.$$

$\gamma_j(t)$ is then given by $\gamma_j(t) = r_0^{b_j}$ where the exponents b_j need to be designed to ensure convergence.

The main result of the paper states sufficient conditions for the observer (17.18) to be *globally convergent*. The conditions in this result are stated in terms of a linear matrix inequality (LMI), whose solution determines the values of the parameters b_j, $1 \leq j \leq n$ used to define the observer.

We now recall the main result of Badri (2022). In the statement below, A is the interaction matrix of the system's dynamics, and c is the vector used to define the output $y(t)$.

Theorem 4 *Suppose that there exists a point $d \in \mathbb{R}^n$ and two positive definite diagonal matrices P, Q in $\mathbb{R}^{n \times n}$ with rank$(I - P^{-1}dc^T) = n$ such that the following LMI is satisfied.*

$$\begin{pmatrix} (P - dc^T)A + A^T(P - dc^T)^T & A^T(P - dc^T) + PA \\ (P - dc^T)^T A + A^T P & (P + Q)A + A^T(P + Q) \end{pmatrix} \prec 0. \tag{17.19}$$

Then setting $b = P^{-1}d$, the observer (17.18) *is globally asymptotically convergent.*

If there is a solution, given by P, Q, d, to the linear matrix inequality (LMI) in (17.19), then the LMI is said to be *feasible*. The feasibility problem for LMIs can be readily solved using numerical solvers such as the LMI toolbox for MATLAB. It is also worth noting that the condition $\text{rank}(I - P^{-1}dc^T)$ is not really restrictive. If the LMI is feasible, the set of solutions will be an open set and $\text{rank}(I - P^{-1}dc^T) = n$ will be satisfied generically (most solutions found by numerical means will satisfy this condition).

Remarks
It is important to note that the matrix A in Theorem 4 is the interaction matrix of the overall system, whereas in Theorem 3, A is the Jacobian evaluated at the positive equilibrium.

A necessary condition for the LMI (17.19) to be feasible is that the matrix A is diagonally stable. In particular, there are no solutions to the LMI (17.19) unless A is a Hurwitz matrix. For example, in the case of a simple predator-prey system without intra-specific competition, the interaction matrix is given by

$$A = \begin{pmatrix} 0 & -b \\ d & 0 \end{pmatrix}.$$

This is not Hurwitz, so Theorem 4 cannot be used to construct a global observer in this case.

As discussed by the author in Badri (2022), an alternative approach to the construction of globally convergent observers has been presented in Boutat and Saif (2016). The approach in this earlier paper relies on differential geometric methods to transform the system to a nonlinear observer normal form, for which observer design is straightforward. The authors demonstrate the technique on a number of low-dimensional systems of predator-prey form. While the method is interesting, powerful, and based on general results, it does require a quite sophisticated calculation to obtain the normal form for a given system, even for low-dimensional examples. On the other hand, the feasibility of the LMI in Theorem 4 can be checked automatically using standard tools such as LMI solvers available in MATLAB.

5 Concluding Remarks

In this chapter, our aim was to introduce the reader to some of the main ideas, concepts, and results that underlie observability and observer design and to illustrate the relevance of these for population monitoring. The field of nonlinear observer design is an active area within control theory, and we have only given a very light introduction here. For a more detailed discussion of recent work on the general problem, the reader should consult any of the references Bernard et al. (2022), Bernard (2019), Boutat and Zheng (2021) mentioned above.

Our discussion has largely focussed on extensions of classical observers to nonlinear systems. It is also worth noting the separate line of work on *interval observers* (Gouzé et al. 2000). Here, a pair of observer systems is constructed such that the true state of the original system is always bounded between the 'lower' and 'upper' trajectories obtained from the observers.

References

Bacaër N (2010) A short history of mathematical population dynamics. Springer, London

Badri V (2022) Global observer for Lotka-Volterra systems. Syst Control Lett 167:105319

Bernard P (2019) Observer design for nonlinear systems. Lecture notes in control and information sciences, vol 479. Springer, Cham

Bernard P, Andrieu V, Astolfi D (2022) Observer design for continuous-time dynamical systems. Annu Rev Control 53:224–248

Boutat D, Saif, M (2016) Observer normal forms for a class of predator-prey models. J Frankl Inst 353:2178–2198

Boutat D, Zheng G (2021) Observer design for nonlinear dynamical systems: differential geometric methods. Lecture notes in control and information sciences, vol 487. Springer

Cushing JM (2024) Matrix models for population, disease, and evolutionary dynamics. American Mathematical Society, Rhode Island

Doak D, Kareiva P, Klepetka B (1994) Modeling population viability for the desert tortoise in the western Mojave Desert. Ecol Appl 4(3):446–460

Ebenman B (1988) Competition between age classes and population dynamics. J Theoret Biol 131(4):389–400

Gámez M, López I, Rodríguez C, Varga Z, Garay J (2017) Ecological monitoring in a discrete-time prey-predator model. J Theoret Biol 429:52–60

Gouzé JL, Rapaport A, Hadj-Sadok MZ (2000) Interval observers for uncertain biological systems. Ecol Model 133(1–2):45–56

Kazantzis N, Kravaris C (2001) Discrete-time nonlinear observer design using functional equations. Syst Control Lett 42:81–94

Krener AJ, Respondek W (1985) Nonlinear observers with linearizable error dynamics. SIAM J Control Optim 23:197–216

Logofet DO, Ulanova NG (2022) From population monitoring to a mathematical model: a new paradigm of population research. Biol Bull Rev 12(3):279–303

López I, Gámez M, Molnár S (2007) Observability and observers in a food web. Appl Math Lett 20:951–957

López I, Luz Puertas M, Rodríguez C, Gámez M, Varga Z (2019) Robust observer design by sign-stability for the monitoring of population systems. J Frankl Inst 356:2240–2257

Luenberger D (1966) Observers for multi-variable linear systems. IEEE Trans Autom Control 2:190–197

Mollet HF, Cailliet GM (2002) Comparative population demography of elasmobranchs using life history tables, Leslie matrices and stage-based matrix models. Marine Freshwater Res 53(2):503–515

Polderman JW, Willems JC (1997) Introduction to mathematical systems theory: a behavioral approach. Texts in applied mathematics, vol 26. Springer

Smith HL, Thieme HR (2011) Dynamical systems and population persistence. American Mathematical Society, Rhode Island

Sundarapandian V (2002) Local observer design for nonlinear systems. Math Comput Model 35:25–36

Sundarapandian V (2004) General observers for discrete-time nonlinear systems. Math Comput Model 39:87–95

Sundarapandian (2004) New results on general observers for discrete-time nonlinear systems. Appl Math Lett 17:1415–1420

Sundarapandian V (2010) Nonlinear observer design for Lotka-Volterra systems. In: IEEE International Conference on Computational Intelligence and Computing Research. https://doi.org/10.1109/ICCIC.2010.5705770

Thau FE (1973) Observing the states of nonlinear dynamical systems. Int J Control 18:471–479

Varga Z (2008) Applications of mathematical systems theory in population biology. Period Math Hungar 56(1):157–168

Part IV
Societal Innovations: Case Studies

Chapter 18
The Influence of Rafting on the Use of Riparian Atlantic Forest by Middle to Large Terrestrial Mammals

Maria J. Robim, João Paulo Villani, Douglas da Silva Menezes, Fernanda Cristina de Barros, Hélio Alexandre de Souza, Maria Cecília Barbosa de Toledo, Carlos I. Piña, Rafael de Andrade Moral, and Luciano Martins Verdade

Abstract Ecotourism has great potential to benefit protected areas. In addition to the economic impact on local communities, it has generated increased political and financial support for conservation. However, tourist activities can negatively impact natural environments. In this study we evaluated the impact of rafting activities on medium to large terrestrial mammals in the Serra do Mar State Park/Santa Virgínia Unit (PESM/NSV). We installed camera-traps along the Paraibuna river in areas with and without rafting and hiking activities. We found that both species richness and abundance of medium to large terrestrial mammals were negatively affected by rafting, but not by hiking. The white-lipped peccary (*Tayassu pecari*) was the species most affected. However, this impact should be monitored in the medium to long term, given the lack of data previous to the implementation of these activities and the possible habituation of the species over time. In such context, future studies should test variable frequencies of rafting activities to estimate its cost-benefit relationship with conservation units of Atlantic Forest.

Keywords Ecotourism · Wildlife management · Human impact · Neotropics

M. J. Robim · J. P. Villani · D. da Silva Menezes · F. C. de Barros · H. A. de Souza
Serra do Mar State Park – Santa Virgínia Unit, São Luís do Paraitinga, SP, Brazil

M. C. B. de Toledo
University of Taubaté/Laboratory of Ecology, Taubaté, SP, Brazil

C. I. Piña
CICyTTP-CONICET Entre Ríos/UADER, España, Diamante, Entre Ríos, Argentina

R. A. Moral
Department of Mathematics and Statistics, Maynooth University, Maynooth, Co. Kildare, Ireland

L. Martins Verdade (✉)
Wildlife Monitoring Consultancy, Campinas, São Paulo, Brazil
e-mail: luciano.verdade@wildlifemgmtco.com

L. Martins Verdade, R. A. Moral (eds.), *Wildlife Monitoring*,
https://doi.org/10.1007/978-3-032-05821-8_18

1 Introduction

The World Tourism Organization (UNWTO) estimates that 10% of the world's tourists seek environment-related activities, which annually generate an estimated 260 billion USD worldwide, approximately 70 million USD of which (approximately 0.03%) is generated in Brazil (Vasconcelos et al. 2012). Accordingly, nature tourism plays an important role in isolated regions with low-income populations and few job opportunities. However, with the increasing political and financial support for conservation, the economic impact could be significant for both local communities and protected areas (Okech 2008). Within the adventure tourism industry, rafting is characterized by developing sustainable activities, adopting mechanisms that seek to increase the ecological awareness of participants and implementing measures to minimize the impact of tourist flow in fragile communities (Brasil 2008).

While ecotourism has the potential to generate benefits for full protection conservation units, the fact that tourist activities in these areas may have negative impacts should not be overlooked (Leung and Farrell 2002; Buckley 2004; Buckley et al. 2016). The impact of recreational activities on the fauna in natural areas can be direct or indirect (Hidinger 1996; DeLong 2002; Newsome et al. 2002; Buckley 2004; Cline et al. 2007). The importance and magnitude of these effects are related to the extent, intensity, and duration of the activity. In such cases, the vulnerability, rarity, and importance of habitat to wildlife should be considered. For instance, a comparative study in the *Serra dos Órgãos* National Park suggests a significant reduction in species richness and abundance of medium to large mammals and birds on the visited trails (Cunha 2010). To evaluate the extent to which species and habitats are affected, specific long-term studies on the impact of recreational activities are recommended (Cole and Landres 1995). In Brazil, studies focusing on the impact of recreational activities on the wildlife are still scarce.

Rafting was invented in 1869, when John Wesley Powel organized the first expedition on the Colorado River, USA, using rowboats. Inflatable boats appeared in the United States in 1936. In 1980, the "self-bailer" raft was introduced, along with new models made of lighter and more resistant materials, which popularized the sport (Marchi and Mezzadri 2003; CBCa 2014). According to the Brazilian Canoe Confederation (*Confederação Brasileira de Canoagem*—CBCa), the history of rafting in Brazil is recent. In 1982, the first rafts were introduced, and the first Brazilian company was created, offering rafting trips down the Paraíba do Sul and Paraibuna Rivers in the municipality of Três Rios (state of Rio de Janeiro). Currently, an estimated 50 companies operate in Brazil, mostly in São Paulo, Santa Catarina, and Rio Grande do Sul (Marchi and Mezzadri 2003; CBCa 2014). In the region of São Luís do Paraitinga, the first implementation phase of rafting activity on the Paraibuna River within the boundaries of Serra do Mar State Park—Santa Virgínia Unit (PESM-NSV) lasted from 1997 to 2006. In this period, carrying capacity studies were conducted, and the recommendations of the minimal impact practice "Leave no Trace" procedures were adopted (Raimundo and Villani 2000; Robim et al. 2011). In 2007, rafting activities were halted to allow for studies and the implementation of regulations (Villani et al. 2009). In 2011, the PESM-NSV was São Paulo State's first conservation unit to obtain authorization for commercial rafting trips in its area. Hiking is usually carried out complementarily to rafting in tourism of adventure, which can not only increase tourists' contact with wilderness along trails, but also increase a possible impact on the wildlife in riparian areas (Enríquez-de-Salamanca and Díaz-Sierra 2023).

PESM-NSV includes patches of Atlantic Forest, which is considered a hot spot for its biodiversity (Mittermeier et al. 2005), including 83 species of non-flying medium to large terrestrial mammals (Reis et al. 2014). However, most of its original area has been replaced by agriculture and urbanization (Dean 1997), which resulted on a great number of species endangered with extinction (Chiarello et al. 2008). As the largest protected area of Atlantic Forest, PESM is crucial for their conservation. For this reason, the possible impacts of ecotourism should be evaluated. In such a context we hypothesized that rafting and hiking activities, as well as their interaction could affect species richness and frequency of occurrence of medium to large terrestrial mammals in riparian forest at PESM/NSV.

2 Study Area

The PESM is the largest remnant of the Atlantic Forest Biome in Brazil with an area of approximately 320,000 ha. The park was created on August 30, 1977, by Decree No. 10,251 and it is divided into the following units: Bertioga, Caraguatatuba, Cunha, Curucutu, Itariru, Itutinga-Pilões, Padre Dória, Picinguaba, Santa Virgínia, and São Sebastião (São Paulo 2006). The Santa Virgínia Unit (NSV) has been established on May 2, 1989, comprising former Ponte Alta and Santa Virgínia farms, where its current headquarters are located (23°24′ S and 45°03′ W) (São Paulo 2006). The NSV is spread over the municipalities of São Luís do Paraitinga, Natividade da Serra, Cunha, and Ubatuba, with approximately 17,500 ha (Fig. 18.1).

The NSV is in the Atlantic Plateau of the state of São Paulo, a geomorphological macro-unit that is located Northwest to the cliffs of the Serra do Mar, predominantly consisting of steep terrain between 840 and 1500 m of altitude. The average slope ranges from 24° to 37°, peaking over 40°. According to Koppen classification, the regional climate is humid temperate without a dry season (Setzer 1996). Precipitation varies from 1800 to 2800 mm/year, with frequent orographic rainfall and fog formation (São Paulo 2006).

The area is covered by montane and upper-montane rainforest (Veloso et al. 1991) with six species of palm trees: *Euterpe edulis*, *Geonoma gamiova*, *G. pohliana*, *G. schottiana*, *Attalea dubia*, and *Syagrus pseudococos*. According to Tabarelli et al. (1994), 60% of the area is composed of pristine forest complemented by *campo limpo* (grassland), *compo sujo* (shrubby grassland), *capoeira* (secondary forest), *capoeirão* (dense secondary forest), and *Eucalyptus saligna* plantations.

The NSV Public Use Program is divided into two subprograms, namely, "Visitation and Sustainable Tourism" and "Environmental Education". The program

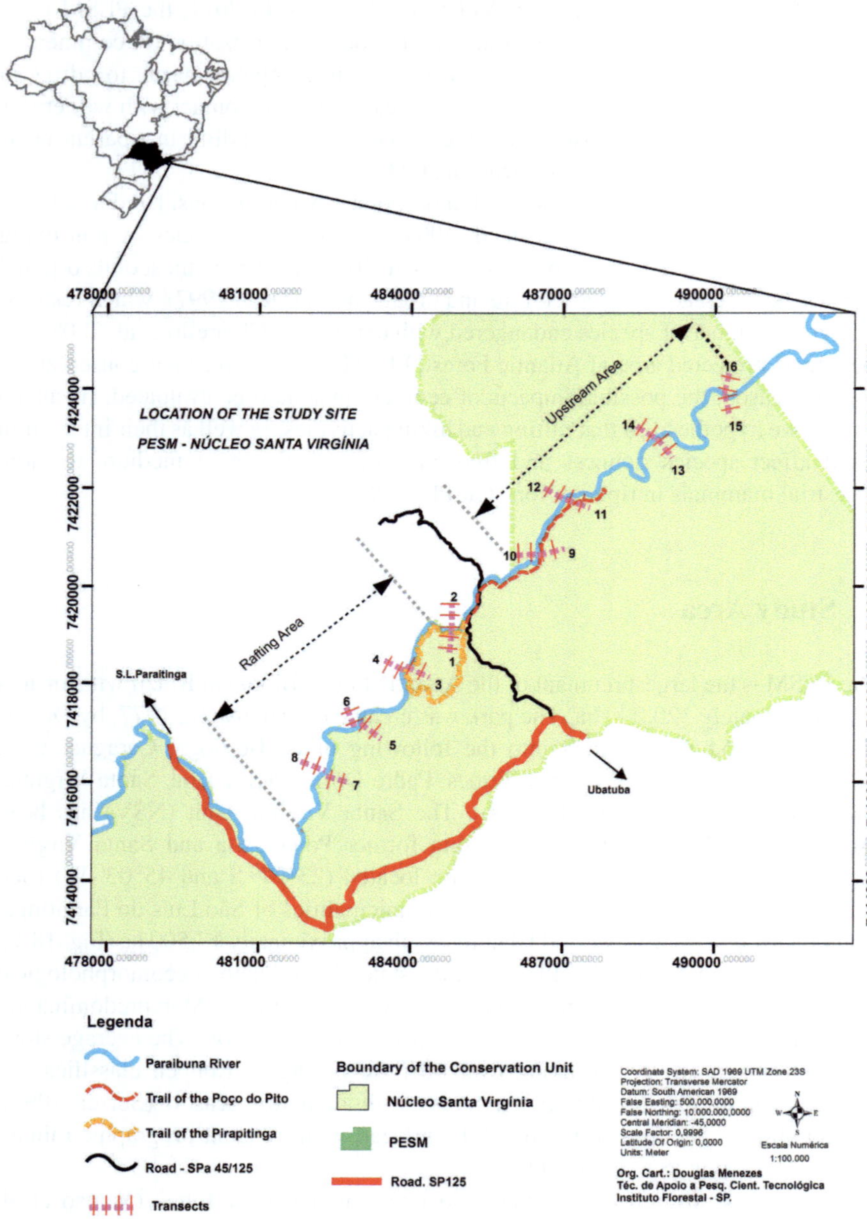

Fig. 18.1 Study area at *Parque Estadual Serra do Mar—Núcleo Santa Virgínia*, state of São Paulo, Brazil

receives 5000 visitors per year on average and develops activities with students, teachers and the general public, including hikes, talks, environmental interpretation and education activities.

A rafting license has been granted for a private company from São Luís do Paraitinga in January 2011, along 10 km of the Paraibuna River (23°20′07″ S–45°08′46″ W, 23°22′58″ S–45°10′46″ W) (Fig. 18.1), for 2 days every other week with up to 20 tourists per day. Rafting trips usually last 6 hours on average, from 9:00 am to 4:00 pm. The inflatable rafts are 3.7 m long.

3 Methodology

3.1 Sampling Methodology

The sampling area of the present study was composed of two parts as follows:

(a) Rafting area: Eight 500-m transects transversal to the Paraibuna river, with four of them crossing the Pirapitinga hiking trail; and,
(b) Non-rafting area: Eight 500-m transects transversal to the Paraibuna river upstream to the rafting area, with four of them crossing the Poço do Pito hiking trail.

We installed six camera-traps in each transect, placed at 0, 100, 200, 300, 400 and 500 m far away from the water on each side of the river bank. The cameras were set up for 10 days in each of 18 campaigns from January through June 2013, totaling 17,280 camera-trap days (i.e., 6 cameras × 16 transects × 18 campaigns × 10 days per campaign). During this study, 13 rafting trips took place, with a total of 243 visitors.

The camera-traps were attached to tree trunks at 50 cm high facing South to avoid excessive sun light. We recorded species richness and frequency of occurrence (i.e., number of photos per species) of medium to large terrestrial mammals. We assume the number of photos per species is correlated with its frequency of occurrence which, in turn, is related to the individuals' use of space (although not necessarily with species abundance). In addition, consecutive records of the same species at the same camera were deemed independent when there was an interval between them of at least 1 hour, unless multiple individuals could be distinguished by appearing simultaneously in the same photo (O'brien et al. 2003).

3.2 Analytical Methodology

We fitted Poisson generalized linear mixed models to the richness data, including the effects of distance from the water, rafting, hiking, and all two-way and three-way interactions between these effects, as well as random intercepts per transect, since observations taken on the same transect are correlated. We assessed the significance of the effects using likelihood-ratio (LR) tests for nested models. Goodness-of-fit was assessed using half-normal plots with a simulated envelope (Moral et al. 2017). We fitted the same models to the data on the number of photos, separately for each species encountered. When a certain species was not found in all transects that included e.g., rafting or hiking, these effects were excluded from the analysis.

We also carried out marginal analyses ignoring the distance covariate, by collapsing the number of photos over distance per transect. This allowed us to fit Poisson log-linear models to analyse the 2 × 2 contingency table for each combination between rafting (yes and no) and hiking (yes and no) for each species, as to whether there was an interaction between rafting and hiking activities and the number of photos taken. Significance of the effects was assessed using chi-squared tests for the differences between deviances.

All analyses were carried out in R (R Core Team 2024). To fit the mixed models, we used package lme4 (Bates et al. 2015), and all visualisations were produced using package ggplot2 (Wickham 2006). All data and code are available at https://www.github.com/rafamoral/rafting_analyses.

4 Results

Twelve species from eight families and six orders of medium to large terrestrial mammals have been detected by a total of 1039 photos (Table 18.1). Five of them represented approximately 85% of the total number of photos taken by camera-traps: White-lipped peccary, Collared peccary, Tapir, Paca and Opossum. Three of them are considered threatened with extinction in the state of São Paulo (São Paulo 2014): Puma, Tapir, and White-lipped peccary. When analysing species richness (Fig. 18.2), there were no significant interactions between distance, rafting and hiking (triple interaction: LR = 1.27, d.f. = 1, p = 0.261; rafting × distance interaction: LR = 0.07, d.f. = 1, p = 0.788; hiking × distance interaction: LR = 0.42, d.f. = 1, p = 0.519; rafting × hiking interaction: LR = 2.97, d.f. = 1, p = 0.085), nor there was a significant main effect of hiking (LR = 2.27, d.f. = 1, p = 0.132). However, the main effects of both distance (LR = 3.38, d.f. = 1, p = 0.066) and rafting (LR = 3.36, d.f. = 1, p = 0.067) were marginally significant at a 5% level, indicating that richness declined slightly in areas with rafting, and with a larger distance from the water.

We found a significant interaction between rafting and hiking on the abundance of *Dasypus* sp., *Didelphis* sp., *Leopardus* sp., *Puma concolor*, and *Tapirus terrestris* (Table 18.2; Fig. 18.3), whereas only the main effect of rafting was significant for *Pecari tajacu*, and only the main effect of hiking was significant for *Monodelphis* sp. The main effect of distance was also significant for *Puma concolor*, whereas the interaction between distance and hiking was significant for *Agouti paca*, and the triple interaction between distance, hiking and rafting was significant for *Sciurus aestuans* (Table 18.2; Fig. 18.3).

Disregarding the distance from the water in the analyses, we found significant interactions between rafting and hiking on the abundance of *Dasypus* sp., *Didelphis*

Table 18.1 Number of photos of medium to large terrestrial mammals detected by camera trapping at the Serra do Mar State Park—Santa Virgínia Unit, in the riparian Atlantic Forest with and without rafting and hiking activities along Paraibuna River

Order	Family	Species	Common names	Food habit	No. of photos	%
Carnivora	Felidae	*Puma concolor*	Puma	Carnivore	12	1.1
		Leopardus sp.	Wild cat	Carnivore	36	3.4
	Mustelidae	*Eira barbara*	Tayra	Omnivore	8	0.8
Artiodactyla	Tayassuidae	*Pecari tajacu*	White-lipped peccary	Frugivore	111	10.8
		Tayassu pecari	Collared peccary	Frugivore	332	32.0
Peryssodactila	Tapiridae	*Tapirus terrestris*	Tapir	Frugivore	152	14.6
Rodentia	Dasyproctidae	*Agouti paca*	Paca	Frugivore	179	17.3
		Dasyprocta aguti	Agouti	Frugivore	24	2.2
	Sciuridae	*Sciurus aestuans*	Brazilian squirrel	Frugivore	38	3.7
Marsupialia	Didelphidae	*Monodelphis sp.*	Opossum	Insectivore	104	10.0
		Didelphis aurita	Big-eared opossum	Omnivore	12	1.1
Cingulata	Dasipodidae	*Dasypus sp.*	Armadillo	Insectivore	31	3.0
Total					1039	100

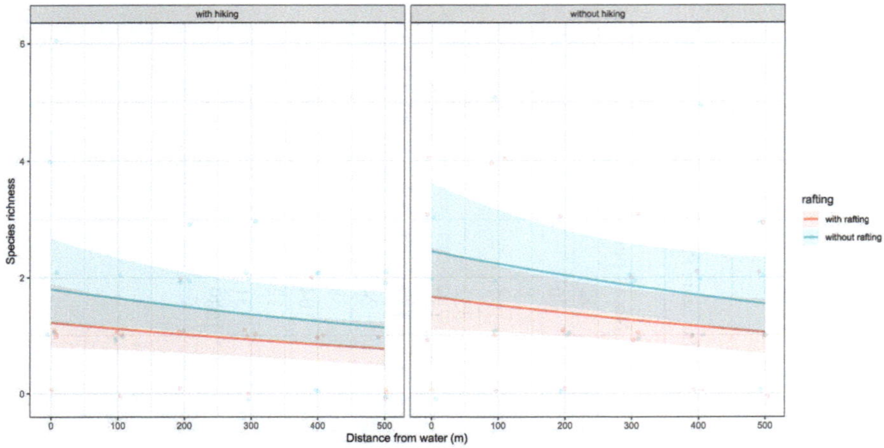

Fig. 18.2 Predicted richness (solid curves) and 95% confidence intervals for the true richness (shaded areas) of medium to large terrestrial mammals, found in areas with or without rafting and hiking activities at the PESM-NSV, at different distances from the Paraibuna river bank. Transparent points represent the actual observations and are jittered horizontally to aid visualization

sp., *Leopardus* sp., *Monodelphis* sp., *Pecari tajacu*, *Puma concolor*, *Sciurus aestuans*, and *Tapirus terrestris* (Table 18.3). For *Agouti paca*, only the main effect of hiking was significant. No *Dasyprocta aguti* individuals were found in areas with hiking, whereas no *Tayassy tajacu* individuals were found in areas with rafting. Finally, *Eira barbara* was only found in areas with rafting and without hiking.

5 Discussion and Conclusion

The observed species richness in this study corresponds to 80% of the medium and large mammal species described by Rocha-Mendes et al. (2015) on a previous camera-trapping study in two areas of the PESM-NSV, with only crab-eating fox (*Cerdocyon thous*), jaguarundi (*Herpailurus yagouaroundi*), and giant anteater (*Tamandua tetradactyla*) not detected in the present study. A relatively high diversity of frugivores is likely related to the high local abundance and diversity of palm trees, which corroborates the relevance of conservation units in the Atlantic Forest in especial for medium to large terrestrial mammals and their trophic network (Chiarello 2000; Araujo et al. 2008; Galetti et al. 2009; Cullen Jr et al. 2001; Norris et al. 2011; Brocardo et al. 2012; Rocha-Mendes et al. 2015).

The reduction of species richness in areas of rafting is predominantly related to game species, which tend to avoid human presence as a response to hunting pressure (Verdade 1996; Chiarello 1999, 2000; Galetti et al. 2009; Cunha 2010). However, if poaching is effectively enforced, as it should be in conservation units, animals can get habituated to human presence during rafting and hiking activities. Such

Table 18.2 Likelihood-ratio test statistics, all associated with 1 degree of freedom, and associated p-values (between parentheses) for the effects of distance, rafting, hiking and interactions, obtained from the Poisson generalized linear mixed models fitted to the data on the number of photos where each species appears

Species	Effect						
	Distance	Rafting	Hiking	Distance × Rafting	Distance × Hiking	Rafting × Hiking	Distance × Rafting × Hiking
Agouti paca	–[a]	2.26 (0.133)	–[a]	1.01 (0.314)	**4.28 (0.039)**	0.00 (0.952)	0.15 (0.700)
Dasyprocta aguti	0.73 (0.394)	2.95 (0.086)	–[b]	0.49 (0.482)	–[b]	–[b]	–[b]
Dasypus sp.	2.96 (0.085)	–[a]	–[a]	1.80 (0.179)	1.22 (0.269)	**6.73 (0.009)**	0.00 (0.999)
Didelphis sp.	0.10 (0.755)	–[a]	–[a]	0.00 (0.999)	0.00 (0.999)	**4.70 (0.030)**	0.00 (0.999)
Eira barbara	0.59 (0.442)	–[c]	–[c]	–[c]	–[c]	–[c]	–[c]
Leopardus sp.	3.25 (0.072)	2.50 (0.114)	2.98 (0.084)	0.25 (0.615)	3.18 (0.074)	**3.83 (0.050)**[d]	0.00 (0.999)
Monodelphis sp.	0.00 (0.994)	2.48 (0.116)	**6.92 (0.009)**	0.27 (0.604)	1.11 (0.292)	0.20 (0.656)	0.23 (0.633)
Pecari tajacu	1.96 (0.161)	**4.35 (0.037)**	0.60 (0.439)	0.52 (0.470)	0.17 (0.684)	0.01 (0.941)	0.18 (0.667)
Puma concolor	**11.75 (0.001)**	–[a]	–[a]	0.00 (0.999)	0.00 (0.999)	**9.75 (0.002)**	–[e]
Sciurus aestuans	–[a]	–[a]	–[a]	–[a]	–[a]	–[a]	**4.33 (0.038)**
Tapirus terrestris	0.14 (0.705)	–[a]	–[a]	0.04 (0.833)	0.03 (0.871)	**5.29 (0.021)**	0.00 (0.999)
Tayassu tajacu	0.00 (0.999)	–[f]	1.04 (0.37)	–[f]	0.35 (0.552)	–[f]	–[f]

Significant effects (at a 5% level) are represented in bold

[a]When higher-order interaction terms are significant, we do not present lower-order interactions and/or main effects

[b]No individuals were found in transects crossing hiking trails

[c]No individuals were found in transects without rafting or crossing hiking trails

[d]Effect was marginally significant at a 5% level, hence why the main effects are still presented

[e]The model including the triple interaction did not converge due to lack of information in the data to estimate it

[f]No individuals were found in transects with rafting

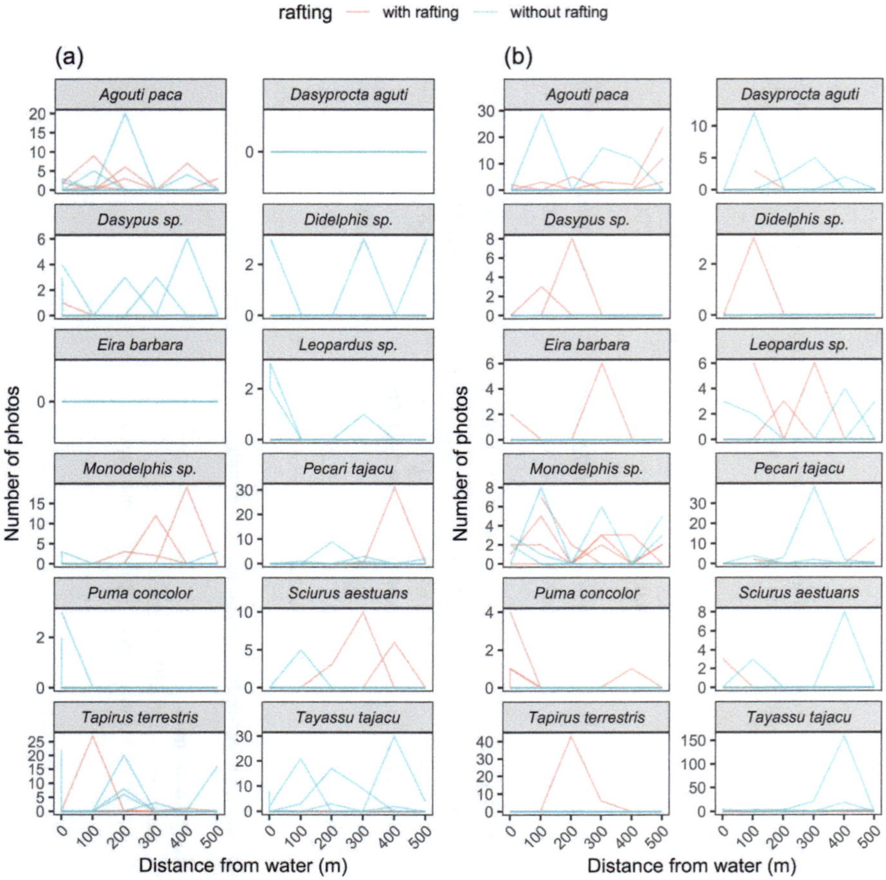

Fig. 18.3 Number of photos taken from 12 different species of medium to large terrestrial mammals, found in areas with and without rafting and hiking activities at the PESM-NSV, at different distances from the Paraibuna riverbank. Each connected line represents one transect where the camera traps were installed. (**a**) With hiking. (**b**) Without hiking

possibility can be tested by the establishment of a medium- to long-term wildlife monitoring program along with touristic operations. This is particularly relevant for the white-lipped peccary, whose use of space at the PESM/NSV appears to be determined by distance to the park border and roads, and amount of primary forest. Considering that this species is an important seed disperser, a possible reduction of its use of space might affect the forest composition in the long-term (Travassos 2011).

White-lipped peccaries, and possibly pumas (not detected in areas with rafting in this study), appear to be affected by rafting activities. However, possible interactions might happen with rafters as the species tend to use mostly areas close to the water. Their sight might be attractive to rafters and hikers. Therefore, proper care should be

Table 18.3 Marginal number of photos taken of 12 different species of medium to large terrestrial mammals at the PESM-NSV in transects with or without rafting and hiking activities, differences of deviances and associated p-values (in between parentheses)

Species	Rafting	Hiking		Hiking (significance)	Rafting (significance)	Interaction (significance)
		With	Without			
Agouti paca	With	33	54	**10.43 (0.001)**	0.14 (0.709)	0.00 (0.988)
	Without	35	57			
Dasyprocta aguti	With	0	3	**33.27 (<0.001)**	**15.19 (<0.001)**	–[a]
	Without	0	21			
Dasypus sp.	With	1	11	–[b]	–[b]	**33.44 (<0.001)**
	Without	19	0			
Didelphis sp.	With	0	3	–[b]	–[b]	**13.50 (<0.001)**
	Without	9	0			
Eira barbara	With	0		**11.09 (<0.001)**	**11.09 (<0.001)**	–[a]
	Without	0	0			
Leopardus sp.	With	0	15	–[b]	–[b]	**11.81 (<0.001)**
	Without	9	12			
Monodelphis sp.	With	36	34	–[b]	–[b]	**11.63 (<0.001)**
	Without	6	28			
Pecari tajacu	With	31	14	–[b]	–[b]	**22.36 (<0.001)**
	Without	16	50			
Puma concolor	With	0	7	–[b]	–[b]	**16.30 (<0.001)**
	Without	5	0			
Sciurus aestuans	With	19	3	–[b]	–[b]	**12.62 (<0.001)**
	Without	5	11			
Tapirus terrestris	With	28	49	–[b]	–[b]	**90.16 (<0.001)**
	Without	75	0			
Tayassu tajacu	With	0	0	**55.66 (<0.001)**	**460.25 (<0.001)**	–[a]
	Without	99	233			

All tests are associated with 1 degree of freedom. Significant effects at a 5% level are highlighted in bold

[a]There is not enough data in the contingency table to estimate this effect reliably

[b]When higher-order interaction terms are significant, we do not present lower-order interactions and/or main effects

taken to mitigate the chance of accidents, which requires environmental education efforts and enforcement by park rangers.

An interaction between rafting and hiking activities affect the use of space by medium to large terrestrial mammals in the present study. As a result, these species might have their dispersion and dispersal (Caughley and Sinclair 1994) patterns, as well as their home range and mating system affected (Rosalino et al. 2014; Pogratz et al. 2024).

Terrestrial mammals tend to use trails themselves to forage and sent marking (Silva 2014). This might increase the chance of wildlife sighting by hikers. However, it can also increase the chance of incidents between animals and tourists. For this reason, education and information to the tourists are necessary.

There is usually a trade-off between the environmental cost and the socioeconomic benefits of ecotourism in conservation units. In such a context, the planning systems for public use of conservation units are based on goals that consider its long-term management, as well as the parameters that can be used as indicators to assess whether those goals are achieved (Robim et al. 2011: 208). As stated above, a long-term wildlife monitoring program is required to help the decision-making process about the cost-benefit of touristic activities in conservation units (Verdade et al. 2022).

We could not evaluate the impact of rafting and hiking frequency in this study as they practically did not vary. However, since both activities are affected by pluviosity, future studies should prioritize the possible impact of seasonality and frequency of these touristic operations on medium to large terrestrial mammals (Raimundo and Villani 2000).

Acknowledgments This study was partially funded by the Biota Program/São Paulo Research Foundation (FAPESP) (Proc. No. 2006/60954-4, 2017/01304-4). The Serra do Mar State Park/Santa Virgínia Unit provided fully supported for field data collection. LMV has a productive fellowship from CNPq (Proc. No. 312049/2015-3). M. Veloso, V. C. Veloso, F. V. C. Veloso, H. Carneiro, A. B. Claro, D. Sbruzzi, H. A. de Souza, L. M. da Silva, J. Briet, F. Miranda Carvalho, E. N. do Prado, D. Alves, A. C. dos Santos, D. A. G. Bolderini, A. C. da Silva, P. S. de Toledo, E. F. Custódio, C. A. das Neves, V. M. dos Santos, J. S. dos Santos, J. P. B. Villani, C. Beraldo, R. Alves, L. Lemes, M. C. da Conceição and G. Souza provided support for field work.

References

Araujo RM, Souza MB, Ruiz-Miranda CR (2008) Densidade e tamanho populacional de mamíferos cinegéticos em duas Unidades de Conservação do Estado do Rio de Janeiro, Brasil. Iheringia. Sér Zool, Porto Alegre 98:391–396. https://doi.org/10.1590/S0073-47212008000300014

Bates D, Maechler M, Bolker B, Walker S (2015) Fitting linear mixed-effects models using lme4. J Stat Softw 67(1):1–48. https://doi.org/10.18637/jss.v067.i01

Brasil/Ministério do Turismo e ABETA-Associação Brasileira das Empresas de Ecoturismo e Turismo de Aventura (2008) [Internet]. Diagnóstico do turismo de aventura no Brasil. [cited 2008 oct 23]. Available from: http://www.abeta.com.br/pt-br/pgn.asp?id_pg=61&nivel=1

Brocardo CR, Rodarte R, Bueno RS, Culot L, Galetti M (2012) Non-volant mammals of Carlos Botelho State Park, Paranapiacaba Forest *Continuum*. Biota Neotrop 12(4):198–208. https://doi.org/10.1590/S1676-06032012000400021

Buckley RC (2004) Impacts of ecotourism on terrestrial wildlife. In: Buckley R (ed) Environmental impacts of ecotourism. CAB International, Wallingford, pp 211–228

Buckley RC, Morrison C, Castely JG (2016) Net effects of ecotourism on threatened species survival. PLoS One 11:e0147988. https://doi.org/10.1371/journal.pone.147988

Caughley G, Sinclair ARE (1994) Wildlife ecology and management. Blackwell Scientific Publications, Boston

CBCa – Confederação Brasileira de Canoagem (2014) Rafting: história. [Internet]. [cited 2014 aug.23]. Available from: https://canoagem.org.br

Chiarello AG (1999) Effects of fragmentation of the Atlantic forest on mammal communities in South-Eastern Brazil. Biol Conserv 89:71–82. https://doi.org/10.1016/S0006-3207(98)00130-X

Chiarello AG (2000) Influência da caça ilegal sobre os mamíferos e aves das matas de Tabuleiro do norte do estado do Espírito Santo. Bol Mus Biol Mello Leitão 11(12):229–247

Chiarello AG, Aguiar LDS, Cerqueira R, Melo FR, Rodrigues FH, Silva VMF (2008) Mamíferos ameaçados de extinção no Brasil. Livro vermelho da fauna brasileira ameaçada de extinção 2:680–880

Cline R, Sexton N, Stewart S (2007) A human-dimensions review of human – wildlife disturbance a literature review of impacts, frameworks, and management solutions. U.S. Geological Survey Open-File Report, 1111, p 88

Cole DN, Landres PB (1995) Indirect effects of recreation on wildlife. In: Knigh RL, Gutzwiller KJ (eds) Wildlife and recreationists: coexistence through management and research. Island Press, Washington, DC, pp 183–202

Cullen L Jr, Bodmer ER, Valladares-Padua C (2001) Ecological consequences of hunting in Atlantic forest patches, São Paulo, Brazil. Oryx 35(2):137–144. https://doi.org/10.1046/j.1365-3008.2001.00163.x

Cunha AA (2010) Negative effects of tourism in a Brazilian Atlantic forest national park. J Nat Conserv 18:291–295. https://doi.org/10.1016/j.jnc.2010.01.001

Dean W (1997) With broadax and firebrand: the destruction of the Brazilian Atlantic Forest. University of California Press, Berkeley

DeLong AK (2002) Managing visitor use & disturbance of waterbirds – a literature review of impacts and mitigation measures – prepared for stillwater National Wildlife Refuge. Dept. of the Interior, U.S. Fish and Wildlife Service, Region 1, Portland, OR, 120 p

Enríquez-de-Salamanca Á, Díaz-Sierra R (2023) Impact assessment of temporary activities and events. Integr Environ Assess Manag 19(5):1320–1332

Galetti M, Giacomini HC, Bueno RS, Bernardo CSS, Marques RM, Bovendorp RS et al (2009) Priority areas for the conservation of Atlantic forest large mammals. Biol Conserv 142:1229–1241. https://doi.org/10.1016/j.biocon.2009.01.023

Hidinger AL (1996) Measuring the impacts of ecotourism on animal populations: a case study of Tikal National Park. Yales For Enivorn Bull 99:49–59

Leung Y-F, Farrell TA (2002) Visitor impact knowledge is basic. [CD-ROM]. In: Portifolio of statements and presentations – the world ecotourism summit, May 19–22, 2002, Quebec City, Canada

Marchi KB, Mezzadri FM (2003) História da canoagem e do rafting. In: ANPUH – XXII Simpósio Nacional de História – João Pessoa, 2003. Anais... João Pessoa, pp 1–7. [cited 2014 aug 23]. Available from: http://anpuh.org/anais/wp-content/uploads/ANPUH.S22.pdf

Mittermeier R, Gil PR, Hoffmann M, Pilgrim J, Brooks T, Mittermeier CG, Lamoreaux J, da Fonseca GAB (2005) Hotspots revisited: earth's biologically richest and most endangered terrestrial ecoregions. CEMEX, Mexico City

Moral RA, Hinde J, Demétrio CGB (2017) Half-normal plots and overdispersed models in R: the hnp package. J Stat Softw 81(10):1–23. https://doi.org/10.18637/jss.v081.i10

Newsome D, Moore SA, Dowling DK (2002) Natural area tourism: ecology, impacts and management. Channel View Publications, Clevedon, 340 p

Norris D, Rocha-Mendes F, de Barros F, Ferraz S, Villani JP, Galetti M (2011) How to not inflate population estimates? Spatial density distribution of white-lipped peccaries in a continuous Atlantic forest. Anim Conserv 14:492–501. https://doi.org/10.1111/j.1469-1795.2011.00459.x

O'Brien TG, Kinnaird MF, Wibisono HT (2003) Crouching tigers, hidden prey: Sumatran tiger and prey populations in a tropical forest landscape. Anim Conserv 6:131–139. https://doi.org/10.1017/S1367943003003172

Okech RN (2008) Ecotourism and the economy: case study of Mara & Amboseli in Kenya. Revista de turism-studii si cercetari in turism/J Tour Stud Res Tour 5(5):9–13

Pogratz EJ, Holt RD, Brown JS (2024) Ecology of fear: acclimation and adaptations to hunting by humans. Sustainability 16(3):1216. https://doi.org/10.3390/su16031216

R Core Team (2024) R: a language and environment for statistical computing. R Foundation for Statistical Computing, Vienna. https://www.R-project.org/

Raimundo S, Villani JP (2000) Estudo de capacidade de carga e proposta de regulamentação do rafting no Núcleo Santa Virgínia – Parque Estadual da Serra do Mar (SP). In: Congresso Brasileiro de Unidades de Conservação, 2, Campo Grande-MS, 2000. Anais... Campo Grande, MS, pp 232–242

Reis NR, Fregonezi MN, Peracchi AL, Shibatta OA, Sartore ER, Rossaneis BK, Santos VR, Ferracioli P (2014) Mamíferos Terrestres de Médio e Grande Porte da Mata Atlântica. Technical Books Editora, Rio de Janeiro

Robim MJ, Gallo H Jr, Magro TC, Villani JP (2011) Monitoramento dos Impactos do rafting na zona ripária do rio Paraibuna, Parque Estadual da Serra do Mar-Núcleo Santa Virgínia. Rev OLAM – Tecnol Ciênc 11:203–227

Rocha-Mendes F, Neves CL, Nobre RDA, Marques RM, Bianconi GV, Galetti M (2015) Non-volant mammals from Núcleo Santa Virgínia, Serra do Mar State Park, São Paulo, Brazil. Biota Neotrop 15:1–9. https://doi.org/10.1590/1676-06032014000814

Rosalino LM, Lyra-Jorge MC, Verdade LM (2014) Adaptation and evolution in changing environments. In: Verdade LM, Lyra-Jorge MC, Piña CI (eds) Applied ecology and human dimensions on biological conservation. Springer, Heidelberg, Germany, pp 53–71. https://doi.org/10.1007/978-3-642-54751-5_4. ISBN 978-3-642-54750-8

São Paulo Leis, Decretos, etc (2014) Decreto N° 60.133, de 7 de Fevereiro de 2014. Declara as espécies da fauna silvestre ameaçadas de extinção, as quase ameaçadas e as deficientes de dados para avaliação no Estado de São Paulo e dá providências correlatas. In: São Paulo, Imprensa Oficial do Estado 124(27):25. [Internet]. [cited 2016 sept 28]. Available from: http://www.al.sp.gov.br/norma/?id=172345

São Paulo/Secretaria do Meio Ambiente/Instituto Florestal (2006) Parque Estadual da Serra do Mar – Plano de Manejo [CD-ROM]. São Paulo, 441 p

Setzer J (1996) Atlas climático e ecológico do Estado de São Paulo. Comissão Interestadual da Bacia Paraná-Uruguai, São Paulo

Silva MX (2014) Efetividade de áreas protegidas para a conservação da biodiversidade: padrões de ocupação de mamíferos no Parque Nacional de Foz do Iguaçu. Dissertação de Mestrado. Instituto de Biociências da Universidade de São Paulo, Departamento de Ecologia, 44 p

Tabarelli M, Villani JP, Mantovani W (1994) Estudo comparativo da vegetação de dois trechos de floresta secundária no Núcleo Santa Virgínia/SP. Rev Inst Florestal São Paulo 6:1–11

Travassos L (2011) Impacto da sobrecarga em populações de mamíferos e suas interações ecológicas nas florestas neotropicais. Oecol Aust 15:380–411

Vasconcelos FP, Silva ACP, Costa LF (2012) Turismo de aventura e ecoturismo: entre práticas e normas no contexto brasileiro. Rev Iberoam Tur – RITUR, Penedo 2:108–138. [cited 2014 sept 5]. Available from: http://www.seer.ufal.br/index.php/ritur

Veloso HP, Rangel-Filho ALR, Lima JCA (1991) Classificação da vegetação brasileira, adaptada a um Sistema Universal. IBGE, Departamento de Recursos Naturais e Estudos Ambientais, Rio de Janeiro

Verdade LM (1996) The influence of hunting pressure on the social behavior of vertebrates. Rev Bras Biol 56:1–13

Verdade LM, Bianchi RC, Galetti PM Jr, Pivello VR, Silva WR, Uezu A (2022) Biodiversity management and research in multifunctional landscapes. Biota Neotrop 22(spe):e20221407. https://doi.org/10.1590/1676-0611-BN-2022-1407

Villani JP, Silva AN, Lorejan SF, Pisciotta K, Robim MJ, Ribeiro MCG et al (2009) Caminos para la implementación del rafting en una unidad de conservación del bioma mata atlántica, São Paulo, Brasil. [Cd-ROM]. In: VII Convención del Médio Ambiente y Desarrollo – VI Congreso de Áreas Protegidas, Anais, Havana, Ministerio de Ciencia, Tecnología y Medio Ambiente de la República de Cuba, pp 191–203

Wickham H (2006) ggplot2: elegant graphics for data analysis. Springer, New York

Chapter 19
Wild Boar Introduction Might Result in the Selection of Larger Pumas in Central-South Brazil

Fernado Cesar C. Azevedo, Rafael de Andrade Moral, and Luciano Martins Verdade

Abstract The introduction of wild boars in Central-South Brazil resulted in many environmental impacts, including damage to biotic and abiotic components of natural environments and agroecosystems, as well as hazards to livestock production, and risks to humans. As adult wild boars can reach up to 200 kg they can alter the local food web, as both the predators of small vertebrates and invertebrates, and the possible prey of large predators. As a matter of fact, as wild boars got widespread, we started to detect pumas' tracks near tracks of wild boar females and offspring in our field station in Central-South Brazil. More recently we started to detect piglets' hairs and bones in pumas' scats, which suggests a learning process by local pumas to such novel prey. The combination of predator skills and body size plasticity might result on a selection of larger pumas in Central-South Brazil, including areas under relatively high human population in suburban areas and intensively managed agricultural landscapes. These questions are discussed in this chapter.

Keywords Evolution · Wildlife management · Trophic ecology · Body mass · Felids · Exotic species · Carnivores · Agricultural landscapes

---

F. C. C. Azevedo
Universidade Federal de São João del Rei, Departamento de Ciências Naturais, São João del Rei, MG, Brazil

Instituto Pró-Carnívoros, Atibaia, SP, Brazil

R. A. Moral
Department of Mathematics and Statistics, Maynooth University, Maynooth, Co. Kildare, Ireland

L. Martins Verdade (✉)
Wildlife Monitoring Consultancy, Campinas, São Paulo, Brazil
e-mail: luciano.verdade@wildlifemgmtco.com

L. Martins Verdade, R. A. Moral (eds.), *Wildlife Monitoring*,
https://doi.org/10.1007/978-3-032-05821-8_19

1 Wild Boar Introduction in Brazil

The introduction and subsequent population booming of wild boars in Central-South Brazil are associated with a plethora of environmental impacts, including damage to biotic (i.e., plants and animals) and abiotic (i.e., soil and water) components of natural environments and agroecosystems, as well as hazards to livestock production (Kmetiuk et al. 2023), and risks to humans (Milda et al. 2023). As adult wild boars can reach up to 200 kg (Macdonald 1985: 502), they can alter the local food web, as both the predators of small vertebrates and invertebrates, and the possible prey of large predators like the jaguar (Verdade et al. 2016a). Pumas, on their turn, have a large distribution range, from Patagonia to Western Canada, with an expressive variation in adult body mass from approximately 50–100 kg (Iriarte et al. 1990; Jansen and Jenks 2011). In certain areas of Central-South Brazil (e.g., São Paulo state) pumas tend to be smaller, usually preying on small rodents and birds (Gheler-Costa et al. 2018). However, as wild boars got widespread, we started to detect pumas tracks near tracks of wild boar females and offspring in our field station in Angatuba municipality, in Central São Paulo state (Fig. 19.1). More recently we started to detect remaining parts of piglets (e.g., hair, bones) in pumas scats (Fig. 19.2), which suggests a learning process by local pumas to such novel prey.

The combination of predator skills and body size plasticity might result on a selection of larger pumas in Central-South Brazil, including areas under relatively

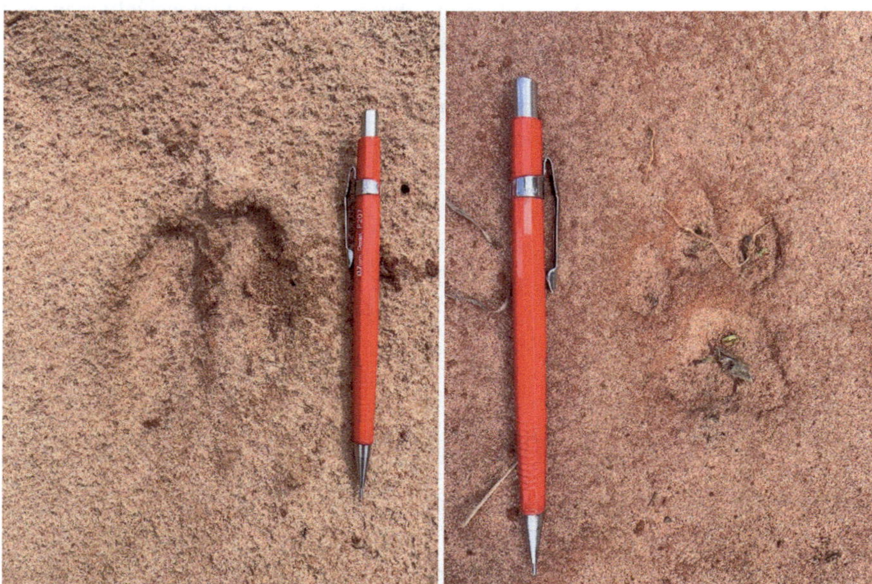

Fig. 19.1 Wild boar and pumas share the same net of dirt roads in agricultural and silvicultural landscapes, Angatuba municipality, southeast Brazil. (Photos: LMV)

Fig. 19.2 Remaining part of a piglet in pumas feces at Fazenda Três Lagoas, Angatuba municipality, southeast Brazil. (Photos: LMV)

high human population in suburban areas and intensively managed agricultural landscapes (Hayward et al. 2016; Zanin et al. 2020). Such a scenario demands proactive wildlife management to deal with conceptual, technological, and societal challenges (Verdade et al. 2014).

2 Prey Size Versus Predator Body Size in Felids

The relationship between prey size and predator body size in felids represents a complex interplay of ecological, physiological, and evolutionary pressures. Such a pattern results in distinct prey specializations across felid species, often correlated with body size and reflected in corresponding prey preferences and hunting strategies. Felid body mass evolution exhibits a pattern of convergent evolution toward two primary optima: a small size (approximately 5 kg) and a large size (approximately 100 kg), reflecting the combined influences of phylogenetic history and prey selection (Cuff et al. 2015). A general correlation exists between felid body size and prey size, with larger felids typically targeting larger prey and smaller felids focusing on smaller prey (Gittleman 1985; Cuff et al. 2015). Craniodental morphology also provides insights into prey size preferences. Large prey specialists exhibit adaptations such as robust canines and broad muzzles, facilitating the capture and subjugation of comparatively large prey items (Meachen-Samuels and Van Valkenburgh 2009). Conversely, small prey specialists possess relatively smaller canines and narrower muzzles, adaptations that enhance speed and agility for pursuing smaller, more elusive prey (Meachen-Samuels and Van Valkenburgh 2009).

Larger felids, such as lions (*Panthera leo*) and tigers (*Panthera tigris*), have evolved to exploit larger prey, a strategy facilitated by their greater gut capacity, which allows for the consumption of substantial food quantities in a single feeding event, thereby reducing hunting frequency (Cuyper et al. 2019). Conversely, smaller felids, frequently target smaller prey. Their limited stomach capacity and comparatively rapid digestion rates require more frequent hunting bouts (Cuyper et al. 2019). However, exceptions and overlaps in such a pattern are well documented (Gittleman 1985), and factors such as ecological conditions, competition, and evolutionary history can lead to variations in prey size preferences among felids. Additionally, some felids exhibit mixed prey strategies, allowing them to adapt to different ecological niches and prey availability (Vissia et al. 2022).

The joint evolution of predator and prey body sizes is influenced by ecological factors such as prey availability and competition, although these factors primarily affect predator size rather than the predator-prey size ratio (Troost et al. 2008). Among carnivores, prey size and diversity tend to increase with predator body mass, highlighting the adaptive significance of body size in relation to prey characteristics (Gittleman 1985).

3 Body Size Variation in Pumas: An Evolutionary Perspective

One of the most enthralling issues in ecological and evolutionary biology is the body size variation in pumas. A complex interaction of geographic, ecological, and biological factors, including biome, prey availability, and individual characteristics such as sex and age influence this variation. The puma is a sexually dimorphic species with males being 30–50% larger than females (Gay and Best 1996). In addition, pumas present significant differences in body size across their geographic range. Generally, geographic variation in size is mostly correlated with latitude where populations from temperate regions have larger body sizes, preying on larger animals (Iriarte et al. 1990). In contrast, populations in tropical habitats tend to be smaller, hunting smaller and more diverse prey, resulting on smaller body size (Iriarte et al. 1990; Gay and Best 1996). Such pattern is often attributed to Bergmann's rule, which suggests that larger body size is advantageous in colder climates (Lindstedt and Boyce 1985).

Puma body mass correlates positively with the mean body mass of vertebrate prey and negatively correlated with food niche breadth, suggesting that larger pumas can exploit fewer larger prey (Iriarte et al. 1990). Seasonal changes also affect prey selection, with pumas opting for smaller prey during warmer months (Bates-Mundell et al. 2024). Puma body mass can fluctuate rapidly, depending on recent food intake. These felids are known to engage in gorging behavior, particularly following successful hunts, resulting in rapid increases in body mass (Bates-Mundell et al. 2024). Such behavior is tightly coupled with ecological factors, including prey size selection and seasonal variations in prey abundance. Additionally, individual sex and age play a role in prey selection. Male pumas often exhibit a tendency to select larger prey items as they mature, whereas females may focus on smaller prey, reflecting distinct foraging strategies potentially related to experience, prey availability and prevailing environmental conditions (Azevedo et al. 2018; Bates-Mundell et al. 2024).

4 How Big Can a Puma Get and How Long Would It Take

Pumas can be considered the largest of small cats. Adult males measure 1.5–2.4 m, whereas females reach around 2.1 m of total length (i.e., including the tail) (Iriarte et al. 1990; Jansen and Jenks 2011). While adult males typically weigh between 53 and 100 kg (average around 80 kg), some exceptionally large individuals can exceed 110 kg. Adult females are generally smaller and lighter, weighing between 29 and 64 kg (average around 50 kg).

Pumas exhibit rapid growth rate, particularly in their first year, They typically reach nearly adult size by approximately 18 months, although they continue to develop and gain muscle mass through their second and third years of age, influenced by genetic background, habitat, and, most importantly, prey availability (Iriarte et al. 1990; Jansen and Jenks 2011; Barceló et al. 2022).

Areas with abundant prey support the development of larger, stronger pumas, as ample food resources allow for optimal growth. Although pumas can attain substantial size, metabolic constraints limit their maximum size, as larger carnivores incur higher energy demands (Whitfield 2007). This balance between size and energy efficiency is crucial for their survival in diverse habitats. While pumas body size is generally adapted to its environment, factors such as habitat fragmentation and human encroachment may negatively impact populations and, consequently, the individuals' body size and health (McRae et al. 2005).

5 Learning Process in Felids as Predators

The learning process in felids as predators is multifaceted, involving observational learning, social interactions, and individual experiences that shape their predatory skills. These processes are crucial for their survival, particularly in environments with variable prey availability (Elbroch and Quigley 2013; Arbon and Thornton 2024; Bogoni et al. 2017). Felids exhibit a range of learning strategies that enhance their hunting efficiency and adaptability. Social interactions and play among kittens are significant predictors of adult predatory behavior. Play activities correlate with future hunting skills, suggesting that social play is a form of practice for real-life predation (Caro 2010). Young felids often learn predatory behaviors by observing their mothers and siblings. Observational learning in domestic cats has been shown to be more effective than conventional training methods. Cats that observed others performing tasks acquired responses faster and with fewer errors (John et al. 1968). Moreover, social facilitation plays a role in problem-solving and learning. For instance, lions that initially failed to solve a task succeeded after observing successful peers and apparently remembered solutions over extended periods, indicating the importance of social learning in complex environments, and advanced cognitive abilities, such as problem-solving and memory, which are necessary for adapting to new challenges (Borrego et al. 2016). Predators also adapt their strategies based on environmental cues, such as distinguishing between noxious and palatable prey, which involves learning from past experiences to optimize food consumption (Tsoularis 2007). While the learning process in felids is primarily driven by observation and social interactions, individual experiences and environmental factors may also play a crucial role (John et al. 1968; Borrego et al. 2016). These elements collectively contribute to the development of effective predatory behaviors, ensuring the survival and success of felids in diverse ecological contexts.

6 Wildlife Management in Agricultural Landscapes

Pumas have already learned to prey on introduced wild boars in southern South America (Zanin et al. 2020). They likely started with piglets as they usually do with calves in domestic cattle (Palmeira et al. 2008). The wild boar population is still booming, despite their legal hunting control in Brazil (Kmetiuk et al. 2023). In addition, jaguars might (re)colonize areas where pastures have been replaced by *Eucalyptus* and sugarcane plantations by a combination of hunting pressure decrease and food resources increase (Verdade et al. 2016a). There is usually a tradeoff between prey size, risk of injury and predation frequency on predator-prey relationships (Caro and Fitzgibbon 1992). Thus, the following hypotheses can be established concerning the relationship between pumas and wild boars in Central-south Brazil:

(a) Prey selection may vary between male and female pumas, with piglets being theoretically preferred by adult females and young wild boars by adult males. If so, larger females could benefit by preying on larger piglets, whereas larger males could benefit by preying on larger young wild boars. Such a pattern would result in a selection for large prey for both sexes;
(b) Mesopredators might benefit from an increase in small prey offer, no longer prioritized by pumas; and,
(c) If present, jaguars might concentrate their predation pressure on females and young adult male wild boars, in theory inaccessible to pumas.

Table 19.1 Possible impacts of top predators' recovery and increase in body size and correspondent management interventions

Impacts	Management intervention	References
Livestock depredation	Guardian dogs	Moral et al. (2016)
	Improve livestock management	Palmeira et al. (2008)
Changes in local trophic ecology	Monitoring trophic ecology	Campos et al. (2026, this volume)
The presence of top predators in human modified environments	Conservation	Azevedo et al. (2018) Verdade and Moral (2026, this volume)
Incidents with humans	Improvement and increase wild boars' hunting control	Marchini (2014) and Verdade et al. (2016a)

The introduction—and the lack of its effective hunting control—of wild boars in Central-South Brazil might cause a relevant impact on local food web, besides all environmental impacts already reported. Top predators are usually considered keystone species for conservation as they mold trophic processes and, consequently, the diversity and abundance of their prey (Natsukawa and Sergio 2022). However, their possible population recovery and body size increase, especially in areas with relatively high human population density require proactive wildlife management to avoid problems related to human-wildlife interactions (Table 19.1).

To be able to test such hypotheses, it is important to design observational studies that can capture the adequate temporal scale. Since the aim is to identify an evolutionary process, the temporal scale should be large enough to cover different generations of pumas and wild boars. Resource availability would be the main determinant of potential sampling methods. For instance, capturing pumas to precisely weigh them over multiple years would be the ideal approach, albeit very expensive both in terms of human and financial capital. However, a competitive precision level could also be attained by using a combination of camera traps and machine vision techniques, which can infer size and body mass of pumas based on images by calculating size and estimating body mass (Jansen and Jenks 2011). Accessory data on puma and wild boar abundances would certainly be helpful to establish the hypothesized causal relationship, and this could be done inexpensively through vestige sampling (Mimnagh et al. 2026; this volume).

Top predators in human-modified environments may result in a conflict of interest between conservationists and landowners (Marchini 2014). On one side, agricultural and silvicultural landscapes provide habitat for a relevant part of wildlife, despite their innumerous negative environmental impacts (Verdade et al. 2014, 2016b). On the other side, their coexistence with humans requires management interventions to avoid incidents with humans and livestock depredation, including the use of guardian dogs (Moral et al. 2016), improvements in livestock management (e.g., Palmeira et al. 2008), and ideally the implementation of a Wildlife Service in Brazil (Verdade et al. 2022). In addition, the likely changes in the local food web should be assessed by a monitoring program which included not only the diversity of ecological patterns (e.g., taxonomic diversity), but also the complexity of ecological and evolutionary processes (e.g., functional and phylogenetic diversity, sensu Faith and Pollock 2014), in special the trophic process (Table 19.1). Finally, improving and increasing wild boars' hunting control tend to decrease the chance of interactions between large felids and humans, as they tend to avoid hunters, in special those with hunting dogs (Weston and Stankowich 2014).

Acknowledgements Fazenda Três Lagoas gave us full support for the field studies that served as basis for this chapter. Fundação de Amparo à Pesquisa do Estado de São Paulo (FAPESP, Grant No. 2017/01304-4) funded the study which served the basis for the perspectives here presented.

References

Arbon JJ, Thornton A (2024) Social learning and culture in mammals. Oxford University Press. https://doi.org/10.1093/oxfordhb/9780198869252.013.25

Azevedo FC, Lemos FG, Freitas-Junior MC, Rocha DG, Azevedo FCC (2018) Puma activity patterns and temporal overlap with prey in a human-modified landscape at Southeastern Brazil. J Zool 305(4):246–255. https://doi.org/10.1111/jzo.12558

Barceló G, Perrig PL, Dharampal P, Donadio E, Steffan SA, Pauli JN (2022) More than just meat: carcass decomposition shapes trophic identities in a terrestrial vertebrate. Funct Ecol 36(6):1473–1482. https://doi.org/10.1111/1365-2435.14041

Bates-Mundell L, Williams SH, Sager-Fradkin K, Wittmer HU, Allen ML, Cristescu B, Wilmers CC, Elbroch LM (2024) Season, prey availability, sex, and age explain prey size selection in a large solitary carnivore. Ecol Evol 14. https://doi.org/10.1002/ece3.11080

Bogoni JA, Graipel ME, Volkmer de Castilho P, Peroni N (2017) Development of predatory behaviours in young southern tigrinas (*Leopardus guttulus*). Mammalia 81(4):421–424. https://doi.org/10.1515/MAMMALIA-2015-0165

Borrego N, Borrego N, Dowling B (2016) Lions (*Panthera leo*) solve, learn, and remember a novel resource acquisition problem. Anim Cogn 19(5):1019–1025. https://doi.org/10.1007/S10071-016-1009-Y

Campos CB, Gheler-Costa C, Esteves CF, Rech-Fernandes TC, Rosalino LM, Piña CI, Moral RA, Verdade LM (2026) Sampling scats in wildlife surveys: trophic approach in wildlife surveys. In: Martins Verdade L, Moral RA (eds) Wildlife monitoring: integrating conservation and innovation in human-altered landscapes. Springer, Cham (this volume)

Caro TM (2010) Relations between kitten behaviour and adult predation. Ethology 51(2):158–168. https://doi.org/10.1111/J.1439-0310.1979.TB00680.X

Caro TM, Fitzgibbon CD (1992) Large carnivores and their prey: the quick and the dead. In: Crawley MJ (ed) Natural enemies: the population biology of predators, parasites and diseases. Blackwell Scientific Publications, Oxford, UK, pp 117–142

Cuff AR, Cuff AR, Randau M, Head JJ, Hutchinson JR, Pierce SE, Goswami A (2015) Big cat, small cat: reconstructing body size evolution in living and extinct Felidae. J Evol Biol 28(8):1516–1525. https://doi.org/10.1111/JEB.12671

Cuyper A, Clauss M, Carbone C, Codron D, Cools A, Hesta M, Janssens G (2019) Predator size and prey size–gut capacity ratios determine kill frequency and carcass production in terrestrial carnivorous mammals. Oikos 128(1):13–22. https://doi.org/10.1111/OIK.05488

Elbroch LM, Quigley H (2013) Observations of wild cougar (*Puma concolor*) kittens with live prey: implications for learning and survival. Can Field-Nat 126(4):333–335. https://doi.org/10.22621/CFN.V126I4.1380

Faith DP, Pollock LJ (2014) Phylogenetic diversity and the sustainable use of biodiversity. In: Applied ecology and human dimensions in biological conservation. Springer Berlin Heidelberg, Berlin, Heidelberg, pp 35–52

Gay SW, Best TL (1996) Age-related variation in skulls of the puma (*Puma concolor*). J Mammal 77(1):191–198. https://doi.org/10.2307/1382720

Gheler-Costa C, Botero GP, Reia L, Gilli LC, Comin FH, Verdade LM (2018) Ecologia trófica de onça-parda (*Puma concolor*) em paisagem agrícola. Rev em Agronegócio e Meio Ambiente 11(1):203–225. https://doi.org/10.17765/2176-9168.2018v11n1p203-225

Gittleman JL (1985) Carnivore body size: ecological and taxonomic correlates. Oecologia 67(4):540–554. https://doi.org/10.1007/BF00790026

Hayward MW, Kamler JF, Montgomery RA, Newlove A, Rostro-García S, Sales LP, Van Valkenburgh B (2016) Prey preferences of the jaguar Panthera onca reflect the post-Pleistocene demise of large prey. Front Ecol Evol 3:148. https://doi.org/10.3389/fevo.2015.00148

Iriarte JA, Franklin WL, Johnson WE, Redford KH (1990) Biogeographic variation of food habits and body size of the America puma. Oecologia 85(2):185–190. https://doi.org/10.1007/BF00319400

Jansen BD, Jenks JA (2011) Estimating body mass of pumas (*Puma concolor*). Wildl Res 38(2):147–151. https://doi.org/10.1071/WR10109

John ER, Chesler P, Bartlett F, Victor I (1968) Observation learning in cats. Science 159(3822):1489–1491. https://doi.org/10.1126/SCIENCE.159.3822.1489

Kmetiuk LB, Biondo LM, Pedrosa F, Favero GM, Biondo AW (2023) One Health at gunpoint: impact of wild boars as exotic species in Brazil-a review. One Health 17:100577. https://doi.org/10.1016/j.onehlt.2023.100577

Lindstedt SL, Boyce MS (1985) Seasonality, fasting endurance, and body size in mammals. Am Nat 125(6):873–878

Macdonald DW (1985) The encyclopedia of mammals. Facts of File, New York

Marchini S (2014) Who's in conflict with whom? Human dimensions of the conflicts involving wildlife. In: Verdade LM, Lyra-Jorge MC, Piña CI (eds) Applied ecology and human dimensions in biological conservation. Springer, Heidelberg. ISBN 978-3-642-54750-8, pp 189–209. https://doi.org/10.1007/978-3-642-54751-5_13

McRae BH, Beier P, Dewald LE, Huynh LY, Keim P (2005) Habitat barriers limit gene flow and illuminate historical events in a wide-ranging carnivore, the American puma. Mol Ecol 14(7):1965–1977. https://doi.org/10.1111/j.1365-294x.2005.02571.x

Meachen-Samuels J, Van Valkenburgh B (2009) Craniodental indicators of prey size preference in the Felidae. Biol J Linn Soc 96(4):784–799. https://doi.org/10.1111/J.1095-8312.2008.01169.X

Milda D, Ramesh T, Kalle R, Gayathri V, Thanikodi M, Ashish K (2023) Factors driving human–wild pig interactions: implications for wildlife conflict management in southern parts of India. Biol Invasions 25(1):221–235. https://doi.org/10.1007/s10530-022-02911-6

Mimnagh N, Ferreira IEP, Verdade LM, Moral RA (2026) Counting animals we can't see: a triple Poisson model for scarce vestige data. In: Martins Verdade L, Moral RA (eds) Wildlife monitoring: integrating conservation and innovation in human-altered landscapes. Springer, Cham (this volume)

Moral RA, Azevedo FCC, Verdade LM (2016) The use of sheepdogs in sheep production in southeastern Brazil. Pastor Res Policy Pract 6(1):1–7. https://doi.org/10.1186/s13570-016-0065-x

Natsukawa H, Sergio F (2022) Top predators as biodiversity indicators: a meta-analysis. Ecol Lett 25(9):2062–2075

Palmeira FBL, Crawshaw PG Jr, Haddad CM, Ferraz KMPMB, Verdade LM (2008) Cattle depredation by puma (*Puma concolor*) and jaguar (*Panthera onca*) in Northern Goiás, Central-Western Brazil. Biol Conserv 141(1):118–125

Troost TA, Kooi BW, Dieckmann U (2008) Joint evolution of predator body size and prey-size preference. Evol Ecol 22(6):771–799. https://doi.org/10.1007/S10682-007-9209-1

Tsoularis A (2007) A learning strategy for predator preying on edible and inedible prey. Acta Biotheor 55(3):283–295. https://doi.org/10.1007/s10441-007-9020-y

Verdade LM, Moral RA (2026) Rewilding agricultural landscapes: conceptual, technological, and societal constraints. In: Martins Verdade L, Moral RA (eds) Wildlife monitoring: integrating conservation and innovation in human-altered landscapes. Springer, Cham (this volume)

Verdade LM, Lyra-Jorge MC, Piña CI (2014) Redirections in conservation biology. In: Verdade LM, Lyra-Jorge MC, Piña CI (eds) Applied ecology and human dimensions in biological conservation. Springer, Heidelberg. ISBN 978-3-642-54750-8, pp 3–17. https://doi.org/10.1007/978-3-642-54751-5_1

Verdade LM, Palomares F, Couto HTZ, Polizel JL (2016a) Recent land use changes and the expansion of an exotic potential prey: a possible redemption for Atlantic forest jaguars? Anim Conserv 19:209–211. https://doi.org/10.1111/acv.12221

Verdade LM, Gheler-Costa C, Lyra-Jorge MC (2016b) The multiple facets of agricultural landscapes. In: Gheler-Costa C, Lyra-Jorge MC, Verdade LM (eds) Biodiversity in agricultural landscapes of southeastern Brazil. De Gruyter, Berlin, pp 2–13. https://doi.org/10.1515/9783110480849-003

Verdade LM, Bianchi RC, Galetti PM Jr, Pivello VR, Silva WR, Uezu A (2022) Biodiversity management and research in multifunctional landscapes. *Biota*. Neotropica 22(spe):e20221407. https://doi.org/10.1590/1676-0611-BN-2022-1407

Vissia S, Fattebert J, van Langevelde F (2022) Leopard density and interspecific spatiotemporal interactions in a hyena-dominated landscape. Ecol Evol 12(10):e9365

Weston MA, Stankowich T (2014) Dogs as agents of disturbance. In: Gomper ME (ed) Free ranging dogs & wildlife conservation. Oxford University Press, Oxford, UK, pp 94–116

Whitfield J (2007) How big can a meat-eater get? Nature. https://doi.org/10.1038/news070115-4

Zanin M, Adrados B, Vargas Foletto VCD, Paula Camargo BD, Porfirio G, Matos Dias DD, Palomares F (2020) What should I eat: feeding behaviour of puma in a Brazilian protected semi-arid area. Hystrix Ital J Mamm 31:5. https://doi.org/10.4404/hystrix-00265-2019

Chapter 20
Rewilding Agricultural Landscapes: Conceptual, Technological, and Societal Constraints

Luciano Martins Verdade and Rafael de Andrade Moral

Abstract Multifunctional agricultural landscapes congregate both biological production and conservation. Their conservation value might be improved by the recovery of depleted populations and the reintroduction of locally extinct species, which could be called rewilding agricultural landscapes. In this chapter we discuss its conceptual, technological, and societal constraints related to planning capacity and decision-making process on the edge of human-wildlife coexistence.

Keywords Wildlife management · Reintroduction · Agroecosystems · Monitoring · Human-wildlife coexistence

1 Introduction

The expansion of humans on the planet has been resulting in mass extinction of other species since the Pleistocene (Martin 1966; Alroy 2001). More recently, agriculture has resulted on a considerable increase in human population, with also a tremendous impact on other species due to direct habitat loss and environmental contamination by agrochemical residues, especially after the so-called "green revolution" (Tripp 1996; Horlings and Marsden 2011). Agriculture *lato sensu* (i.e, including livestock production and forestry) can be considered as biological production, whereas the prevention or remediation of wild species decline, or extinction, can be considered as biological conservation (Verdade et al. 2014a). On the one hand, agriculture has become the basis of what we usually call civilization but, on the other, it is one of the major human threats to wildlife and the wilderness (Verdade et al. 2014a, 2016).

L. Martins Verdade (✉)
Wildlife Monitoring Consultancy, Campinas, São Paulo, Brazil
e-mail: luciano.verdade@wildlifemgmtco.com

R. A. Moral
Department of Mathematics and Statistics, Maynooth University, Maynooth, Co. Kildare, Ireland

L. Martins Verdade, R. A. Moral (eds.), *Wildlife Monitoring*,
https://doi.org/10.1007/978-3-032-05821-8_20

Conservation units, with reduced to no direct human activities, have been created in most countries to prevent the loss of wilderness and wildlife extinction or defaunation (Dirzo et al. 2014; Galetti et al. 2016). However, it would be impossible to promote conservation of all biota only in conservation units, such as national parks and biological reserves, even if they worked perfectly—which they do not—in any country of the world (Knight et al. 2006). Despite all the problems wildlife faces in current agricultural landscapes, such areas are fundamental to their conservation as a relevant number of species dwell there (Verdade et al. 2014b). However, the wild strains of domesticated animals and plants are among the most endangered taxa on the planet (Cassidy and Mullin 2007). Notwithstanding, they are fundamental to the long-term agriculture sustainability as they furnish wild genes to promote resistance or tolerance to environmental changes and new diseases to their domesticated relatives. Such wild strains desperately need the wilderness and consequently conservation units to survive. For these reasons, the frequently assumed-as-inevitable conflict between biological production and biological conservation should be rather seen as an interdependent relationship. In other words, agricultural landscapes are as fundamental to biological conservation as conservation units are fundamental to agriculture sustainability (Verdade et al. 2014a).

The effectiveness of conservation units to promote biological conservation is just a matter of following the contract, or simply doing whatever they are supposed to do, even though this is not easy. Contrastingly, to effectively promote biological conservation, agricultural landscapes need to add it as a secondary but fundamental mission to their primary mission of biological production. This is only possible in multifunctional agricultural landscapes (Martinelli et al. 2010). Biological conservation includes not only mitigating extinction risk, but also recovering depleted populations or locally extinct species (Caughley 1994; Verdade et al. 2014a). Rewilding agricultural landscapes would be a relevant step to improving multifunctional agricultural landscapes. However, such a challenge brings conceptual, technological, and societal constraints (Table 20.1). In this chapter, we discuss them in the nexus of planning and decision making in a human-wildlife coexistence context. Usually, conceptual, and technological constraints limit our planning capacity, whereas societal constraints limit the decision-making process.

2 Conceptual Constraints

The terms defaunation and rewilding replaced—at least partially—terms like local extinction and reintroduction of wildlife (Dirzo et al. 2014). Ecological restoration usually refers to the recovery of ecological patterns and processes (Martin 2017). In such context, the concept of resilience from Physics has been adapted to ecosystems under anthropic pressures (Holling 1996). Agricultural landscapes are composed by novel ecosystems, usually called agroecosystems, which partially or totally replace pristine ecosystems with a consequent population decline or local extinction of

Table 20.1 References on conceptual, technological, and societal constraints related to multifunctional agricultural landscapes

Level	Nexus	Theme	References
Conceptual	Planning	Multifunctional agricultural landscapes	Martinelli et al. (2010), Merckx and Pereira (2014), Verdade et al. (2014b), and Song et al. (2020)
		Ecological restoration and rewilding	Schadt et al. (2002), Caro (2007), Sandon et al. (2013), Lorimer et al. (2015), Jorgensen (2015), Corlett (2016), Nogués-Bravo et al. (2016), Martin (2017), and Walpole et al. (2020)
		Land sharing vs. land sparing	Green et al. (2005), Phalan et al. (2013, 2016), and Ruffell et al. (2017)
		Agricultural landscape matrix as habitat: Permeability vs. carrying capacity	Gascon et al. (1999), Ricketts (2001), Vandermeer and Carvajal (2001), Kupfer et al. (2008), Prugh et al. (2008), Franklin and Lindenmayer (2009), Fahrig et al. (2011), and Millan et al. (2015)
		Spatial vs. temporal-spatial heterogeneity	Verdade et al. (2014b), Martin et al. (2012), Timo et al. (2015), Lopes et al. (2016), and Verdade et al. (2020)
		Rapid evolution	Thompson (1998), Hairston et al. (2005), and Rosalino et al. (2014)
Technological	Application	Wildlife-friendly management techniques	Gheler-Costa et al. (2013), Athayde et al. (2014), and Millan et al. (2015)
		Monitoring: Patterns vs. processes	Dunning et al. (1992), Levin (1992), Rouget et al. (2003), Soulé et al. (2003), Bennett et al. (2009), Lindenmayer and Likens (2010), Cadotte et al. (2011), Estes et al. (2011), Verdade et al. (2014b), Valiente-Banuet et al. (2015), and Souza et al. (2017)
Societal	Decision making	Reintroduction: Priority criteria and procedures	Seddon (1999), Fischer and Lindenmayer (2000), Seddon et al. (2005, 2007), Enserink and Vogel (2006), Hayward et al. (2007), Seddon et al. (2014), Houde et al. (2015), and Watson and Watson (2015)
		Stakeholders' perceptions	Riley and Decker (2000), Gigliotti et al. (2000), and Worthington et al. (2010)
		Capacity building	Verdade (2004) and König et al. (2020)
		Land use regulation: Public policy vs. certification	Endres et al. (2015)

a relevant part of its original community, and the introduction of exotic invasive species and environmental contaminants. However, a significant proportion of the wildlife can surprisingly resist, tolerate, or adapt to such impacts, which gives agricultural landscapes some conservation value (Verdade et al. 2014b). To improve it we need to deal with the following conceptual constraints: ecological restoration, matrix as habitat, land sharing/sparing approach, spatiotemporal heterogeneity, and rapid evolution.

Although we may restore some ecological patterns, we might not be able to restore some ecological and evolutionary processes in agricultural landscapes. Since biological production requires agricultural management to be sustainable, biological conservation would require wildlife management to be feasible. In such circumstance we might not be doing ecological restoration per se, but some kind of ecological management at the landscape scale.

A relevant portion of the wildlife present in agricultural landscapes use the matrix as habitat (e.g., Millan et al. 2015; Timo et al. 2015; Rodrigues et al. 2017; Verdade et al. 2020). Therefore, we should estimate its habitat quality instead of its permeability. However, to avoid the tricky concept of habitat (see, for instance, Mitchell 2005), it might be worth measuring species resource use, such as food and shelter (Garshelis 2000). At this point, it is necessary to understand that most species are unable to survive only at the landscape matrix (Timo et al. 2016; Martin et al. 2012; Gheler-Costa et al. 2012; Dotta and Verdade 2011). Thus, it might be wise to estimate species-specific carrying capacity for the whole landscape, although the concept of carrying capacity may also be a tricky one (Dhondt 1988).

Although the debate land sharing vs. land sparing has been cooling down (Kremen 2015), it is still relevant to the biological conservation in agricultural landscapes because the spatial heterogeneity is directly related to species richness and functional diversity at the landscape level (Magioli et al. 2021). However, the concept of sharing and sparing is a matter of scale. For instance, a land sparing property can be part of a land sharing landscape. Nevertheless, multifunctional agricultural landscapes would only be functional on land sharing landscapes.

Most of the landscape ecology field operates in two spatial dimensions and a smashed temporal dimension (e.g., Forman 1995). However, in agricultural landscapes the biomass pulse dramatically varies between the matrix and the native vegetation (see Holt 2008) with still unknown consequences for wildlife behavioral ecology and demography (Verdade et al. 2014b). Spatiotemporal heterogeneity of agricultural landscapes can be particularly relevant for the evolutionary processes, in special for species with shorter lifespan. For instance, after no more than 10 generations the Montane grass mouse (*Akodon montensis*) has longer feet in *Eucalyptus* plantations than on the native vegetation nearby in Southern Brazil (Rosalino et al. 2013). Other anthropogenic pressures, such as hunting, have been long known to affect vertebrates' behavioral ecology (Verdade 1996; Potratz et al. 2024). Relatively fast evolutionary responses to the strong anthropogenic pressure, usually called rapid evolution (Thompson 1998) might be relevant in agricultural landscapes due to their high spatial-temporal heterogeneity (Levin 1968).

3 Technological Constraints

The development of wildlife-friendly management techniques is paramount for the effectiveness of biological conservation in agricultural landscapes by both decreasing mortality rate (e.g., Gheler-Costa et al. 2022) and increasing carrying capacity (e.g., Millan et al. 2015). The reintroduction of locally extinction species will require research and development programs for new agricultural management techniques which are friendly to the reintroduced species as well.

Rewilding agricultural landscapes will require more than anything else new technologies to establish long-term scale-crossing monitoring programs of both agricultural sustainability and biodiversity conservation (Verdade et al. 2014a; 2026, this volume). Such program would form the basis for planning the human coexistence with wildlife in the nexus of the interaction between biological production and biological conservation. Previous experiences (e.g., Lindenmayer and Likens 2010) can be useful to avoid waste of resources on reinventing the wheel. However, it would be prudent to assume we still need to learn how to monitor the planet's biota, including the species living in widespread agricultural landscapes (Verdade and Moral 2026, this volume).

From technical purposes rewilding agricultural landscapes would imply in both increasing the remaining species diversity and reintroducing locally extinct species. The former would require concrete measures to increase agricultural landscapes' carrying capacity. The latter would require the identification of the causes of species extinction and their inactivation, and to define priority criteria for species reintroduction.

The following criteria could be tentatively used to prioritize species to reintroduction programs:

(a) Species whose local extinction causes are no longer active;
(b) Species that effectively contribute to environmental quality (e.g., keystone species, engineer species);
(c) Species that effectively contribute to local food webs; and,
(d) Species that effectively contribute to ecosystem services.

4 Societal Constraints

The stakeholders' perceptions on the coexistence between humans and wildlife would also be paramount for the success of a rewilding program of agricultural landscapes as it is one of the main drivers of conflicts that may arise among distinct parts of society (Marchini 2014). The decision-making process about possible interventions should be balanced between political and technical personnel from both the biological production sectors (e.g., agro-industry and consumers) and the biological conservation sectors (e.g., NGOs and environmental agencies). Their technical support should come predominantly but possibly not exclusively from the monitoring program mentioned above. However, such program would require an effective governmental Wildlife Service, which would operate at both federal and state/provincial levels. Such a Service would require a consistent investment on capacity building in some countries (Verdade 2004).

The decision-making process concerning rewilding multifunctional agricultural landscapes would most of all require a consistent public policy on land use regulation. The Brazilian Forest Code (BFC) is likely one of the best examples in the world. It was created in the early 1930s based on the land sharing approach almost seven decades before such debate and taking into consideration ecological, geographical, and historical use of original biomes (Metzger et al. 2010).

Complementarily to public policy on land use regulations, the certification process of biological production and their commodities might be an effective tool to sustain biodiversity on agricultural landscapes (Endres et al. 2015). The Forest Stewardship Council (FSC), related to forestry sector, is likely the most rigorous certification process of any agro-industrial sector in the world (Burger et al. 2005). Thanks to it, the forestry sector has been attaining consistent results, both economically and environmentally (IBA 2019).

5 Final Considerations

Biological production and biological conservation can be interdependent processes instead of representing an inevitable conflict in agricultural landscapes. The land sharing approach is more appropriate to such nexus than land sparing. In such a context, rewilding agricultural landscapes is an effective possibility. However, to do it we need to deal with the conceptual, technological, and societal constraints discussed above which could limit our planning capacity and decision-making process on the edge of human-wildlife coexistence.

Future conceptual studies should prioritize ecological restoration, land-sharing in distinct scales, spatial-temporal heterogeneity, and rapid evolution. In addition, research and technology programs should prioritize the development of wildlife-friendly management techniques, habitat and carrying capacity measurement procedures, and long-term crossing-scale biodiversity monitoring procedures. Finally, the decision-making process concerning the relationship between biological production and biological conservation should include priority criteria for species reintroduction, stakeholders' perception, capacity building, and land use regulations by public policy and certification.

Acknowledgements This study has been funded by *Fundação de Amparo à Pesquisa do Estado de São Paulo* (FAPESP, Proc. No. 2017/01304-4) and *Agência Nacional de Energia Elétrica* (ANEEL, Proc. No. PD-0747-0225/2025. LMV held a Productivity Fellowship from *Conselho Nacional de Desenvolvimento Científico e Tecnológico* (CNPq, Proc. No. 312049/2015-3).

References

Alroy J (2001) A multispecies overkill simulation of the end-Pleistocene megafaunal mass extinction. Science 292:1892–1896

Athayde EA, Cancian LF, Verdade LM, Morellato LPC (2014) Functional and phylogenetic diversity of scattered trees in agricultural landscape: implications for conservation. Agric Ecosyst Ecol 199:272–281

Bennett AF, Haslem A, Cheal DC, Clarke MF, Jones RN, Koehn JD, Lake PS, Lumsden LF, Lunt ID, Mackey BG, Nally RM, Menkhorst PW, New TR, Newell GR, O'Hara T, Quinn PG, Radford J, Robinson D, Watson JEM, Yen AL (2009) Ecological processes: a key element in strategies for nature conservation. Ecol Manage Restor 10(3):192–199

Burger D, Hess J, Lang B (2005) Forest certification: an innovative instrument in the service of sustainable development? Eschborn, Germany

Cadotte MW, Carscadden K, Mirotchnick N (2011) Beyond species: functional diversity and the maintenance of ecological processes and services. J Appl Ecol 48:1079–1087

Caro T (2007) The Pleistocene re-wilding gambit. Trends Ecol Evol 22(6):281–283

Cassidy R, Mullin M (2007) Where the wild things are now: domestication reconsidered. Routledge, New York

Caughley G (1994) Directions in conservation biology. J Anim Ecol 63(2):215–244

Corlett RT (2016) Restoration, reintroduction and rewilding in a changing world. Trends Ecol Evol 31(6):453–462

Dhondt AA (1988) Carrying capacity: a confusing concept. Acta Oecol (Oecol Gen) 9(4):337–346

Dirzo R, Young HS, Galetti M, Ceballos G, Isaac NJB, Collen B (2014) Defaunation in the Anthropocene. Science 345(6195):401–406

Dotta G, Verdade LM (2011) Medium to large-sized mammals in agricultural landscapes of South-Eastern Brazil. Mammalia 75:345–352. https://doi.org/10.1515/MAMM.2011.049

Dunning JB, Danielson BJ, Pulliam HR (1992) Ecological processes that affect populations in complex landscapes. Oikos 65(1):169–175

Endres J, Diaz-Chavez R, Kafka SR, Pelkmans L, Seabra JEA, Walter A (2015) Sustainability certification. In: Souza GM, Victoria RL, Joly CA, Verdade LM (eds) Bioenergy & sustainability: bridging the gaps, vol 72. SCOPE, Paris, pp 136–178, 660–680, 779 p. ISBN 978-2-9545557-0-6

Enserink M, Vogel G (2006) The carnivore comeback. Science 314:746–749

Estes JA, Terborgh J, Brashares JS, Power ME, Berger J, Bond WJ, Carpenter SR, Essington TE, Holt RD, Jackson JBC, Marquis RJ, Oksanen L, Oksanen T, Paine RT, Pikitch EK, Ripple WJ, Sandin SA, Scheffer M, Schoener TW, Shurin JB, Sinclair ARE, Soulé ME, Virtanen R, Wardle D (2011) Trophic downgrading of Planet Earth. Science 333:301–306

Fahrig L, Baudry J, Brotons L, Burel F, Crist T, Fuller R, Sirami C, Siriwardena G, Martin J (2011) Functional landscape heterogeneity and animal biodiversity in agricultural landscapes. Ecol Lett 14:101–112

Fischer J, Lindenmayer DB (2000) An assessment of the published results of animal relocations. Biol Conserv 96:1–11

Forman RTT (1995) Land mosaics: the ecology of landscapes and regions. Cambridge University Press, Cambridge, UK

Franklin JF, Lindenmayer DB (2009) Importance of matrix habitats in maintaining biological diversity. PNAS 106(2):349–350

Galetti M, Brocardo CR, Begotti RA, Hortenci L, Rocha-Mendes F, Bernardo CSS, Bueno RS, Nobre R, Bovendorp RS, Marques RM, Meirelles F, Gobbo SK, Beca G, Schmaedecke G, Siqueira T (2016) Defaunation and biomass collapse of mammals in the largest Atlantic forest remnant. Anim Conserv 20(3):270–281

Garshelis DL (2000) Delusions in habitat evaluation: measuring use, selection, and importance. In: Boitani L, Fuller TK (eds) Research techniques in animal ecology: controversies and consequences. Columbia University Press, New York, pp 111–164

Gascon C, Lovejoy TE, Bierregaard RO Jr, Malcolm JR, Stouffer PC, Vasconcelos HL, Laurance WF, Zimmerman B, Tocher M, BorgeS S (1999) Matrix habitat and species richness in tropical forest remnants. Biol Conserv 91:223–229

Gheler-Costa C, Vettorazzi CA, Pardini R, Verdade LM (2012) The distribution and abundance of small mammals in agroecosystems of Southeastern Brazil. Mammalia 76:185–191. https://doi.org/10.1515/mammalia-2011-0109

Gheler-Costa C, Sabino-Santos G Jr, Amorim LS, Rosalino LM, Figueiredo LTM, Verdade LM (2013) The effect of pre-harvest fire on the small mammal assemblage in sugarcane fields. Agric Ecosyst Environ 171:85–89. https://doi.org/10.1016/j.agee.2013.03.016

Gheler-Costa C, Sabino-Santos G Jr, Rosalino LM, Amorim LS, Maia FGM, Moral RA, Ferreira IEP, Figueiredo LTM, Piña CI, Verdade LM (2022) The influence of sugarcane pre-harvest fire on hantavirus prevalence in Neotropical small mammals. Ecosphere 13(9):e4241. https://doi.org/10.1002/ecs2.4241

Gigliotti L, Decker DJ, Carpenter LH (2000) Developing the wildlife stakeholder acceptance capacity concept: research needed. Hum Dimens Wildl 5(76):82

Green RE, Cornell SJ, Scharlemann JPW, Balmford A (2005) Farming and the fate of the wild nature. Science 307:550–555

Hairston NG, Ellner SP, Geber MA, Yoshida T, Fox JA (2005) Rapid evolution and convergence of ecological and evolutionary time. Ecol Lett 8:1114–1127

Hayward MW, Adendorff J, O'Brien J, Sholto-Douglas A, Bissett C, Moolman LC, Bean P, Fogarty A, Howarth D, Slater R, Kerley GIH (2007) Practical considerations for the reintroduction of large, terrestrial, mammalian predators based on reintroductions to South Africa's Eastern Cape Province. Open Conserv Biol J 1:1–11

Holling CS (1996) Engineering resilience versus ecological resilience. In: Schulze P (ed) Engineering within ecological constraints. National Academy of Engineering, Washington, DC, pp 31–44

Holt RD (2008) Theoretical perspectives on resource pulses. Ecology 89(3):671–681

Horlings LG, Marsden TK (2011) Towards the real green revolution? Exploring the conceptual dimensions of a new ecological modernisation of agriculture that could 'feed the world'. Glob Environ Chang 21(2):441–452

Houde ALS, Garner SR, Neff BD (2015) Restoring species through reintroductions: strategies for source population selection. Restor Ecol 23(6):746–753

IBA (2019) Relatório 2019. Indústria Brasileira de Árvores, São Paulo, 80 p

Jorgensen D (2015) Rethinking rewilding. Geoforum 65:482–488

Knight AT, Driver A, Cowling RM, Maze K, Desmet PG, Lombard AT, Rouget M, Botha MA, Boshoff AF, Castley JG, Goodman PS, Mackinnon K, Pierce SM, Sims-Castley R, Stewart WI, Von Hase A (2006) Designing systematic conservation assessments that promote effective implementation: best practice from South Africa. Conserv Biol 20(3):739–750

König HJ, Kiffner C, Kramer-Schadt S, Fürst C, Keuling O, Ford AT (2020) Human-wildlife coexistence in a changing world. Conserv Biol 34(4):786–794

Kremen C (2015) Reframing the land-sparing/land-sharing debate for biodiversity conservation. Ann N Y Acad Sci 1355(1):52–76

Kupfer JA, Malanson GP, Franklin SB (2008) Not seeing the ocean for the islands: the mediating influence of matrix-based processes on forest fragmentation effects. Glob Ecol Biogeogr 15:8–20

Levin R (1968) Evolution in changing environments. Princeton University Press, Princeton

Levin SA (1992) The problem of pattern and scale in ecology. Ecology 73(6):1943–1967

Lindenmayer DB, Likens GE (2010) The science and application of ecological monitoring. Biol Conserv 143:1317–1328

Lopes PC, Marques TS, Martin PS, Gheler-Costa C, Verdade LM (2016) Patterns of leaf-litter amphibian diversity in a silvicultural landscape of southeastern Brazil. In: Gheler-Costa C, Lyra-Jorge MC, Verdade LM (eds) Biodiversity in agricultural landscapes of southeastern Brazil. De Gruyter, Berlin, pp 226–242

Lorimer J, Sandom C, Jepson P, Doughty C, Barua M, Kirby K (2015) Rewilding: science, practice and politics. Annu Rev Environ Resour 40:39–62
Magioli M, Ferraz KMPMB, Chiarello AG, Galetti M, Setz EZF, Paglia AP, Abrego N, Ribeiro MC, Ovaskainen O (2021) Land-use changes lead to functional loss of terrestrial mammals in a Neotropical rainforest. Perspect Ecol Conserv 19(2):161–170. https://doi.org/10.1016/j.pecon.2021.02.006
Marchini S (2014) Who is in conflict with whom? Human dimensions of the conflicts involving wildlife. In: Verdade LM, Lyra-Jorge MC, Piña CI (eds) Applied ecology and human dimensions in biological conservation. Springer, Heidelberg. ISBN 978-3-642-54750-8, pp 189–209. https://doi.org/10.1007/978-3-642-54751-5_13
Martin PS (1966) Africa and Pleistocene overkill. Nature 212:339–342
Martin DM (2017) Ecological restoration should be redefined for the twenty-first century. Restor Ecol 25(5):668–673
Martin PS, Gheler-Costa C, Lopes PC, Rosalino LM, Verdade LM (2012) Terrestrial non-volant small mammals in agro-silvicultural landscapes of Southeastern Brazil. For Ecol Manag 282:185–195. https://doi.org/10.1016/j.foreco.2012.07.002
Martinelli LA, Joly CA, Nobre CA, Sparovek G (2010) A falsa dicotomia entre a preservação da vegetação natural e a produção Agropecuária. Biota Neotrop 10:323–330
Merckx T, Pereira HM (2014) Reshaping agri-environmental subsidies: from marginal farming to large-scale rewilding. Basic Appl Ecol 16:95–103
Metzger JP, Lewinsohn TM, Joly CA, Verdade LM, Martinelli LA, Rodrigues RR (2010) Brazilian law: full speed in reverse? Science 329:276–277
Millan CH, Develey PF, Verdade LM (2015) Stand-level management practices increase occupancy by birds in exotic *Eucalyptus* plantations. For Ecol Manag 336:174–182. https://doi.org/10.1016/j.foreco.2014.10.005
Mitchell SC (2005) How useful is the concept of habitat? A critique. Oikos 110(3):634–638
Nogués-Bravo D, Simberloff D, Rahbek C, Sanders NJ (2016) Rewilding is the new Pandora's box in conservation. Curr Biol 26:87–91
Phalan B, Bertzky M, Butchart SH, Donald PF, Scharlemann JP, Stattersfield AJ, Balmford A (2013) Crop expansion and conservation priorities in tropical countries. PLoS One 8(1):e51759
Phalan B, Green RE, Dicks LV, Dotta G, Feniuk C, Lamb A, Strassburg BBN, Williams DR, zu Ermgassen EKHJ, Balmford A (2016) How can higher-yield farming help to spare nature? Science 351:450–451
Potratz EJ, Holt RD, Brown JS (2024) Ecology of fear: acclimation and adaptations to hunting by humans. Sustainability 16:1216. https://doi.org/10.3390/su16031216
Prugh LR, Hodges KE, Sinclair ARE, Brashares JS (2008) Effect of habitat area and isolation on fragmented animal populations. Proc Natl Acad Sci USA 105:20770–20775
Ricketts TH (2001) The matrix matters: effective isolation in fragmented landscapes. Am Nat 158:87–99
Riley S, Decker DJ (2000) Wildlife stakeholder acceptance capacity for cougars in Montana. Wildl Soc Bull 28(4):931–939
Rodrigues TF, Kays R, Parsons A, Versiani NF, Paolino RM, Pasqualotto N, Krepschi VG, Chiarello AG (2017) Managed forest as habitat for gray brocket deer (Mazama gouazoubira) in agricultural landscapes of southeastern Brazil. J Mammal 98(5):1301–1309
Rosalino LM, Martin PS, Gheler-Costa C, Lopes PC, Verdade LM (2013) Allometric relations of Neotropical small rodents (Sigmodontinae) in anthropogenic environments. Zool Sci 30:585–590. https://doi.org/10.2108/zsj.30.585
Rosalino LM, Martin PS, Gheler-Costa C, Lopes PC, Verdade LM (2014) Neotropical small mammals' diversity in the early cycle of commercial Eucalyptus plantations. Agrofor Syst 88(3):427–436. https://doi.org/10.1007/s10457-014-9702-9
Rouget M, Cowling RM, Pressey RL, Richardson DM (2003) Identifying spatial components of ecological and evolutionary processes for regional conservation planning in the Cape Floristic Region, South Africa. Divers Distrib 9:191–210

Ruffell J, Clout MN, Didham RK (2017) The matrix matters, but how should we manage it? Estimating the amount of high-quality matrix required to maintain biodiversity in fragmented landscapes. Ecography 40(1):171–178

Sandon C, Donlan CJ, Svenning JC, Hansen D (2013) Chapter 24: Rewilding. In: Macdonald DW, Willis KJ (eds) Key topics in conservation biology 2. Blackwell Publishing, pp 430–451

Schadt S, Revilla E, Wiegand T, Knauer F, Kaczensky P, Breitenmoser U, Bufka L, Červený J, Koubek P, Huber T, Staniša C, Trepl L (2002) Assessing the suitability of central European landscapes for the reintroduction of Eurasian lynx. J Appl Ecol 39(2):189–203

Seddon PJ (1999) Persistence without intervention: assessing success in wildlife reintroductions. Trends Ecol Evol 14(12):503

Seddon PJ, Soorae PS, Launay F (2005) Taxonomic bias in reintroduction projects. Anim Conserv 8:51–58

Seddon PJ, Armstrong DP, Maloney RF (2007) Developing the science of reintroduction biology. Conserv Biol 21(2):303–312

Seddon PJ et al (2014) Reversing defaunation: restoring species in a changing world. Science 345. https://doi.org/10.1126/science.1251818

Song B, Robinson GM, Bardsley DK (2020) Measuring multifunctional agricultural landscapes. Land 9(8):260

Soulé ME, Estes JA, Berger J, Del Rio MC (2003) Ecological effectiveness: conservation goals for interactive species. Conserv Biol 17(5):1238–1250

Souza GM, Ballester MVR, Cruz CHB, Chum H, Dale B, Dale VH, Fernandes ECF, Foust T, Karp A, Lynd L, Maciel R, Milanez A, Nigro F, Osseweijer P, Verdade LM, Victoria RL, Van der Wielen L (2017) The role of bioenergy in a climate-changing world. Environ Dev 23:57–64. https://doi.org/10.1016/j.envdev.2017.02.008

Thompson JN (1998) Rapid evolution as an ecological process. Trends Ecol Evol 13(8):329–332

Timo TP, Lyra-Jorge MC, Gheler-Costa C, Verdade LM (2015) Effect of the plantation age on the use of Eucalyptus stands by medium to large-sized wild mammals in south-eastern Brazil. iForest-Biogeosciences and Forestry 8(2):108

Timo TP d C, Costa JC, Gheler-Costa C, Lyra-Jorge MC, Verdade LM (2016) Wildlife and landscape geometry in silvicultural landscapes. In: Gheler-Costa C, Lyra-Jorge MC, Verdade LM (eds) Biodiversity in agricultural landscapes of southeastern Brazil. De Gruyter, Berlin, pp 293–327. https://doi.org/10.1515/9783110480849-019

Tripp R (1996) Biodiversity and modern crop varieties: sharpening the debate. Agric Hum Values 13(4):48–63

Valiente-Banuet A, Aizen MA, Alcántara JM, Arroyo J, Cocucci A, Galetti M, García MB, García D, Gómez JM, Jordano P, Medel R, Navarro L, Obeso JR, Oviedo R, Ramírez N, Rey PJ, Traveset A, Verdú M, Zamora R (2015) Beyond species loss: the extinction of ecological interactions in a changing world. Funct Ecol 29:299–307

Vandermeer J, Carvajal R (2001) Metapopulation dynamics and the quality of the matrix. Am Nat 3:211–220

Verdade LM (1996) The influence of hunting pressure on the social behavior of vertebrates. Rev Bras Biol 56(1):1–13

Verdade LM (2004) A exploração da fauna silvestre no Brasil: jacarés, sistemas e recursos humanos. Biota Neotrop 4(2):1–12

Verdade LM, Lyra-Jorge MC, Piña CI (2014a) Redirections in conservation biology. In: Verdade LM, Lyra-Jorge MC, Piña CI (eds) Applied ecology and human dimensions in biological conservation. Springer, Heidelberg. ISBN 978-3-642-54750-8, pp 3–17. https://doi.org/10.1007/978-3-642-54751-5_1

Verdade LM, Penteado M, Gheler-Costa C, Dotta G, Rosalino LM, Pivello VR, Piña CI, Lyra-Jorge MC (2014b) The conservation value of agricultural landscapes. In: Verdade LM, Lyra-Jorge MC, Piña CI (eds) Applied ecology and human dimensions in biological conservation. Springer, Heidelberg. ISBN 978-3-642-54750-8, pp 91–102. https://doi.org/10.1007/978-3-642-54751-5_6

Verdade LM, Palomares F, Couto HTZ, Polizel JL (2016) Recent land use changes and the expansion of an exotic potential prey: a possible redemption for Atlantic forest jaguars? Anim Conserv 19:209–211. https://doi.org/10.1111/acv.12221

Verdade LM, Moral RA, Calaboni A, Amaral MVSG, Martin PS, Amorim LS, Gheler-Costa C, Piña CI (2020) Temporal dynamics of small mammals' in *Eucalyptus* plantations in Southeast Brazil. Global Ecol Conserv 24:e01217. https://doi.org/10.1016/j.gecco.2020.e01217

Verdade LM, Moral RA (2026) Why should we monitor fauna and flora in multifunctional landscapes? In: Verdade LM, Moral RA (eds) Wildlife monitoring: integrating conservation and innovation in human-altered landscapes. Springer, Amsterdam, pp 3–15. https://doi.org/10.1007/978-3-032-05821-8_1

Verdade LM, Toledo RM, Pivello VR, Jacobus AP, Moral RA (2026) Long-term, cross-spatial-scale wildlife monitoring: a protocol for human-modified environments in Central-South Brazil. In: Verdade LM, Moral RA (eds) Wildlife monitoring: integrating conservation and innovation in human-altered landscapes. Springer, Amsterdam. pp 431–444, 350 p. https://doi.org/10.1007/978-3-032-05821-8_21

Walpole EH, Toman E, Stidham M, Wilson R (2020) The science and practice of ecological restoration: a mental models analysis of restoration practitioners. Environ Syst Decis 40(4):588–604

Watson DM, Watson MJ (2015) Wildlife restoration: mainstreaming translocations to keep common species common. Biol Conserv 191:830–838

Worthington T, Tisdale J, Kemp P, Williams I, Osborne PE (2010) Public and stakeholder attitudes to the reintroduction of the burbot, *Lota lota*. Fish Manag Ecol 17:465–472

Chapter 21
Long-Term, Cross-Spatial-Scale Wildlife Monitoring: A Protocol for Human-Modified Environments in Central-South Brazil

Luciano Martins Verdade, Vânia R. Pivello, Renato M. Toledo, Ana P. Jacobus, and Rafael de Andrade Moral

Abstract Agricultural and silvicultural landscapes provide complementary conservation to a relevant part of the biota, which demands a standardized monitoring protocol to be adequately managed. In this chapter we present a standardized monitoring protocol of non-flying terrestrial mammals and the native vegetation present in agricultural and silvicultural landscapes of Central-South Brazil. A total of 11 indicators from five taxonomic groups are divided into three categories: (1) the diversity of biological patterns, (2) the complexity of eco-evolutionary processes, and (3) their interactions, from individual to landscape level. Such a monitoring protocol might be useful for the effectiveness of multifunctionality of agricultural landscapes as well as for their environmental certification process by quantifying a relevant part of their biota over time.

Keywords Standardized monitoring protocol · Biodiversity · Biocomplexity · Wildlife management · Agricultural landscapes

L. Martins Verdade (✉)
Wildlife Monitoring Consultancy, Campinas, São Paulo, Brazil
e-mail: luciano.verdade@wildlifemgmtco.com

V. R. Pivello · R. M. Toledo
Universidade de São Paulo/IB/Departamento de Ecologia, São Paulo, SP, Brazil

A. P. Jacobus
Institute of Bioenergy Research-São Paulo State University, São Paulo, SP, Brazil

R. A. Moral
Department of Mathematics and Statistics, Maynooth University, Maynooth, Co. Kildare, Ireland

L. Martins Verdade, R. A. Moral (eds.), *Wildlife Monitoring*,
https://doi.org/10.1007/978-3-032-05821-8_21

1 Scope

In this chapter we present a standardized monitoring protocol of non-flying terrestrial mammals and the native vegetation present in agricultural and silvicultural landscapes of Central-South Brazil. First, non-flying terrestrial mammals are not necessarily the best indicators of anthropic impacts on biodiversity. However, they are usually more familiar to the public (Deere et al. 2024)—and consequently, to the market—than other vertebrates and invertebrates, which makes them a good model for laypeople. In addition, Central-South Brazil is not a model either of agricultural or silvicultural systems for the whole world. However, this region has an intensive biological production with a consistent presence of remnants of native vegetation thanks to the Brazilian Forest Code (BFC), which makes it a good model for the combination of intensive agricultural/silvicultural production and wildlife conservation under a land-sharing approach.

The present monitoring protocol aims at improving the decision-making process regarding biological conservation, damage control and sustainable use of wildlife in human-modified environments, distributed over areas previously occupied by semideciduous Atlantic Forest (Caughley 1994; Verdade et al. 2014a). It should be applied ideally for as long as agricultural or silvicultural operations last (Verdade et al. 2014b, 2016, 2020, 2022, Chaps. 1, 10, and 11, this book).

As the diversity of biological patterns are usually molded by the complexity of eco-evolutionary processes, this protocol is based on three components: biodiversity, biocomplexity, and their interactions. The following innovations make it possible: population density estimate of wild species based on their vestiges (Chaps. 2, 4, and 12, this book) and DNA analysis (Chap. 5, this book), the use of native vegetation biomass as an indicator of both biodiversity and biocomplexity (Chaps. 6 and 7, this book), and the use of Bayesian hierarchical modelling that allows us to account for observation error, latent ecological states, and complex data dependencies (Chap. 8, this book). In the following sections we describe indicators, methodology (both sampling and analytical) and governance aspects, i.e., what to measure, how to measure/estimate, and who/when/where to measure, respectively (see Chap. 1, this book).

2 Indicators

A total of 11 indicators from five taxonomic groups are divided into three categories: the diversity of biological patterns (hereafter called biodiversity), the complexity of eco-evolutionary processes (hereafter called biocomplexity), and their interactions, from individual to landscape level. The number of indicators per taxon varies from three to canids and felids to six to arboreal and arborescent plants. The indicators

Table 21.1 Indicators to be used in the monitoring protocol of wildlife and its habitats in agricultural and silvicultural landscapes of Central-South Brazil

Type	Ecological level	Indicator	Taxonomic group
Biodiversity	Species	Population density (ind./km^2)	Arboreal/arborescent plants
			Medium to large mammals
			Canids
			Felids
		Population sex ratio (m:f)	Canids
			Felids
		Relative abundance	Small mammals (preyed by canids and felids)
	Community	Species richness & composition	Arboreal/arborescent plants
			Medium to large mammals
			Canids
			Felids
			Small mammals (preyed by canids and felids)
Biocomplexity	Community	Functional diversity	Arboreal/arborescent plants
			Medium to large mammals
		Phylogenetic diversity	Arboreal/arborescent plants
			Medium to large mammals
		AGC[a] (kg/ha)	Arboreal/arborescent plants
		AGB[b] (kg/ha)	Arboreal/arborescent plants
Interactions	Landscape	AGC (kg/ha)	Estimated by AGB
		Population	AGB: Population of medium/large mammal species
			AGC: Population of medium/large mammal species
		Trophic complexity	AGB: Trophic complexity

[a] Above-ground carbon
[b] Above-ground biomass of native vegetation

of biodiversity are at the species and community levels, whereas the indicators of biocomplexity are used at community level only, and their interactions are used at the landscape level (Table 21.1). Such a combination of indicators at distinct levels allows us to extrapolate data from the local to the regional scale.

3 Methodology

3.1 *Monitoring Sites*

3.1.1 Local Scale

In general, the decision-making process regarding biodiversity monitoring tends to be site-specific, which results in relatively high costs. In addition, sampling methodology tends to be based on relatively expensive field methods, like camera-trapping and passive sound-recording devices. The present protocol is based on a network of long-term monitoring sites (LTMS), which can be considered representative of the whole area of interest (e.g., the area covered by landscapes dominated by a specific plantation). In addition, it uses non-invasive and inexpensive sampling methods based on vestiges, which results in a significant reduction in monitoring costs, which can facilitate its application by agro-industrial sectors under certification.

In terms of number of sites to be monitored, power studies based on simulations can be used, where the uncertainty should be accommodated via a proper distributional assumption (e.g., Poisson and extensions for count data, binomial and extensions for discrete proportions, beta for continuous proportions, normal for continuous and symmetrical data, and skewed distributions with positive support for strictly continuous data; see Chap. 9, this book). According to the expected degree of variation of the indicator being monitored, a different number of sites may be proposed. However, based on experiential and empirical knowledge from our group, as a rule of thumb, 15 sites per focused sector when dealing with landscapes in Central-South Brazil tends to work well in terms of cost-benefit and pragmatism, while allowing for the detection of discrepancies with acceptable power. Naturally, this number should be tweaked accordingly depending on the area to be surveyed. If there is a very fast shift in a temporal trend, then fewer sites could be used; conversely, if the shift is very mild, a very high number of sites and/or time points would be necessary to differentiate this signal from random variation.

Each sampling site should be composed of a grid of 25 plots (20 $\times$ 20 m each, adapted from Toledo et al. 2024) and a trail of approximately 30 km. The plots should be distributed over the vertices of a 5 $\times$ 5 km grid (adapted from Magnusson et al. 2013), 1 km far from each other. However, the grid can be adapted to fit a specific property. The trail can be built over the lines of the grid or locally adapted to topography, plantation areas, and local trails for local traffic and transportation of agricultural commodities (Fig. 21.1). Each plot can be considered a sampling unit of arboreal and arborescent plants, whereas the trail can be considered the sampling unit of medium to large terrestrial mammals.

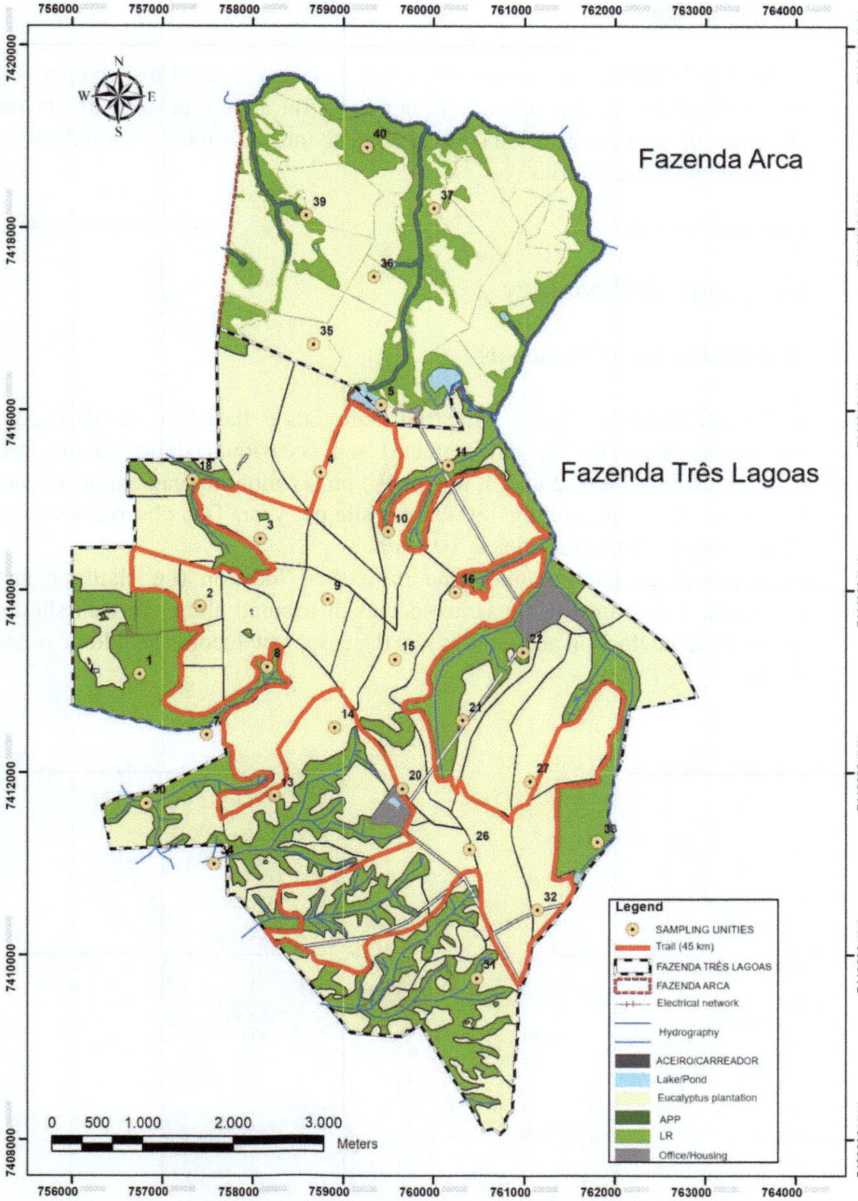

Fig. 21.1 Spatial distribution of parcels and trails locally adapted to topography, plantation areas, and local trails for logging transportation, locally called "aceiros" (i.e., those located on the border between the plantations and areas of native vegetation) and "carreadores" (i.e., those which cross the plantations). Fazendas Três Lagoas and Arca, Angatuba municipality, state of São Paulo, Brazil. Conservation áreas: APP (Area of Permanent Protection) and LR (Legal Reserve)

3.1.2 Regional Scale

Remote sensing should be used to determine land use (i.e., spatial distribution and respective distribution of plantations and conservation areas), as well as above-ground biomass of native vegetation (Fig. 21.2). The interaction between indicators should be based on these metrics.

3.2 Sampling Methodology

3.2.1 Medium to Large Mammals

The sampling of medium-to-large terrestrial mammals is based on vestiges (e.g., footprints, feces, marks on the environment) and occasional sights on the trail (Verdade et al. 2013, Chaps. 2 and 3, this book) on six monthly campaigns during the wet season every year, totaling 180 km per site per year. Two observers should do it simultaneously on motorcycles at 10 km/h.

Vestiges should be photographed and have their location (i.e., latitude and longitude) taken. Footprints of the same species distributed along the trail should be registered at first detection and then, if it is the case, new records should be made every 100 m.

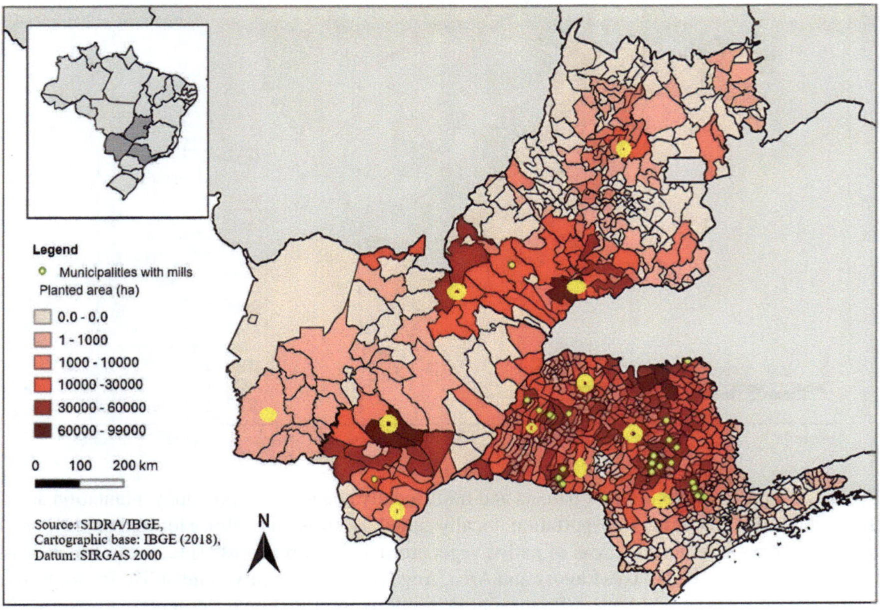

Fig. 21.2 Standardized monitoring protocol for fauna and flora. Hypothetical long-term study sites: sugarcane. (Adapted from Silva and Pereira (2023))

3.2.2 Canids and Felids

Besides the location and photograph of their vestiges and occasional sights, as described above (i.e., photographed and located), canids and felids should have 20–50 g samples of their fecal material collected and preserved in 50 ml sterile screw-cap plastic with silica beads. The tubes should be kept refrigerated at 3 °C during fieldwork activities (up to 2 months) and afterwards kept at −20 °C in the laboratory until DNA extraction (Chap. 5, this book).

3.2.3 Arboreal and Arborescent Plants

Every individual of arboreal and arborescent plants with diameter at breast height (DBH) $\geq$3 cm (equivalent to basal area $\geq$7.0 cm^2) should be identified to the lowest possible taxonomic level and have its DBH and height measured (Toledo et al. 2024), as follows:

(a) DBH (cm): It should be measured by caliper, dendrometric tape or regular tape of proper size. Multi-stemmed trees should have each stem's DBH measured (FAO 2004).
(b) Height (m): It should be estimated by laser hypsometer, a metric tape with a clinometer, or by using a graded scale (Bettinger et al. 2020).

3.3 *Analytical Methodology*

3.3.1 Species Richness

Species' richness should be determined for arboreal and arborescent plants and medium to large terrestrial mammals for each sampling site. Two measures of species' richness should be considered:

(a) Detected species richness: The number of species detected.
(b) Estimated species richness: The estimate of the asymptote of the rarefied species incidence curve using a Bootstrap procedure (Krebs 1999: 422–423), or nonlinear (Bayesian) modeling.

3.3.2 Species Composition

The species composition is the list of species detected. It should be determined for arboreal and arborescent plants, medium to large terrestrial mammals, and small mammals for each sampling site. Amplicon-based metabarcoding of fecal samples or soil residues using universal primers provide a comprehensive inventory of mammalian species present in the area. In addition, individual genotyping should be performed using nuclear microsatellites (STRs) or single nucleotide polymorphism (SNP) panels developed for key target species (Deiner et al. 2017). These nuclear markers allow for individual identification within species, enabling further estimates of population size, sex ratio, and genetic diversity from non-invasively collected samples.

3.3.3 Phylogenetic Diversity (PD)

The number of families detected for the species above should be used as a proxy of PD (adapted from Faith and Pollock 2014). Ideally, non-native species should be accounted for separately.

3.3.4 Functional Diversity (FD)

For arboreal and arborescent plants, the functional diversity should be considered number of arboreal strata plant life forms, as a proxy of FD proposed by Faith and Pollock (2014). Similarly to PD, non-native species should be accounted for separately also for FD. For medium-to-large terrestrial mammals the following functional groups should be considered (Dotta and Verdade 2007): insectivore/omnivores, frugivore/omnivores, carnivores, frugivore/herbivores, and herbivore/grazers.

3.3.5 Population Density

Population density should be estimated for each species of arboreal and arborescent plants, medium to large terrestrial mammals, canids, felids, and small mammals for each sampling site, as follows:

(a) Arboreal and arborescent plants: Total number of individuals of each species detected at all sampling units of each sampling site divided by its area (km^2).
(b) Medium to large terrestrial mammals: Estimated by the number of individuals detected based on the minimum and maximum home range sizes found in literature for the species (adapted from Fragoso et al. 2016, Chaps. 2 and 3, this book).
(c) Canids and felids: Estimated by the number of individuals detected by vestiges (as stated above for medium-to-large terrestrial mammals) spatialized by the minimum and maximum home range sizes found in literature for the species (adapted from Fragoso et al. 2016, Chaps. 2 and 3, this book). In addition, identification of distinct individuals genotyping in a population should be taken by nuclear microsatellites or single-nucleotide polymorphisms (SNPs) panels (Waits and Paetkau 2005, Chap. 5, this book). This allows for extrapolation of the number of individuals per unit area, considering habitat type and species-

specific behavior. Then, the number of individuals detected by molecular analysis should be used to establish a collector curve rarefied by Bootstrap, whose asymptote is an estimate of the total number of individuals. Such a procedure is analogous to the estimate of species richness described above.

(d) Small mammals: The rodents and marsupials preyed by canids and felids should be identified through DNA metabarcoding of fecal samples using vertebrate-specific primers targeting mitochondrial markers, such as 12S rRNA or cytochrome b. This non-invasive approach enables high-resolution identification of prey species and has been widely applied in trophic ecology and dietary studies of elusive predators. Metabarcoding allows for the detection of both dominant and rare prey taxa, including cryptic or soft-bodied species that are often underrepresented in traditional morphological diet analysis (De Barba et al. 2014; Shehzad et al. 2012; Kartzinel et al. 2015). It also improves comparability across studies and ecosystems by standardizing taxonomic resolution and reducing observer bias.

3.3.6 Sex Ratio per Species

Sex ratio should be determined for each canid and felid species. Molecular sexing can be conducted by targeting sex-linked genetic markers, such as the *CHD1* gene in birds or the *amelogenin* gene in mammals, using PCR amplification from non-invasive samples including feces, feathers, hair, or saliva. These markers allow reliable differentiation between male and female genotypes due to sex-specific sequence variants or length polymorphisms (Fridolfsson and Ellegren 1999; Arandjelovic et al. 2010). In cases where more than one sex-specific locus is required—such as in species with degraded DNA or complex sex chromosome systems—multiplex PCR or sequencing-based methods (e.g., amplicon sequencing or hybrid capture of sex-linked regions) can be applied (Morin et al. 2020). High-throughput sequencing can also enhance the resolution of sex assignment in samples with low DNA yield or where traditional markers may fail, enabling accurate determination of sex ratios in wild populations under natural conditions.

3.3.7 Relative Abundance

The relative abundance of each species of small mammals preyed by canids and felids should be estimated based on sequencing data derived from DNA extracted from non-invasive samples such as feces or soil. The number of sequence readings assigned to each species through metabarcoding can serve as a proxy for their relative abundance within the sampled community. Although reading counts are subject to amplification biases and primer efficiencies, they have been shown to correlate with biomass or relative frequency in several ecological contexts (Deagle et al. 2019; Lamb et al. 2019). This approach allows for robust comparative assessments of species' presence and relative dominance across different locations or sampling units. To minimize distortions caused by uneven sequencing depth or technical variation, abundance of readings should be pre-processed through methods such as rarefaction, relative scaling, or variance-stabilizing transformations (Mcmurdie and Holmes 2014).

3.3.8 Above-Ground Biomass (AGB)

Although plant biomass can be considered a measure of relative abundance (Bullock 2006), here we take it separately as an indicator for the sake of simplicity. AGB should be calculated for each sampled tree by allometric equation for pantropical tree communities, using specific wood density (ρ, g/cm^3), DBH (cm) and tree height (H, m): $AGB = 0.063(\rho\ DBH^2\ H)^{0.976}$ (Chave et al. 2014). Subsequently, individual biomass values should be summed and standardized as tons per hectare (ton/ha). The species wood density can be obtained from Chave et al. (2009) and Zanne et al. (2009).

3.3.9 Above-Ground Carbon (AGC)

AGC should be estimated by AGB multiplied by 0.46 (Doraisami et al. 2024).

3.3.10 Population

The population of each species of arboreal/arborescent plants can be estimated by their AGB and AGC at local scales throughout the whole area covered by landscapes dominated by the agroindustrial sector in question (Toledo et al. 2024). The population of each species of medium-to-large mammals should be estimated by their range of population density among sampling sites extrapolated to the whole area of landscapes dominated by the agroindustrial sector in question. For canids and felids there should be two estimates, one based on vestiges and another based on DNA analyses, as described above.

3.3.11 Trophic Complexity

An analogous collector curve should be assembled for each sampling site based on the predator-prey relationship (see Fig. 4.4, Campos et al. (2026), Chap. 4 of this book). The β-coefficient of the linearized mathematical model can be considered a direct measure of its diversity, or the chance that the next pair of predator-prey is a new one (analogous to the measurement of species diversity, e.g., Maurer and McGill 2011). However, since such diversity index refers to the complexity of a process—not a diversity of a pattern—it can be considered a biocomplexity measure. The relationship between it and AGB and AGC might be used to estimate trophic complexity at local scale throughout the whole area covered by landscapes dominated by the agroindustrial sector in question.

4 Governance

The monitoring protocol of wildlife and its habitats can be useful in two complementary levels regarding wildlife management in agricultural landscapes: public policy and environmental certification. In both levels the detection of possible impacts of changes in land use and agricultural management might be faster. Consequently, the decision-making process about possible interventions could be improved. However, it would be wise to keep in mind that monitoring programs tend to be more effective to detect discrepancies than their causes. Therefore, the decision-making process associated with them usually includes the following steps: monitoring, detection of discrepancies, empirical vs. experimental approaches, interventions, and their evaluation (Fig. 21.3). Such an approach is based on adaptive management (Chap. 1, this book), which might result on the improvement of wildlife management as well as the development of wildlife-friendly management techniques.

The monitoring protocol presented here might be particularly useful for the effectiveness of multifunctionality of agricultural landscapes by quantifying a relevant part of their biota over time. In addition, it might also be useful for environmental certification processes, which are already consolidated in some agroindustrial sectors for the same reason. The standardization of a monitoring protocol for distinct agroindustrial sectors might simplify the comparison of their conservation values as well as their eventual improvement. Finally, the present protocol might be particularly useful for the certification process of Biodiversity-

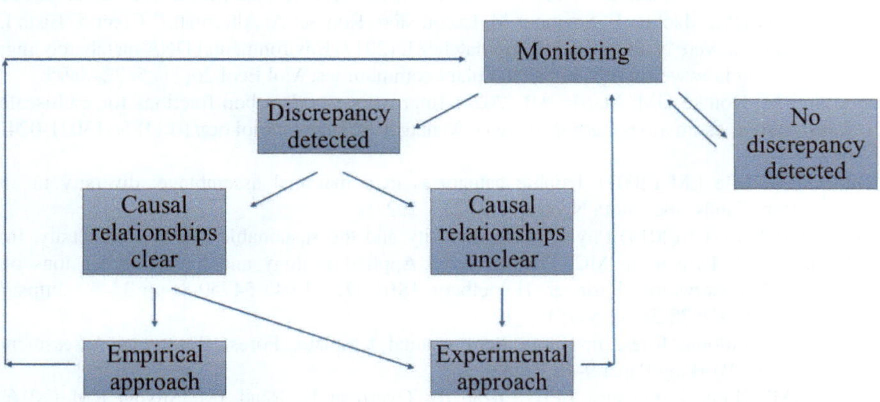

Fig. 21.3 *Modus operandi* of a biodiversity monitoring protocol

credit (Wauchope et al. 2024) and even Carbon-credit, given its applicability to agricultural landscapes.

Acknowledgements This study was partially funded by Fundação de Amparo à Pesquisa do Estado de São Paulo (FAPESP, Proc. No. 2017/01304-4) and Agência Nacional de Energia Elétrica (ANEEL, Proc. No.)

References

Arandjelovic M, Head J, Kühl H, Boesch C, Robbins MM, Maisels F, Vigilant L (2010) Effective non-invasive genetic monitoring of multiple wild western gorilla groups. Biol Conserv 143(7):1780–1791. https://doi.org/10.1016/j.biocon.2010.04.030

Bettinger P, Izlar B, Harris T, Cieszewski C, Conrad J, Greene D, Mech A, Shelton J, Siry J, Kane M, Merry K, Baldwin S, Smith J (2020) Handbook of land and tree measurements. Center for Forest Business, University of Georgia, Athens. ISBN 978-0-9821139-1-2

Bullock JM (2006) Plants. In: Sutherland WJ (ed) Ecological census techniques: a handbook, 2nd edn. Cambridge University Press, Cambridge, UK, pp 186–213

Caughley G (1994) Directions in conservation biology. J Anim Ecol 63:215–244

Chave J, Coomes D, Jansen S, Lewis SL, Swenson NG, Zanne AE (2009) Towards a worldwide wood economics spectrum. Ecol Lett 12(4):351–366

Chave J, Réjou-Méchain M, Búrquez A, Chidumayo E, Colgan MS, Delitti WBC, Duque A, Eid T, Fearnside PM, Goodman RC, Henry M, Martínez-Yrízar A, Mugasha WA, Muller-Landau HC, Mencuccini M, Nelson BW, Ngomanda A, Nogueira EM, Ortiz-Malavassi E, Pelissier R, Ploton P, Ryan CM, Saldarriaga JG, Vieilledent G (2014) Improved allometric models to estimate the aboveground biomass of tropical trees. Glob Chang Biol 20(10):3177–3190. https://doi.org/10.1111/gcb.12629

De Barba M et al (2014) DNA metabarcoding multiplexing and validation of data accuracy for diet assessment: application to omnivorous diet. Mol Ecol Resour 14(2):306–323

Deagle BE et al (2019) Counting with DNA in metabarcoding studies: how should we convert sequence reads to dietary data? Mol Ecol 28(2):391–406

Deere NJ, Ramadiyanta E, Sibarani MC, Hadi AN, Andayani N, Ginting Y et al (2024) Selecting umbrella species as mammal biodiversity indicators in tropical forest. Biol Conserv 292:110511

Deiner K, Bik HM, Mächler E, Seymour M, Lacoursière-Roussel A, Altermatt F, Creer S, Bista I, Lodge DM, de Vere N, Pfrender ME, Bernatchez L (2017) Environmental DNA metabarcoding: transforming how we survey animal and plant communities. Mol Ecol 26(21):5872–5895

Doraisami M, Domke GM, Martin AR (2024) Improving wood carbon fractions for multiscale forest carbon estimation. Carbon Balance Manag 19:25. https://doi.org/10.1186/s13021-024-00272-2

Dotta G, Verdade LM (2007) Trophic categories in a mammal assemblage: diversity in an agricultural landscape. Biota Neotrop 7(2):287–292

Faith DP, Pollock LJ (2014) Phylogenetic diversity and the sustainable use of biodiversity. In: Verdade LM, Lyra-Jorge MC, Piña CI (eds) Applied ecology and human dimensions on biological conservation. Springer, Heidelberg. ISBN 978-3-642-54750-8, pp 35–52. https://doi.org/10.1007/978-3-642-54751-5_12

FAO (2004) National forest inventory field manual template. Forest Resources Assessment Programme, Working Paper 94/E Rome

Fragoso JMV, Levi T, Oliveira LFB, Luzar JB, Overman H, Read JM, Silvius KM (2016) Line transect surveys underdetect terrestrial mammals: implications for the sustainability of subsistence hunting. PLoS One 11(4):e0152659

Fridolfsson AK, Ellegren H (1999) A simple and universal method for molecular sexing of non-ratite birds. J Avian Biol 30(1):116–121
Kartzinel TR et al (2015) DNA metabarcoding illuminates dietary niche partitioning by African large herbivores. Proc Natl Acad Sci 112(26):8019–8024
Krebs CJ (1999) Ecological methodology, 2nd edn. Addison Wesley Longman, Menlo Park
Lamb PD et al (2019) How quantitative is metabarcoding: a meta-analytical approach. Mol Ecol 28(2):420–430
Magnusson WE, Braga-Neto R, Pezzini F, Baccaro F, Bergallo H, Penha J, Rodrigues D, Verdade LM, Lima A, Albernaz AL, Hero J-M, Lawson B, Castilho C, Drucker D, Franklin E, Mendonça F, Costa F, Galdino G, Castley G, Zuanon J, do Vale J, Campos L, Luizão R, Cintra R, Barbosa RI, Lisboa A, Koblitz RV, Mendes-Pontes AR, da Cunha CN (2013) Biodiversidade e monitoramento ambiental integrado. Áttema Editorial, Santo André. ISBN 978-85-65551-04-5, 352 p
Maurer BA, McGill BJ (2011) Measurement of species diversity. In: Magurran AE, McGill BJ (eds) Biological diversity: Frontiers in measurement and assessment. Oxford University Press, Oxford, UK, pp 55–65
Mcmurdie PJ, Holmes S (2014) Waste not, want not: why rarefying microbiome data is inadmissible. PLoS Comput Biol 10(4):e1003531
Morin PA, Rykiel EJ et al (2020) SNP-based molecular sexing for non-model organisms using next-generation sequencing. Methods Ecol Evol 11(1):95–103
Shehzad W et al (2012) Carnivore diet analysis based on next-generation sequencing: application to the leopard cat (Prionailurus bengalensis) in Pakistan. Mol Ecol 21(8):1951–1965
Silva LR, Pereira MFV (2023) Use of territory and financial valuation: the Cosan Group and the Brazilian Sugar-Energy Sector. Soc Nat 35:e68172
Toledo RM, Pivello VR, Perring MP, Verdade LM (2024) Natural vegetation biomass and the dimension of forest quality in tropical agricultural landscapes. Ecol Appl 34(3):e2950. https://doi.org/10.1002/eap.2950
Verdade LM, Moreira JR, Ferraz KMPMB (2013) Counting capybaras. In: Moreira JR, Ferraz KMPMB, Herrera EA, Macdonald DW (eds) Capybara: biology, use and conservation of an exceptional neotropical species. Springer, New York. ISBN 978-1-4614-3999-8, pp 357–370. https://doi.org/10.1007/978-1-4614-4000-0_22
Verdade LM, Penteado M, Gheler-Costa C, Dotta G, Rosalino LM, Pivello VR, Piña CI, Lyra-Jorge MC (2014a) The conservation value of agricultural landscapes. In: Verdade LM, Lyra-Jorge MC, Piña CI (eds) Applied ecology and human dimensions in biological conservation. Springer, Heidelberg. ISBN 978-3-642-54750-8, pp 91–102. https://doi.org/10.1007/978-3-642-54751-5_6
Verdade LM, Lyra-Jorge MC, Piña CI (2014b) Redirections in conservation biology. In: Verdade LM, Lyra-Jorge MC, Piña CI (eds) Applied ecology and human dimensions in biological conservation. Springer, Heidelberg. ISBN 978-3-642-54750-8, pp 3–17. https://doi.org/10.1007/978-3-642-54751-5_1
Verdade LM, Gheler-Costa C, Lyra-Jorge MC (2016) The multiple facets of agricultural landscapes. In: Gheler-Costa C, Lyra-Jorge MC, Verdade LM (eds) Biodiversity in agricultural landscapes of southeastern Brazil. De Gruyter, Berlin, pp 2–13. https://doi.org/10.1515/9783110480849-003
Verdade LM, Moral RA, Calaboni A, Amaral MVSG, Martin PS, Amorim LS, Gheler-Costa C, Piña CI (2020) Temporal dynamics of small mammals' in *Eucalyptus* plantations in Southeast Brazil. Global Ecol Conserv 24:e01217. https://doi.org/10.1016/j.gecco.2020.e01217
Verdade LM, Bianchi RC, Galetti PM Jr, Pivello VR, Silva WR, Uezu A (2022) Biodiversity management and research in multifunctional landscapes. Biota Neotrop 22(spe):e20221407. https://doi.org/10.1590/1676-0611-BN-2022-1407
Waits LP, Paetkau D (2005) Noninvasive genetic sampling tools for wildlife biologists: a review of applications and recommendations for accurate data collection. J Wildlife Manag 69:1419–1433. https://doi.org/10.2193/0022-541X(2005)69[1419:NGSTFW]2.0.CO;2

Wauchope HS, Zu Ermgassen SO, Jones JP, Carter H, Schulte to Bühne H, Milner-Gulland EJ (2024) What is a unit of nature? Measurement challenges in the emerging biodiversity credit market. Proc R Soc B 291(2036):20242353. https://doi.org/10.1098/rspb.2024.2353

Zanne AE, Lopez-Gonzalez G, Coomes DA, Ilic J, Jansen S, Lewis SL, Miller RB, Swenson NG, Wiemann MC, Chave J (2009) Global wood density database. Dryad. Identifier: http://datadryad.org/handle/10255/dryad.235

Index

L. Martins Verdade, R. A. Moral (eds.), *Wildlife Monitoring*,
https://doi.org/10.1007/978-3-032-05821-8

MIX
Papier aus verantwortungsvollen Quellen
Paper from responsible sources
FSC® C105338

If you have any concerns about our products,
you can contact us on
ProductSafety@springernature.com

In case Publisher is established outside the EU,
the EU authorized representative is:
Springer Nature Customer Service Center GmbH
Europaplatz 3, 69115 Heidelberg, Germany

Printed by Libri Plureos GmbH
in Hamburg, Germany